Smart Innovation, Systems and Technologies

Volume 267

D1795686

Series Editors

Robert J. Howlett, Bournemouth University and KES International,
Shoreham-by-Sea, UK

Lakhmi C. Jain, KES International, Shoreham-by-Sea, UK

The Smart Innovation, Systems and Technologies book series encompasses the topics of knowledge, intelligence, innovation and sustainability. The aim of the series is to make available a platform for the publication of books on all aspects of single and multi-disciplinary research on these themes in order to make the latest results available in a readily-accessible form. Volumes on interdisciplinary research combining two or more of these areas is particularly sought.

The series covers systems and paradigms that employ knowledge and intelligence in a broad sense. Its scope is systems having embedded knowledge and intelligence, which may be applied to the solution of world problems in industry, the environment and the community. It also focusses on the knowledge-transfer methodologies and innovation strategies employed to make this happen effectively. The combination of intelligent systems tools and a broad range of applications introduces a need for a synergy of disciplines from science, technology, business and the humanities. The series will include conference proceedings, edited collections, monographs, handbooks, reference books, and other relevant types of book in areas of science and technology where smart systems and technologies can offer innovative solutions.

High quality content is an essential feature for all book proposals accepted for the series. It is expected that editors of all accepted volumes will ensure that contributions are subjected to an appropriate level of reviewing process and adhere to KES quality principles.

Indexed by SCOPUS, EI Compendex, INSPEC, WTI Frankfurt eG, zbMATH, Japanese Science and Technology Agency (JST), SCImago, DBLP.

All books published in the series are submitted for consideration in Web of Science.

More information about this series at https://link.springer.com/bookseries/8767

Vikrant Bhateja · Jinshan Tang ·
Suresh Chandra Satapathy · Peter Peer · Ranjita Das
Editors

Evolution in Computational Intelligence

Proceedings of the 9th International
Conference on Frontiers in Intelligent
Computing: Theory and Applications
(FICTA 2021)

 Springer

Editors
Vikrant Bhateja
Department of Electronics
and Communication Engineering
Shri Ramswaroop Memorial College
of Engineering and Management
(SRMCEM)
Lucknow, Uttar Pradesh, India

Dr. A.P.J. Abdul Kalam Technical
University
Lucknow, Uttar Pradesh, India

Suresh Chandra Satapathy
School of Computer Engineering
Kalinga Institute of Industrial Technology
(KIIT)
Bhubaneswar, India

Ranjita Das
Department of Computer Science
and Engineering
National Institute of Technology (NIT)
Mizoram
Aizawl, India

Jinshan Tang
College of Computing
Michigan Technological University
Michigan, MI, USA

Peter Peer
Faculty of Computer and Information
Science
University of Ljubljana
Ljubljana, Slovenia

ISSN 2190-3018   ISSN 2190-3026   (electronic)
Smart Innovation, Systems and Technologies
ISBN 978-981-16-6618-6   ISBN 978-981-16-6616-2   (eBook)
https://doi.org/10.1007/978-981-16-6616-2

Organisation

Chief Patron

Prof. Rajat Gupta, Director, NIT Mizoram

Patrons

Prof. Saibal Chatterjee, Dean (Academics), NIT Mizoram
Dr. Alok Shukla, Dean (Students' Welfare), NIT Mizoram
Dr. P. Ajmal Koya, Dean (Research & Consultancy), NIT Mizoram
Dr. K. Gyanendra Singh, Dean (Faculty Welfare), NIT Mizoram

General Chair

Dr. Jinshan Tang, College of Computing, Michigan Technological University, Michigan, US

Publication Chairs

Dr. Yu-Dong Zhang, Department of Informatics, University of Leicester, Leicester, UK
Dr. Peter Peer, Faculty of Computer & Information Science, University of Ljubljana, Slovenia
Dr. Suresh Chandra Satapathy, KIIT, Bhubaneshwar

Conveners

Dr. Ranjita Das, Head, Department of CSE, NIT Mizoram
Dr. Anumoy Ghosh, Head, Department of ECE, NIT Mizoram

Organising Chairs

Dr. Ranjita Das, Head, Department of CSE, NIT Mizoram
Dr. Anumoy Ghosh, Head, Department of ECE, NIT Mizoram
Dr. Rudra Sankar Dhar, Asst. Professor, Department of ECE, NIT Mizoram
Dr. Chaitali Koley, Assistant Professor, Department of ECE, NIT Mizoram
Mr. Sandeep Kumar Dash, Assistant Professor, Department of CSE, NIT Mizoram

Publicity Chairs

Dr. Chaitali Koley, Assistant Professor, Department of ECE, NIT Mizoram
Mr. Sushanta Bordoloi, Trainee Teacher, Department of ECE, NIT Mizoram
Mr. Sandeep Kumar Dash, Assistant Professor, Department of ECE, NIT Mizoram
Mr. Lenin Laitonjam, Trainee Teacher, Department of CSE, NIT Mizoram

Advisory Committee

Aime' Lay-Ekuakille, University of Salento, Lecce, Italy
Annappa Basava, Department of CSE, NIT Karnataka
Amira Ashour, Tanta University, Egypt
Aynur Unal, Standford University, USA
Bansidhar Majhi, IIIT Kancheepuram, Tamil Nadu, India
Dariusz Jacek Jakobczak, Koszalin University of Technology, Koszalin, Poland
Dilip Kumar Sharma, IEEE U.P. Section
Ganpati Panda, IIT Bhubaneswar, Odisha, India
Jagdish Chand Bansal, South Asian University, New Delhi, India
João Manuel R. S. Tavares, Universidade do Porto (FEUP), Porto, Portugal
Jyotsana Kumar Mandal, University of Kalyani, West Bengal, India
K. C. Santosh, University of South Dakota, USA
Le Hoang Son, Vietnam National University, Hanoi, Vietnam
Naeem Hanoon, Multimedia University, Cyberjaya, Malaysia
Nilanjan Dey, TIET, Kolkata, India
Noor Zaman, Universiti Tecknologi, PETRONAS, Malaysia

Pradip Kumar Das, Professor, Department of CSE, IIT Guwahati
Roman Senkerik, Tomas Bata University in Zlin, Czech Republic
Sriparna Saha, Associate Professor, Department of CSE, IIT Patna
Sukumar Nandi, Department of CSE, IIT Guwahati
Swagatam Das, Indian Statistical Institute, Kolkata, India
Siba K. Udgata, University of Hyderabad, Telangana, India
Tai Kang, Nanyang Technological University, Singapore
Ujjawl Maulic, Department of CSE, Jadavpur University
Valentina Balas, Aurel Vlaicu University of Arad, Romania
Yu-Dong Zhang, University of Leicester, UK

Technical Program Committee Chairs

Dr. Steven L. Fernandes, Creighton University, USA
Dr. Vikrant Bhateja, Shri Ramswaroop Memorial College of Engineering and Management (SRMCEM), Lucknow, U.P., India

Technical Program Committee

A. K. Chaturvedi, Department of Electrical Engineering, IIT Kanpur, India
Abdul Rajak A. R., Department of Electronics and Communication Engineering
Birla Institute of Dr. Nitika Vats Doohan, Indore, India
Ahmad Al- Khasawneh, The Hashemite University, Jordan
Alexander christea, University of Warwick, London UK
Amioy Kumar, Biometrics Research Lab, Department of Electrical Engineering, IIT Delhi, India
Anand Paul, The School of Computer Science and Engineering, South Korea
Anish Saha, NIT Silchar
Apurva A. Desai, Veer Narmad South Gujarat University, Surat, India
Avdesh Sharma, Jodhpur, India
Bharat Singh Deora, JRNRV University, India
Bhavesh Joshi, Advent College, Udaipur, India
Brent Waters, University of Texas, Austin, Texas, United States
Chhaya Dalela, Associate Professor, JSSATE, Noida, Uttar Pradesh, India
Dan Boneh, Computer Science Dept, Stanford University, California, USA
Dipankar Das, Jadavpur University
Feng Jiang, Harbin Institute of Technology, China
Gengshen Zhong, Jinan, Shandong, China
Harshal Arolkar, Immd. Past Chairman, CSI Ahmedabad Chapter, India
H. R. Vishwakarma, Professor, VIT, Vellore, India
Jayanti Dansana, KIIT University, Bhubaneswar, Odisha, India

Preface

This book is a collection of high-quality peer-reviewed research papers presented at the 9th International Conference on Frontiers in Intelligent Computing: Theory and Applications (FICTA-2021) held at National Institute of Technology, Mizoram, Aizawl, India, during 25–26 June 2021.

The idea of this conference series was conceived by few eminent professors and researchers from premier institutions of India. The first three editions of this conference: FICTA-2012, 2013 & 2014 were organized by Bhubaneswar Engineering College (BEC), Bhubaneswar, Odisha, India. The fourth edition FICTA-2015 was held at NIT, Durgapur, W.B., India. The fifth and sixth editions FICTA-2016 and FICTA-2017 were consecutively organized by KIIT University, Bhubaneswar, Odisha, India. FICTA-2018 was hosted by Duy Tan University, Da Nang City, Viet Nam. The eighth edition FICTA-2020 was held at NIT, Karnataka, Surathkal, India. All past eight editions of the FICTA conference proceedings are published in Springer AISC Series. Presently, FICTA-2021 is the ninth edition of this conference series which aims to bring together researchers, scientists, engineers, and practitioners to exchange and share their theories, methodologies, new ideas, experiences, applications in all areas of intelligent computing theories and applications to various engineering disciplines like Computer Science, Electronics, Electrical, Mechanical, Bio-Medical Engineering, etc.

FICTA-2021 had received a good number of submissions from the different areas relating to computational intelligence, intelligent data engineering, data analytics, decision sciences, and associated applications in the arena of intelligent computing. These papers have undergone a rigorous peer-review process with the help of our technical program committee members (from the country as well as abroad). The review process has been very crucial with minimum 02 reviews each; and in many cases 3–5 reviews along with due checks on similarity and content overlap as well. This conference witnessed more than 400+ submissions including the main track as well as special sessions. The conference featured five special sessions in various cutting-edge technologies of specialized focus which were organized and chaired by eminent professors. The total toll of papers included submissions received cross country along with 10 overseas countries. Out of this pool, only 108 papers were given acceptance and segregated as two different volumes for publication under the proceedings. This volume consists of 54 papers from diverse areas of Evolution in Computational Intelligence.

The conference featured many distinguished keynote addresses in different spheres of intelligent computing by eminent speakers like: Dr. Jinshan Tang (Professor in College of computing at Michigan Technological University) and Prof. Sukumar Nandi (Department of Computer Science & Engineering, Indian Institute of Technology, Guwahati, Assam, India). Dr. Jinshan Tang keynote lecture on "Automatic segmentation of COVID-19 infections from medical images with Deep convolutional neural network" give an idea on the recent research trends for segmenting COVID-19 infections in CT slices. The technique only requires scribble supervision, with uncertainty aware self ensembling and transformation consistent techniques. Also Prof. Sukumar' talk on the use and challenges of federated learning received ample applause from the vast audience of delegates, budding researchers, faculty, and students.

We thank the advisory chairs and steering committees for rendering mentor support to the conference. An extreme note of gratitude to Dr. Ranjita Das (Head, Department of CSE, NIT Mizoram, Aizawl, India) and Dr. Anumoy Ghosh (Head, Department of ECE, NIT Mizoram, Aizawl, India) for providing valuable guidelines and being an inspiration in the entire process of organizing this conference. We would also like to thank Department of Computer Science and Engineering and Department of Electronics and Communication Engineering, NIT Mizoram, Aizawl, India, who jointly came forward and provided their support to organize the ninth edition of this conference series.

We take this opportunity to thank authors of all submitted papers for their hard work, adherence to the deadlines and patience with the review process. The quality of a refereed volume depends mainly on the expertise and dedication of the reviewers. We are indebted to the technical program committee members who not only produced excellent reviews, but also did these in short time frames. We would also like to thank the participants of this conference, who have participated the conference above all hardships.

Lucknow, Uttar Pradesh, India	Dr. Vikrant Bhateja
Michigan, USA	Dr. Jinshan Tang
Bhubaneswar, Odisha, India	Dr. Suresh Chandra Satapathy
Ljubljana, Slovenia	Dr. Peter Peer
Aizawl, Mizoram, India	Dr. Ranjita Das

Contents

About the Editors

Dr. Vikrant Bhateja is associate professor in Department of Electronics & Communication Engineering (ECE), Shri Ramswaroop Memorial College of Engineering and Management (SRMCEM), Lucknow (Affiliated to AKTU) and also the Dean (Academics) in the same college. His areas of research include digital image and video processing, computer vision, medical imaging, machine learning, pattern analysis, and recognition. He has around 160 quality publications in various international journals and conference proceedings. He is a associate editor of IJSE and IJACI. He has edited more than 30 volumes of conference proceedings with Springer Nature and is presently EiC of IGI Global: IJNCR journal.

Dr. Jinshan Tang is currently a professor in the College of Computing at Michigan Technological University. He received his Ph.D. degree from Beijing University of Posts and Telecommunications and postdoctoral training at Harvard Medical School and the National Institute of Health. His research covers wide areas related to image processing and imaging technologies. His specific research interests include machine learning, biomedical image analysis and biomedical imaging, biometrics, computer vision, and image understanding. He has obtained more than three million US dollars grants as a PI or Co-PI. He has published more than 110 refereed journals and conference papers. He has also served as a committee member at various international conferences. He is a senior member of IEEE and a co-chair of the Technical Committee on Information Assurance and Intelligent Multimedia-Mobile Communications, IEEE SMC Society. He serves/served as a editors or guest editors of more than 10 journals.

Suresh Chandra Satapathy is Ph.D. in Computer Science, currently working as Professor and at KIIT (Deemed to be University), Bhubaneshwar, Odisha, India. He held the position of the National Chairman Div-V (Educational and Research) of Computer Society of India and is also a senior member of IEEE. He has been instrumental in organizing more than 20 International Conferences in India as Organizing Chair and edited more than 30 book volumes from Springer LNCS, AISC, LNEE, and SIST Series as Corresponding Editor. He is quite active in research in the areas of swarm intelligence, machine learning, data mining. He has developed a new optimization algorithm known as social group optimization (SGO) published in Springer Journal. He has delivered a number of Keynote address and Tutorials in his areas of expertise in various events in India. He has more than 100 publications in reputed journals and conference proceedings. He is in Editorial Board of IGI Global, Inderscience, Growing Science journals and also Guest Editor for Arabian Journal of Science and Engineering published by Springer.

Peter Peer is a full professor of computer science at the University of Ljubljana, Slovenia, where he heads the Computer Vision Laboratory, coordinates the double degree study program with the Kyungpook National University, South Korea, and serves as a vice-dean for economic affairs. He received his doctoral degree in computer science from the University of Ljubljana in 2003. Within his post-doctorate, he was an invited researcher at CEIT, San Sebastian, Spain. His research interests focus on biometrics and computer vision. He participated in several national and EU-funded R&D projects and published more than 100 research papers in leading international peer reviewed journals and conferences. He is co-organizer of the Unconstrained Ear Recognition Challenge and Sclera Segmentation Benchmarking Competition. He serves as Associated Editor of IEEE Access and IET Biometrics. He is a member of the EAB, IAPR, and IEEE.

Dr. Ranjita Das is currently serving as Head and Assistant Professor, Department of Computer Science and Engineering, National Institute of Technology Mizoram. She has joined the National Institute of Technology Mizoram in the year 2011. She did her Ph.D. from NIT Mizoram, M. Tech from Tezpur University, and B. Tech. from NIT Agartala. She has over 10 years of teaching experience. Her research was in the areas of pattern recognition, information retrieval, computational biology, and machine learning. She has published 20 journal and international conference papers in various journals with SCI impact factors, SCOPUS index, and also in conference proceedings of Springer, IEEE, etc. She has two ongoing sponsored projects funded by DBT and SERB. Under her supervision, presently ten research scholars are doing research work. She was recipient of best paper awards in the conferences IEEE-INDICON-2017, ICACCP-2019, IC4E-2020.

Chapter 1
A Comprehensive Study of Page-Rank Algorithm

Surabhi Solanki, Seema Verma, and Kishore Chahar

Abstract Web-mining played an important role in the internet. Web-mining approaches are applied to search relevant information from WWW. The process of web-mining mainly depends on the page rank. The immense pool of data related to the user's queries and this data is stored in the various webpages and increases rapidly. Yet, all data is not relevant to the user. Whenever a user writes a query in the search engine, it should be able to give the relevant information to the user. To provide documents according to the relevance page-ranking algorithm have been applied which used the web-mining methods to order the documents according to their content score, relevance, and importance. There are some algorithms that use Web-Structure-Mining in order to search structure of web or some algorithms that use Web-Content-Mining in order to search the content of documents, whereas some algorithm uses the combination of both algorithms. In this paper, the comparative study on different page-rank algorithms is discussed and found.

1.1 Introduction

Internet is a huge resource of data like raw data, text, picture, sound, and video. In-formation retrieval is used to search relevant information from the unstructured-data [1], page-ranking is a most important step in process of information-retrieval. In internet data is growing rapidly, development of data on the WWW and the client's reliance on web, there trouble in overseeing web data and fulfilling the user's query. In the process of ranking [2], first the user submitted the query then using information retrieval techniques, as per the data relevance it filters the web age and after that it creates the index. This index is identified according to the rank of the web.

S. Solanki (✉) · S. Verma · K. Chahar
Banasthali Vidyapith, Rajasthan Tonk, India

© The Author(s), under exclusive license to Springer Nature Singapore Pte Ltd. 2022   1
V. Bhateja et al. (eds.), *Evolution in Computational Intelligence*,
Smart Innovation, Systems and Technologies 267,
https://doi.org/10.1007/978-981-16-6616-2_1

1.1.1 Web-Mining

Retrieval of information that will be present in the form of previously unknown non-trivial and necessary for large collection of data is called data mining and when it is applied on internet is called web-mining [3]. The primary goal of web-mining is to find out the necessary pattern in web data in the following ways collect and analyze the information. Web-mining is categorized into three sub-categories. These three sub-categories of web-mining are shown in Fig. 1.1.

1.1.1.1 Web-Content-Mining

Web-Content-Mining is the useful procedure toward mining data that are generally text, picture, audio, and visual documents from the substances of webpage. It incorporates extraction of structured data, compatibility, and integration of webpages in which the data has similar meaning. Some approaches which are used in Web-Content-Mining are shown in Fig. 1.2.

1.1.1.2 Web-Structure-Mining

Web structure mining is used to create summary of websites and web age in structured form. To generalize and describe it used the structure in the form of tree. Web-structure-mining depends on the link-structure with or there is no linked description. Some algorithms like Hits, Page Rank have been used for modeling web topologies. The primary reason for structure mining is to find relation between webpages which is previously unknown.

Fig. 1.1 Sub-categories of Web-Mining

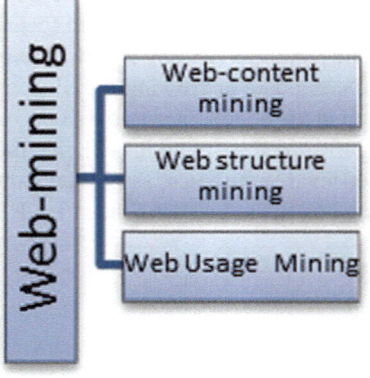

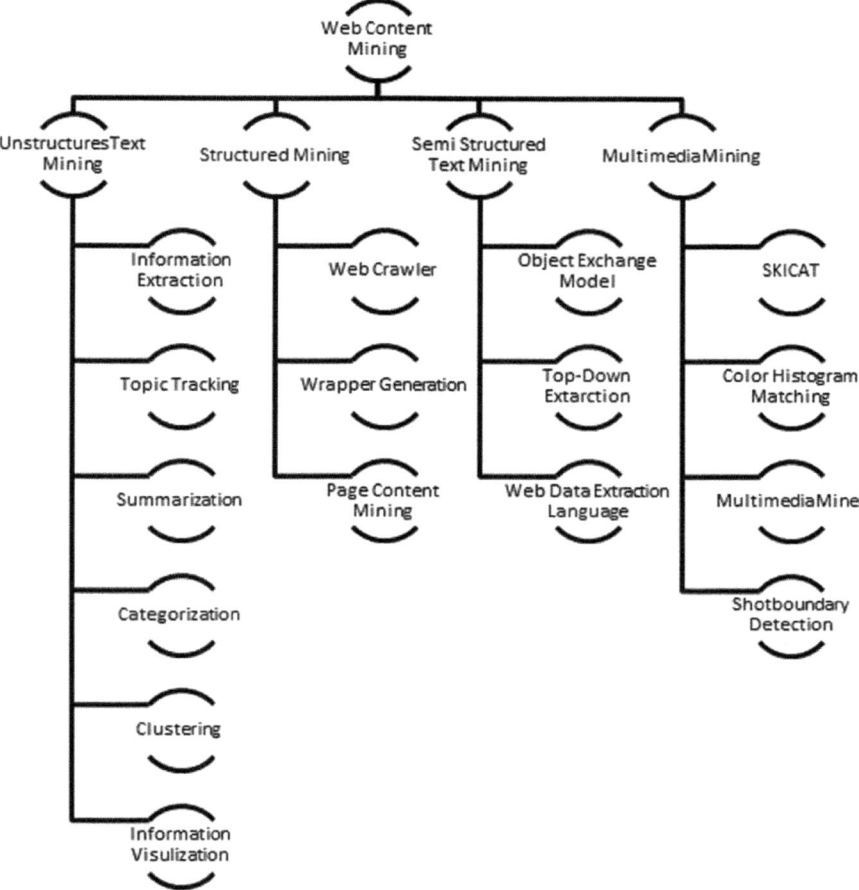

Fig. 1.2   Web-content mining

1.1.1.3   Web-Usage Mining

Web-Usage-Mining is also known as web-log-mining that is utilized to know the user's behavior for website. While user interacts with network it predicts the behavior of user. It includes the four stages of processing shown in Fig. 1.3.

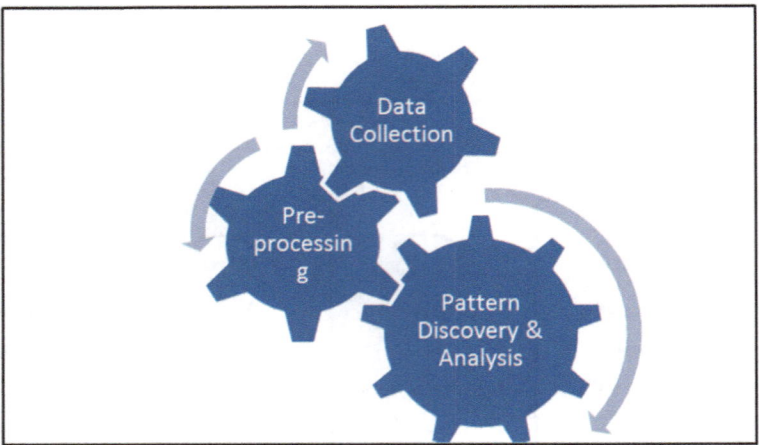

Fig. 1.3 Stages of web-usage mining

1.2 Related Work

Page rank is a method of giving page rating for find out the importance of webpages in terms of relevance using the Web-Structure-Mining. Lots of researchers have been developing the page-rank algorithm.

Xing et al. [4] proposed an idea of weighted page-rank (WPRA) algorithm, which relied on traditional page-rank algorithm. It used the concept of large rank value to popular (relevant) pages in the place of giving equal score to all out links. WPR used both in links as well as out links to evaluate the page rank of the webpage instead of using in links used by traditional page-rank algorithm. Efficiency of WPRA is good as contrast to conventional page-rank algorithm.

Le et al. [5] developed method which used the TS-local rank. In this method rank has been calculated using the interconnectivity of the webpages. This method used a Meta search engine name of search engine is TS-Meta Search. Zhang et al. [6] used new-page-rank to accelerate the page rank score of webpages.

Yang et al. [7] proposed a modified page rank method to design the internal link by taking this access to the mining user's patterns and also again giving the rank to the webpages. It applied for small area for web search. Aung et al. [8] used refined page rank to solve the problem of differentiating the web links and the different issues of the user's access-routine. Qiao et al. [9] developed concept of Sim-Rank to refine the simple page-rank algorithm by assigning the probability of searching a page by giving initial value to page rank of each page. This approach is based on similarity measure of webpages and used it to divide the database of web into several social networks.

Dubey et al. [10] developed a modified-page-rank algorithm, which used the normalization-approach. Normalization approach calculated the mean-value to decrease the no. of iterations of page-rank algorithm. Kumar et al. [11] have developed page-rank algorithm used the approach of visit-of-links. Mostly page-rank algorithm is using the web-structure/Web-Content-Mining. In this algorithm, consider a factor by no. of visits to the in links of webpages by using the concept of larger the page-rank score increases the importance of the webpages. Yen et al. [12] have been proposed associated-page-rank algorithm that used relevance measure of webpages. Associated page-rank algorithm is used to increase the accuracy of the documents. This algorithm considered the efficiency and decreases the difficulty of topic-sensitive web-page-rank. Haveliwala et al. [13] developed Topic sensitive page-rank algorithm that can handle the problem of theme draft. Some pages are not considered relevant in other fields but those pages get a higher score in some field. Sharma et al. [14] proposed a weighted-page-content-rank-algorithm which used Web-Content and also the Web-Structure-Mining and also uses the weighted-page-rank algorithm. Amin et al. [15] have developed a score base page-rank algorithm. This algorithm involves web -content and Web-Usage-Mining. So it involves both syntactic and semantic approach. Syntactic score calculation is based on total no. of similarity of words matched in the webpages and semantic score calculation is based on the use of synonym words matched to the webpages.

Tuteja [16] has proposed weighted-page-rank algorithm that rely on visit-of-links. Finally, the addition of the both scores used to search the relevance of each webpage.

1.3 Page Rank

Page rank is a method of giving page rating for find out the importance of webpages. Initially, there existed Google page-rank algorithm after that so many modifications applied on conventional-page-rank algorithm like weighted page-rank algorithm, Hits algorithm etc. Page ranking can be computed by using in links, out links and topic sensitivity. There are some page-ranking-algorithms discussed as follows.

1.3.1 Page-Rank Algorithm

Larry Page and Sergly Brin have developed the algorithm for page rank. It is a part of research project began in 1995 and prompted a practical model in1998. Shortly, Page and Brin established Google.

Webpages are organized as a network. Each page is denoted as a node and each link is denoted as directed edge. The entire web defined as a directed graph. Page rank mainly represents the necessity of the webpages in a form of numeric value. One page connects to other page is defined as Page-A, vote for other page. Page-A connects from Page-A to Page-B that means vote on Page-A to Page-B. If Page-A is important for itself it means that the vote from Page-A to Page-B ought to carry a lot

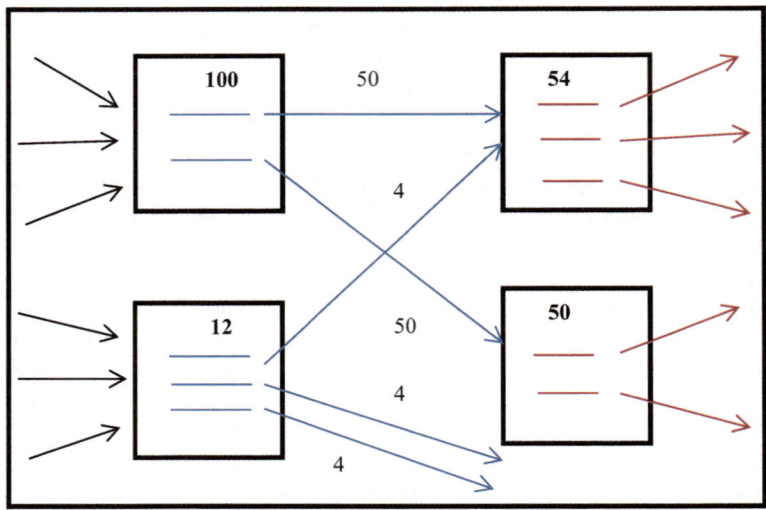

Fig. 1.4 Example of page-rank algorithm

of weight. So, webpage importance is defined as more votes means the page must be more important, here an example of Page-rank algorithm shown in Fig. 1.4.

Let page rank as "PR". So formula of page-rank algorithm is as follows:

$$PR(A) = (1 - d) + d\left(\frac{PR(T_1)}{C(T_1)} + \cdots + \frac{PR(T_n)}{C(T_n)}\right) \tag{1.1}$$

where, PR (A) defined as rank of Page-A. PR (Ti) defined as rank of Page Ti that connects to Page-A. C (Ti) defined as no. of out-links of Page-Ti. D defined as damping-factor the value should be defined in-between 0 and 1 (usually set to 0.85).

Dangling Links

Page with no outbound links is called dangling links. Until the value of page-rank has been computed, pages that have no outbound links removed from the database to prevent the page rank from negative effect of dangling links.

Search-engines used page-rank algorithm to display the webpage in order to given page-rank. Page-rank algorithm is mainly turn on link structure. Some disadvantages of page-rank algorithm are as follows:

- Ranking position of search engine rely on the current status of the website, while page rank is formulated by page-rank algorithm, which used the stored database, that modified rarely once in a duration of 2–3 months.
- Page rank is calculated as well as stored duration of indexing not at the duration of submission of query.

- Page-rank score of page-rank algorithm equally divide the values of page rank to all outbound link webpages but there are some out-links are not relevant to the query of the user.

To overcome these problems one additional weighted factor is introduced for page-rank calculation and new algorithm has come into the use known as weighted-page-rank algorithm.

1.3.2   Weighted Page-Rank Algorithm

WPR algorithm defined a large page-rank value as popular page in place of split the page-rank value equally among the out-links pages. Every out-link page receives the score proportional to the importance of page. Popularity of page can be defined as $w^{in}_{(v,u)}$ for no. of in-links and $w^{out}_{(v,u)}$ for no. of out-links. $w^{in}_{(v,u)}$ is defined as weight of link(v, u)

$$w^{in}_{(v,u)} = \frac{I_U}{\sum_{p \in R(v)} I_p} \tag{1.2}$$

where I_U defined as no. of in-links of Page-(u). I_p: defined as no. of in-links of Page-(p) and R (v) defined as reference-page-list of Page-(v).
   $w^{out}_{(v,u)}$ is defined as weight of link (v, u).

$$w^{out}_{(v,u)} \frac{O_U}{\sum_{p \in R(v)} O_p} \tag{1.3}$$

where, O_U: defined as no. of out-links of the page-(u). O_p: defined as no. of out-links of the page-(p) and R (v): defined as reference-page-list of page (v).
   So, the procedure of modified-page-rank is as follows:

$$PR(u) = (1-d) + \sum_{v \in B(u)} PR(v) w^{in}_{(v,u)} w^{out}_{(v,u)} \tag{1.4}$$

Weighted-page-rank algorithm resolves the ranking problem according to their relevance. But there still remains the problem of calculation of page-rank value at the time of indexing and also the issue of user's independence. To overcome these problems a new algorithm has been developed called HITS Algorithm.

1.3.3   HITS Algorithm

Jon Kleinberg has been introduced hyperlink-induced topic-search-algorithm. This algorithm removes the problem of evaluation of page rank at indexing time. In this algorithm calculate the page rank at user's query time. Hits algorithm is used for the link structure algorithm. Rank of webpages depends on two values Authority and Hub in place of depends on a single value. Authority assigns the pages having important information and Hub assign the webpage which works as resource lists. Mostly the webpages at the same time work as hubs or authorities. The HITS algorithm can work on two steps as follows.

1.3.3.1   Sampling Step

In the process of sampling collected the collection of relevant pages from the given user's query.

1.3.3.2   Iterative Step

In the process of iteration, find out the hub and authorities by using the result of sampling step.

The score of Hub (H_p) and the score of Authority (A_p) can be formulated as:

$$H_p = \sum_{q \in I_{(p)}} A_q \tag{1.5}$$

$$A_p = \sum_{q \in B_{(p)}} H_q \tag{1.6}$$

where H_p, A_p defined as the score of Hub and Authority, respectively. $I_{(p)}$: defined as of the reference-pages of page- p. $B_{(p)}$: defined as collection of the referrer-pages of page-p.

Hub-score of the webpage is equal to the summation of authority weight of all the pages which connects to the webpage. Authority value of the webpage is equivalent to the summation of the hub weights of all the webpages that connects to it.

HITS used the back links and the forward links. HITS also solve the problem of calculation of page rank at indexing time. But HITS algorithm faced the problems of topic drift, generation of automatic links, efficiency.

Table 1.1 Comparison of Page-Ranking Algorithms

Algorithm	Context Considered	Parameters	Relevance	Method
Page ranking	Web-structure-mining	Backlinks	Less	Computation of page rank at the time of indexing
Page ranking using Visit Of Links (VOL)	Web-usage-mining, Web-structure–mining	Visit of links, Back-links	Less	Define larger score to out-link
Weighted page rank	Web-structure-mining	Backlinks, Forward links	Less	Computation of page rank at the query time
Weighted-page rank (VOL)	Web-usage-mining, Web-structure–mining	Visit of links, Back-links	More	Define larger score to out-link
Page Content Ranking	Web-content-mining	Content	More	Compute new score of top n pages and return the relevant document to the users

1.4 Gap Analysis

From the comprehensive literature studied out for the various page-ranking algorithm of information retrieval, comparison has been carried out. Table 1.1 has shown the comparison of page-ranking-algorithm that is relying on some methods, parameters, relevance etc.

1.5 Conclusion

This paper concludes different concepts of web-mining, page-ranking and its different techniques then studied the conventional-page-rank algorithm; study the uses and limitation of it. To overcome the limitation of conventional-page-rank algorithm so many modified page-rank algorithms have been introduced like weighted page-rank algorithm, HITS algorithm etc. Page rank provides relevant results as per the user's necessity. Currently page-rank algorithm is used by many social networking algorithms like twitter to suggest follow to the user's account and so many fields.

References

1. Solanki, S., Verma, S., Chahar, K.: A comparative study of information retrieval using machine learning. Adv. Comput. Intell. Syst. 35–42 (2020)
2. Duhan, N., Sharma, A.K., Bhatia, K.K.: Page ranking algorithms: a survey. In: 2009 IEEE International Advance Computing Conference, pp. 1530–1537. IEEE (2009)
3. Kumar, A., Singh, R.K.: Web mining overview, techniques, tools and applica-tions: a survey. Int. Res. J. Eng. Technol. (IRJET) 3(12), 1543–1547 (2016)
4. Xing, W., Ghorbani, A.: Weighted pagerank algorithm. In: Proceedings Second Annual Conference on Communication Networks and Services Research, pp. 305–314. IEEE (2004)
5. Le, D.B., Prasad, S.: TS-LocalRank: A topic similarity local ranking algorithm for re-ranking web search results. In: 2009 International Conference on Advanced Technologies for Communications, pp. 197–200. IEEE (2009)
6. Zhang, L., Ma, F.: Accelerated ranking: a new method to improve web structure mining quality. J. Comput. Res. Dev. 41(1), 98–103 (2004)
7. Yang, Q., Zhang, H.H., Li, T.: Mining web logs for prediction models in WWW caching and prefetching. In: Proceedings of the Seventh ACM SIGKDD International Conference on Knowledge Discovery and Data Mining, pp. 473–478 (2001)
8. Aung, M.T., Tun, K.N.N.: To construct implicit link structure by using frequent sequence miner (fs-miner). In: 2009 International Conference on Computer Engineering and Technology, vol. 1, pp. 549–553. IEEE (2009)
9. Qiao, S., Li, T., Li, H., Zhu, Y., Peng, J., Qiu, J.: SimRank: a page rank approach based on similarity measure. In: 2010 IEEE International Conference on Intelligent Systems and Knowledge Engineering, pp. 390–395 (2010)
10. Dubey, H., Roy, B.N.: An improved page rank algorithm based on optimized normalization technique. Int. J. Comput. Sci. Inf. Technol. 2(5), 2183–2188 (2011)
11. Kumar, G., Duhan, N., Sharma, A.K.: Page ranking based on number of visits of links of web page. In: 2011 2nd International Conference on Computer and Communication Technology (ICCCT-2011), pp. 11–14. IEEE (2011)
12. Yen, C.-C. et al.: Pagerank algorithm improvement by page relevance measurement. In: IEEE International Conference on Fuzzy Systems, FUZZ-IEEE. IEEE (2009)
13. Haveliwala, T.H.: Topic-sensitive pagerank: a context-sensitive ranking algorithm for web search. IEEE Trans. Knowl. Data Eng. 15(4), 784–796 (2003)
14. Sharma, P., Bhadana, P., Tyagi, D.E.E.P.A.K.: Weighted page content rank for ordering web search result. Int. J. Eng. Sci. Technol. 2(12), 7301–7310 (2010)
15. Amin, M.S., Kabir, S., Kabir, R.: A score based web page ranking algorithm. Int. J. Comput. Appl. 110(12) (2015)
16. Tuteja, S.: Enhancement in weighted page rank algorithm using VOL. J. Comput. Eng. 14(5), 135–141 (2013)

Chapter 2
Live Emotion Verifier for Chat Applications Using Emotional Intelligence

Nirmalkumar Patel, Falguni Patel, and Santosh Kumar Bharti

Abstract Accuracy is essential in developing and researching innovative products. For example, in a live customer helpline chat, an honest dialogue between its users can be achieved by a live emotion analysis and verification to corroborate messages of both ends which ultimately nullifies the deceit made to complainers on the live chat. This emotion artificial intelligent verifier works on the concept of licensing or declining authenticity of message by comparing emotions found in text messaging and facial expressions. In this paper, we proposed an artificial intelligence-based live emotion verifier that acts as an honest arbitrator which first, recognizes live facial expressions under four labels namely, 'Happiness', 'Surprise', 'Sadness' and 'Hate' using Convolutional Neural Network (CNN) using a miniXception model. Simultaneously, it predicts the same labels of emotions in live text messages using these text classifiers—Support Vector Machine (SVM), Random Forest (RF), Naive bales (NB) and Logistic Regression (LR). We observed that among all four classifiers, SVM attained the highest accuracy for text prediction.

2.1 Introduction

Messaging is an intrinsic part of human interaction through which we transit our casuals, secrets and emotions. Today it mostly happens through human–computer interaction in the form of messaging applications. It is rolling in all genres like compliant filing, social chatting, interactive surfing, emergency guidance, counseling and so on. However, do messaging applications possess the 'observation' factor? That means, can these chat applications dissect true facial emotions held by their users through 'meticulous observations' of those on going chat conversations? Which is the only factor that differentiates humans from machines and at present, intellectuals are working on this direction of making machines more human-like prudent through machine learning. Currently, messaging apps does not sense the true intention of the

---

N. Patel · F. Patel (✉) · S. Kumar Bharti
Pandit Deendayal Energy University, Gandhinagar 382007, Gujarat, India

© The Author(s), under exclusive license to Springer Nature Singapore Pte Ltd. 2022
V. Bhateja et al. (eds.), *Evolution in Computational Intelligence*,
Smart Innovation, Systems and Technologies 267,
https://doi.org/10.1007/978-981-16-6616-2_2

11

user behind each message just by reading it. Therefore, comparison is required to be analyzed between emotions acknowledged in message contexts and facial expressions during live chatting and the emotion contradiction must be justified that would give true credentials of each message.

In this way, chat applications may work more like human and help receivers to identify fraud or deceit if done by senders. But we have found no research on simultaneous emotion tracking in the text and the face of the same user during text conversation to verify the assumed emotions and therefore, we have proposed and researched it in this paper to introduce the idea of finding conflict between predicted emotions in the research areas of emotion recognition and transparent communication.

2.1.1 Motivation

Faster computations and complex algorithms have made results more promising in the field of machine learning. However, in the real-world application, it is not necessary that we always have high-quality input. Therefore, in the case of honesty or transparency, we also need another way to confirm the predicted results (emotions). Emotion recognition may increase possibilities for Artificial Intelligence (AI) to correctly predict human motives respective to their apparent ones. However, Are the humans always predictable? and how can a modest AI predict the true intentions of deceitful ones particularly? Therefore, there is a requirement of automatic emotion verifier to detect human true intentions.

Emotion verification can be extremely helpful in for following applications:

– Social media apps—To find any fraud or evil intention of messenger
– Complaints handling chatbot—To find if the complaints are valid
– Companion chatbot—To find if users are actually needed any help

The rest of this article is propagated as follows: Sect. 2.2 exhibits literature survey. Preliminary is given in Sect. 2.3. The proposed architecture is detailed in Sect. 2.4. The experimental results are drawn in Sect. 2.5. Finally, Sect. 2.6 concludes the article.

2.2   Literature Survey

In 1967, Albert Mehrabian presented the '3V law' as 7–38–55% of the communications is verbal, vocal and visual respectively [1] which itself represents importance of non-verbal conversations (emotions namely) in human lives. In the present, most communications are done through social media like Twitter which is an exemplary source to study multiple emotions of its users through their short text messages. Even though multiple emotion classification of Twitter dataset has been already achieved with about 72% accuracy by Balabantaray et al. [2], it is still to be remained to classify through celebrated NLP classifiers like random forest, logistic regression and Naive Base which we have successfully conducted in this paper. Besides from traditional key-board feed texts, a survey [3] suggests that today most of emotion recognition is carried through multi-modal emotion recognition comprising different roots like audio, video, text et al. which further improves to recognize traditional basic emotions such as happiness, sadness, disgust, surprise, hatred or scare. Also there are researches [4–7] in the field of automatic/live facial emotion detection which we have studied as for emulating them in a specific part in this paper.

2.3 Preliminaries

In this section, we describe the existing tools and technology used in this work based on literature survey.

2.3.1 Live Facial Emotion Recognition

Dataset We have used FER—2013 dataset [8] which comprises of 35,888 images labeled in 7 emotions as 'Happiness', 'Disgust', 'Sadness', 'Angry', 'Surprise', 'Fear' and 'Neutral'. In this work, only four emotions are considered namely 'Happiness', 'Sadness', 'Angry', 'Surprise' where 'Angry' is revisited as 'Hate'. For live emotion capturing, images are taken from a live webcam while chatting.

Preprocessing In the preprocessing step, we convert all the images into 48*48 pixel grey scale. After conversion, we write it back into CSV file with two columns namely 'emotion' and 'pixel'. Each emotion is encoded with a numeric code from 0 to 6 representing predefined emotions in the dataset and each pixel comprises each numeric value separated by space in a single row. We followed a standard method to pre-process the dataset by scaling them in a range of $[-1, 1]$ as a superior range for neural system models. Again scaled to the range $[0, 1]$, divide by 255, minus 0.5 and add 2 to convert into the range $[-1, 1]$.

2.3.2 Text Emotion Prediction

Dataset For textual emotion analysis and prediction, we have used Twitter dataset [9] taken from Kaggle for text classification competitions. It includes 40,000 tweets labeled to thirteen different human emotions written in columns of tweet ID, the author, the text content of tweets and emotion depicted by tweet. For the live emotion extract, we have used an input message field for new text which is considered as a tweet similar to the Twitter dataset by our model.

Preprocessing Lemmatization is used for preprocessing of dataset. For better but not accurate corrections, we reverted repetition of letters in a word assuming that no word has more than twice consecutively repeating letters. By which we accumulated rare words having minor repetition which are mostly proper nouns and other insignificant words. Thus, we can easily remove them by following the idea that these words have very little role in deciding the sentiment of the text. Spelling mistakes are ignored due to avoid complexity in algorithms.

2.4 Proposed Work

The proposed live emotion verification is achieved in two fold namely Facial expression recognition and Text emotion prediction as shown in Fig. 2.1. Both the task performed simultaneously by getting two results and then these results are compared with each other to investigate contradiction between the emotion labels.

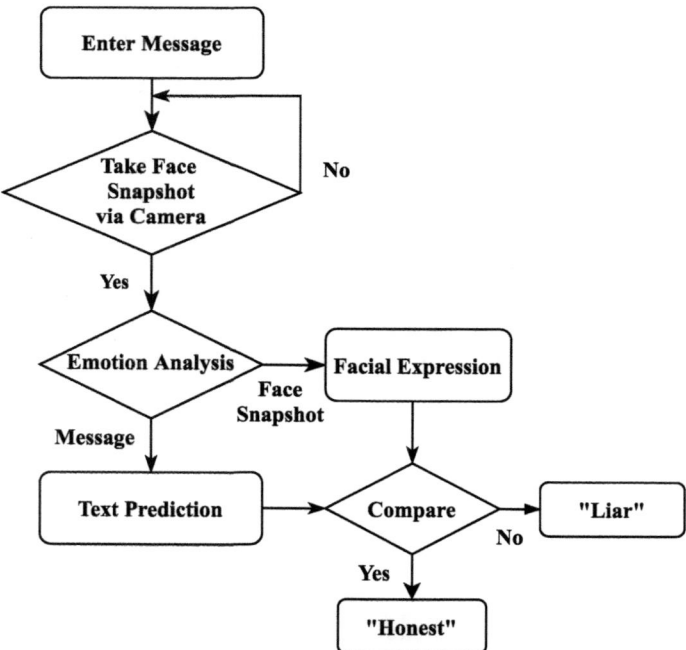

Fig. 2.1 Flowchart for live emotion verification

2.4.1 Live Face Emotion Detection

Live Face Emotion recognition algorithm works in two steps as shown in Fig. 2.2. In the first step it recognizes the face in the given image and marks the boundary. To recognize the face in given image, we train our model using following classifiers:

- Haar-feature based cascading classifier [10] : It captures front-side faces in a given image better than other available face detectors. It works in real-time, i.e., it can detect a face in a live camera also. It flags faces in images by bounding boxes on the face.
- miniXception CNN Model [11] : A pre-trained architecture that only takes input in pixel size of 48*48 and predicts seven labels of emotions. In proposed approach, we restricted to only four labels, i.e., Happiness, Sadness, Surprise and Hate in the output layer. We selected this architecture because it is an improvised form of the Xception CNN model [12] which gives the best accuracy while training it on the validation set. This CNN model promisingly learns the characterized features of emotions from the provided training dataset. Present-day CNN designs, for example, Xception influence from the blend of two of the best test predictions in CNNs: the utilization of residual modules and separable depth wise convolutions.

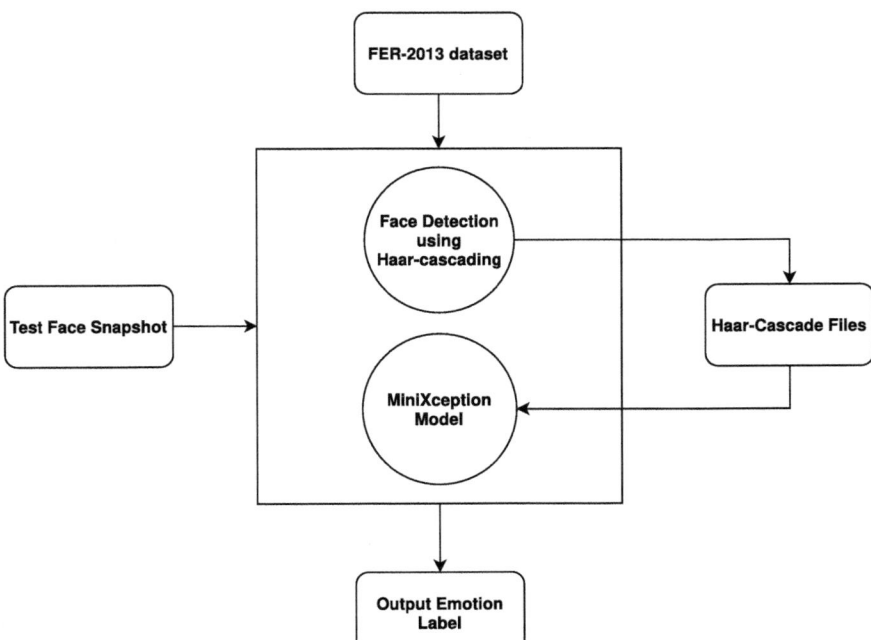

Fig. 2.2 Live face detection algorithm

In second step we detect the emotion of the bounded face through testing the trained model over live face snapshot using keras library and accordingly label those faces.

2.4.2   Text Emotion Prediction

Emotion prediction from chat message can also be achieved in two phase. Firstly, selecting best accurate model for any given text dataset. Secondly, testing live message on selected best prediction model.

Classifiers We have built a model comprising four different probability classifiers used for text emotion prediction widely used in natural language processing (NLP) such as Naive Bayes, Logistic Regression, Linear Support Vector Machine and Random Forest. Further, elaborating the text emotion prediction model step by step as follows:

– Feed preprocessed given dataset to training model comprising above 4 mentioned classifiers
– Best accurate classifier is selected for future text prediction by comparing accuracy of all classifiers.
– Now text entered by a user is taken as a test tweet for this model with the best accurate classifier and Twitter dataset for emotion prediction.
– After that, the trained model detects emotion in a text and for the entire chat conversation, emotions of entered texts are repeatedly predicted with each message.

Firstly in order to select best accurate model for any given test dataset we carefully considered above all classifiers in our model instead of picking one because text messages in real-world applications are complex in terms of grammar structure, slang, acronyms and abbreviations which can clearly be seen in the Twitter dataset also. This complexity can vary the result of each classifier for different dataset irrespective of its architecture due to that we can select the best promising classifier only after running a given training dataset On every one of them. In the case of a Twitter dataset, linear SVM has given the best accuracy so our text emotion prediction model will take linear SVM for all new messages entered by users. These all classifiers label text into four human emotions such as Happiness, Sadness, Surprise and Hate. Secondly we have used feature extraction in order to test a live message through the chosen model at the end of the above first step.

Feature Extraction Every word parameter from the data is numerically represented for the text classification process. We considered widely known two different features mainly TF-IDF and Count Vectors [13] to construct the required numeric input base for the above classifiers. The Term Frequency- Inverse Document Frequency (TF-IDF) features the importance of data or word to its documents and vice-versa in terms of frequency and rarity. The Count Vectors is a direct approach to analyze a particular sentiment for given data by counting total repetitions of each word representing each sentiment in a word vector array made of given data.

2.4.3 Comparing Both Analysis and Justify Emotion Contradiction

Both abstracted emotion labels of face and text are compared and if they contradict then output is 'Liar' otherwise 'Honest'. In equal case, i.e., message sent has honest or fraud account respective to its output. For example, if a complainer has sent 'sad' characteristic message but in reality, he is trying to deceive the receiver side which can only be judged by his 'happy' face then live emotion verifier also captures his face snapshot through camera and compares both emotions retrieved in face and text and alerts receiver by sending output result as 'Liar' with message sent by complainer.

2.5 Experimental Results

After conducting an experiment of total 12 repetitions/epochs comprising all 4 emotion cases of 3 persons, the accuracy of each probability classifier for Twitter dataset is shown in Table 2.1.

Thus the best model in the case of a Twitter dataset is linear Support Vector Machine (SVM) which has achieved up to 71% accuracy due to the nature of this specific dataset where an emotion of the text is heavily dependent on the presence of some significant adjective and this is proven that the accuracy obtained for predicting the emotions from face using miniXception model is 95.60% and corresponding precision and recall rates are 93 and 90% [14]. So final accuracy to correctly predict the labels—'honest' and 'liar' by comparing emotions verified in text prediction through SVM and facial expressions using miniXception is achieved up to 83.30 having precision and recall rates 82 and 80 respectively.

Table 2.1 Accuracy comparison of various existing ML techniques

Classification algorithm	Accuracy (%)
Support vector machine	71
Logistic regression	70
Naive Bayes	69
Random forest	70

2.6 Conclusion and Future Work

This paper has highlighted the importance of a real-time social lie detector especially for the professional world where everyone relies on trust for better judgment and future actions. In this work, we proposed an intelligent emotion verifier for the identification of fraud or evil intentions and licensing their validation. In order to identify emotion of users' chat text, we deployed four classifiers namely, Naive Bayes Classifier, Logistic Regression, Linear Support Vector Machine (SVM) and Random Forest. With the help of emotion from facial expression, it also identifies the real intention of the user for communication or conversation. We observe that the CNN architecture-based miniXception model for image detection was performed best. Chat applications train their model of each thread on the same face because every chat groups have defined users each time in conversation chat of particular thread so same user is predicted in model during a single chat session which can increase accuracy of predicting emotions of the same face with the number of messages entered and because of that our emotion verifier does not rely on the accuracy of those training dataset every time as we use them just to build up model for the first time only and after that it can predict promising result with increasing accuracy respect to increasing messages transferred in each chat.

In the future, we will increase accuracy for text and face emotion classification methods. Currently, we have focused more on facial expression and ignored the text complexity like spelling mistakes or acronyms. However, the longer conversations in a single chat will definitely improve output. We could also make it more user-friendly by expanding this model for multi-users in a single chat but for that models must be reinvented and tested for newly produced test cases of multi-users. Till now this model is very promising for two users lie detection through emotion prediction.

References

1. Mehrabian, A., Ferris, S.R.: Inference of attitudes from nonverbal communication in two channels. J. Consult. Psychol. 31(3), 248–252 (1967)
2. Balabantaray, R.C., Mohammad, M., Sharma, N.: Multi-class twitter emotion classification: a new approach. Int. J. Appl. Inf. Syst. 4(1), 48–53 (2012)
3. Marechal, C., Mikolajewski, D., Tyburek, K., Prokopowicz, P., Bougueroua, L., Ancourt, C., Wegrzyn-Wolska, K.: Survey on AI-Based Multimodal Methods for Emotion Detection, pp. 307–314. Springer, Berlin (2019)
4. Azcarate, A., Hageloh, F., Van de Sande, K., Valenti, R.: Automatic Facial Emotion Recognition, pp. 1–6. Universiteit van Amsterdam (2005)
5. Bindu, M.H., Gupta, P., Tiwary, U.S.: Cognitive model-based emotion recognition from facial expressions for live human computer interaction. In: 2007 IEEE Symposium on Computational Intelligence in Image and Signal Processing, pp. 351–356. IEEE (2007)
6. Duncan, D., Shine, G., English, C.: Facial Emotion Recognition in Real Time, pp. 1–7. Stanford University (2016)
7. Dagar, D., Hudait, A., Tripathy, H.K., Das, M.N.: Automatic emotion detection model from facial expression. In: 2016 International Conference on Advanced Communication Control and Computing Technologies (ICACCCT), pp. 77–85. IEEE (2016)

8. Goodfellow, I.J., Erhan, D., Carrier, P.L., Courville, A., Mirza, M., Hamner, B., Cukierski, W., Tang, Y., Thaler, D., Lee, D.H., Zhou, Y.: Challenges in representation learning: a report on three machine learning contests. In: International Conference on Neural Information Processing, pp. 117–124. Springer, Berlin (2013)

9. Bouazizi, M., Ohtsuki, T.: Sentiment analysis: from binary to multi-class classification: a pattern-based approach for multi-class sentiment analysis in Twitter. In: 2016 IEEE International Conference on Communications (ICC), pp. 1–6. IEEE (2016)

10. Viola, P., Jones, M.: Robust real-time face detection. In: Null, pp. 747–751. IEEE (2001)

11. Arriaga, O., Valdenegro-Toro, M., Plöger, P.: Real-time convolutional neural networks for emotion and gender classification (2017). arXiv:1710.07557

12. Chollet, F.: Xception: Deep learning with depthwise separable convolutions. In: Proceedings of the IEEE Conference on Computer Vision and Pattern Recognition, pp. 1251–1258 (2017)

13. Manning, C.D., Raghavan, P., Schütze, H.: Scoring, term weighting and the vector space model. Introd. Inf. Retr. **100**, 2–4 (2008)

14. Fatima, S.A., Kumar, A., Raoof, S.S.: Real time emotion detection of humans using Mini-Xception algorithm. In: IOP Conference Series: Materials Science and Engineering, vol. 1042, No. 1, p. 012027. IOP Publishing (2021)

Chapter 3
Text to Speech Conversion of Handwritten Kannada Words Using Various Machine Learning Models

Chandravva Hebbi, J. S. Sooraj, and H. R. Mamatha

Abstract Recognition of handwritten characters and words is challenging due to the presence of complex character sets and the complexity of the words. The machine learning models with feature extraction methods will help us to solve the problem of recognizing handwritten words. The various preprocessing techniques applied to the word are Bilateral filters, resizing the images to find the Region of Interest (ROI) by contour detection and cropping the images. After resizing the image, it is further deskewed for better results. The recognition of handwritten Kannada words by extracting histogram of oriented gradients (HOG) features from the word image using various Machine Learning (ML) techniques are presented in this paper. Then the recognized word is converted to speech using the Google Text-to-Speech (gTTS) API. The dataset consists of 54,742 handwritten word images. Various machine learning models like Support Vector Machine (SVM), k-nearest neighbors (KNN), and random forest were applied to the dataset. Average accuracy of 88% is obtained using the SVM classifier with Radial Basis Function (RBF) kernel.

3.1 Introduction

Character or word recognition has been a topic of immense interest in the field of image processing for the last two decades. Based on the application and the type of input required for the system, these recognition systems are classified into two categories namely, offline word recognition systems where the input is readily available to us (it can be an image, document, a video containing the word to be recognized) and in online word recognition system, the input is taken at real-time i.e., the input is read as the word is being written. The online recognition system works on the temporal information of the word that is available to it. The offline system on

C. Hebbi · J. S. Sooraj (✉) · H. R. Mamatha
Department of Computer Science and Engineering, PES University, Bengaluru, Karnataka 85, India

H. R. Mamatha
e-mail: mamathahr@pes.edu

© The Author(s), under exclusive license to Springer Nature Singapore Pte Ltd. 2022   21
V. Bhateja et al. (eds.), *Evolution in Computational Intelligence*,
Smart Innovation, Systems and Technologies 267,
https://doi.org/10.1007/978-981-16-6616-2_3

the other hand banks on information that can be interpreted from the text itself, like the input image pixel intensity values. In the case of offline recognition systems, the word to be recognized can be printed or handwritten. While both handwritten and printed documents share common strategies to recognize the word, but handwritten word recognition systems are harder to solve as handwritings tend to differ and noise reduction becomes harder to resolve due to the presence of occlusion.

Primarily there are two approaches to recognize offline words. In the first approach, character segmentation is done and each character is recognised individually. In the second approach, the entire word is treated as one entity to recognize it. While splitting the word may provide flexibility to the program, for applications that require the meaning of the word to stay intact, such as a dictionary, it is better to treat the word as a whole for recognition as we do away with extra computation such as segmentation. The shear number of languages present in the world today makes word recognition a daunting task as each individual language possesses a unique hurdle to overcome.

A lot of work has been reported in building an OCR system for languages such as English, French, Chinese, etc. There are also a few works in Indian languages such as Hindi, Bangla, Kannada, Tamil, and Telugu. When we considered OCR works for Kannada scripts, we found that majority of them focused on character recognition. The works that did exist on Kannada word recognition were few and most of them considered smaller datasets. Our work focuses on recognizing handwritten Kannada words with a dataset consisting of 210 different words collected from different people and providing an audio output for the same. The use cases or the applications of the work are word spotting from the document images, recognition of words, and converting the recognized words to machine editable form, or the output of the model can be a voice. This can act as a teaching aid for the parents to make their children learn about the name of the places, animals, plants, birds, and things.

3.1.1 Kannada Language

Kannada is the official language of Karnataka which is derived from Kadamba script. The character set consists of 13 vowels and 34 consonants, some of the characters in the set are similar in structure. Due to the similarity in the structure and complexity of the characters, recognition of the characters is challenging. Kannada letters are a combination of a Kannada consonant (C) and vowel (V) and so the number of possible combinations would be 34 x 13. The possible combinations will be $34 \times 34 \times 13$. These combinations of characters are called complex conjugates (Vattaksharagalu). When the combination of CCV is considered the complexity of words increases due to varying writing styles, the location of the characters in the words. Sample examples of the words are shown in Fig. 3.1.

The words in Fig. 3.1a, are a combination of consonants and vowels and the words in Fig. 3.1b are the combination of consonants and vowels. The words that are considered in this work are combinations of consonants and vowels. The rest of the paper

(a)   (b)

Fig. 3.1 Sample handwritten words

is organized into a literature review that presents the existing work, a methodology which is followed by results and discussions, and finally the conclusions.

3.2   Related Work

The works in the field of building OCR for foreign languages and also for some of the Indian languages are enormous. But the works in the field of word recognition for English, Arabic are more compared to the works in the recognition of handwritten words for Indian languages mainly the South Indian languages like Kannada, Tamil, Telugu, and Malayalam. The work carried out in the field of Handwritten word recognition for English, Arabic, and Indic scripts are presented below.

Recognition of printed and handwritten words is discussed in [1]. Six-features of the words namely mean, standard deviation pixel intensity value, Otsu's threshold value, number of local maxima, percentage of pixels belonging to upper quarter, percentage of pixels belonging to the lower quarter of the pixel intensities were extracted. The extracted features were sorted and a decision tree was constructed using these values to classify the words. Dataset consists of 1000 handwritten and printed words each. An accuracy of 96.80% was reported.

Recognition of online handwritten Bangla words using the different feature vectors and SVM classifiers with Radial Basis Function (RBF) kernel is presented in [2]. Dataset consists of 27,798 Bangla words written by 298 people. For lexicons of sizes 50 and 110, an accuracy of 89.92% and 87.73% respectively were claimed.

In [3] the database building for middle eastern languages Arabic, Dari, Farsi, Pashto, and Urdu has been presented. The data is written by 1600 people from 4 countries and the words are related to finance. The gradient features of the word are extracted and given to the classifiers SVM, Modified Quadratic Discriminant Function (MQDF), and Multi-Layer Perceptron (MLP). The authors claim a very good accuracy.

Recognition of Devanagari handwritten words is presented in [4]. Word is segmented into matras and consonants. A headline removal algorithm was used to extract individual characters. The middle regions are extracted using the chain code. For each of the zones, different feature extraction methods and classifiers were used. The Multi-Layer Perceptron (MLP) classifier is used for the recognition. The authors have reported an accuracy of 93.4% and 93.5% for the words with two characters and six characters respectively with the dataset of 3600 words.

In [5] recognition of handwritten English words with and without segmentation of the words is discussed. The authors use the markov random field (MRF) model and Pseudo-2D Bi-Moment Normalization (P2DBMN) models as their classifiers. Recognition of handwritten Arabic words using structural features like loops, ascenders, descenders, and diacritic is discussed.

In [6] Variations of Naive Bayes classifier is used to recognize the handwritten words present in the IFN-ENIT dataset. The dataset consists of Tunisian city names. Authors have claimed accuracy of 90.02% with the Horizontal and Vertical Hidden Markov Model (VH-HMM). The dataset consists of 7881 words.

The focus of work in [7] is to recognize Kagunita in Kannada scripts and classify them accurately. The image document was segmented to extract each word and in turn, each word is segmented to extract characters. Different feature extraction methods like Fourier–Wavelet coefficients, Zonal features, Zernike moments have been used. The recognition process was carried out using Back Propagation Neural Networks (BPNN). Authors have claimed accuracy of 85–90% for vowels and 80–85% for consonants.

The objective of research work in [8] was to identify the best feature extraction method and classifiers for Kannada word recognition. The image was enhanced for easier feature extraction. The features considered were the Gray Level Co-occurrence Matrix (GLCM), Shape-based, and graph-based features. The classifiers considered were SVM and K-means. The average recognition rate of the SVM classifier is better than K-means. The reported recognition rate for the SVM classifier was 88.96%.

Different preprocessing techniques like noise reduction, binarization, greyscale conversion, Gaussian filtering have been discussed in [9]. Geometric, regional, gradient features are used for feature extraction for the handwritten Devanagari words. Neural Networks is used as a classifier, where a single hidden layer and 10 neurons gave the most required output. The overall accuracy reported is 94% using Neural Networks. The objective of the work in [10] was to recognize offline Kannada words. Data is collected from 60 different writers. The number of words in the dataset was 1200 words. Binarization, median filtering, skew detection, thinning, edge-detection were used for preprocessing. FCC (Freeman's Chain Code) is one of the techniques for pattern recognition and shape analysis. Minimum distance classifiers like Manhattan's, Euclidean distance are used for classification. An accuracy of 92% was claimed.

In [11], the feature analysis methods for handwritten Kannada Kagunitha were discussed. For classification, Neural Network, Multi-Layer Perceptron (MLP) with backpropagation is used. Here two new ways of recognition are carried out: one is the concept of the cutting image in which vowels and consonants are recognized separately. The other one is merging moments and Gabor transforms, and statistical features for the recognition. Handwritten word recognition using MLP based Classifier with a holistic approach was discussed in [12]. The MLP based classifier is built to reach the prediction of precision/accuracy. Dataset consists of 50 handwritten document images written in Bangla mixed with few English words. Binarization of the dataset and then features are computed from the image. The features are extracted and fed to MLP classifier for training and testing.

In [13], offline handwritten Kannada text recognition using Support Vector Machine, and Zernike Moments for feature extraction were described. If the input data has some skewness in it, then the correction using Wigner-Ville distribution of horizontal projections was used in the preprocessing step, Projection profiles were used for segmentation. SVM with the Kernel trick was useful for pattern classification and recognition. The recognition rate reported was nearly 94%. In [14] the methods for word spotting and recognition using various deep learning models are presented. The feature extraction methods used are Local Binary Pattern (LBP), HOG, and Gabor. An accuracy of 82% is claimed by the authors with a dataset of 11,000 words written by 100 people using the CNN model with spatial transformation.

From the literature survey, it becomes evident that a lot of work has been done for English, Arabic and few works have been reported on Devanagari and Bangla word recognition. Very few works have been reported in the recognition of Kannada words. Some of the works have been presented in the recognition of handwritten words for other languages using the standard datasets like IAM-onDB, IFN-ENIT. The work in the field of handwritten Kannada word recognition is minimal, the size of the dataset used by the author is smaller and the dataset is not available publicly. Hence there is a need to build a recognition system for handwritten words.

3.3 Proposed Methodology

3.3.1 Overview

The various steps involved in the proposed methodology are preprocessing, feature extraction, training, and testing the model.

Various models are used to get better results by reducing the overfitting and time taken to predict words using dimensionality reduction techniques like Principal Component Analysis (PCA) and Latent Dirichlet allocation (LDA). The predicted word is converted to audio by using gTT's API. Figure 3.2 shows the proposed system architecture.

3.3.2 Dataset

The dataset consists of 54,742-word images. Deskewing is applied to reduce the skew present in the image. Figure 3.3 shows the effect of the deskewing the word.

Deskewing is followed by resizing all the images to 100×200 pixels. Resizing the images is necessary as HOG features will be used for word recognition. If the images are left to be of variable size, the HOG features extracted will be of variable length, making comparisons between two sets of features to be difficult. Figure 3.4. provides a sample of the dataset.

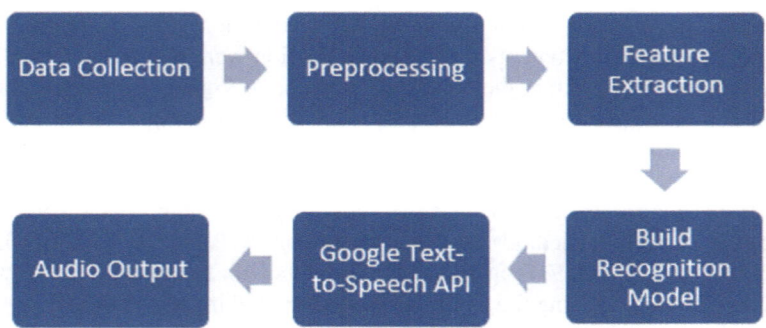

Fig. 3.2 Proposed methodology

Fig. 3.3 Image deskewed by 12 degrees

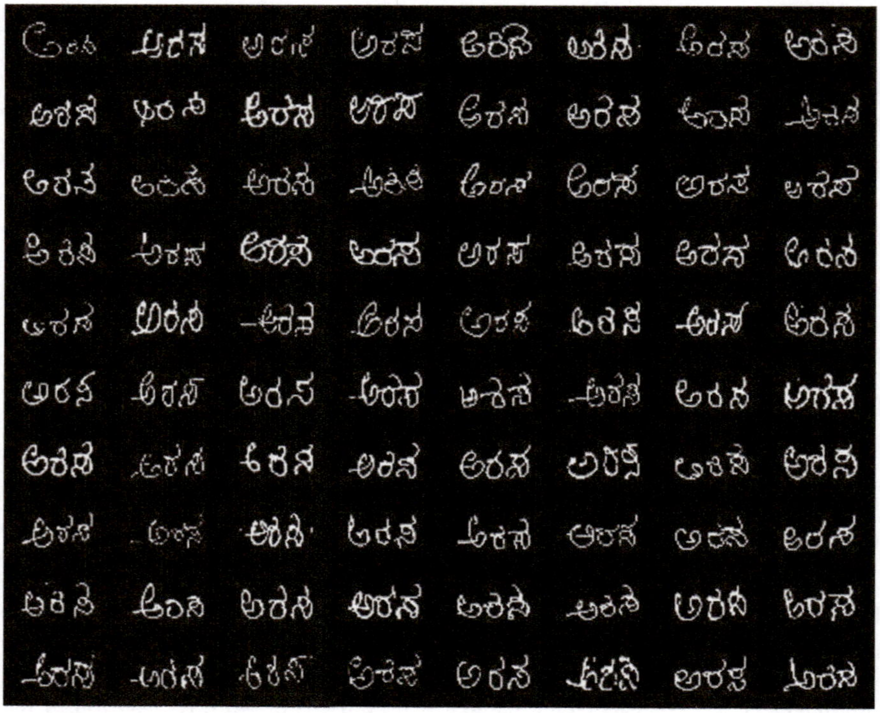

Fig. 3.4 Sample dataset

3.3.3   Preprocessing

The Pre-processing step is started with converting a color image to a grayscale image. Then an adaptive thresholding method is applied to convert the image to a binarized image as shown in Fig. 3.5.

After binarization, the Gaussian noise present in the image is removed using a bilateral filter. Bilateral filters help in smoothing and removal of noise while preserving the edges of an image. Morphological operations such as opening and closing are applied to enhance the images. Figure 3.6 shows the output image after noise removal.

In Fig. 3.6 (right) still, there is a noise in the edges of the image. Canny edge detection method with appropriate threshold value is used to remove the noise present at the edges. This was done to retain only the word and eliminate the noise that the morphological operations failed to remove. After the edge detection, the image is cropped. Figure 3.7 shows the preprocessed word after applying various preprocessing techniques like binarization noise removal, edge detection, and deskewing. These preprocessing steps are applied only to the new word image input by the user and not to the dataset.

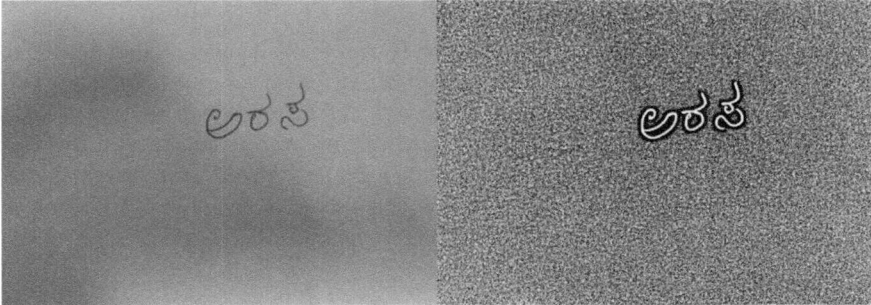

Fig. 3.5 Binarization of image

Fig. 3.6 Filtering to reduce noise

Fig. 3.7 Final processed
image

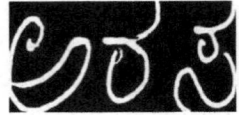

3.3.4   Feature Extraction

Images are represented as a 2D array with each value representing the intensities of
that pixel. Since grayscale images are considered, there is only one channel present,
and the intensity values in the array range from 0 to 255. Since the machine learning
model used was SVM, it is not feasible to directly input these raw pixel intensities
for fitting the model. Therefore, it is necessary to extract valuable information from
the input image using feature extraction techniques. In this work, the HOG feature
extraction method is being considered. The HOG features count the occurrence of
gradient orientation of localized parts of the image. This ensures that the direction
of edges present in the image is retained.

The method is broken down into three steps, the **first step** begins by dividing the
image into blocks, and further dividing the blocks into cells which is a grid of pixels.
Next, the gradients of each pixel in the cell in the x and y-direction are calculated
using the formulae (3.1) and (3.2).

The **second step** is to build the histogram (rather than a 1−d array that represents
the histogram) out of gradients and orientations calculated for each pixel in the cell.
We take the orientation (direction) as the x-axis and the total gradient as the y-axis.
We will consider a bin size of 20 degrees and therefore, we will end up with $180°/20°$
= 9 bins in our histogram i.e., we can classify each pixel in the cell to belong to one
of these 9 bins based on its orientation. The magnitude of each bin will depend on
the total gradient of all the pixels that belong to the bin.

$$\text{Slope in } x = S_x = P(x+1, y) - P(x-1, y) \qquad (3.1)$$

$$\text{Slope in } y = S_y = P(x, y+1) - P(x, y-1) \qquad (3.2)$$

$$\text{Gradient} = \sqrt{(S_x)^2 + (S_y)^2} \qquad (3.3)$$

$$\text{Orientation} = \arctan(S_y/S_x) \qquad (3.4)$$

For the case presented in Fig. 3.8, let's consider the pixel P (x, y),

Using the formulae (3.1) (3.2) (3.3), we can calculate the total gradient and
orientation,

$$\text{Gradient} = \sqrt{8} = 2.828$$

For example,

	P (x, y-1) =4	
P (x-1, y) =4	P (x, y) =3	P (x+1, y) =2
	P (x, y+1) =2	

Fig. 3.8 Representation of cell from the image where P (x, y) represents the pixel value at (x, y)

$$\text{Orientation} = \arctan{(2/2)} = 45 \text{ degrees}$$

Now if we consider the bin size to be 20 degrees, then the 9 bins will be 0, 20, 40, 60, 80, 100, 120, 140, 160.

As 45 lies between 40 and 60, its gradient will be split and assigned to both 40 and 60 bins. This is done in such a way that the bin closer to the orientation receives a higher fraction of the gradient. In our case,

$$\text{Bin 40 is assigned } (60 - 45)/20 * 2.828 = 2.121$$

$$\text{Bin 60 is assigned } (45 - 40)/20 * 2.828 = 0.707$$

By repeating the above steps for all the pixels in a cell, we can reduce a $2-$d cell into a $1-$d array of size 9.

The **third step** is to normalize the gradients. The gradients are sensitive to lighting conditions and therefore normalizing the gradients becomes a necessary step. Each block consisting of m-rows and n-columns i.e., mxn cells are considered for normalization. In this case, 14×14 cells have been considered for block normalization. This would give the final HOG features for building our model. Figure 3.9 shows a visualization of the HOG features.

The number of HOG features extracted is significantly more. Training and testing models will take more time as the number of features extracted are more. To avoid this, dimensionality reduction techniques are used to scale down the number of features. PCA and LDA dimensionality reduction methods have been used. HOG features with PCA gave better results than HOG features with LDA.

Fig. 3.9 HOG features
visualization

3.3.5 Recognition Model

To know the best classifier for the dataset, different classification models like Random Forest, Support vector machines, and K-Nearest Neighbors were tested and hyperparameters were tuned using cross-validation techniques. Variations in SVM such as Linear SVC and RBF-SVM have been used. Weighted KNN's were also built to improve the accuracy of KNN. The best results were obtained when HOG-PCA features with the RBF-SVM classifier. An accuracy of 88.7% is obtained with the method. These models were then stored in memory using object serialization which was carried out by a pickle module in python.

3.3.6 The Output of the Model

Each word in the dataset was first converted to its equivalent Unicode using a English to Kannada Unicode converter. This converter takes the transliteration of the Kannada word in English as input and provides the corresponding Unicode value.

The transliteration of ಅರಸis 'arasa'. The word in English is converted to its corresponding Unicode. The Unicode representation of each character is a unique set of digits with a prefix of '&#' and the last character is a semicolon indicating the end of the sequence. In the case of the consonants, if the consonant is not modified, the consonant's Unicode is appended with '್' to indicate the absence of a modifier. If the consonant is modified by any other consonant or vowel, '್' is replaced with the modifier's Unicode value.

The Unicode forming the entire word is found by concatenating the Unicode of all the letters in the word.

Unicode conversion of the word 'arasa'.

ಅ(a): ಅ   ರ(r): ರ್   ಸ(s): ಸ್

The characters (r) and (s) end with ್ to indicate the absence of a modifier.

Table 3.1. ಅರಸinto the corresponding letters,

The characters 'ra' and 'sa' are being affected by the modifier, ್ is being replaced with the Unicode for 'a'.

These Unicode values are input to Google's gTTS API to convert to the audio output. These output files are stored in the local system. Each time a prediction is made, this pre-generated audio file is played as an output.

Table 3.1 Breakdown for the word arasa

ಅರಸ (arasa)	ಅ (a)	ರ ಅ (ra)	ಸಅ (sa)
ಅರಅಸಅ	&#3205;	ರಅ	ಸಅ

3.4   Results and Discussions

The work was conducted using a dataset of 54,742-word images consisting of 210 different classes. Stratified sampling was used to select the data to train the model to include each class in almost equal proportion.

Some words were very well classified and gave an accuracy of 100%. There were words like 'hogu' whose accuracy was 56% because the writing styles in the dataset varied drastically. The final accuracy was obtained with the SVM using the RBF kernel. There was not much increase in the accuracy when regularization parameter C of the SVM method was increased. The parameter C controls the tradeoff between smooth decision boundary and classifying the training point correctly. Hence, C was set to 1 to avoid overfitting of the data. The average accuracy of the proposed system is 88%. Figure 3.10 shows the results obtained from various classification models.

The accuracy of the model is decreased by 1% when the dimensions of the features were reduced using LDA and PCA as the small amount of information was lost as the last n−k columns are rejected. But there was a significant improvement in the time required for the recognition. Table 3.2 gives the performance effect of PCA and LDA on RBF-SVM.

Hence it was necessary to reduce some of the features by using the dimensionality reduction methods to improve the performance of the proposed method. When the dimension of the features was reduced there was a drop in the performance by 1% with PCA-HOG and 4% with LDA-HOG. But there was a huge reduction in the time.

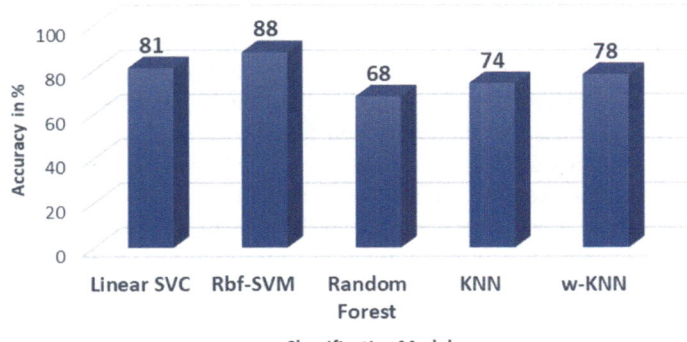

Fig. 3.10   Comparative study of classification models

Table 3.2   Performance improvements gained by using PCS and LDA

Feature extraction (for SVM model)	Accuracy (%)	The time before (s)	Time after (s)
PCA-HOG	87	10,852.44	3463.50
LDA-HOG	84	10,852.44	6237.62

Table 3.3 A comparative study of Handwritten word recognition methods

Authors	Feature extraction method	Classifier	Dataset size	Accuracy
Patel et al. [10]	Freeman's Chain Code (FCC)	Euclidean distance classifier, dynamic time wrapping	1200	92%
Patel et al. [8]	Locality Preserving Projections (LPP)	Support Vector Machine (SVM)	600	80%
Tulika Sureka et al. [14]	KAZE, HoG, LBP, gabor wavelets	Convolutional Neural Network (CNN)	11,000	82%
Proposed method	HOG	SVM-RBF	54,742	88%

From Table 3.3, the proposed method performed better than the existing methods presented by the other authors. In [8] and in the proposed method, the SVM classifier used is the same, but feature extraction methods are different, the dataset, size of the dataset are different and the dataset is versatile as it is collected from a larger number of people. The overall performance of the system is better than the existing methods. The dataset used by each of the authors is different.

3.5   Conclusion

The objective of this work is to recognize the 210 different handwritten Kannada words collected from 370 people and provide the audio output of the recognized words. Some of the machine learning models like Random Forest, SVM, KNN, and weighted KNN were used as classifiers, and features were obtained from the HOG feature extraction method. The average accuracy of 88% was obtained using an SVM classifier with an RBF kernel. The proposed model helps people to search the data online by giving their input in handwritten word image form. The future enhancement of the given work would be to provide additional information regarding the word. And also, adding additional words to the dataset to make it more versatile and can be used as a Kannada dictionary that could help teach meaning along with pronunciation.

References

1. Malakara, S., Dasa, R.K., Sarkarb, R., Basub, S., Nasipurib, M.: Handwritten and printed word identification using gray-scale feature vector and decision tree classifier. Procedia Technol. **10**, 831–839 (2013). https://doi.org/10.1016/j.protcy.2013.12.428
2. Srimany, A., Chowdhuri, S.D., Bhattacharya, U., Parui, S.K.: Holistic recognition of online handwritten words based on an ensemble of svm classifiers. In: 11th IAPR International

Workshop on Document Analysis Systems, pp. 86–90 (2014). https://doi.org/10.1109/DAS.2014.67

3. Nobile, N., Khayyat, M., Lam, L., Suen, C.Y.: Novel handwritten words and documents databases of five middle eastern languages. In: 2014 14th International Conference on Frontiers in Handwriting Recognition, pp. 152–157 (2014)

4. Kumar, S.: A study for handwritten devanagari word recognition. In: 2016 International Conference on Communication and Signal Processing, pp. 1009–1014. (2016). https://doi.org/10.1109/ICCSP.2016.7754301.

5. Zhu, B., Shivram, A., Govindaraju, V., Nakagawa, M.: Online handwritten cursive word recognition by combining segmentation-free and segmentation-based methods. In: 15th International Conference on Frontiers in Handwriting Recognition, pp. 416–422 (2016)

6. Khemiri, A., Echi, A.K., Bela˙ıd, A., Elloumi, M.: A system for off-line arabic handwritten word recognition based on bayesian approach. In: 15th International Conference on Frontiers in Handwriting Recognition 2016, pp. 560–565 (2016). https://doi.org/10.1109/ICFHR.2016.0108.

7. Aravinda, C.V, Prakash, H.N., Lavanya, S.: Kannada handwritten character recognition using multi feature extraction techniques. Int. J. Sci. Res. (IJSR) 3(10), 911–916 (2014)

8. Patel, M.S., Kumar, R.: Offline kannada handwritten word recognition using support vector machines (SVM). Int. J. Comput. Sci. Inf. Technol. Res. 936–942 (2015)

9. Ansari, S., Bhavani, S., Sutar, U.S.: Devanagari handwritten word recognition using efficient and fast feed forward neural network classifier. Int. J. Adv. Res. (IJAR) 2034–2042 (2017)

10. Sandyal, K.S., Patel, M.S.: Offline handwritten kannada word recognition. In: Proceedings of 07th IRF International Conference, pp. 19–22 (2014)

11. Ragha, L.R., Sasikumar, M.: Feature analysis for handwritten kannada kagunita recognition. Int. J. Comput. Theory Eng. 3(1), 94–102 (2011)

12. Acharya, A., Rakshit, S., Sarkar, R., Basu, S., Nasipuri, M.: Handwritten word recognition using MLP based classifier: a holistic approach. IJCSI Int. J. Comput. Sci. 422–526 (2013)

13. Thungamani, M., Ramakhanth Kumar, P., Prasanna, K., Rau, S.K.: Offline handwritten kannada text recognition using support vector machine using zernike moments. IJCSNS Int. J. Comput. Sci. Netw. Sec. 128–134 (2011)

14. Tulika Sureka, K.S.N., Swetha, I.A., Mamatha, H.R.: Word recognition techniques for kannada handwritten documents. In: 2019 10th International Conference on Computing, Communication and Networking Technologies (ICCCNT), pp. 1–7. Kanpur, India (2019)

Chapter 4
An Improved Approach for Automated Essay Scoring with LSTM and Word Embedding

Dadi Ramesh and Suresh Kumar Sanampudi

Abstract Automatic essay scoring has been shown to be an effective mechanism for quickly assessing student responses in the education system. It has already a wide variety of applications to solve, but there are evaluating the essays based on statistical features like Bag of Words (BoG), Term Frequency-Inverse Document Frequency (TF-IDF). Some of the evaluating approaches are considering the features like Word embedding with Glove, Word2Vec, One hot encoding. Both types of approaches are not fulfilling essay evaluation and not able to retrieve semantic information from essays. Here we are evaluating the essay with Word2Vec and Long Short-Term Memory (LSTM) with K-Fold cross-validation and we got an accuracy of 85.35.

4.1 Introduction

A manually scoring student response like essays and long answers is a daunting task for evaluators in the education system. In extensive exams where the number of students is more than thousands, manual scoring requires a lot of time and no reliability guarantee. But automated essay scoring (AES) system evolved to assess student responses on a large scale based. Thereby computers reduced the human effort in the assessment process. Advancement in artificial intelligence and natural language processing has improved the progress in AES. So many researches concentrated on AES to improve the accuracy and reliability of automation systems.

However, in earlier AES was implemented on statistical features like a number of words, a number of sentences, length of sentences and the average length of sentences, etc., to prepare vector and trained machine learning model, later researchers tried to

D. Ramesh (✉)
Research Scholar in JNTU Hyderabad, School of Computer Science and Artificial Intelligence, SR Univrsity, Warangal, India

S. K. Sanampudi
JNTUH College of Engineering Jagitial, Nachupally Jagtial dist, Telangana, India
e-mail: sureshsanampudi@jntuh.ac.in

© The Author(s), under exclusive license to Springer Nature Singapore Pte Ltd. 2022   35
V. Bhateja et al. (eds.), *Evolution in Computational Intelligence*,
Smart Innovation, Systems and Technologies 267,
https://doi.org/10.1007/978-981-16-6616-2_4

retrieve style-based features like parts of speech, grammar, etc. For vector preparation. From the last decade, 2010–2020, researchers focused on content-based features like coherence, cohesion, and completeness to assess the student responses. These features generally say student writing genre and explanation of content to the prompt.

There are two challenges in AES majorly: the first is preparing the feature vector from the essay set, second is training the feature vector with a proper machine learning model. Preparing feature vectors from essay sets, including semantic information, is a significant challenge in AES [1]. Many researchers have used many types of text embedding (vector) [2, 3] methods, but no one fulfilled the essay's semantic vector. The embedding techniques are numbers from the essay, Bag of Words [4], word frequency vector like Tf-Idf [4], Word2vec [5], and GloVec. But no technique is preparing content-based feature vectors.

In machine learning models [6–8], the researchers have used the regression model, support vector machine, and random forest, but these models are not sequenced to sequence learning and don't find the essay's coherence. The deep learning models convolutional neural network (CNN), recurrent neural network (RNN) in that CNN models also do not learn sequence to sequence but are used for N-gram feature extraction. RNN models are sequence to sequence learning models that will assess the student response.

The rest of the paper is organized as follows: In Sect. 4.2, we explained the related work, and in Sect. 4.3, we described our approach for AES, we explained results and analysis in Sect. 4.4, and the conclusion in Sect. 4.5.

4.2 Related Work

The first AES system developed by Ajay et al. [9], the project essay grading extracts the features like character count, word count, and sentence length to grade essays. Foltz et al. [10] introduced an Intelligent Essay Assessor (IEA) by evaluating content using latent semantic analysis to produce an overall score, but these systems failed to retrieve the content-based features from essays the models trained on statistical features.

Dasgupta et al. [11] implemented a sequence to sequence learning model to train feature vectors. They used the glove to prepare feature vectors, and the vectors are transferred to the CNN layer to retrieve local features. After CNN, they stacked the RNN layer for the actual sequence to sequence training they embedded an activation layer to predict the score of the essay.

Wang et al. [12] implemented bi-LSTM to train feature vectors and prepared a feature by word-level encoding with Word2vec [5] library, which works on one-hot encoding to prepare word vectors.

Kumar et al. [13] implemented an autosas for short answer scoring. They stacked CNN and RNN layers in autosas. Retrieved various features like context, POS, prompt overlap, etc.; based on these features, they classified an essay whether to recommend or not. They assigned a score for individual features with that, and they classified the essay.

Liu et al. [14] developed two stage learning frameworks for essay scoring. In the first stage they found essay semantic score, coherence score, and prompt relevant score. They found semantic score with BERT it is word-level encoding library and by LSTM they calculated coherence and prompt relevant score. In the second stage they concatenated all three scores and trained a xgboost to assign score for the essay.

Darwish et al. [15] predicted essay scores based on syntax and semantic features. The syntax features were found with lexical analysis and parsing method and semantic features predicted with Tf-IDF vector and predicted final score of the essay.

Zhu and Sun [16] used Glove for word embedding, trained the LSTM model to give final scores, and retrieved some statistical features for the essay's final score. Wang et al. (2020). Implemented regression and classification layers both in Bi-LSTM for essay scoring. They classified the essays based on the ground truth value.

Uto [17] implemented a stacked deep neural network with CNN and RNN. They extracted the features lookup table layer, which will convert all words to vector, and pulled N-gram level features with CNN with zero padding—finally trained LSTM with sigmoid activation function to give the final score.

4.3 Method

We proposed single-dimensional LSTM [18] with K-fold cross-validation on the ASAP dataset. In the AES system, one should assess the essay based on the content and relevance to the prompt. For that, we used LSTM [18] recurrent neural network model to train the essay word vector. LSTM [4] is a sequence to sequence learning model which learns complex patterns from vectors. The Word vector is prepared with the Word2Vec [5] NLP library. The word2vec will prepare a vector from essay words into a minimum of 32 dimensions.

4.3.1 Data Set

We used ASAP publicly available dataset from kaggle competition which is largest corpus for AES and used by maximum number of researches. It contains 8 prompts each prompt has on an average on above 1500 essays. The human rater scores each essay is in between 0 and 60. A detailed description of essay dataset is given in Table 4.1.

Table 4.1 ASAP kaggle data set

Essay set	No. of essays	Average length of essays	Rating range
1	1783	350	2–12
2	1800	350	1–6
3	1726	150	0–3
4	1772	150	0–3
5	1805	150	0–4
6	1800	150	0–4
7	1569	250	0–30
8	723	650	0–60

4.3.2 Proposed Model

Typically, AES system will have two components, namely, essay representation into a vector, essay scoring. For vector representation of essays, first divided the entire essay into sentence-wise and removed stop words with NLTK python library's help from the essay. From that, we prepared a list with all bow of each essay. Later converted bow to vector form with Word2Vec [5] NLP library. Word2Vec converts the words into a vector with a minimum of 32 dimensions for each word. Figure 4.1 will illustrate the complete AES scoring approach with LSTM [18].

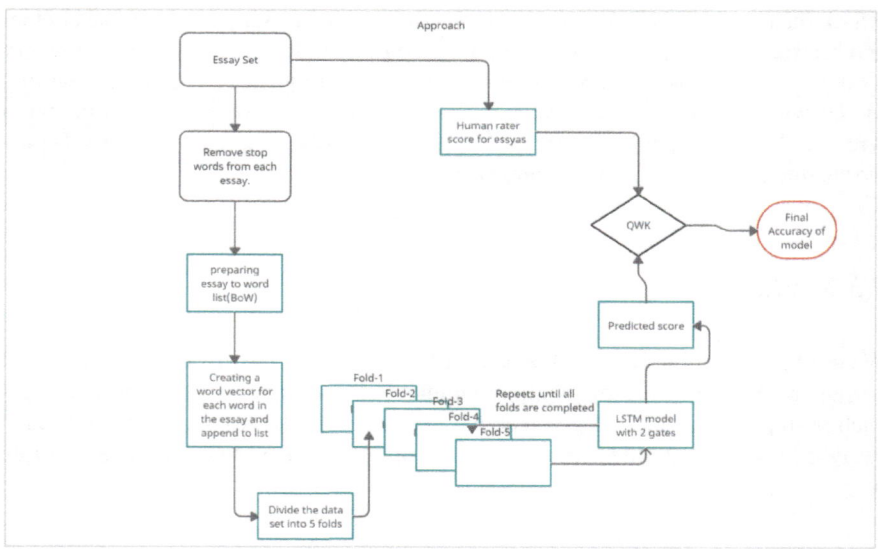

Fig. 4.1 Proposed approach

Table 4.2 K-fold cross-validation QWK score

K-Fold cross Validation	QWK score
Fold-1	0.8489
Fold-2	0.8576
Fold-3	0.8628
Fold-4	0.8451
Fold-5	0.8529
Average	85.35

We developed LSTM [18] with two layers to train the essay vector. The LSTM [18] is a deep learning network quite good for sequence to sequence learning. In an essay all the sentences are connected to each other, so Essay scoring needs sequence to sequence learning to predict the score of the essay. We prepared LSTM [18] with pretrained hyper parameters like learning rate as 0.1, batch size is 64. We used a recurrent dropout rate of 0.4, and the activation function is Relu. Calculated loss with a mean square error, and we used rmsprop as the optimizer.

To train LSTM [18] with feature vectors first we divided the total essay set into five groups (fold). Each fold contains a subset of the essay's features vector. Out of 5 folds [19, 20], four folds for training and the remaining one-fold for testing. The training process will repeat until all the folds are completed as testing and training. Each time one-fold is for testing, and the remaining are for training. Hear each fold trained for six iterations on the LSTM model [18].

Quadratic weighted Kappa (QWK) is used for comparing the predicted score with the actual score. That is to find the mutual agreement between two raters: human rater and system. The final QWK score we got after all iterations is 85.35.

4.4 Results and Analysis

The results of fivefold cross-validation with each fold QWK score are shown in Table 4.2. The mean QWK score we achieved with our AES system is 85.35. Our model LSTM with K-fold cross-validation [19, 20] outperformed better than other models: Tirthankar Dasgupta, Yaman Kumar, and all, which used various neural network models like CNN, LSTM from 2018 to 2020, for AES on the Kaggle dataset. The comparisons of our model with various models with results are shown in Table 4.3.

4.5 Conclusion and Future Work

In this work, we proposed LSTM with K-Fold cross-validation and retrieved features with word2vec for scoring essays. The experiments on the publicly available ASAP dataset and showed that our model outperformed the current approaches. According

Table 4.3 Comparison of results

System	Approach	Dataset	Features applied	Results
Tirthankar Dasgupta et al. [11]	CNN -bidirectional LSTMs neural network	ASAP Kaggle	Content and physiological features	QWK 0.786
Wang et al. [12]	Bi-LSTM	ASAP Kaggle	Word embedding sequence	QWK 0.724
Yaman Kumarm et al. (2019)	Random Forest CNN, RNN neural network	ASAP Kaggle short Answer	Style and content-based features	QWK 82%
Jiawei Liu et al. (2019)	CNN, LSTM, BERT	ASAP Kaggle	semantic data, handcrafted features like grammar correction, essay length, number of sentences, etc	QWK 0.709
Darwish et al. [15]	Multiple Linear Regression	ASAP Kaggle	Style and content-based features	QWK 0.77
Zhu and Sun [16]	RNN (LSTM, Bi-LSTM)	ASAP Kaggle	Word embedding, grammar count, word count	QWK 0.70
Uto(B) and Okano [17]	Item Response Theory Models (CNN-LSTM, BERT)	ASAP Kaggle		QWK 0.749
Proposed model	LSTM with multiple layers	ASAP Kaggle	Word2Vec, Sentence vector, BoW	QWK 0.8535

to the agreement between the machine learning model and human graders, compared to other models, it is 5−7% improved, and we achieved the mean average QWK score of 85.35.

The accuracy has increased with our model, and we extracted features from the essay with word2vec, but it extracts word-level features, so there is a chance of missing semantics of the essay in our model, which is an essential feature in essay grading. We want to work on it in the future and assess the essay by considering features like cohesion, coherence, and completeness.

References

1. Dadi R. et al.: Iop Conf. Ser.: Mater. Sci. Eng. **981**, 022016 (2020)
2. Mikolov, T., Sutskever, I., Chen, K., Corrado, G., Dean, J.: Distributed representations of words

and phrases and their compositionality. In: Proceedings of the 26th International Conference on Neural Information Processing Systems, vol. 2, pp. 3111–3119 (NIPS'13). Curran Associates Inc., Red Hook, NY, USA (2013)

3. Sheshikala, M. et al.: Natural language processing and machine learning classifier used for detecting the author of the sentence. Int. J. Recent Technol. Eng. (IJRTE) (2019)

4. Dolamic, L., Savoy, J.: When stopword lists make the difference. J. Am. Soc. Inf. Sci. Technol. **61**(1), 200–203 (2010)

5. Mikolov, T. et al.: Efficient Estimation of Word Representations in Vector Space. ICLR (2013)

6. Ramesh, D.: Enhancements of artificial intelligence and machine learning. Int. J. Adv. Sci. Technol. **28**(17), 16–23 (2019). Accessed from http://sersc.org/journals/index.php/ijast/article/view/2223

7. Al, S.M. et al.: A comprehensive study on traditional ai and ann architecture. Int. J. Adv. Sci. Technol. **28**, no. 17, 479–87 (2019)

8. Al, S.N.P. et al.: Variation analysis of artificial intelligence, machine learning and advantages of deep architectures. Int. J. Adv. Sci. Technol. **28**(17), 488–95 (2019)

9. Ajay, H.B., Tillett, P.I., Page, E.B.: Analysis of essays by computer (AEC-ii) (no. 8–0102). Washington, DC: U.S. Department of Health, Education, and Welfare, Office of Education, National Center for Educational Research and Development (1973)

10. Foltz, P.W., Laham, D., Landauer, T.K.: The intelligent essay assessor: applications to educational technology. Interact. Multimed. Electron. J. Comput.-Enhanc. Learn. **1**(2) (1999). http://imej.wfu.edu/articles/1999/2/04/ index.asp.

11. Dasgupta, T., Naskar, A., Dey, L., Saha, R.: Augmenting textual qualitative features in deep convolution recurrent neural network for automatic essay scoring. In: Proceedings of the 5th Workshop on Natural Language Processing Techniques for Educational Applications, pp. 93–102 (2018)

12. Wang, Y. et al.: Automatic essay scoring incorporating rating schema via reinforcement learning. EMNLP (2018)

13. Kumar, Y., Aggarwal, S., Mahata, D., Shah, R.R., Kumaraguru, P., Zimmermann, R.: Get it scored using autosas - an automated system for scoring short answers. AAAI (2019)

14. Liu, J., Xu, Y., Zhao, L.: Automated Essay Scoring Based on Two-Stage Learning (2019). abs/1901.07744

15. Darwish S.M., Mohamed S.K.: Automated essay evaluation based on fusion of fuzzy ontology and latent semantic analysis. In: Hassanien, A., Azar, A., Gaber, T., Bhatnagar, R.F., Tolba, M. (eds.) The International Conference on Advanced Machine Learning Technologies and Applications (2020)

16. Zhu W., Sun Y.: Automated essay scoring system using multi-model machine learning. In: Wyld D.C. et al. (eds.) MLNLP, BDIOT, ITCCMA, CSITY, DTMN, AIFZ, SIGPRO (2020)

17. Uto M., Okano M.: Robust neural automated essay scoring using item response theory. In: Bittencourt, I., Cukurova, M., Muldner, K., Luckin, R., Millán, E. (eds.) Artificial Intelligence in Education. Aied 2020. Lecture Notes in Computer Science, vol 12163. Springer, Cham (2020)

18. Hochreiter, S., Schmidhuber, J.: Long short-term memory. Neural Comput. **9**(8), 1735–1780 (1997). https://doi.org/10.1162/neco.1997.9.8.1735

19. Refaeilzadeh, P., Tang, L., Liu, H.: Cross-validation. In: Liu, L., Özsu, M.T. (eds.) Encyclopedia of Database Systems. Springer, Boston, MA (2009). https://doi.org/10.1007/978-0-387-39940-9_565

20. Yadav, S., Shukla, S.: Analysis of k-fold cross-validation over hold-out validation on colossal datasets for quality classification. In: 2016 IEEE 6th International Conference on Advanced Computing (IACC), pp. 78–83. Bhimavaram, India (2016). https://doi.org/10.1109/iacc.2016.25.

Chapter 5
RoMaPla: Using t-Test for Evaluating Robustness of Marathi Plagiarism

Jatinderkumar R. Saini and Prafulla B. Bafna

Abstract Identifying plagiarism of the document is a mandatory task in the academic domain. Generally online available tools are used to check plagiarism. These tools calculate similarity between the documents using a sequence of the tokens/words present in the documents which are to be compared. A semantic relationship between the words for eg., word and its synonym are treated as different, while calculating the similarity between the documents. Few tools may be available for checking the similarity of English documents. But checking the plagiarism of Marathi documents is comparatively untouched field. Information present in the Marathi language is growing due to multilingual processing. The existing MaPla (Marathi Plagiarism checker) proved that Document synset matrix for Marathi (DSMM) similarity results are near to readings observed using cognitive ability of humans and it was performed on 4 documents. To further confirm robustness of MaPla, we experimented with 24 documents to calculate the similarity between all pairs of documents using cosine measure. Thus two, 24×24 matrices are formulated using DSMM and manual readings. Paired t-test, which was not carried out in MaPla, proves that there is no significant difference between two matrices and hence proves the robustness of the proposed technique.

5.1 Introduction

Referring and using other researchers' or persons' work, ideas like poetry, cinema, etc. output without citing or acknowledging these researchers is known as plagiarism[8]. An application like detection of a replica of web pages, etc. will benefit due to plagiarism checking technique [1–3]. Though the importance of natural languages [6, 8] is increased for eg., research carried out in Hindi language and so on, very

J. R. Saini · P. B. Bafna (✉)
Symbiosis International (Deemed University), Symbiosis Institute of Computer Studies and Research, Pune, India
e-mail: prafulla.bafna@sicsr.ac

© The Author(s), under exclusive license to Springer Nature Singapore Pte Ltd. 2022   43
V. Bhateja et al. (eds.), *Evolution in Computational Intelligence*,
Smart Innovation, Systems and Technologies 267,
https://doi.org/10.1007/978-981-16-6616-2_5

few plagiarism tools are present in these languages. Marathi is used as commercial language in some states of India for eg., Goa, Maharashtra, etc. The Marathi language exists in more than 20 dialects. Online text data in Marathi is growing exponentially [4, 5] eg., Charolya (small poems), play scripts, and so on. Plagiarism of this text could be identified by identifying similarity between texts also known as a mapping between texts. To perform any operation on the text there are several NLP steps [11, 12] like tokenization that is separating text into smaller meaningful chunks, removal of noise from the text also known as stopwords [7–9] thus stopwords removal increases the efficiency of an algorithm. Dimensions are reduced due to the removal of stip words. POS tagger means classifying prepositions, nouns, adverbs, adjectives, and so on from the text. One of the methods to normalize text is known as lemmatization. Also stemming is another method to normalize the text, but lemmatization is more correct as it produces grammatically correct root words known as lemmas. Stemming just discards the inflectional end. 'जैन धर्माची सर्व तत्त्वे ही अहिंसावर आधारित आहेत.'In the first phase of tokenization, all terms present in the sentence are split. Eg., 'जैन', 'धर्माची', and so on. Nouns, adjectives, adverbs are identified. Then stopwords like 'ही'are removed. In the lemmatization phase, words like 'धर्माची'are converted into 'धर्म'and its stem is 'धर्मा'. To find out plagiarism between the text data, the row-column format is necessary. In this Row-column form documents and significant terms/features are placed, respectively. It is termed as DTMM. The significance of the term is decided by the frequency of the term. The specific threshold frequency is decided, the terms having more frequency than threshold frequency are termed as significant terms. To involve semantics of the term Document synset matrix for Marathi (DSMM) [4] was developed. Many times instead of repeating the term, its synonyms are used. It means that term is significant in the document but term and synonyms are considered separately thus their frequency is also not counted together, as a result, due to less frequency, these terms are considered as insignificant terms and incorrect DTMM. Instead of considering synonyms separately, a group of synonyms with their frequency are used in DSMM. Thus in DSMM significant synset are placed in columns and documents occupy rows. Matrix entry represents feature weight of corresponding synset with respect to text document. After formation of a Matrix, Cosine similarity measure is applied which depicts the similarity between the documents' pairs.

Robustness is the capacity of a system to survive and produce correct output in case of variation in inputs in different forms eg., a number of inputs, quality of inputs, and so on [11].

The paired sample t-test is a statistical measure to decide about the average difference between two groups of input sets is zero. In this, each observation is measured two times, which results in observations' pairs. [11].

Order of the different sections in the paper is as mentioned. Research carried out by the different researchers is presented in the form of a literature review, in the next, that is the second section. Next, that is the third section details the research methodology. Retrieved results and discussions are stated as a fourth section. Next, the fifth section includes conclusions to complete the paper.

5.2 Literature Review

Several deep learning (DL) based approaches are contributing to NLP. Social networking is benefited by the DL approach. Sentiment analysis is used in social computing. There are several research opportunities in the field of pharmacovigilance. The evolution of DL techniques to support NLP operations is stated. The Arabic language is not much explored by researchers, it is a resource-scarce language. There are several challenges which are faced while performing ANLP (Arabic NLP) tasks. Optical character recognition problems are solved using DL. Machine translation is one more area to apply NLP techniques. Several areas where NLP techniques can contribute to making correct decisions are explored [1].

Spell checker tools are needed to develop in different Indian languages. Different spelling mistakes present in the text are identified and corrected by the spell checker. It can be also combined with various applications. Various approaches of spell checking and their applications are discussed [8].

The ambiguity present in language and speech is focused. It also provides linguistic knowledge with their categories. Different classification algorithms allow having different categories of this knowledge. In the last decade, several machine learning techniques were developed to solve the problem of disambiguation of linguistic knowledge [9].

Different supervised techniques are reviewed and applied to concepts of linguistic knowledge, deep learning, maxentropy techniques are discussed as a classification problem. Different datasets and online corpus are stated.

NLP includes not only linguistic computation but statistics too. Identifying and processing language that human use is the primary objective of NLP. Text classification, semantic analysis are different activities of NLP and these are still evolving. Unlike different algorithms that are already in a stable state, different steps of NLP along with the techniques are still in the research phase. Statistics along with Probability theory are parts of NLP [10].

Semantic analysis of Marathi text is done to detect plagiarism between documents. Checking plagiarism is required in academic fields, which is achieved by using different tools. These tools are based on the similarity between the words and do not consider similarity between the words. Several NLP steps like removal stopwords and others are used to find frequency of the terms. To identify the plagiarism similar words are identified, for this purpose, wordnet is referred [5].

As availability of literature is growing, gradually plagiarism is also increasing. Detecting plagiarism is a critical but significant need in academic and other fields too. N-gram approach is used to detect Marathi text plagiarism. Word-wise checking is done to identify plagiarism on Marathi corpus. It can detect paraphrasing as well as copy-paste plagiarism. The technique is easy but covers depth by processing sentence and paragraph level text [11].

Restructuring the sentence by swapping words or their meanings but keeping the meaning same is called paraphrasing. Paraphrase identification is the process which identifies sentence with similar meaning but different words. The semantic and structural relationships between the words are used to identify paraphrasing of Marathi text. Word-vector, sumo metric, word-order are used to measure statistical similarity. UNL graphs drawn for each sentence identify the semanticity of the sentences and semantic equivalence is expressed.

Statistical and semantic scores are added up to get the final similarity score. The applications like plagiarism detection, question answering, text summarization are benefited due to paraphrasing detection [12].

5.3 Research Methodology

It has been already proved that MaPla [4] which uses DSMM produces the best results compared to DTMM and context-based plagiarism checking tool. MaPla checks Plagiarism between two input text documents using synsets and involves context in the similarity checking process. The performance of MaPla is evaluated using four documents. But to prove the robustness of the DSMM for plagiarism checking, the existing approach is extended to more documents and more authors. Robustness assures the quality of the system for increased and different types of input. Figure 5.1 shows the research methodology steps used to perform experimental analysis.

1. Choose topics and assign the topics to different authors to submit a write-up
    Total of 6 topics are chosen. Four authors are required to write on one topic, so 6 (topics)* 4 (authors per topic), that is total of 24 authors are asked to submit a write-up. The topics chosen are 'Bhagwan Mahaveer', 'Ahimsa param dharam', 'Corona', 'Woman empowerment', 'Technology : developes or kills', 'Green revolution.

2. Compare all the generated documents using human intelligence and DSMM
    Matching between the pair of documents is found. For manual comparison, Marathi linguists are chosen. Authors are indicated as A1 to A24 and topics are shown as T1 to T6, A1_T1 indicates author1 writing on topic 1, thus for 24th author submitting the 6th topic, it is represented as A24_T6. A comparison between write-ups submitted by all authors is done by linguists and the percentage of similarity is recorded in the form of a matrix. This is called as a manual comparison. DSMM is used which identifies the semantics and context of the documents while performing its comparison.

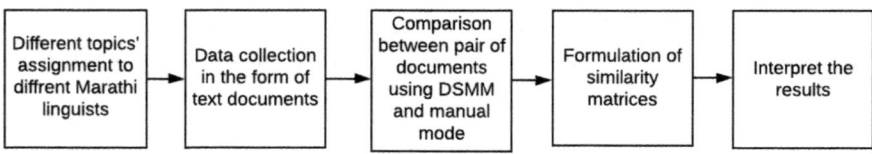

Fig. 5.1 Steps in research methodology

3. Formulation of similarity matrices

   Square matrices (24x24) are formulated to store similarity measure between every pair of all documents using DSMM and manual comparison. A1_T1, A2_T1, A3_T1, A4_T1, A5_T2…A24_T6 are placed in rows and columns of a matrix and similarity score is recorded in the form of percentage in the corresponding entry of the matrix. The plagiarism between documents A1_T1 and A2_T1 will be the same as between A2_T1 and A1_T1.

4. Interprete results

To compare whether there is a significant difference between similarity measures obtained using manual reading and DSMM, a paired t-test is used assuming the hypothesis that there is no significant difference between DSMM and manual similarity measurements.

5.4 Results and Discussions

Table 5.1 depicts the document samples written on the topic 'Ahimsa' (non-violence). Total 4 authors submitted write-up. One sample handwritten write-up and its typed version are shown in Table 5.2. Total number of tokens present in the write-up is 356, after removing stopwords the number is reduced to 200. After carrying out lemmatization total number of unique tokens left are 96.

Table 5.2 shows that topic 1 (T1) that is 'Bhagwan Mahaveera' is assigned to 4 authors A1, A2, A3, A4. The next topic is assigned to the next author that is A5. The cosine similarity measure obtained for the mapping between the same documents that are A1_T1 and A1_T1 is always 1 for both of the methods. Also, the similarity between A1_T1 and A2_T1 is marked with '*', which indicates the same as between

Table 5.1 Write up submitted by an author

	जैन धर्मात अहिंसा हा सर्व सजीवांबद्दल प्रतिबंधित वर्तन आहे. अहिंसाचा शाब्दिक अर्थ हिंसाचार न करणे. याचा तांत्रिक अर्थ पद्धतशीर आणि निषिद्ध आहे. भावनिक प्रवृत्ती बाळगणे, मारणे किंवा फक्त प्रवृत्तीचा प्रतिकार करणे ही निषेधात्मक अहिंसा आहे; स्वत: ची धार्मिकता, स्वत: ची नीतिमत्त्व, आत्मशिक्षण, अध्यात्मिक सेवा, उपदेश, ज्ञान चर्चा इत्यादी पद्धतशीर अहिंसा आहेत.
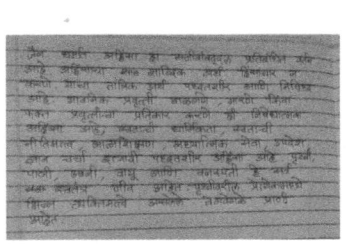 Sample paragraph of the document submitted by author	A typed version of the same paragraph

Table 5.2 Preassumptions of a matrix by five authors

	A1_T1	A2_T1	A3_T1	A4_T1	A5_T2
A1_T1	1	*	#		
A2_T1	*	1			
A3_T1	#		1		
A4_T1				1	
A5_T2					1

A2_T1 and A1_T1. '#' also indicates the same similarity between the corresponding document pair.

Table 5.3 shows terms/synset groups along with their frequencies. Two write-ups are processed to extract terms. A total of 378 unique lemmas/terms are extracted. These terms are then reduced to 324 synset groups. To find out frequent terms, a 60% threshold is chosen. The maximum frequency is 12, so only those terms having frequency of more than 7 are chosen to formulate DSMM. Thus tokens till 'आगम' having frequency 7 will be included in DSMM and rest tokens are ignored eg., 'केवल', 'प्रसार' as their frequencies are 3 and 2, respectively, which is less than 7.

Table 5.4 shows the construction of DSMM using significant synsets retrieved after processing the write-ups. The entries in the matrix represent the frequency of the synset group/term for the corresponding document. For example, in the first topic written by the first author (A1_T1), a number of times the word 'ज्ञान ' has occurred is 7. Total column heads are 91.

Table 5.3 Synset groups and frequency

Synset groups/terms	ज्ञान	दिप, दिवा		आगम	केवल	प्रसार
Frequency	12	8		7	3	2

Table 5.4 DSMM using significant synsets

Author_Topic	ज्ञान	दिप, दिवा		केवल	आगम
A1_T1	7	3		2	4
A2_T1	5	5		1	3

Table 5.5 Cosine similarity measure using DSMM

Author_Topic	A1_T1	A2_T1
A1_T1	1	0.72
A2_T1	0.72	1

Table 5.6 Manual commonality in the form of %

Author_Topic	A1_T1 (%)	A2_T1 (%)	…	A23_T6 (%)	A24_T6 (%)
A1_T1	100	73	…	0	0
A2_T1	73	100	…	10	0
…	…	…	…	…	…
A23_T6	0	10	…	100	85
A24_T6	0	0	…	85	100

A cosine measure was applied on DSMM. Table 5.5 shows the similarity score in the form of a cosine measure. The similarity score between the same documents is always 1. The similarity score between A1_T1 and A1_T1 is 1. Thus all the diagonal elements represent the same entry of row and the column has value 1. Also, similarity score of A1_T1 and A2_T1 is the same as between A2_T1 and A1_T1 which is 0.72 that is 72%. The threshold considered for the similarity between the documents is 0.70, so all the document pairs having a similarity score of less than 0.70 will be considered as dissimilar.

Table 5.6 shows the similarity between document pairs which is expressed manually. Marathi linguists are asked to record the similarity between document pairs. The similarity between the write-ups written by two different authors A1 and A2 on the first topic (T1) 'Bhagwan Mahaveer' is represented as A1_T1 and A1_T2 and the value is 73%.

Figure 5.1 shows similarity measure scores for all 24 documents using DSMM and manual mode X-axis represents the documents and the Y-axis represents the similarity score. A paired T-test was used to prove that there is no significant difference between the similarity values obtained manually and using DSMM. The p-value calculated for two matrices is $p = 0.01$ which is much lower than the usual significance level of alpha $= 0.05$. This strongly establishes the efficacy of the proposed approach and supports the hypothesis, that there is no significant difference between values of DSMM and Manual readings (Fig. 5.2).

5.5 Conclusions

The proposed approach extends MaPla to prove the robustness that is RoMaPla. 24 different authors and 6 different topics are chosen. One topic is assigned to 4 authors to submit a write-up. The retrieved text documents were input to DSMM in pair form

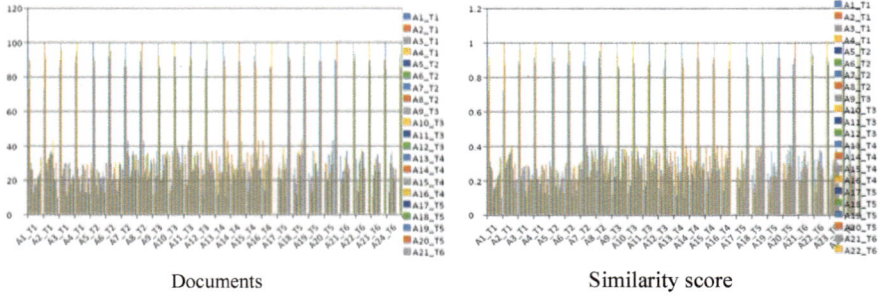

Documents   Similarity score

Fig. 5.2 Similarity measures using DSMM and Manual Mode

to find out the similarity using cosine measure. Also, manual reading of similarity measurements are recorded in the form of a percentage. Two 24 × 24 matrices are formulated based on the similarity readings. The paired t-test is applied. The obtained p-value is 0.01 less than alpha (0.05) shows that there is no significant difference between the values of the two matrices (DSMM and manual similarity measurement matrix) means they are almost similar. Thus the robustness of the proposed technique is proved.

References

1. Naik, R.R., Landge, M.B., Mahender, C.N.: Development of marathi text corpus for plagiarism detection in the marathi language. Corpus **6**, 340 (2011)
2. Lamba, H., Govilkar, S.: A survey on plagiarism detection techniques for indian regional languages. Int. J. Comput. Appl. **975**, 8887 (2017)
3. Shenoy, N., Potey, M.A.: Semantic similarity search model for obfuscated plagiarism detection in Marathi language using Fuzzy and Naïve Bayes approaches IOSR. J. Comput. Eng. **18**(3), 83–88 (2016)
4. Bafna P.B., Saini J.R.: MaPla: a marathi plagiarism checker using document synset matrix. Int. J. Adv. Sci. Technol. (2020). in press
5. Bafna P.B., Saini J.R.: Marathi text analysis using unsupervised learning and word cloud. Int. J. Eng. Adv. Technol. **9**(3) (2020)
6. Naik, R.R., Landge, M.B.: Plagiarism detection in marathi language using semantic analysis. In: Scholarly Ethics and Publishing: Breakthroughs in Research and Practice, pp. 473–482. IGI Global (2019)
7. Al-Ayyoub, M., Nuseir, A., Alsmearat, K., Jararweh, Y., Gupta, B.: Deep learning for Arabic NLP: a survey. J. Comput. Sci. **26**, 522–531 (2018)
8. Gupta, N., Mathur, P.: Spell Checking Techniques in NLP: A Survey (2012)
9. Khan, W., Daud, A., Nasir, J.A., Amjad, T.: A survey on the state-of-the-art machine learning models in the context of NLP. Kuwait J. Sci. **43**(4) (2016)
10. Ranjan, N., Mundada, K., Phaltane, K., Ahmad, S.: A survey on techniques in NLP. Int. J. Comput. Appl. **134**(8), 6–9 (2016). odelling, pa

11. Naik, R.R., Landge, M.B., Mahender, C.N.: Word level plagiarism detection of marathi text using N-Gram approach. In: International Conference on Recent Trends in Image Processing and Pattern Recognition, pp. 14–23. Springer, Singapore (2018)
12. Srivastava, S., Govilkar, S.: Paraphrase identification of marathi sentences. In: International Conference on Intelligent Data Communication Technologies and Internet of Things, pp. 534–544. Springer, Cham (2018); Intelligent Computing: Theory and Applications, pp. 797–806. Springer, Singapore (2018)

Chapter 6
To Analyse the Impact of Water Scarcity in Developing Countries Using Machine Learning

Kiran S. Raj and Priyanka Kumar

Abstract It is quite common to hear the phrase 'Water is the elixir of life'. In the current scenario, water has become an alarmingly depleted quantity. Overconsumption of resources has affected the water table and humans have a big cost to pay. The notion of 'Water Consumption' has grown to form stronger and deeper roots, awareness regarding the same has thrust upon an individual. Despite all the moves in the right direction, access to drinking water is a huge cause of concern in developing nations. Water Pumps are considered as one of the most important innovations of all time. In developing nations, it is quite a common sight to spot a water pump for it strives to be a single point of destination to usable water. In recent times, most of the water pumps are not functioning, some partially functional and some functioning, this is a cause of concern. With the aid of Data Mining and Machine Learning concepts, being able to create a model to predict the functionality of the same can prove to be fruitful. To analyse and understand the impact of the availability of water in Tanzania, by studying the operational status of water pumps around the country, and the number of water-related deaths in the region. Through solid inference, the knowledge regarding the same can be extrapolated to different regions around the globe suffering from the same crisis.

6.1 Introduction

Around 70% of the surface of the Earth is covered by water. Despite the enormous amount of water present, only 3% of it is suitable for human consumption. In the current world with a staggering total population of 7.6 billion individuals, almost 1 billion face issues of admittance to water. The interest for water would surpass an

K. S. Raj (✉) · P. Kumar
Department of Computer Science and Engineering,
Amrita School of Engineering, Amrita Vishwa Vidyapeetham, Coimbatore, India
e-mail: cb.en.u4cse17430@cb.students.amrita.edu

P. Kumar
e-mail: k_priyanka@cb.amrita.edu

© The Author(s), under exclusive license to Springer Nature Singapore Pte Ltd. 2022   53
V. Bhateja et al. (eds.), *Evolution in Computational Intelligence*,
Smart Innovation, Systems and Technologies 267,
https://doi.org/10.1007/978-981-16-6616-2_6

amount of 40% by 2020, say the United Nations. This is a cause of concern, individuals/communities would thereby be struggling for a glass of water [10]. In certain spots, this emergency may introduce more emotional shapes. In certain nations, the water emergency can have its belongings potentiated by the interminable money-related emergencies, compassionate emergencies, wars, and so forth [9]. The roots of 'Water Conservation' have become stronger as the awareness regarding the same has to spread.

The problem of water scarcity in Tanzania The United Republic of Tanzania is an East African country. It is a unitary state consisting of 26 regions, and in addition to its continental part, includes the archipelago of Zanzibar in the Indian Ocean. Tanzania is surrounded by great central plains, tropical savanna, forest highlands and an equatorial shoreline. Mount Kilimanjaro is also situated in Tanzania. Tanzania is not known for its picturesque location but rather as a place adversely affected by accessibility to drinking/safe water. An average estimate of 2.46 million people in Tanzania doesn't have access to drinking water. On surveying the situation, a fundamental reason aggravating the problem revolves around the functionality of water pumps. Most of the pumps are often non-functional or those that require immediate repair. Water pumps are an essential source of drinking water to the general public. This paves an important point in Tanzania's water security, water table. One can drill on to the water pumps as a conversation for why people in Tanzania are struggling without clean water. The goal of the case study is to devise a Machine Learning model capable of predicting the functionality of water pumps and finding relevant inferences which could be used as crucial information in solving related problems in different regions around the world.

6.2   Formulation of Problem Statement

Our task is to identify the functional status of a water pump, given various details about it. This establishes our problem to be a classification problem [7], and as such, we will use several approaches to tackle it. Beyond the scope of simple classification [8], we also explore the dataset using numerous statistical techniques to understand which factors affect the lifespan of water pumps the most. We then perform clustering upon the data to uncover latent patterns in the data, that will allow us to preemptively diagnose the health of a water pump, and service it ahead of time, before a potential failure. The expense of establishment of a standard water siphon goes from 100 to 2000, and introducing this siphon requires penetrating which can be anything between 1000 and 3000 [2]. On the other hand, maintaining the pump would only cost tens of dollars, which is vastly less expensive than installing a new one. Hence, this analysis will save a lot of money for the Government/NGO organization [5].

6.3 Background Work and Literature Survey

Internationally, 20% of the water points are not functional while an extra 10% are functional yet with issues. Different examinations show that in Sub-Saharan Africa (primarily in Tanzania), it is assessed that around 30–36% of water focuses are not practical with nations, for example, Côte d'Ivoire (65 per cent) and Sierra Leone (65%) encountering high paces of water points failures and nations, for example, Madagascar (10%) and Benin (22%) encountering lower levels.

Another issue is that there is no broadly agreed-upon definition of functionality. This is a significant part of data accuracy because the error of the information gathered—particularly when the meaning of functionality isn't characterized—could unfavourably influence programmes that are planned dependent on defective information. While most studies utilize the parallel differentiation of functioning or not functioning, which on occasion is viewed as deficient, different overviews incorporate the idea of partial functionality.

Based on the above study we decided to choose the dataset which includes the concept of partial functionality.

Geographic and hydrological factors such as the different geographical zones and groundwater characteristics are valuable to future dynamic cycles in planning, developing or dealing with the water points at areas with changing hydro-geological qualities [1]. Truth be told, the Shapely disintegration shows that the hydrological factors are a higher priority than those of area and geographic zone types. While intercessions can't be intended to entirely change these components, with proper and precise hydrological information they can be intended to moderate or decrease the unfriendly consequences for water focuses. At last, the absence of hydrological information is a significant obstruction to understanding the water point disappointments particularly during the main year after their establishment.

So, for getting accurate results we have taken a dataset which includes both hydrological and geological factors.

In previous works, the empirical results from the logistic regressions confirm the evidence provided in the Tanzania context as well as those found in the literature. Water points managed by village committees had a much higher likelihood of failure than those managed by private operators or water authorities [4, 6].

So we tried to build the model with different algorithms to check the best model which can be suitable for our functionality.

6.4 Architecture Diagram

Figure 6.1 illustrates the architecture diagram involved in the entire system.

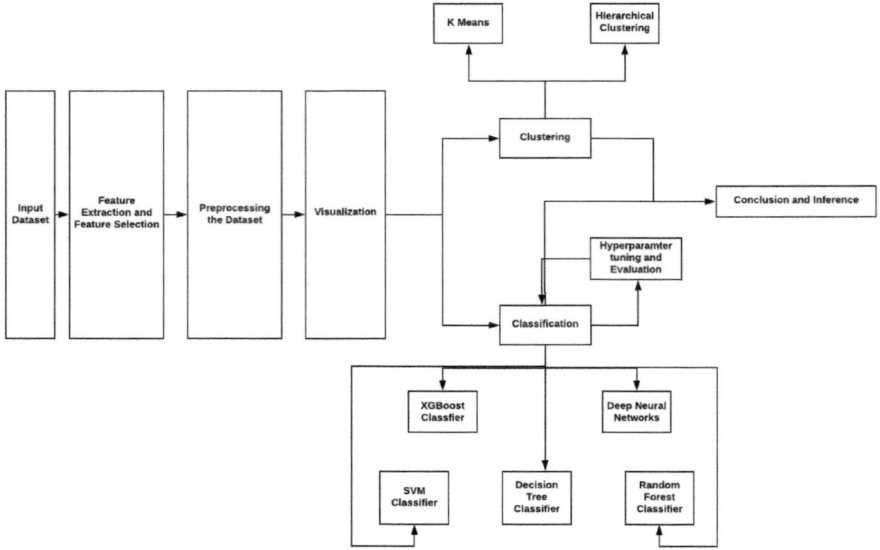

Fig. 6.1 Architecture diagram of the system

6.5 Data Collection and Preparation

The dataset used for the system is one dataset produced by the Taarifa and Tanzanian Ministry of Water, which contains at a glance:

- Size of the dataset = (59400, 40); 59400 examples (rows) for training.
- The number of Features = 40.
- Number of Labels = 3, namely functional, functional needs repairs, and non-functional for the respective water pumps.

6.5.1 Feature Selection and Extraction

The dataset that is obtained for the process has 40 feature vectors. It is essential for Machine Learning [3] to learn the appropriate and relevant features to predict with high accuracy. Feature Selection is a dimensionality reduction technique where the total number of features is reduced. With feature selection, only a key subset of the original features are retained and the rest of the irrelevant/redundant features are dropped during the process. Feature Selection in this situation proves to be a crucial point in determining the accuracy of the effective model. From manual inspection, it is found out that certain features like 'id' and 'payment_type' were irrelevant. On further inspection features such as : (source, source_class), (extraction_type, extraction_type_group), extraction_type_class), (payment, payment_type), (water_quality, quality_group), (subvillage, region, region_code, district_code, lga, ward) and (waterpoint_type, waterpoint_type_group). All these features exhibit similar representation/redundant values. Hence, the redundant features were removed to prevent the model from overfitting.

To further analyse the extent of the impact of the features, a method of Backward Elimination is used. Backward Elimination is a technique which is used to select those features which would serve as important features to the model and remove those that are unnecessary. This technique is used to remove features that would not have a major/significant impact on the performance of the model. The hypothesis made on on the feature vectors are:

– Null Hypothesis H_o = Features vectors are dependent on one another.
– Alternate Hypothesis H_a = Feature vectors are independent of one another.
– The significance level chosen is = 0.05.

Hence on performing backwards elimination for all the feature values for which the computed P-value is less than that of the Significance Level (i.e. 0.05) null hypothesis can be rejected. Hence, those values are retained. For those values above the significance level, the feature with the highest P-value above the significance level is removed from the feature list and this process is iterated again until only those features with P-values less than the significance level are obtained.

Feature Extraction refers to the creation of smaller new features that would retain most of the useful information. This is a technique that would add a new feature which would encompass the information retained by one or more features. The results of feature extraction: A new feature called 'Age' which denotes the age of the water pump. It is obtained as df['date_recorded'] - df['construction_year'] (That is the last recorded date subtracted from the construction year) A new feature called 'Population/Year' which is the ratio of total population by its age. A new called 'Water/Person' which is the ratio of the total amount of water and the population. These newly created features provide useful information contained in more than one of the parent features.

6.5.2 Data Visualization

It is essential to visualize the way different features are distributed in vector space. It is through this visualization meaningful inferences regarding the significance of various parameters are observed. In Figure 6.2, the visualization focuses on the count of each scheme management present to each of the status_group (that is the functionality of the pump). It is evident from the data that VMC is the most frequent scheme management group with the highest amount of functional, non-functional and functional needs to repair water pumps. Since most of the water pumps belong to the scheme, it is essential to detailedly focus more on this group as they also have the most non-functional pumps. Another important observation is regarding the SWC group, though their count of water pumps is less, the count of non-functional water pumps

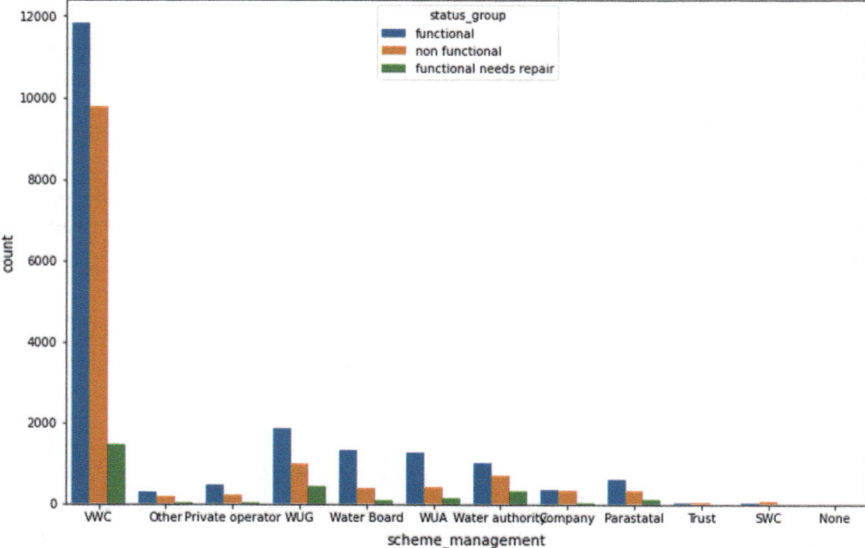

Fig. 6.2 Count wrt status group

exceeds that of the functional ones. With the help of geopandas in Fig. 6.3, plotting the various latitude and longitudes on the African map. This visualization gives an understanding of where the various water pumps are located and the functionality of the water pumps. Green indicates the functional water pumps, Yellow indicates the functional but needs repair water pumps and the Red indicates the non-functional water pumps. An interesting observation can be made about the non-functional water pumps, they occur in clusters (dense, neighbours of non-functional water pumps are also non-functional). Figure 6.4 focuses on the probabilistic distribution of the construction year. Though the distribution does not follow any traditional mathematical distribution, an important observation to note is regarding the skewness of the graph. The depicted distribution seems to be highly negatively skewed, denoting that a lot of water pumps were only recently constructed (since the 2005's), hence it is important to closely monitor the status of those water pumps in that year owing to the count of them.

6.5.3 Data Cleaning and Pre-Processing

Null Values

Due to loss of data or the mishaps that happen during data entry, there are null values present in the dataset which have to be imputed or filled (or even dropped). On analysing the state of the null values present in the feature vectors. The features

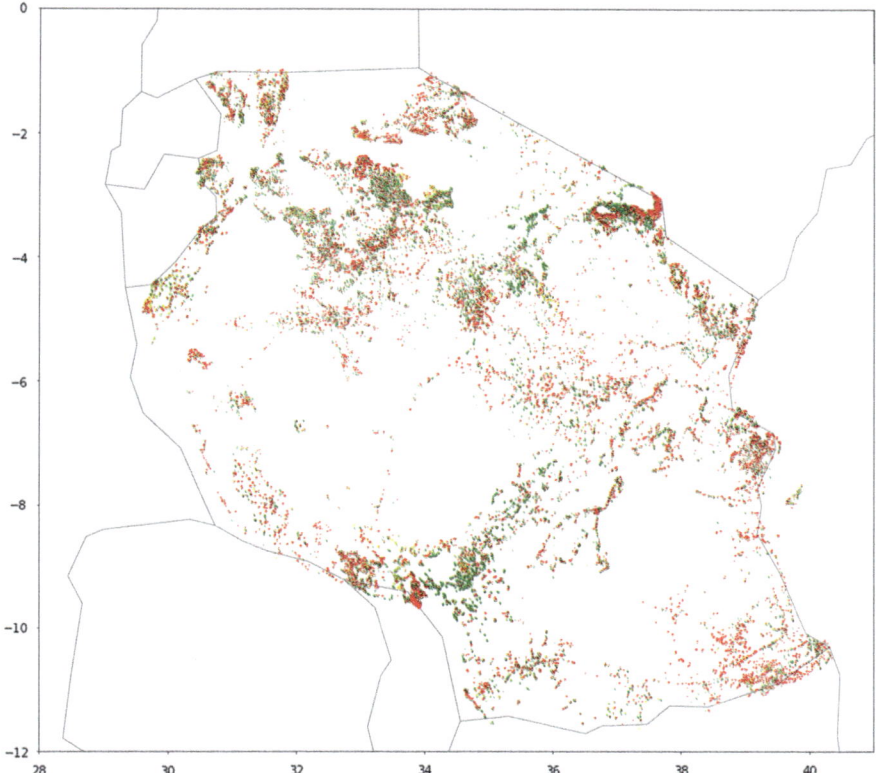

Fig. 6.3 Geographical visualization

funder, installer, subvillage, public_meeting, scheme_management, scheme_name and permit have null values. The common way to fill in these values is to understand the impact of the features on the target variable. For all the null values features, the central tendency value 'mode' is used to fill the null values. The reason behind this logic is to obey the coherence between the data. The frequently occurring value would play an apt association when it comes to categorical data (instead of mean/median), the null values would be filled by the most frequently occurring category. (Another thing to note is that none of the records could be dropped as the count of the null values is sufficiently large—would lose a considerable amount of data if dropped).

Categorical Features
The features present in the dataset such as ['basin', 'scheme_management', 'extraction_type', 'management_group', 'payment_type', 'water_quality', 'quantity_group', 'source_type', 'source_class', 'waterpoint_type', 'waterpoint_type_group'] are all categorical variables. These variables seem important to identify the specificity about a particular water pump (Hence, cannot be removed). Machine Learning

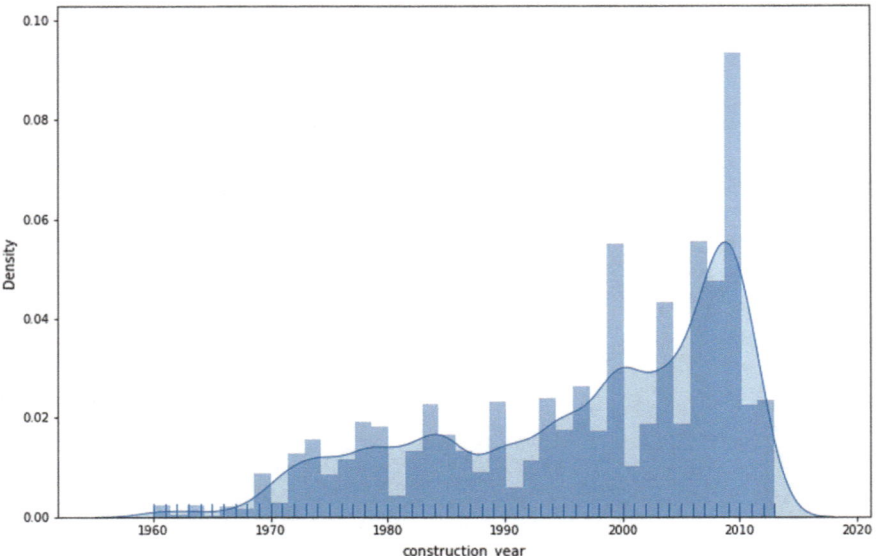

Fig. 6.4 Histogram wrt construction year

Models don't understand strings/text as such, hence it is important to convert it into a particular way to mathematically model an equation that satisfies the criteria to predict the target accurately. The way the categorical variables are converted to such a format is as follows:

- Label Encode each of the categorical variables. That is, put an integer to represent each of the unique values present in the categorical variable.
- Since label encoding would bring about an order to the nominal variables it is essential to make sure that there isn't any bias between the data points in the categorical feature. Hence, One Hot Encoding is performed to create a representation of the categorical feature in the specified format that would be mathematically modelled.
- Care is taken to prevent the dummy variable trap. Dummy variable trap occurs when there is a very high correlation between the one-hot encoded values, i.e, a particular value can be predicted by the other values. This would lead to inaccuracy in the prediction and inefficient for the model to generalize the same. Hence, dummy variable traps are avoided by removing a featured one hot encoded column.
- The pandas function of get_dummies() is used to encode the categorical variable into dummy variables for processing.

Scaling is done to normalize the angle of different values present in the dataframe to the same scale. This would help in speeding up computation as well as avoid weight-related problems because of the abnormality of the scale between the features. StandardScaler is used to normalize the features vectors in the process.

6.6   Performance Evaluation and Result

The different models tried out for the classification are Decision Tree Classifier, Random Forest Classifier, Mini Batch KMeans Clustering, Extreme Gradient Boosting (XGBoost), SVC (Support Vector Machines Classifier), Deep Neural Networks (DenseNets)

Model Tweaking, Regularization, Hyper Parameter Tuning It is through careful consideration of the different hyperparameters, that the classification model is tuned and the hyperparameters modified. With the help of SciKit Learn's GridSearchCV, the perfect blend of hyperparameters was searched and the models were retrained with the new set of hyperparameters (Table 6.1).

For the Deep Learning Model, the various combinations of the number of neurons were modified to find the best possible fit. On various iterations, the following were the model configurations used to obtain the best results. The configurations are as follows:

– 5 Dense layers with 64, 128, 128, 64, 3 neurons respectively. Rectified Linear Unit (ReLU) is the activation function used in the intermediate hidden layers.
– The output layer has 3 nodes reflecting on the number of classification labels. Softmax as the activation function used in the final layer.
– The model was trained for 200 epochs with a batch size of 10, the loss function used for the same is categorical_crossentropy and the optimizer function used is Adaptive Moment Estimation (Adam).
– The final output for the model gave a training accuracy of 84.2% and a test accuracy of 78.2%.

It's safe to say that accuracy is not the only criteria that have to be taken into consideration from the model's result. To evaluate classifiers different parameters like Precision, Recall, F1 score. All these parameters have to be in check to decide the best classifier out of the lot that we've tested. Out of the lot, one of the best ways to compare the performance of the classifier is by using the ROC/AUC curve. The Area

Table 6.1 Performance comparison

Performance comparison		
Model	Train accuracy	Test accuracy
Decision tree classifier	93.7%	74.3%
Random forest classifier	93.7%	78.19%
Support Vector Machine Classifier (SVC)	77.13%	76.15%
XGBoost classifier	78.39%	76.7%
Deep neural network	84.2%	78.2%

Under the Curve for the model with the highest magnitude is a good way to judge the performance of multi-label classifiers. (ROC details out on True Positive Rate and the False Positive Rate) On further analysing the classifier based on Receiver Operating Characteristics and Area Under the Curve, the best classifier out of the lot was the Deep Neural Networks with a close second of Random Forest Classifier (Deep Neural Networks = 0.82, Random Forest Classifier = 0.80).

6.7 Conclusion and Future Enhancements

Data mining classification models, for example, Decision tree, Neural Network, XGBoost Classifier, Support Vector Machine classifier and Random Forest were utilized to anticipate if a Water pump is functional, non-functional or needs fix. Given the misclassification rate, Deep Neural Networks with custom layers were the best model in grouping the Water Pumps. From the model, the main considerations that contribute in deciding the usefulness of the Water pump are recognized as are longitude, locale code, area code, source, gps stature, amount, extraction type, instalment and Water-point type. The Water Pumps towards the eastern locale of Tanzania have acquired more support issues than that of Water Pumps in different districts of the nation. The cost of a standard pump ranges from 100 to 2000. Installing this pump requires drilling which can be anything between 1000 and 3000. On the other hand, maintaining the pump would only cost tens of dollars, which is vastly less expensive than installing a new one. This project aids in efficiently identifying pumps which require maintenance, so that the Government/NGOs can focus their aid to the regions that require them the most. Other than the monetary cost, there is a more pressing issue. If a water pump is non-functional, for that duration, water would be scarce in the region. The time a water pump is non-functional must be minimized, optimally 0.

Future Enhancements:

– Overlay network graphs on the geological guide of non-useful water pumps to speak to the most limited separation between all the Water Pumps.
– Collect more data about the impact of water supply in the region—population, birth rate, death rate, waterborne diseases, etc.
– Identify historical data about wear and tear in water pumps to estimate the lifespan of the pump, and identify the best points in time to service them.
– Predict the time duration a pump will be out of service so that alternate arrangements can be made for water supply

References

1. Dhanush, G.A., Raj, K.S., Kumar, P.: Blockchain aided predictive time series analysis in supply chain system. In: Mekhilef, S., Favorskaya, M., Pandey, R.K., Shaw, R.N. (eds.) Innovations in Electrical and Electronic Engineering, pp. 913–925. Springer, Singapore (2021)
2. Mittal, S., Mittal, M., Khan, M.S.A.: Ground-Level Water Predication Using Time Series Statistical Model, pp. 427–437 (2021). https://doi.org/10.1007/978-981-15-5421-6_43
3. Moleda, M., Momot, A., Mrozek, D.: Predictive maintenance of boiler feed water pumps using scada data. Sensors **20**(2) (2020). https://doi.org/10.3390/s20020571, https://www.mdpi.com/1424-8220/20/2/571
4. Pandey, S., Muthuraman, S., Shrivastava, A.: Data Classification Using Machine Learning Approach, pp. 112–122 (2018). https://doi.org/10.1007/978-3-319-68385-0_10
5. Qin, T.L., Yan, D.H., Wang, G., Yin, J.: Water demand forecast in the baiyangdian basin with the extensive and low-carbon economic modes. J. Appl. Math. **2014**, 673485 (2014)
6. Raghavendra Prasad, J.E., Senthil, M., Yadav, A., Gupta, P., Anusha, K.S.: A comparative study of machine learning algorithms for gas leak detection. In: Ranganathan, G., Chen, J., Rocha, Á. (eds.) Inventive Communication and Computational Technologies, pp. 81–90. Springer, Singapore (2021)
7. Raj, K.S., Nishanth, M., Jeyakumar, G.: Design of Binary Neurons with Supervised Learning for Linearly Separable Boolean Operations, pp. 480–487 (2020). https://doi.org/10.1007/978-3-030-37218-7_54
8. Saleth, R.: Water scarcity and climatic change in india: The need for water demand and supply management. Hydrol. Sci. J.—J. Sci. Hydrol. **56**, 671–686 (2011). https://doi.org/10.1080/02626667.2011.572074
9. Ununiversity: Monitoring sustainability of rural water supplies in sub-saharan africa. United Nations University https://unu.edu/projects/monitoring-sustainability-of-rural-water-supplies-in-sub-saharan-africa-ph-d-sekela-twisa.html#outline
10. Wilson, D.L., Coyle, J.R., Thomas, E.A.: Ensemble machine learning and forecasting can achieve 99% uptime for rural handpumps. PLOS ONE **12**(11), 1–13 (2017). https://doi.org/10.1371/journal.pone.0188808, https://doi.org/10.1371/journal.pone.0188808

Chapter 7
Deep Learning Algorithms for Object Detection—A Study

A. Naveenkumar and J. Akilandeswari

Abstract In the field of computer vision, creating an effective object detection model is a difficult task. Object detection is the process of predicting objects and locating their instances in a given image. It is used in a wide range of applications, including self-driving cars, navigating visually impaired people in an indoor/outdoor environment, counting crowds, detecting vehicles, tracking objects, etc. Traditionally, the opencv's feature extraction and feature detecting algorithms are employed to object detection. However, in the real world, the performance of those algorithms is unsatisfactory. In recent days, different deep learning models are available to extract and learn the objects' features to detect the objects and also the performance of these algorithms is acceptable in the real environment. This paper compares the SSD, YOLOv3 and YOLOv4 for detecting objects in the indoor environment and discusses the aspects of sparse and dense prediction for the various deep learning algorithms used for object detection.

7.1 Introduction

Object detection is the process of classifying and locating the instance of an object in the image. Object detection provides the solution to numerous essential vision challenges such as labeling, scene perception, object recognition, image captioning, movement or behavioral recognition, etc. Also, it plays an important role in autonomous navigation, e-commerce, insurance agencies, health care, content-based image recovery, smart video surveillance, etc. Problems with object detection have gained considerable attention in the current decade. Deep learning algorithms are the most powerful extraction techniques and find patterns automatically in the given data. The latest advances in the methodology of deep learning have given the

A. Naveenkumar (✉) · J. Akilandeswari
Department of IT, Sona College of Technology (Autonomous), Salem, Tamil Nadu, India

J. Akilandeswari
e-mail: akilandeswari@sonatech.ac.in

© The Author(s), under exclusive license to Springer Nature Singapore Pte Ltd. 2022   65
V. Bhateja et al. (eds.), *Evolution in Computational Intelligence*,
Smart Innovation, Systems and Technologies 267,
https://doi.org/10.1007/978-981-16-6616-2_7

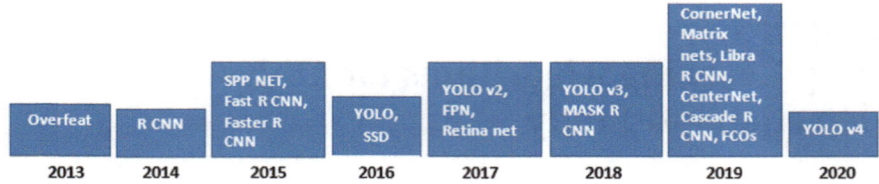

Fig. 7.1 HYPERLINK "sps:id::fig1∥locator::gr1∥MediaObject::0" Evolution of Deep learning object detection algorithm

best results in object detection. The deep learning object detection algorithms are Over feat, RCNN, Fast RCNN, Faster RCNN, Libra RCNN, Cascade RCNN, Retina Net, SPPnet, FPN, SSDnet, YOLO V1, YOLO V2, YOLO V3 and YOLO V4 [2, 3, 4] that operate on the advantages of CNN [1] for object detection. And also, these algorithms extract the features from either bounding boxes or key points to detect the objects. Figure 7.1 shows the evolution of the Deep learning object detection algorithm. The algorithms are broadly categorized into single-stage object detector and two-stage detector based on the feature detection techniques. Compared to the two-stage detector performance, the one-stage detector gives good results. In this paper, the performance of object detection algorithms namely, SSD, YOLOv3 and YOLOv4 are compared.

The rest of the paper is structured as Sects. 7.2–7.5. Section 7.2 explains the litereature survey of object detection techniques. The summarization of object detection techniques is defined in Sect. 7.3. The experimental findings are investigated in Sect. 7.4. The conclusion is given in Sect. 7.5.

7.2   Related Work

Overfeat et al. [5], the first single step deep learning object detector, was introduced by Pierre Sermanet et al. For classification, localization and identification, the multi-scale and sliding window techniques were incorporated with CNN. The bounding boxes are accumulated in order to maximize the detection confidence. On the ILSVRC2013, overfeat is assessed. Girshick et al. [6] presented the methods for the regional proposal to detect objects on the PASCAL VOC 2010–2012 and ILSVRC2013 datasets named as region with CNN (RCNN). Input images were split into almost 2000 regions and then each region was passed to the CNN to detect the features followed by applied SVM to identify the objects.

He et al. [7] have developed an alternative approach called Spatial Pyramid Pooling Network (SPPnet) for CNN-based object detection. The SPP layer was located between the convolutional layer and the fully connected layer to solve the fixed size/scale problem. The fixed length representation is created by SPP layers and then passed to the fully connected layers to detect the objects. On the ImageNet and PASCAL VOC datasets, SPPnet demonstrated. Another approach Fast RCNN is an alternative to RCNN which had been implemented by Ross Girshick [8]. In this process, Region of Interest (RoI) principles were implemented. The input image was fed to convolutional layers, then RoI was applied to create a feature vector and softmax was used for object classification. On data sets VGG16 and PASCAL VOC 2010–2012, the Fast RCNN principle was trained and tested. One modification to the FAST RCNN was made and introduced as Faster RCNN by Ren et al. [9] to increase the detection speed. Two modules are used in this article. Regions are selected in the first module based on the performance of the convolutional layer. To detect the objects in the regions, the detected regions are then transferred to the FAST RCNN detector. The Regional Proposal network was the method of selecting regions from the input image. On the PASCAL VOC 2007 and 2012 and MS COCO dataset, the concepts were evaluated. Redmon et al. [10] implemented the You Look only once to detect human-like objects. This technique divided the image of the input into the S × S grid, and grids were given as the Convolutional layer input. The output of the convolutional layer was the confidence score with the label that was used to determine whether or not the object was present on that grid. If the score is zero, it implies that no entity existed. On the PASCAL VOC data set of S value 7, YOLO was evaluated.

The alternative methods for YOLO, called single shot detectors (SSD), were introduced by Liu et al. [12]. In this analysis, the area proposal techniques were eliminated, and the fixed collection of bounding boxes was used to detect the feature map. Then the scores in the boxes for the presence of objects are calculated. The model was evaluated on datasets of PASCAL VOC, MS COCO and ILSVRC. The previous work of YOLO was expanded by Redmon et al. [11], and the YOLO 900 was introduced (version 2). In this work, boundary boxes were detected using anchor boxes instead of selecting boundary boxes on fully connected layers, and batch normalization, high-resolution classifiers, and dimension clusters are also used to improve accuracy and speed. Object classification is achieved by the newly implemented darknet-19 model. The PASCAL VOC and MS COCO datasets were assessed for this work.

The Feature Pyramid Network (FPN) was introduced by Lin et al. [13] in deep learning for object detection. FPN is built for designing high-level semantic feature maps for all sizes of images. The FPN was evaluated on the COCO dataset with the Fast RCNN detector. The MASK RCNN definition had been developed by He et al. [14], which was helpful for segmentation detection and object detection. The same two steps of Faster RCNN were followed here. However, the second layer, Fast RCNN's ROI definition, was used to segment objects into regions and detect objects through regression. For implementing the MASK RCNN, the MS COCO dataset was used. The new one-stage object detector, which gave similar precision to the two-stage object detector RCNN named as RetinaNet, was presented by Lin et al. [15]. The Focal loss for handling the class imbalance was introduced in this work using the scalable method of cross-entropy loss. Method of focal loss was tested on the MS COCO dataset. Another YOLO version 3 (v3) definition, which was the extension work of YOLOv2, was introduced by Redmon et al. [16]. The darknet-53 was used for classification and predicted the object score using logistic regression for each boundary box and was evaluated on the MS COCO dataset.

Law et al. [17] proposed CornerNet, a new one-stage object detector based on the pair of key points for detecting the object. The pair of corners is the top-left corners and bottom-right corner of the bounding boxes. The hourglass network was used as the backbone of CornerNet. Hourglass networks are placed to classify the objects. CornerNet was demonstrated on MS COCO dataset. The new deep object detection architecture, called Matrix Nets (xNets), has been introduced by Rashwan et al. [18]. It detects object corners of various sizes using various matrix layers, and aspect ratios are calculated using CNN-aware components. Simplify the matching process by removing the fully linked layers and directly pointing out the object center. The efficient method for balanced learning for object detection called Libra RCNN was introduced by Pang et al. [19]. This method overcomes the problems of imbalance when training the object detector. Sample level imbalance, feature level imbalance and objective level imbalance are calculated to maintain balanced training learning. Detect the region by combining balanced IoU sampling with a balanced pyramid for region selection and a softmax with a balanced L1 loss system. Another key point-based one-stage object detector named CenterNet has been introduced by Duan et al. [20], which uses the triplets (three) key points to detect objects. The cascade corner pooling and center pooling modules have been developed to detect the corners and center points. The cornet on the MS COCO dataset was evaluated. A new object detector, Cascade RCNN, is proposed by Cai et al. [21] to solve the false positive recognition problem and assign the precise bounding boxes for objects, i.e., object localization problem, by increasing the value of the intersection over union thresholds. Instance segmentation issue is also solved by Cascade RCNN. The MS COCO dataset is used for evaluating the Cascade RCNN. A Fully Convolutional one-stage (FCOS) object detector has been proposed by Tian et al. [22] to eliminate the region proposals and modules of anchor boxes, which prevents complicated intersection over union (IOU) computation and uses the per-pixel prediction for object detection. FCOS was tested on the MS COCO dataset. The new modern detector, called YOLO version 4(v4), was introduced by Bochkovskiy et al. [23]. There are two parts of YOLOv4; a base model that is trained on the ImageNet and the head component to define the object and identify object boundary boxes. Weighted residual connections (WRC), cross-stage partial connections (CSP), self-adversarial training (SAT), activation of Mish, augmentation of a mosaic image, and normalisation of cross mini-batch normalisation (CmBN) are used to classify objects.

7.3   Summarization of Object Detection Techniques

Collecting the data sets, designing the neural networks, training the model and validating the model are the steps to designing an efficient deep learning algorithm. The detector can use either customized datasets or predefined datasets for training. More time is required for customized data set collection. For training and validating the neural networks, a deep learning algorithm requires high computational hardware such as the Graphical Processing Unit (GPU) and the Tensor Processing Unit

(TPU). The mean average precision metric has been used to measure the accuracy of the object detector. Precision means how accurately the model predicts or the percentage of successful predictions. IOU measures the overlap between the real boundary boxes (ground truth) and the predicted boundary boxes. IOU threshold values are predefined in a few datasets to classify whether the prediction is true or false. Frames per second measure the speed of the object detector. The following Table 7.1 shows the summarization of deep learning object detection techniques.

From the literature review, we inferred that the following deep learning algorithms, such as SSD, YOLOv3, and YOLOv4, detect objects at 59fps, 67fps, and 165fps, respectively. SSD, YOLOv3 and YOLOv4 algorithms are used for this research since these algorithms have good accuracy and speed.

7.4 Experimental Setup and Analysis

7.4.1 Environmental Setup

Google Colab [28] has been used for the execution of YOLOv3 and YOLOv4 for this research work. The TensorFlow [27] framework has been used in this work. The GPU and TPU environment for training the model is given by Google Colab. The cushioned chair, telephone, S style chair and office table are taken from the indoor items in this study. Teslav100 16 GB GPU is used for training the model.

7.4.2 Data Collection and Labelling

360° video of cushion chair, telephone, S type chair and office table was taken for this study work of comparison of object detection in the indoor environment. After that, frames were collected, and then the labeling [26] of data was carried out. The collected input image size is 224×224. One of the familiar data labeling strategies is bounding boxes. In bounding boxes, rectangular boxes are used to describe the position of objects, and bounding boxes are usually defined by either two coordinates $(\times 1, y1)$ and $(\times 2, y2)$ or by one bounding box coordinate $(\times 1, y1)$ and width (w) and height (h). The sample image labeling is shown in Fig. 7.2.

7.4.3 Training

After data labeling and downloads the weights of mobilenet of SSD [29] and darknet weights of YOLOv3 [24] and YOLOv4 [25] are downloaded. Then, the two algorithms are trained by setting classes $= 4$ (because four objects), the number of training

Table 7.1 Summarization of deep learning object detection techniques

Object Detection Techniques	Authors	Type-based on a stage	Dataset Used for evaluation	Accuracy (%)	Speed
Overfeat	Sermanet et al.	Single-stage object detector	ILSVRC2013	24.3	–
RCNN	Girshick et al.	Two-stage object detector	PASCAL VOC 2010–12	53.7	–
			ILSVRC2013	31.4	
SPPNet	He et al.	Two-Stage Object detector	PASCAL VOC 2007	59.2	–
Fast RCNN	Girshick	Two-Stage Object detector	PASCAL VOC 2007	70.0	5 fps
			PASCAL VOC 2012	68.4	
			VGG16	61.4	
Faster RCNN	Ren et al.	Two-Stage Object detector	PASCAL VOC 2007	73.2	17 fps
			PASCAL VOC 2012	70.4	
			MS COCO	75.9	
YOLO	Redmon et al.	Single-stage object detector	PASCAL VOC 2007	59.2	45 fps
			PASCAL VOC 2012	57.9	
SSD	Liu et al.	Two-Stage Object detector	PASCAL VOC 2007	74.3	59 fps
			PASCAL VOC 2012	72.4	
			COCO	31.2	
			ILSVRC2013	43.4	
YOLO900(v2)	Redmon et al.	Single-stage object detector	PASCAL VOC 2007	78.6	40 fps
			PASCAL VOC 2012	73.4	
FPN	Lin et al.	Single-stage object detector	MS COCO	36.2	–
MASK RCNN	He et al.	Two-Stage Object detector	MS COCO	35.7	11 fps
RetinaNet	Lin et al.	Single-stage object detector	MS COCO	39.1	5.4fps
YOLOv3	Redmon et al.	Single-stage object detector	MS COCO	33.0	67 fps

(continued)

Table 7.1 (continued)

Object Detection Techniques	Authors	Type-based on a stage	Dataset Used for evaluation	Accuracy (%)	Speed
CornerNet	Law et al.	Single-stage object detector	MS COCO	42.1	–
Matrix Nets	Rashwan et al.	Single-stage object detector	MS COCO	47.8	–
Libra RCNN	Pang et al.	Two-Stage Object detector	MS COCO	43.0	–
CenterNet	Duan et al.	Single-stage object detector	MS COCO	47.0	28 fps
Cascade RCNN	Cai et al.	Two-Stage Object detector	MS COCO	42.8	7 fps
FCOS	Tian et al.	Single-stage object detector	MS COCO	44.7	7.7fps
YOLOv4	Bochkovskiy et al.	Single-stage object detector	MS COCO	43.5	165 fps

Cushion Chair   Telephone   Rolling Chair   S type Chair   Office Table

Fig. 7.2 Sample Image labeling

iterations $= 10{,}000$, learning rate $= 0.001$, batch size $= 64$. Figure 7.3 shows the loss and accuracy of SSD, YOLOv3 and YOLOv4 object detection algorithms. From this Fig. 7.3, it is pointed out, YOLOv4 has approximately 200 times faster than SSD and YOLOv3.

7.4.4 Results Discussion

In this experiment, evaluate the performance of object detection algorithms in an indoor environment. The test data set is utilized to validate the object detection

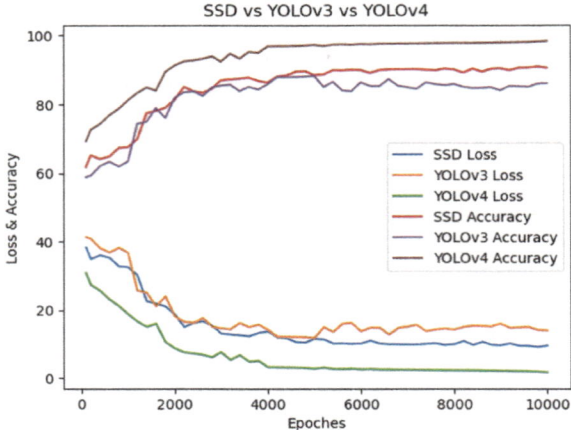

Fig. 7.3 Loss and accuracy of SSD, YOLOv3 and YOLOv4

model such as YOLOv3 and YOLOv4. The sample result output is demonstrated in Fig. 7.4.

In Fig. 7.4, it can be noted that the SSD and YOLOv3 algorithm doesn't detect few objects in the test image. The reason for not detecting an object is the size of the objects, and also there is overlapping of objects. YOLOv4 algorithm produces better results when compared to SSD and YOLOv3. It is also believed that the YOLOv4 algorithm detects almost all the objects in the test image. The accuracy of object

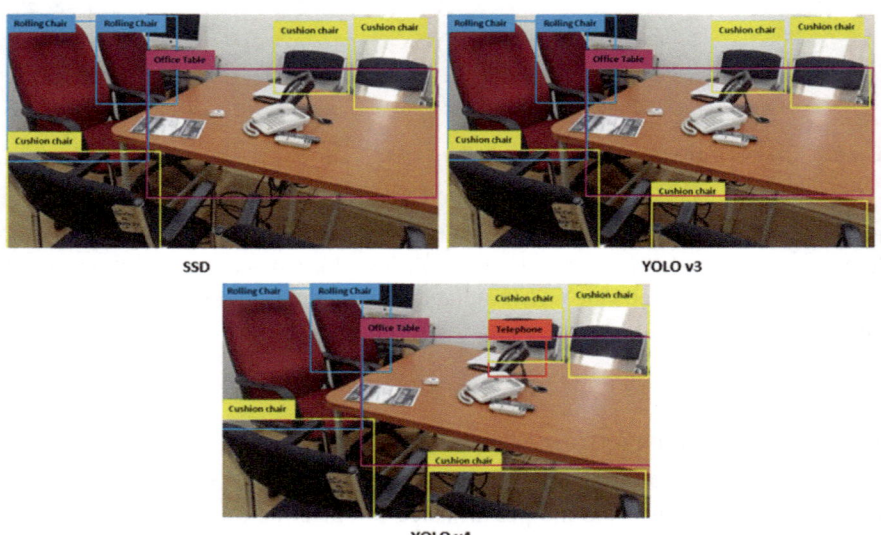

Fig. 7.4 Result of algorithm SSD, YOLO v3 and YOLO v4

Table 7.2 Object detection accuracy

Test image file name	Total number of trained objects in the test image	Number of detected objects			Accuracy based on detection		
		SSD	YOLOv3	YOLOv4	SSD (%)	YOLOv3 (%)	YOLOv4 (%)
Test1.jpg	13	11	10	12	84.61	76.92	92.30
Test2.jpg	12	10	8	11	83.33	66.66	91.66
Test3.jpg	6	5	5	6	83.33	83.33	100

detection models is presented in Table 7.2. The table consists of test images, the number of actual objects in the test image, the number of detected objects by SSD, YOLOv3 and YOLOv4 algorithms. In respect of test images Test1.jpg and Test2.jpg, the YOLOv4 algorithm performs better than the SSD and YOLOv3 algorithm. It detects almost all the objects in the actual input test image.

7.5 Conclusion

In this paper, various deep learning-based object detection algorithms are briefly analyzed. The analysis is categorized based on dense and sparse features of the image dataset. When compared to other objected detection algorithms, the SSD, YOLOv3 and YOLOv4 algorithms provide better accuracy and speed. In this research, SSD, YOLOv3 and YOLOv4 object detection algorithms are evaluated to detect the objects in an indoor environment. The experimental results show that the YOLOv4 algorithm efficiently detects the objects better than SSD and YOLOv3 in an indoor environment. SSD, YOLOv3 algorithm does not properly detect the smaller size of objects and overlapping objects in the test image. It is also noticed that YOLOv4 takes the least amount of time to train the model. In the future, this work can be extended to detect untrained objects in a specific indoor environment.

Declaration The competent authorities allowed us to use the images/data as provided in the study. We shall be entirely responsible for any dispute in the future.

References

1. Krizhevsky, A., Sutskever, I., Hinton, G.E.: Imagenet classification with deep convolutional neural networks. Commun. ACM **60**(6), 84–90 (2017)
2. Zhao, Z.Q., Zheng, P., Xu, S.T., Wu, X.: Object detection with deep learning: A review. IEEE transactions on neural networks and learning systems **30**(11), 3212–3232 (2019)

3. Liu, L., Ouyang, W., Wang, X., Fieguth, P., Chen, J., Liu, X., Pietikäinen, M.: Deep learning for generic object detection: A survey. Int. J. Comput. Vision **128**(2), 261–318 (2020)
4. Han, J., Zhang, D., Cheng, G., Liu, N., Xu, D.: Advanced deep-learning techniques for salient and category-specific object detection: a survey. IEEE Signal Process. Mag. **35**(1), 84–100 (2018)
5. Sermanet, P., Eigen, D., Zhang, X., Mathieu, M., Fergus, R., LeCun, Y.: Overfeat: Integrated recognition, localization and detection using convolutional networks (2013). arXiv:1312.6229
6. Girshick, R., Donahue, J., Darrell, T., Malik, J.: Rich feature hierarchies for accurate object detection and semantic segmentation. In: Proceedings of the IEEE Conference on Computer Vision and Pattern Recognition, pp. 580–587 (2014)
7. He, K., Zhang, X., Ren, S., Sun, J.: Spatial pyramid pooling in deep convolutional networks for visual recognition. IEEE Trans. Pattern Anal. Mach. Intell. **37**(9), 1904–1916 (2015)
8. Girshick, R.: Fast RCNN. In: Proceedings of the IEEE International Conference on CComputer Vision. pp. 1440–1448 (2015)
9. Ren, S., He, K., Girshick, R., Sun, J.: Faster RCNN: Towards real-time object detection with region proposal networks. In: Advances in Neural Information Processing Systems, pp. 91–99 (2015)
10. Redmon, J., Divvala, S., Girshick, R., Farhadi, A.: You only look once: Unified, real-time object detection. In: Proceedings of the IEEE Conference on Computer Vision and Pattern Recognition, pp. 779–788 (2016)
11. Liu, W., Anguelov, D., Erhan, D., Szegedy, C., Reed, S., Fu, C. Y., Berg, A.C.: Ssd: Single shot multibox detector. In: European Conference on Computer Vision, pp. 21–37. Springer, Cham (2016)
12. Redmon, J., & Farhadi, A.: YOLO9000: better, faster, stronger. In: Proceedings of the IEEE Conference on Computer Vision and Pattern Recognition, pp. 7263–7271 (2017)
13. Lin, T.Y., Dollár, P., Girshick, R., He, K., Hariharan, B., elongie, S.: Feature pyramid networks for object detection. In: Proceedings of the IEEE Conference on Computer Vision and Pattern Recognition, pp. 2117–2125 (2017)
14. He, K., Gkioxari, G., Dollár, P., Girshick, R.: Mask RCNN (2017). arXiv:1703.06870.
15. Lin, T.Y., Goyal, P., Girshick, R., He, K., Dollár, P.: Focal loss for dense object detection. In: Proceedings of the IEEE International Conference on Computer Vision, pp. 2980–2988 (2017)
16. Redmon, J., Farhadi, A.: YOLOv3: An incremental improvement (2018). arXiv:1804.02767.
17. Law, H., Deng, J.: Cornernet: Detecting objects as paired keypoints. In: Proceedings of the European Conference on Computer Vision (ECCV), pp. 734–750 (2018)
18. Rashwan, A., Kalra, A., Poupart, P.: Matrix nets: A new deep architecture for object detection. In: Proceedings of the IEEE International Conference on Computer Vision Workshops (2019)
19. Pang, J., Chen, K., Shi, J., Feng, H., Ouyang, W., Lin, D.: Libra RCNN: Towards balanced learning for object detection. In: Proceedings of the IEEE Conference on Computer Vision and Pattern Recognition, pp. 821–830 (2019)
20. Duan, K., Bai, S., Xie, L., Qi, H., Huang, Q., Tian, Q.: Centernet: Keypoint triplets for object detection. In: Proceedings of the IEEE International Conference on Computer Vision, pp. 6569–6578 (2019)
21. Cai, Z., Vasconcelos, N.: Cascade RCNN: high quality object detection and instance segmentation. IEEE Trans. Pattern Anal. Mach. Intell. (2019)
22. Tian, Z., Shen, C., Chen, H., He, T.: Fcos: Fully convolutional one-stage object detection. In: Proceedings of the IEEE International Conference on Computer Vision, pp. 9627–9636 (2019)
23. Bochkovskiy, A., Wang, C.Y., Liao, H.Y.M.: YOLOv4: Optimal Speed and Accuracy of Object Detection (2020). arXiv:2004.10934.
24. https://pjreddie.com/darknet/yolo/
25. https://github.com/pjreddie/darknet
26. GitHub: LabelImg: A graphical image annotation tool (2015)

27. Abadi, M., Barham, P., Chen, J., Chen, Z., Davis, A., Dean, J., Kudlur, M.: Tensorflow: A system for large-scale machine learning. In: 12th {USENIX} Symposium on Operating Systems Design and Implementation ({OSDI} 16), pp. 265–283 (2016)
28. Bisong, E.: Google colaboratory. In: Building Machine Learning and Deep Learning Models on Google Cloud Platform, pp. 59–64. Apress, Berkeley, CA (2019)
29. http://download.tensorflow.org/models/object_detection/ssd_mobilenet_v2_coco_2018_03_29.tar.gz

Chapter 8
A Novel Multiblock Region-Based Arnold Transformation for Image Watermarking Combined with DWT-PSO Technique

Kumari Rinki, Pushpneel Verma, and Ranjeet Kumar Singh

Abstract To ensure safe transmission of data over the internet, digital watermarking is considered to be most promising technique without compromising any duplication, tampering and copyright infringement of data. Unlike conventional Arnold scrambling method, the proposed method is bounded with two-level security in which blocks are scrambled with one key and simultaneously each block is scrambled at pixel level with second key for the same image which increases the security degree of watermark. Further, the encrypted image is used for digital watermarking which is based on DWT-PSO technique. The system also utilizes PSO algorithm to find out the optimal scaling factor to embed the watermark to obtain the highest possible robustness without failing the transparency. The proposed method's performance against imperceptibility and system robustness is evaluated using PSNR and NCC metrices. The proposed method offers a high PSNR value (54.619 dB), which is sufficient to meet any watermarking scheme's imperceptibility requirements. Furthermore, as compared to the NCC values of other international journals, the proposed method's NCC values are nearly 1 for the majority of the attacks. MATLAB tool is used to simulate the results and it generates better result with respect to imperceptibly and more robust under certain attacks as compared with existing systems.

K. Rinki (✉)
Bhagwant University, Ajmer 305004, India

P. Verma
Department of Computer Science and Engineering, Bhagwant University, Ajmer 305004, India

R. K. Singh
Sarla Birla University, Rachi, India

© The Author(s), under exclusive license to Springer Nature Singapore Pte Ltd. 2022   77
V. Bhateja et al. (eds.), *Evolution in Computational Intelligence*,
Smart Innovation, Systems and Technologies 267,
https://doi.org/10.1007/978-981-16-6616-2_8

8.1 Introduction

In today's cyberworld, the conventional copyright law is not enough to protect secret information transfer over the world via Internet. Therefore, it has become very critical to acquire protection for digital data over the network and between individuals. Among various protection mechanism digital watermarking mechanism overcomes the pitfalls of conventional intellectual property rights to protect the multimedia content [1]. The process of digital watermarking technique is subjected to two consecutive module such as embedding module and extracting module. The embedding module begins with embedding the confidential information such as logo, text, image in carrier signal such as image, audio or video object. Here the confidential information is known as watermark and the output of this module is watermarked object. During embedding process, one must take care of tradeoff between robustness and perceptibility quality of watermarked object [2]. Various traditional schemes are developed to design a watermarking system which are categorized on the basis of domain in which they are implemented such as spatial domain and frequency domain approach. Spatial domain approach deals with substitution or alteration of pixels whereas frequency domain works on transforming the original image into frequency domain coefficient before embedding process [14]. Some widely used algorithm of frequency domains are DFT (Discrete Fourier Transform), DWT (Discrete Cosine Transform), DCT (Discrete Wavelet Transform), SVD (Singular Value Decomposition), etc. [1].

This paper introduces a robust hybrid approach of multi-level data security applying on multiblock region-based Arnold transformation encryption algorithm followed by multi-level DWT and PSO techniques. As compared with conventional Arnold transform, the use of improved Arnold transform increases the degree of security due to increase of keys. Next, the different RGB component of multi-level DWT's approximation detail sub band is used to embed the watermark successively, which improves the imperceptibly and robustness of watermark. Thirdly, it makes an effective use of optimization algorithm PSO for estimation of optimal embedding factor, which undoubtedly enhance the robustness of the algorithm without losing its transparency.

The outline of the paper is as follows: Sect. 8.2 briefly describes all mathematical primaries used in the proposed work. Section 8.3 illustrates the proposed techniques and also explains the evaluation of fitness function. Section 8.4 reports the experimental results and gives the performance analysis and comparisons with the other methods. Finally, the conclusion is drawn in Sect. 8.5.

8.2 Mathematical Preliminaries

8.2.1 Conventional Arnold Pixel Scrambling Method

Arnold tranformation is one of the most widely used techniques to encrypt and decrypt the image due to its simplicity, periodicity and invertibility in nature. The mathematical model of Arnold transformation is expressed as [3, 4]:

$$\begin{bmatrix} x_{n+1} \\ y_{n+1} \end{bmatrix} = \begin{bmatrix} 1 & p \\ q & pq+1 \end{bmatrix} \begin{bmatrix} x_n \\ y_n \end{bmatrix} (mod\, N) \tag{8.1}$$

Since Arnold transformation is iterative process, the original image (N × N) can be resorted again, when applying it repletely on transformed image with certain iteration T. But it has to wait for long period of time to restore the image. T is called the period of the transformation and it depends on parameters p, q, and N as shown above in Eq. (8.1). These parameters are also used as secret keys. Here, (x_n, y_n) and (x_{n+1}, y_{n+1}) are the positions of the pixel points in the original image and the transformed scrambled image, respectively.

The mathematical model of inverse Arnold transformation is expressed as [11]:

$$\begin{bmatrix} x_n \\ y_n \end{bmatrix} = \begin{bmatrix} pq+1 & -q \\ -p & 1 \end{bmatrix} \begin{bmatrix} x_{n+1} \\ y_{n+1} \end{bmatrix} (mod\, N) \tag{8.2}$$

8.2.2 Improved Multiblock Region-Based Arnold Transformations Method

The special feature of Arnold transformation is that it uses periodicity. Sometimes it makes the system not robust against the brute force attacks. To improve the security and robustness of the image, the proposed Arnold transformation is not restricted only to the pixels level of an image but also applied on the multiblock region of an image as shown in Fig. 8.1. Unlike conventional Arnold scrambling method, the proposed

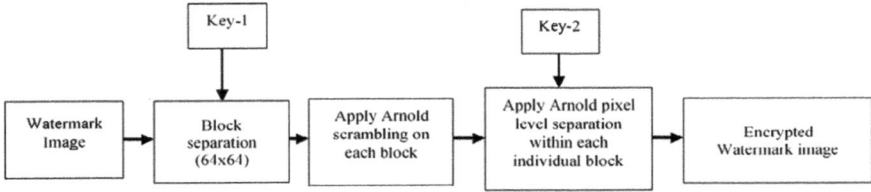

Fig. 8.1 A Multiblock region-based scrambling process

method is bounded with two-level security in which blocks are scrambled with one key and simultaneously each block is scrambled at pixel level with second key. If the first level of descrambling is obtainable then only the second level of descrambling is workable. This improves the complexity of malicious and unauthorized descrambling of an image. Block scrambling through Arnold transform is given as follows.

$$\begin{bmatrix} \{B(x_{i+1})\} \\ \{B(y_{i+1})\} \end{bmatrix} = \begin{bmatrix} 1 & u \\ v & uv+1 \end{bmatrix} \begin{bmatrix} \{B(x_i)\} \\ \{B(y_i)\} \end{bmatrix} (mod M) \tag{8.3}$$

Here, $(B(x_i, y_i)$ represent the block coordinates of an image and M denotes the size of each block.

Scrambling Process. In this process, the color image size of 256×256 is divided into size of 32×32 blocks. Hence, each block is assigned a spatial coordinate in the set of $\{(1, 1), (1, 2) \ldots, (1, 8) \ldots, (8, 1), (8, 2) \ldots, (8, 8)\}$. Now, Arnold transformation is applied onto these blocks with certain iteration (key1) to get a new spatial coordinate according to Eq. (8.3). Simultaneously Arnold pixel scrambling is applied within each block with certain number of iteration (key2) to get the final scrambled image applying Eq. (8.1) [4].

Example of block scrambling process is illustrated in Fig. 8.2. For simplicity consider image of size 512×512 is divided into 16 blocks of size 128×128. After applying Arnold transformation on each block, the coordinates of each block get transferred to new spatial coordinates. For example, after scrambling each block gets transferred from one position to new position. Like block 10 get shifted from its original position (3, 2) to new position i.e., (3, 4) as shown in Fig. 8.2.

Descrambling Process. The Arnold inverse transformation is applied on to scrambled image is known as descrambling process. To get the original image first the scrambled image is divided into blocks and Arnold inverse pixel transformation is carried out to descramble all the pixels in each block of the image based on key2. Again, Arnold inverse transformation is applied to blocks based on key 1 and it reverses back to its correct position to get the original image.

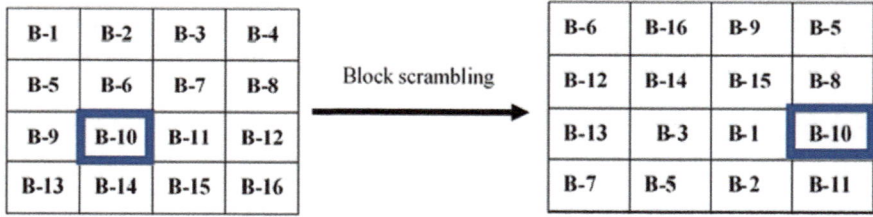

Fig. 8.2 Example of block scrambling process

8.2.3   Discrete Wavelet Transforms (DWT)

The DWT is one of the essential tools of image processing for transforming a signal from spatial domain to frequency domain. Its multiresolution capability of representing an image dragged attention to research helps in feature extraction process in pattern recognition [8]. That's why DWT is more suitable to model the behaviors of Human Visual System (HVS) than other mathematical tools like cosine transform and Fourier transform. It separates the image into four sub band which are different from each other on the basis of their different resolution assets, these are LL, LH, HL and HH (L-Low and H-High) [7]. The first sub band (LL) has low resolution component part, concentrates more energy of an image, means its possess approximation details of an image. The other three bands represent (LH, HL, HH) high-resolution frequency components that have concentrated in the locality of areas related to edges activities of an image such as its texture, edges, contour, etc.

8.2.4   Theoretical Concept of PSO

PSO is biological inspired metaheuristic optimization technique which simulated the behavior of flock of birds, honey bees, school of fishes, where each particle moves around the search space to fill their need and obtain the optimal solution [12].

Mathematical model of PSO. Let there are m particles as a potential solution in N-dimensional space, where each ith particle's position and velocity can be represented as $x_i = (x_{i1}, x_{i2}, \ldots, x_{iN})$ and $v_i = (v_{i1}, v_{i2}, \ldots, v_{iN})$. Each particle also memorized its personal best position (the particle has visited so far) and also respect toward the global best position found by the team. At each iteration, the particles update its velocity and position by the given formula [5, 6, 9, 10].

$$v_{in}(t + 1) = wv_{in}(t) + r_1 c_1 (p_{in}(t) - x_{in}(t)) + r_2 c_2 (g_{in}(t) - x_{in}(t)) \tag{8.4}$$

$$x_{in}(t + 1) = x_{in}(t) + v_{in}(t + 1)$$

where $v_{in}(t)$ and $x_{in}(t)$ denotes velocity and position vector of particles in the n-dimensional component at tth iteration particle. The constant terms c_1 and c_2 are positive acceleration coefficients; r_1 and r_2 are random numbers in range [0, 1] and w is the inertia weight.

8.3 Proposed Work

The proposed algorithm comprises of watermark embedding and watermark extracting process. The details of the two processes are given below.

Watermark Embedding Process. The working steps of watermark embedding are described in following steps and also shown more intuitively in Fig. 8.3.

Input: Cover image (I) of size M x M and watermark image (W) of size (p × p).

Output: Watermarked image (I_w).

1. Apply two-level DWT on cover image using Haar wavelet to obtain a series of multiresolution fine sub-shapes LL_2, HL_2, LH_2 and HH_2. Select approximation coefficient LL_2 for further processing.

$$LL_2 = \text{Two level DWT (I, Haar)}.$$

2. Next, LL_2 sub band are decomposed into read, green and blue (RGB) components (LL_{R2}, LL_{G2} and LL_{B2}) and one component at a time is selected.
3. Encrypt the watermark image with improved AT (Arnold transformation) as described in Sect. 8.2.
4. At step 4, encrypted watermark w_e is decomposed into four subsequent level $LL_{w2}, HL_{w2}, LH_{w2}$ and HH_{w2} after applying two-level DWT on it. Select approximation coefficientLL_{w2}.

$$LL_{w2} = \text{Two level DWT } (w_e, \text{Haar}).$$

5. Now step 2 is repeated for LL_{w2} sub band of encrypted watermark to get three different channels: red, green and blue, respectively (LL_{wR2}, LL_{wG2} and LL_{wB2}).

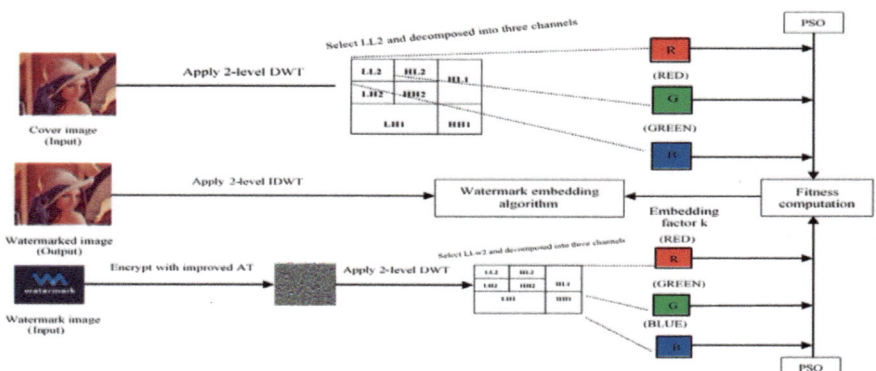

Fig. 8.3 Watermark embedding process

6.   Apply PSO process to compute the fitness function as described in Eq. (8.4).
7.   The output of fitness function will provide optimal embedding strength factor to embed the watermark in cover image.
8.   Select red, green and blue components of LL_2 sub band of cover image to embed the red, green and blue components of watermark such as

$$LL'_{R2} = LL_{R2} + k * LL_{wR2} \text{(Embedding Red component)}$$
$$LL'_{G2} = LL_{G2} + k * LL_{wG2} \text{(Embedding Green component)}$$
$$LL'_{B2} = LL_{B2} + k * LL_{wB2} \text{(Embedding Blue component)}$$

where k is the optimal embedding strength factor optimized by PSO. Two-level inverse DWT is used to apply on all three embedding components to get the image containing watermark information.

$$R_w = \text{Two level IDWT } (LL'_{R2}, Haar)$$
$$G_w = \text{Two level IDWT } (LL'_{G2}, Haar)$$
$$B_w = \text{Two level IDWT } (LL'_{B2}, Haar)$$

9.   Combine all three components together to get final watermarked image (I_w).

$$I_w = \text{RGB component}(R_w, G_w, B_w)$$

Watermark Extraction Process. The extraction of watermark is the reverse process of watermark insertion. The watermark extraction steps are stated below:

Input: Watermarked image (I_w).

Output: Extracted watermark (w').

1.   Perform two-level DWT on I_w to get four sub band $LL'_2, LL'_2, LL'_2 and LL_2$. Like the embedding algorithm select approximation coefficient LL'_2 for further processing.

$$LL'_2 = \text{Two level DWT } (\boldsymbol{I_w, Haar})$$

2.   Now split LL'_2 into RGB components LL'_{R2}, LL'_{G2} and LL'_{B2} accordingly and select one component at a time for extraction.
3.   The extraction is done in following manner from every RGB component.

$$LL'_{wR2} = (LL'_{R2} - LL_{R2})/k$$
$$LL'_{wG2} = (LL'_{G2} - LL_{G2})/k$$
$$LL'_{wB2} = (LL'_{B2} - LL_{B2})/k$$

4.   Two-level inverse DWT is used to apply on all three embedding components to get the image containing encrypted watermark information.

$$w'_R = \text{Two level IDWT } (LL'_{wR2}, \text{Haar})$$
$$w'_G = \text{Two level IDWT } (LL'_{wG2}, \text{Haar})$$
$$w'_B = \text{Two level IDWT } (LL'_{wB2}, \text{Haar})$$

5. Combine all three components together to get final watermark image (w').

$$w' = \text{RGB component}(w'_R, w'_G, w'_B)$$

Optimal Embedding Strength Factor (k) Assessment. The proposed method employs PSO as preprocessing step in order to find out the best (optimal) embedding strength factor to embed the watermark [7], such that the watermarking algorithm obtained the highest possible robustness without losing the transparency. To evaluate the optimal scaling factor the fitness computation formula is depicted as [8]:

$$fitness_i = \left[\frac{1}{\left(\frac{1}{t}\sum_{i=1}^{t} \max(corr_{wi}(ww'_i))\right)} - corr_I(I, I_w) \right]^{-1} \quad (8.5)$$

The fitness function is calculated in the presence of well-known four attacks such as applying median filter(3×3), rotation (0.5), adding salt and pepper noise (0.002) and resizing an image, although other attacking schemes can also be considered or replaced with those used in the PSO optimization process. The terms $corr_I$ and $corr_{wi}$ represent the transparency and robustness, respectively. The term $corr_I$ find out the correlation value between the original and watermarked image, i.e., $corr_I = corr(I, I_w)$, whereas $corr_{wi}$ parameter estimates the correlation value between original watermark and extracted watermark, i.e. $corr_{wi} = corr(w, w'_i)$ under the certain attacks as mentioned above. The $corr_I$ and $corr_{wi}$ are feed back to the PSO repletely until a predefined stopping criterion is satisfied to find out the optimal scaling factor (k). While simulating PSO the set of parameters values were identified for number of particles m is 80, c1 and c2 are chosen as 1.4 and 1.8, respectively, inertia weight is identified as 0.7 and the number of iteration T is considered as 150.

8.4 Comparative Analysis on Experimental Results

Two different metrices PSNR and NCC are considered to evaluate the performance of the proposed method against imperceptibly and system robustness [1]. The proposed method is implemented on MATLAB tool under Window 10 operating system. PSNR is used to evaluate the visible quality of image where NCC figures out the quality of extracted watermark from attacked watermark image. Two standard color images Lena (512×512) and logo (256×256) are taken as cover image and watermark image as shown in Figs. 8.4 and 8.5. The proposed method provides high PSNR value (54.619 dB) which is good enough to maintain the imperceptibility criteria

Fig. 8.4 Cover image

Fig. 8.5 Watermark and
encrypted watermark

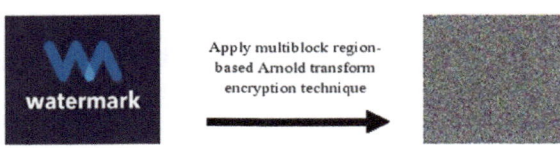

of any watermarking scheme. Table 8.1 shows the different common attacks like rectangular cropping, rotation, salt and paper noise, Gaussian noise, resizing and JPEG compression with quality factor (QF) 10, which are applied on watermarked image to evaluate its robustness. In addition, Fig. 8.6 represents the attacked watermark images and the extracted watermark images from it. While comparing its NCC values with other international journals [3, 7, 13], for most of the attacks the proposed method NCC values are nearly 1. In a summarized way, we can conclude that hybrid approach of AT-DWT-PSO provides better robustness, and can extract watermark with good visual effect under most attacks, so it can be used in copyright protection and content authentication applications.

Table 8.1 NCC comparison results with several existing schemes

Attack-scheme	Proposed scheme	Reference [3] scheme	Reference [7] scheme	Reference [13] scheme
No attack	1	1	1	1
Gaussian noise (0.01)	0.9921	0.9875	0.8071	0.9985
Cropping	0.9895	0.9988	0.7894	0.9865
Rotation	0.9618	0.9381	0.7721	0.9872
Resizing	0.9756	0.9955	0.8321	0.9885
Salt and pepper noise (0.01)	0.9998		0.8214	0.9917
JPEG compression (QF = 10)	0.9973	0.9993	0.9567	0.9898

Types of Attack	Attacked Watermarked Image	Extracted Watermark	NCC Value
Unattacked			1
Gaussian noise (0.01)			0.9921
Rectangular cropping (50 pixels per side)			0.9895
Rotation at 45 degree	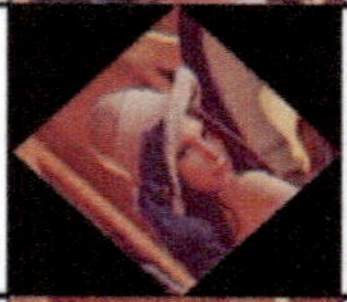		0.9898
Salt & Pepper noise (0.01)			0.9998
JPEG Compression (QF=10)			0.9973
Resizing			0.9756

Fig. 8.6 Several watermarked attacks and respective extracted watermark

8.5 Conclusion

In order to embed the watermark, two-level DWT decomposition is applied over host image and watermark image. This scheme utilizes the PSO search mechanism for determining the best position for concealing the watermark. It also helps to balance both the watermarked image imperceptibility quality and robustness of watermarking. The watermark is gradually embedded using the distinct RGB components of multi-level DWT's approximation detail sub band, which increases the watermark's imperceptibility and robustness. Various attacks are considered to test the robustness of proposed system. The comparison research also reveals that the proposed method is capable of extracting watermarks with good visual effect in the majority of attacks.

References

1. Rinki, K., Verma, P., Singh, R.K.: A novel matrix multiplication based LSB substitution mechanism for data security and authentication. J. King Saud Univ. Comput. Inf. Sci. (2021)
2. Agarwal, N., Singh, A.K., Singh, P.K.: Survey of robust and imperceptible watermarking. Multimedia Tools Appl. 78(7), 8603–8633 (2019)
3. Elayan, M.A., Ahmad, M.O.: Digital watermarking scheme based on Arnold and anti-Arnold transforms. In: International Conference on Image and Signal Processing, pp. 317–327. Springer, Cham (2016)
4. Kaur, S., Talwar, R.: Arnold transforms-based security enhancement using digital image watermarking with complex wavelet transform. Int. J. Electron. Eng. Res. 9, 677–693 (2017)
5. Singh, R., Ashok, A., Saraswat, M.: Optimized robust watermarking technique using CKGSA in DCT-SVD domain. IET Image Proc. 14(10), 2052–2063 (2020)
6. Aslantas, V., Dogan, A.L., Ozturk, S.: DWT-SVD based image watermarking using particle swarm optimizer. In: 2008 IEEE International Conference on Multimedia and Expo, pp. 241–244. IEEE (2008)
7. Tao, H., Zain, J.M., Ahmed, M.M., Abdalla, A.N., Jing, W.: A Wavelet-Based Particle Swarm Optimization Algorithm Applied to Digital Image Watermarking.
8. Lavanya, A., Natarajan, V.: Analyzing the performance of watermarking based on swarm optimization methods. In: Advances in Computing and Information Technology, pp. 167–176. Springer, Berlin, Heidelberg (2013)
9. Wang, Z., Sun, X., Zhang, D.: A novel watermarking scheme based on PSO algorithm. In: International Conference on Life System Modeling and Simulation, pp. 307–314. Springer, Berlin, Heidelberg (2007)
10. Saxena, A., Badal, N.: A deterministic digital watermarking approach based on bird swarm optimization. Int. J. Comput. Sci. Inf. Secur. 14(12), 103 (2016)
11. Rengarajan, A.: Data hiding in encrypted images using Arnold transform. ICTACT J. Image Video Process. 7(1) (2016)
12. Singh, R., Ashok, A.: An optimized robust watermarking technique using CKGSA in frequency domain. J. Inf. Secur. Appl. 58, 102734 (2021)
13. Kamble, S., Maheshkar, V., Agarwal, S., Srivastava, V.K.: DWT-SVD based robust image watermarking using Arnold map. Int. J. Inf. Technol. Knowl. Manage. 5(1), 101–105 (2012)
14. Wu, M., Liu, B.: Watermarking for image authentication. In: Proceedings 1998 International Conference on Image Processing. ICIP98 (Cat. No. 98CB36269), Vol. 2, pp. 437–441. IEEE (1998

Chapter 9
Automated Evaluation of SQL Queries: Eval_SQL

Bhumika Shah and Jyoti Pareek

Abstract The assessment of SQL queries is a time-consuming task for the teacher, as each query needs customized feedback. Automation of such a task can prove beneficial for students as well as teachers. Some of the semi-automated evaluation tools for SQL queries are reported in the literature though none of them provides Quantitative as well as Qualitative feedback. All the evaluation tools available for SQL queries provide a binary type of feedback, which results in the query being right or wrong. However, evaluation could be more meaningful if customized self-explanatory feedback is provided to the student stating the level of correctness of the query along with the description of the mistake committed (if any). Authors have developed "An Automated Assessment tool for SQL Queries: Eval_SQL" which provides the marks even for partially correct query (Quantitative) and the feedback on what went wrong in the query (Qualitative). This can improve the student's learning experience in the virtual world. Eval_SQL also helps to reduce teacher workload, allowing them to focus more on learning-centric tasks.

9.1 Introduction

Computer science courses demand extensive laboratory work, and proficiency in practical implementation is an essential skill required for computer science professionals. Implementation of concepts learned is one of the important criteria for any practical subject in computer science courses. However, the assessment of the said implementation is even more important. DBMS is an important subject taught in higher education in computer science. The practical concepts of DBMS are been implemented in Structured Query Language (SQL). Learning SQL syntax is not a difficult task, but converting a requirement given in natural language into an appropriate query is challenging. Students require a lot of practice to master the art of querying the data. The students are often not sure whether the query been written by them is correct or not. SQL has a very logical syntax, and though the SQL

B. Shah (✉) · J. Pareek
Department of Computer Science, Gujarat University, Ahmedabad, Gujarat, India

© The Author(s), under exclusive license to Springer Nature Singapore Pte Ltd. 2022
V. Bhateja et al. (eds.), *Evolution in Computational Intelligence*,
Smart Innovation, Systems and Technologies 267,
https://doi.org/10.1007/978-981-16-6616-2_9

queries look very simple to understand, translating them from simple English statement to a semantic query proves difficult for the students. For each query written, a teacher/instructor needs to provide customized feedback to the students on the correctness of the query. However, the different variants available in SQL queries demand higher practice for each variety of Query Set. For each such variant, a teacher provides the students with a set of practice questions and assignments to master the SQL queries. However, the assessment of such queries becomes a time-consuming task as each and every query is to be provided with the appropriate and customized feedback and at times the teacher is not able to provide timely feedback to each one of them as students are large in number. Hence, there is a need for a proper assessment system that can automate the evaluation of SQL queries resulting in timely feedback to students and a reduced workload of teachers. The automatic assessment tools can help reduce the burden from teachers by allowing them to focus on student-teacher interactions and other learning-centric tasks.

9.2 Background

There have been various systems proposed on Automated Evaluation of SQL Queries. Each of them differs in its functionality. Some of the systems reviewed evaluate the query just by showing a binary type of feedback like correct or incorrect. One of the Systems provides the table structure to the user and lets the student select the attributes and Auto Generates the Query. One such system is able to provide feedback, by taking peer review and eventually getting reviewed by the teacher. However, there is manual intervention; the system is not evaluating the queries (Table 9.1).

This rightly says that there are hardly any systems available that provide Automated Assessment of SQL Queries. The various tools/systems reviewed for SQL Learning and assessment are described in detail as follows:

In 1997, Kearns et al. [1] proposed a system for Learning SQL that displayed a sequence of images to depict query processing stages step by step on how the query result is determined. The authors have emphasized on visual representation to enhance the semantic understanding of students in the area of SQL.

In 2004, Sadiq et al. [2] proposed an Online SQL Learning workbench "SQLator," which is a web-based interactive tool for learning SQL. The author's claim is to have achieved a high rate of success in determining user queries as correct or incorrect. The authors have used heuristic algorithms for comparing the SQL Queries. However, the student is allowed to execute the query before submitting it for evaluation. Moreover, if the learner is unsuccessful in writing the query, they can access the correct solution.

In 2006, de Raadt et al. [3] proposed SQL tutoring and assessment tool "SQLify." They have combined many features of existing systems like the visualization of the database schema, Query Processing, Semantic feedback, and assessment. de Raadt et al. [3] proposed a pedagogical perspective of the said system in **2007**. They compare and review the existing tools concerning database perspective and pedagogical approach. Relational algebra support was added in the enhanced version.

Table 9.1 Comparison of the existing online assessment tools

Features	Tools SQL learning	Assessment of SQL queries available	Feedback on queries available	Partial grading for syntactically correct queries	Partial grading for syntactically incorrect queries2	Laboratory practice available in the same system	Database used
Esql [4]	✓	–	–	–	–	–	NA
SQLator [7]	✓	✓	–	–	–	✓	SQL Server 2000
Sqlify [6]	✓	✓	✓	–	–	–	MySQL
Assessql [5]	✓	✓	–	–	–	✓	PostgreSQL
ACME [8]	✓	✓	–	–	–	✓	Oracle, SQL Server
AutoGen SQL queries [2]	✓	–	–	–	–	–	PostgreSQL
Partial marking system [1]	✓	✓	✓	✓	–	✓	PostgreSQL
Eval_SQL (our proposed system)	✓	✓	✓	✓	✓	✓	GU_DB (our own DB prototype)

In 2006, another system was proposed by Soler et al. [5] named ACME which maintains a repository of the problems submitted by the teacher and a workbook module for students which consists of different exercises. The third module is the correction module, which compares student's solutions with the teacher's solution or with the output and gives the result to the student, whether correct or incorrect. Different pear libraries were integrated to support Oracle and SQL Server.

In 2014, Cruces et al. [6] proposed a system for the Automatic generation of SQL Queries. The system generates the queries automatically based on the selection. The user selects the schema, attributes, and functions that are used. The system generates the query based on the selection.

In 2016, Chandra et al. [7] proposed partial marking for SQL Queries. The system was developed at IIT Bombay and provided an interactive and automated platform for learning and assessment of SQL Queries. The system provides partial grading to the queries. The generated datasets are compared to determine the correctness of the query, and if the query is incorrect, partial assessment criteria are followed based on how close the student query is with the instructor query. However, the queries are considered only after they are syntactically correct.

As Raadt [4] rightly says "Relational query languages are not Turing complete, and because important subsets of these languages allow decidability of query equivalence, tools can be constructed that provide immediate syntactic and semantic feedback".

The two important considerations about syntactic and semantic feedback play a vital role in the development of the assessment system. All the tools reviewed can provide **semantic** feedback, but when it comes to **syntax checking**, the systems need **to rely on the database** they are using and provide with the error message of the database itself. In addition, they **also fail to grade** queries **partially** in case of **syntactic errors**. **Our system** is able to provide exact and **accurate feedback** as it uses our own Database System Prototype in the back-end named **GU_DB** [8].

The authors have developed the complete learning management system in the form of a virtual laboratory for database systems [8]. It consists of our own database management systems prototype **GU_DB**, which allows the users to implement the concepts learned. As discussed in the previous section, our Database prototype (GU_DB) [8] helps us to provide specific hints to the users when the query is syntactically wrong. It is a complete learning management system (LMS) that involves the components like Tutorial of the subject, Procedure, Simulator, Assessment (Theory and Practical), and Feedback.

9.2.1 Automated Assessment of SQL Queries

Assessment is an important criterion for learning. Once the student has learnt the concepts, there is a need for assessment. During the manual assessment, teachers partially evaluate the student's task. The evaluation is not necessarily zero or full marks. Most of the time, students are awarded partial marks based on the logic applied. Even if the syntax is wrong, the teacher may provide some marks looking at the level of incorrectness. We aim to build an **automated assessment tool which can award partial marks to the students for their task submitted**. Most of the evaluation or auto-grading systems evaluate in the binary whetherthe result is either correct or incorrect. The result is the student does not get to know his/her mistake and do not remain motivated to perform other tasks. Moreover, none of the systems reviewed are able to evaluate the syntactic incorrectness of the Query in terms of the level of incorrectness.

9.3   Proposed Work

Authors have developed a **Framework** for **Automated Partial Marking System for SQL Queries**. The system would award the marks based on the level of correctness of the query. The correctness would be determined as per the rules laid down in the system, which will integrate a proper assessment structure. The marks would be awarded based on the rules defined in the system. Additionally, the system provides rich and constructive feedback to the students on the query submitted.

9.3.1   Objectives of Eval_SQL

Automated assessment systems help the learner(s) by providing the result in terms of marks and feedback instantly. The goal of the proposed system is to make students test their SQL knowledge by applying the queries in the assessment system and getting the result instantly.

The Eval_SQL has been developed to achieve the following objectives:

- Provide **Partial Marking** for SQL Queries.
- Provide **constructive feedback** for student queries to analyse their own mistakes.
- Give a complete learning environment to the student, wherein he/she can learn and immediately apply his/her knowledge and at the same time gets the advantage of validating the knowledge applied by auto evaluation of submitted queries.
- To automate the evaluation to provide immediate feedback/result in order to keep students motivated to keep practicing more.
- Reduce the workload of teacher(s) by automating the task of evaluating a large number of exercises.

9.3.2   Architecture of Eval_SQL

Eval_SQL consists of the following modules.

The diagram in Fig. 9.1 shows the communication between the assessment rule engine and GU_DB to fetch the correctness of the query. The user submits the query

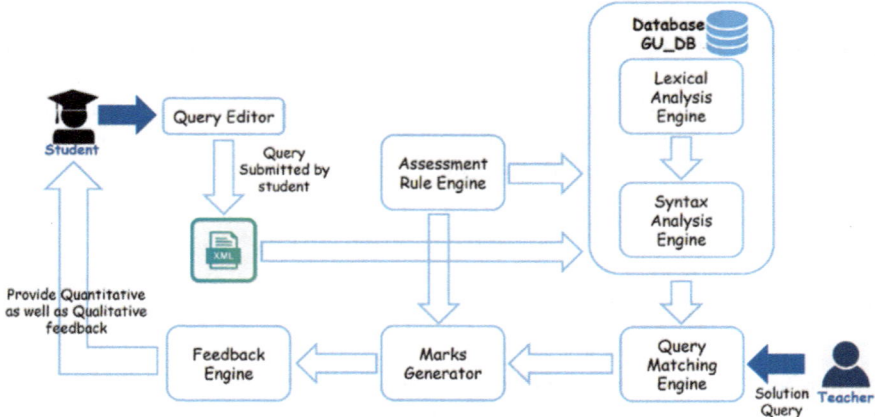

Fig. 9.1 Architecture of Eval_SQL

in the Query editor, which passes the query to the back-end. Eval_SQL integrates GU_DB in the back-end. Hence, the query submitted by the user passes through various phases like Lexical analysis, Syntactical analysis, semantic checking, and finally reaching the Assessment Rule Engine. The diagram highlights how Marks generator and feedback engine display the final result to the user for the query submitted.

9.3.3 Assessment Rule Engine

The Assessment rules are used to generate the final marking for the query, which consists of the following modules.

9.3.3.1 Assessment Module

The Assessment module is one of the core components in the development of an assessment system that provides partial marking for SQL queries along with the feedback.

The image in Fig. 9.2 displays the flow of query being sent for assessment. The query passes through various phases like Lexical, syntactical, and component matching. Any query submitted first needs to communicate through GU_DB, and GU_DB returns the status of the query which is sent for further assessment to Eval_SQL. Eval_SQL then compares the User Query with the Expert query, follows different comparison techniques discussed in the algorithm, and generates the Result. The Result displays the Marks received in the form of percentage and the constructive textual feedback received as displayed in the diagram.

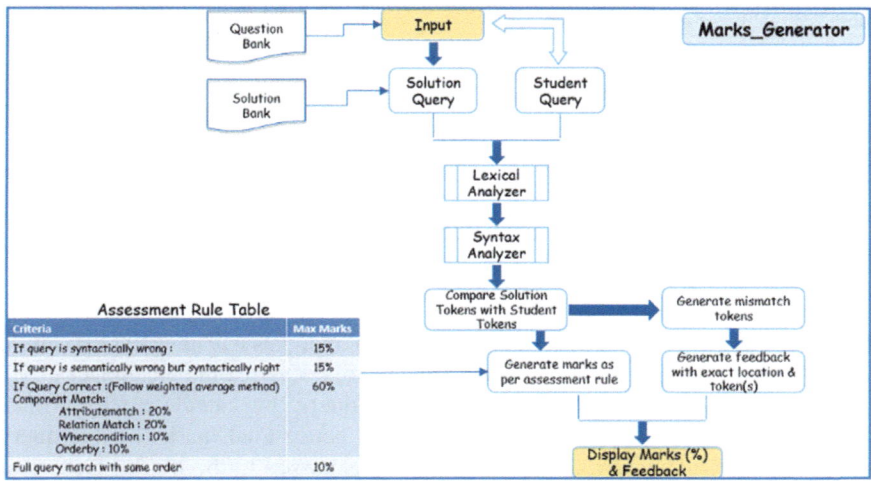

Fig. 9.2 Architecture of Marks_Generator and feedback

The assessment module consists of various modules which are listed as below.

9.3.3.2 Awarding Marks for Different Components of the Query (Marks_Generator)

The system aims at providing partial marks on the query written. However, the partial marks should be awarded only after assessing the validity of the query. The total marks need to be calculated after receiving marks from each of the evaluation stages the query has traversed through and final marks are generated. The Assessment system (Eval_SQL) needs to communicate with the Database (GU_DB) for the validity of the query. GU_DB in turn returns the validity of the query syntactically and semantically. Once the result from GU_DB is fetched, the query is compared with the Solution Query provided by the faculty for the logical analysis. The comparison of a query involves matching the components of the query and having their average. The system is able to provide accurate syntactic error and appropriate feedback to the user as there is GU_DB in the back-end which provides us complete flexibility on the database.

The assessment rules are laid out to have a proper marking scheme for the query. The authors have used the state table for implementation, which informally constructs a finite-state machine. The marking scheme in Marks_Generator is defined as follows.

Marking Scheme

Each component of the query is assigned marks as per the weightage defined for each of them. Maximum marks of the query may vary depending upon the complexity/difficulty level. A Query requiring more components will have more

Table 9.2 Table for marking scheme

Component	Weightage
Keywords (Select, Insert, Update, Delete)	5
Connecting keywords like (values, from, like, group by, where, set, order by, having, etc.)	2
Table name	2
Attribute name	2
Expressions	2

maximum marks whereas a Query requiring less components will have less maximum marks. Later for Normalization, marks obtained are converted in percentage.

The Maximum marks of Solution (Expert) query are calculated first. Since different queries have different components and hence total marks for the query are calculated dynamically. The solution Query is passed to the Marks_Generator for calculation of total marks as described in Table 9.2.

9.3.3.3 Providing Accurate Feedback for the Query (Feedback Engine)

Unlike other systems, our system is able to provide constructive and accurate feedback as we are using our own DBMS Prototype GU_DB. The feedback is provided with an exact component mistake, which helps the learner self-evaluate their mistake. Moreover, all the systems discussed use some or the other database in the back-end like MySQL, PostgreSQL, and so on, hence they need to rely on the said database for the error/message. Our system uses our own database prototype (GU_DB), hence, we have complete flexibility with the system. Our system can fetch the exact mistake including accurate syntactic errors, which the user has made, and accordingly, marks are calculated and feedback is provided.

The distinguishing feature of the Assessment Rule engine is as follows: The system provides two types of feedbacks to the learner:

1.   Scores in terms of Percentage (Quantitative Feedback).
2.   Formative Constructive feedback (Qualitative Feedback).

In the evaluation strategy, providing relevant and timely feedback is an essential criterion. Online assessment systems provide this advantage by giving immediate feedback. There are very few systems in Computer science which provide textual feedbacks, and in Database systems, they are even rare.

Algorithm for Assessment Module

For example, consider the following Queries.

Q.1 List Names and Salary information for employees

Select Name, Salary from Emp; (Expert Query).

In the above query, total marks are calculated as below.

Select [keyword]	5
Name, salary	2 + 2
Emp [table name]	2
From [keyword]	2
Total Marks	13

The total marks calculated for the above query is 13.

This is a simpler query with fewer components

The following is a query with little more complexity.

Q.2. Display names of Clients whose name has the second letter as "A"

Select Name from ClientMaster113 where Name Like "_A%"; (Expert Query).
In the above query, total marks are calculated as below:

Select [keyword]	5	Where [connecting keyword]	2
Name [attribute]	2	Name [attribute]	2
From [keyword]	2	Like [connecting keyword]	2
Clientmaster113 [tablename]	2	"_A%" [expression]	2
		Semicolon	1
Total marks 20			

The total marks calculated for the above query is 20.
The student writes the following query for Q.2
Select Name From Client_Master113 where name like "A%".

This **User Query** will obtain 17 marks (2 marks of expression and 1 mark of semicolon are deducted from total of 20 marks hence user receives 17 marks) as per the expert query and marking scheme. In order to normalize marks, the marks obtained are converted to percentage:

Obtmarksinperc $= obtmarks/maxmarks*100$: e.g. $17/20*100 = 85$.
*Obtained maximum marks are computed by the system (e.g. 85%).

Eval_SQL Algorithm

This section describes the algorithm for calculating the marks and generating the feedback for the query written by the user. The student query is first parsed to **GU_DB** to check for its syntactic and semantic correctness and the status from GU_DB is returned to **Eval_SQL** for component matching, and accordingly, appropriate marks and feedback are recorded.

Input: User Query.

Output: Total Marks received in percentage and Feedback.

Algorithm

1. **Initialization:**
   Establish connection with GU_DB, Parse query and initialise FA
2. **Read the questions from the CSV file (sqlqueries.csv) (goto readfile subroutine)**
   Return with tokenised query, Databasepath, TableMetadata and Tableinfo.Assign MaxMarks for Expert Query [Refer Table1]
      Display qno, question to user
      Accept student Query

3. **Evaluation of student answer**
   Send the student query to GU_DB and return with user answer as Valid or invalid tokens with their exact position from Syntax_validator (For providing feedback to user about exact mistake), Databasepath, TableMetadata and Tableinfo.

4. **Compare User List Tokens with Solution List Tokens**
   Gu_DB returns the syntactical analysis, but to fetch any logical errors, we need to compare it with the Expert query, So, Compare User Token list with Solution Token list,

      Store the Mismatched tokens in "MissingTokens" list
5. **Marks & Feedback Generator**
   Fetch the Token wise marking scheme, assign marks as per their definition and calculate obtained marks. Return with obtained marks and missing tokens and mismatch tokens for providing appropriate feedback obtained marks
6. **Display Result**
   Generate Result in Terms of Feedback displaying the reason for marks deducted and Marks in terms of Percentage.

9.4 Implementation (Eval_SQL)

The following screenshots display the GUI for Automated Assessment of SQL Queries: Eval_SQL (Figs. 9.3 and 9.4).

Once the User clicks on the feedback, the detailed feedback showing the query written by the user and the marks obtained by the user are displayed as follows (Figs. 9.5).

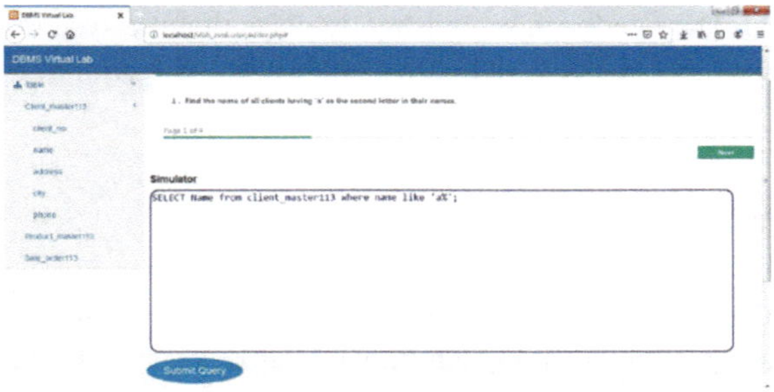
DBMS Virtual Lab
Simulator
Submit Query

Fig. 9.3 User writes first

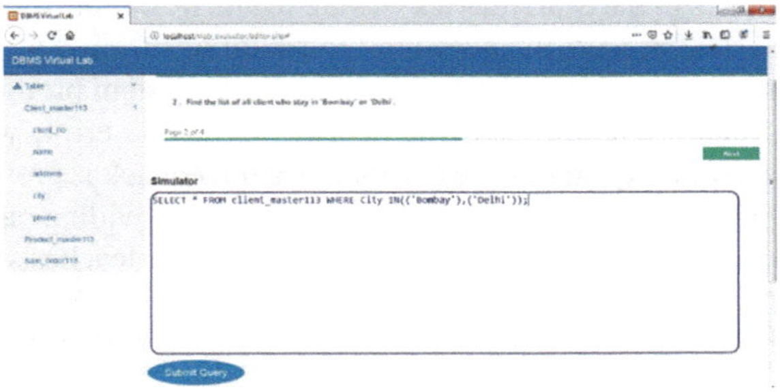
DBMS Virtual Lab
Simulator
Submit Query

Fig. 9.4 User iterates through next queries

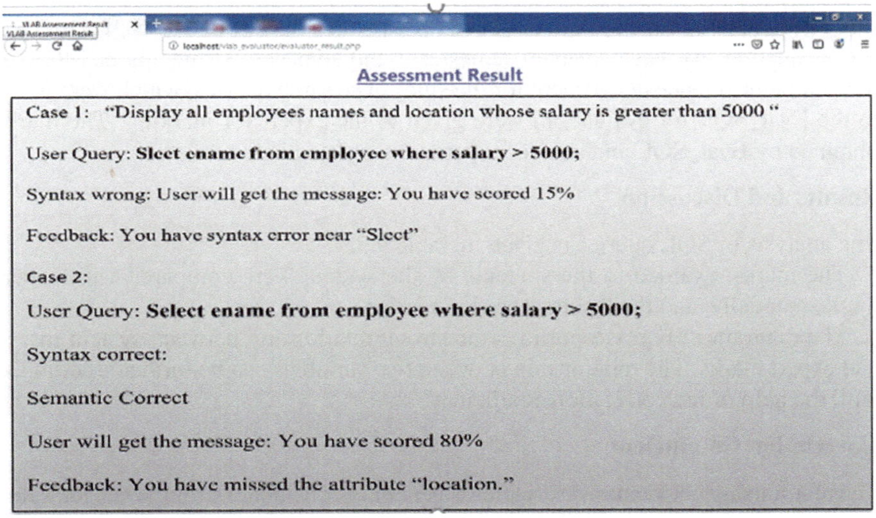
Assessment Result
Case 1: "Display all employees names and location whose salary is greater than 5000 "
User Query: Slect ename from employee where salary > 5000;
Syntax wrong: User will get the message: You have scored 15%
Feedback: You have syntax error near "Slect"
Case 2:
User Query: Select ename from employee where salary > 5000;
Syntax correct:
Semantic Correct
User will get the message: You have scored 80%
Feedback: You have missed the attribute "location."

Fig. 9.5 Feedback displayed to the user

The Feedback screenshot displays that the system provides the exact mistake the user has made, which guides him towards learning.

9.5 Results

Our Automated assessment system: EVAL_SQL provides partial marking as well as Qualitative and Quantitative feedback for SQL Queries to the users. It provides a fully automated assessment of SQL queries and has the ability to partially award the marks to the SQL queries written by the user if the query is incorrect. The evaluation system accepts the questions and solutions from the teacher, and Evaluation system is using our own Database management system prototype GU_DB [10]. Hence, the system is able to provide accurate syntactic error, which the user has made, and accordingly, marks are calculated and feedback is provided. The Eval_SQL provides the advantage of teacher-like marking by providing partial marks to partially correct queries, and at the same time reduces the teacher's workload by automating the process of assessment. The pedagogical consideration in the development of the LMS makes it beneficial to all the students with different learning abilities.

Evaluation Process

The effectiveness of the system has been tested with a dataset of **107 SQL Queries**. For evaluation, we have selected semester 2 and semester 3 students of MCA of the computer science department for the study. The same queries, which were given to the Eval_SQL for evaluation, were given to the expert for marking. The marks obtained by Eval_SQL and subject expert are tabulated below.

Results and Discussion

The analysis of SQL queries is given in Table 9.3.

The marks awarded to the students by the system were compared with expert marks manually, and the results were impressive.

The diagram in Fig. 9.6 portrays the strong relationship between system marks and expert marks. The relationship between two variables can be further ascertained with the help of the correlation coefficient.

Correlation Co-efficient

The relationship between two variables is generally considered strong when the value of r is greater than 0.7. The following scattered plot displays the relationship between two variables namely system marks and expert marks (Fig. 9.7).

The authors have compared the results given by the system with the results provided by an expert. The correlation coefficient was computed to ascertain a positive relationship between the two. The correlation coefficient between the marks given by the system and the expert is **0.8764**, which shows that there is a **highly positive** correlation between the results given by the system developed and the results provided by the expert.

Table 9.3 Marks provided by Eval_SQL and subject expert

Query no.	System marks	Faculty marks	Query no.	System marks	Faculty marks	Query no.	System marks	Faculty marks
1	10	10	41	9	8	81	10	10
2	3	6	42	10	10	82	10	10
3	3	0	43	10	10	83	10	10
4	10	10	44	10	10	84	10	8
5	10	10	45	10	10	85	10	10
6	9	7	46	10	10	86	10	10
7	10	9	47	10	10	87	10	10
8	10	10	48	10	10	88	10	9
9	4	2	49	8	7	89	7	10
10	10	10	50	8	7	90	10	10
11	7	10	51	4	1	91	10	10
12	10	10	52	7	5	92	7	10
13	10	10	53	9	10	93	10	10
14	7	10	54	10	10	94	3	4
15	10	10	55	10	8	95	10	10
16	3	4	56	5	1	96	10	10
17	10	10	57	7	9	97	10	10
18	10	10	58	10	10	98	10	9
19	10	10	59	10	10	99	10	10
20	10	9	60	10	10	100	8	8
21	10	7	61	10	10	101	10	10
22	10	10	62	10	10	102	10	10
23	10	10	63	10	10	103	7	5
24	10	10	64	10	10	104	10	10
25	10	8	65	10	8	105	10	10
26	7	8	66	10	10	106	10	10
27	3	2	67	10	10	107	5	4
28	3	0	68	10	10			
29	7	6	69	10	9			
30	7	6	70	4	7			
31	10	10	71	7	10			
32	8	8	72	9	10			
33	10	10	73	10	10			
34	10	10	74	10	8			
35	7	5	75	5	10			

(continued)

Table 9.3 (continued)

Query no.	System marks	Faculty marks	Query no.	System marks	Faculty marks	Query no.	System marks	Faculty marks
36	10	10	76	7	9			
37	10	10	77	10	10			
38	10	10	78	10	10			
39	5	4	79	10	10			
40	10	10	80	10	10			

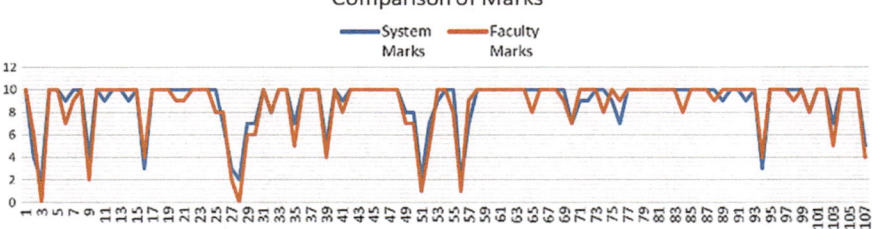

Fig. 9.6 Comparison of system marks and expert marks

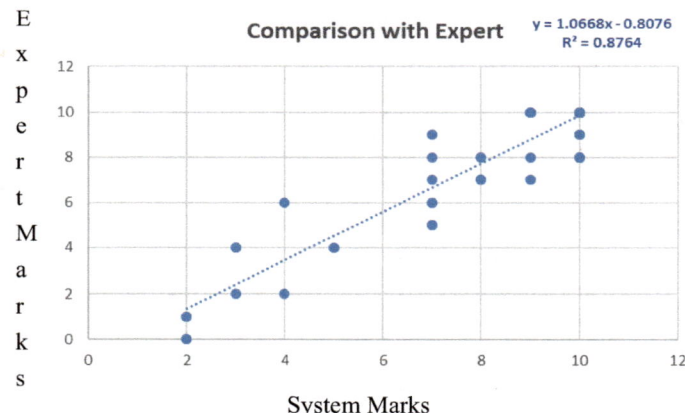

Fig. 9.7 Correlation results

9.6 Conclusion and Future Scope

Automated Evaluation systems are limited in number for SQL Queries. Moreover, there is hardly any system that provides partial marking for the query written along with Qualitative as well as Quantitative feedback for the Queries. The paper presents a partial marking automated system "Eval_SQL," which gives a fully automated experience for the assessment of SQL queries and has the ability to partially award the marks to the SQL queries written by the user. As the evaluation system (Eval_SQL) accepts the questions and solutions from the teacher, the teacher involvement instils confidence in the learner in the evaluation system. Eval_SQL is using the DBMS prototype GU_DB [9] developed by the authors. Hence, the system is able to provide appropriate feedback including accurate syntactic error, which the user has made, and accordingly, marks are calculated and feedback is provided. The Eval_SQL gives the advantage of Teacher like marking by assigning marks to partially correct queries, and at the same time reduces the teacher's workload by automating the process of assessment. In the future, the system will be integrated as part of the Virtual laboratory of Database systems, for automated evaluation of SQL Queries. This will help the students to self-learn and self-assess the progress of the topic learned.

References

1. Kearns, R., Shead, S., Fekete, A.: A Teaching System for SQL, pp. 224–31 (2004)
2. Sadiq, S., Orlowska, M., Sadiq, W., Lin, J.: SQLator: An online SQL learning workbench. ACM SIGCSE Bull. 1–5. http://www.dl.acm.org/citation.cfm?id=1008055
3. de Raadt, M., Dekeyser, S., Lee, T.Y.: Do Students SQLify ? Improving Learning Outcomes with Peer Review and Enhanced Computer Assisted Assessment of Querying Skills, p. 101 (2007)
4. de Raadt, M.: Computer Assisted Assessment of SQL Query Skills, p. 63 (2007)
5. Soler, J., Prados, F., Boada, I., Poch, J.: A web-based tool for teaching and learning SQL. In: International Conference on Information Technology Based Higher Education and Training, ITHET. http://www.acme.udg.es/articles/ithet2006.pdf
6. Cruces, L.: In health informatics, and computer studies. Automatic Generation of SQL Queries Automatic Generation of SQL Queries (2006)
7. Chandra, B., et al.: Partial marking for automated grading of SQL queries. Proc. VLDB Endowm. 9(13), 1541–1544 (2016)
8. Fuller, U., et al.: Developing a computer science-specific learning taxonomy. In: Working Group Reports on ITiCSE on Innovation and Technology in Computer Science Education— ITiCSE-WGR '07, p. 152 (2007). http://www.portal.acm.org/citation.cfm?doid=1345443.1345438

A Publication in Process

9. Bhumika, S., Jyoti, P.: GU_DB: a database management system prototype for academia. Int J Adv Comput Res 11.55(2021):67

Published Doctoral Dissertation or Master's Thesis

10. Shah, B.: An Innovative Framework for Remote Database Experimentation. Department of Computer Science, Gujarat University (2020). http://www.hdl.handle.net/10603/307860ser tation

Chapter 10
Unsupervised Feature Selection Approaches for Medical Dataset Using Soft Computing Techniques

G. Jothi, J. Akilandeswari, S. David Samuel Azariya, and A. Naveenkumar

Abstract An electrocardiogram (ECG) measures the heart's electrical function and has been widely used to detect heart disorders because of their simplicity and non-obtrusive nature. Nowadays, in medical decision support systems, soft computing methods are commonly used. Medical diagnosis helps to acquire various characteristics reflecting the disease's various variations. It is likely to have important, irrelevant, and redundant features to reflect disease, with the aid of various diagnostic procedures. It is a non-trivial job to define a good feature subset for efficient classification. The objective of this research is to assess three different soft computing-based unsupervised feature reduction approaches namely Soft Set-Based Unsupervised Feature Reduction, Unsupervised Quick Reduction, and Unsupervised Relative Reduction, to evaluate the best subset of features with enhanced diagnostic classification accuracy for cardiovascular disease. The output of feature selection algorithms is evaluated with the heart disease dataset from the UCI repository. The experimental findings indicate that the Soft Set-based Unsupervised Feature Redact algorithm achieved good classification accuracy compared to other feature reduction approaches.

G. Jothi (✉)
Department of BCA, Sona College of Arts and Science, Salem, Tamil Nadu, India
e-mail: jothi.g@sonacas.edu.in

J. Akilandeswari · S. David Samuel Azariya · A. Naveenkumar
Department of IT, Sona College of Technology (Autonomous), Salem, Tamil Nadu, India
e-mail: akilandeswari@sonatech.ac.in

S. David Samuel Azariya
e-mail: david@sonatech.ac.in

A. Naveenkumar
e-mail: naveenkumar@sonatech.ac.in

© The Author(s), under exclusive license to Springer Nature Singapore Pte Ltd. 2022   105
V. Bhateja et al. (eds.), *Evolution in Computational Intelligence*,
Smart Innovation, Systems and Technologies 267,
https://doi.org/10.1007/978-981-16-6616-2_10

10.1   Introduction

Heart disease is a type of vascular cardiovascular condition affecting men and women alike. In India, cardiovascular disease is an epidemic and a disease burden and fatality. There is, therefore, a need to tackle the cardiac disease. Several lives are resolved by an early diagnosis of heart disease [1]. The main task for the study of ECG signals to determine the heartbeat case is to determine the required set of characteristics. A core component of information is Feature Selection (FS).

FS is employed to enhance the precision of classification and decrease the computational time of approaches to classification. Where class labels of abstracts are obtainable, we utilized a supervised approach, an alternate approach that is not supervised is sufficient. In certain implementations of soft computing, class marks are not identified, thus declaring recognition of the alternative to unsupervised affection [2, 3]. The soft set theory was implemented by Molodtsov in the year 1999, for grappling with the complexities [4]. Data analysis and decision support systems were applied to the Soft Set Theory. In the Boolean value knowledge process, the soft collection is easy-to-reduce attributes. In this research, the roughest theory-based unsupervised methods are assessed to find the feature subset. In this research, the efficiency of the unsupervised feature reduction approaches focused on soft computing is studied, namely Soft Set-based Unsupervised Feature Reduction (SSUSQR), Unsupervised Quick Reduction (USQR), and Unsupervised Relative Reduction (USRR) for feature reduction. The outcomes of experiments indicate that the enactment of the feature selection/reduction approaches.

The rest of the paper is structured into Sects. 10.2–10.6. Section 10.2 explains the survey of the current studies contained in the classification methods for ECG-based arrhythmia. The preliminaries of soft set theory are defined in Sect. 10.3. The unsupervised algorithms for feature selection are elucidated in Sect. 10.4. The experimental findings are investigated in Sect. 10.5. The conclusion is given in Sect. 10.6.

10.2   Related Work

Table 10.1 includes a literature assessment of current studies on ECG-based arrhythmias. Recently, a lot of algorithms are introduced for feature selection based on Artificial Bees Colony [5], least-square support vector machine (LSSVM) [6], modified rough set [7.8], artificial intelligence, KNN [7], and PCA [8]. Das et al. [9] suggested a group incremental selection algorithm based on rough set theory (RST) for the selection of the optimized subset of features. Genetic algorithms have been used in this method to determine the positive region of RST's target. The efficiency of this system is assessed by the number of selected features, computational times, and classification accuracies. In [10], a new fugitive rough neighborhood (FDNRS)-based feature selection algorithm is proposed. The empirical results show that the FDNRS

Table 10.1 Related work

Name of the authors	Algorithm utilized	The objective of the research
Subanya et al. [5]	ABC, SVM	Feature selection, classification
Acır et al. [6]	Least-square support vector machine (LSSVM)	Classification
Kumar et al. [12]	Modified soft rough set	Classification
Jothi et al. [13]	Tolerance rough set theory	Feature selection, classification
Deekshatulu et al. [1]	Artificial neural network (ANN)	Classification
Ritaban et al. [7]	Optimal feature selection algorithm, K-nearest neighbor (KNN)	Feature selection, classification
Liuh-Chii Lin et al. [8]	Principal component analysis (PCA), fuzzy logic	Feature selection, classification
Das et al. [9]	Rough set theory-based genetic algorithm	Feature selection, classification
Sang et al. [10]	Fuzzy dominance neighborhood rough sets	Feature selection
Zhao et al. [11]	Classified nested equivalence class based entropy method	Feature selection

algorithm can update dynamically controlled data efficiently. Zhao et al. [11] introduced an entropy-based feature selection method for different UCI machine learning datasets. In this method, the classified CNEC approach is applied for the calculation of the information entropy-based importance for the selection of features with rough set theory. In this CNEC approach, the experimental analysis, 31 datasets, including large and high-dimensional datasets, from various sources, such as the repository for the UCI and KDD Cup competitions are employed to assess the effectiveness of the implemented approach.

10.3   Preliminaries

10.3.1   ECG Signal

The heart is a muscle that contracts rhythmically and pumps blood across the body. This contraction begins in the atrial sinus, which functions as a natural pacemaker, and spreads throughout the remainder of the muscle. This spread of the electric signal is modeled [14–16]. As a consequence of this activity, electrical currents are created on the surface of the body generating changes in the electrical potential of the skin surface. These signals can be recorded or quantified using electrodes and appropriate tools. The difference in the electrical potential between points indicated by the skin electrodes is often enhanced with the help of an instrumentation (operational) amplifier with optical insulation. The signal is then sent to the high-pass and the low-pass alienation filter as a second stage. In a digital-analog converter, it finally emerges.

10.3.2 Soft Set Theory

In soft set theory, U denotes a universal set, and E is a set of parameters/attributes, $P(U)$ is the power set of U, and $A \subseteq E$ [4]. A couple (F, A) is referred to as soft set above U, where F is a mapping given by

$$F : A \rightarrow P(U) \tag{10.1}$$

10.3.3 Multi-Soft Set

The 'multi-soft set' is based on the analysis of the multi-valued information system of $S = (U, A, V, f)$ in $|A|$ $S = (U, A, V(\{0.1\}), f)$, in which the cardinality of A is indicated $|A|$ [17].

$$(F, E) = \{(F, a_i) : i = 1, 2, \ldots .|A|\} \tag{10.2}$$

10.3.4 AND Operation in Multi-Soft Sets

Let $(G, H) = ((G, x_i) : i = 1, 2, \ldots .|A|)$ be a multi-soft set over U and multi-valued information system $S = (U, A, V, f)$. The AND operation among (G, x_i) and (G, x_j) is described as [17], $(G, x_i) \text{AND}(G, x_j) = (G, x_i \times x_j)$, where

$$I(Vx_i, Vx_j) = F(Vx_i) \cap F(Vx_j), \forall (Vx_i, Vx_j) \in x_i \times x_j, \text{for} 1 \leq i, j \leq |A| \tag{10.3}$$

10.3.5 Attribute Reduction

Definition 1 Let $(G, H) = ((G, x_i) : i = 1, 2, \ldots .|A|)$ and $S = (U, S, V, f)$. S is a set of attributes; $C \subseteq S$ is called a reduct for S if $R_{F(y_1 \times \ldots \ldots \times y_{|B|})} = R_{F(x_1 \times \ldots \ldots \times x_{|A|})}$ [18].

Definition 2 Let us assume that $C \subseteq S$ is an attribute subset, a $\in$ S is an attribute, $|A| = |IND(A)|$, the importance of a for A is denoted by $\text{SIGNI}_A(a)$, the definition is [18]

$$\text{SIGNI}_A(a) = 1 - |A \cup \{a\}|/|A| \qquad (10.4)$$

Definition 3 $COR(CA)$ denotes all of the independent attribute sets of the conditional attributes (CA), namely $COR(CA) = REDUCT(CA)$, where the reduction of $CA is REDUCT(CA)$ [18].

10.4   Unsupervised Feature Selection Approaches

In several data mining applications, class labels are not identified, and therefore, Unsupervised Feature Selection (UFS) has been accorded the importance to present research on the UFS. Different unsupervised algorithms such as Unsupervised Quick Reduct (USQR), Unsupervised Relative Reduct (USRR), and Soft Set-based Unsupervised Quick Reduct (SSUSQR) methods are being studied in this study.

10.4.1   Unsupervised Quick Reduct (USQR) Algorithm

Without producing any subsets, the USQR algorithm quantifies a decline. The average dependence of each sub-attribute is specified by the algorithm (Eq. 10.5) and the best feature set is selected [19].

$$\gamma_P(a) = \frac{|POS_P(a)|}{|U|}, \forall a| \in A \qquad (10.5)$$

10.4.2   Unsupervised Relative Reduction (URR) Algorithm

The Unsupervised Relative Reduction (URR) algorithm starts by taking all functions of the dataset into account. It iteratively assesses each function and decides the relative degree of dependency. You can eliminate the feature if your relative dependency is equal to one. This approach continues until all features are examined [19].

10.4.3 Soft Set-Based Unsupervised Quick Reduct (SSUSQR) Algorithm

The dimensionality reduction is performed with the use of AND operation in the soft set based on an unsupervised, quick reduction technique. It starts with a blank set and tries to discern the cardinality of the universal set. The cardinality of insight is established for each conditional property. The key attribute is the attribute with the highest cardinality value in the reduction set. If a maximum cardinality value is available for more than one attribute, $SIGNI(x)$ is found and this value is the core one. In the next step, it takes a subset of properties and the subset of features that have the upper indiscernibility cardinality to combine other attributes with the $COR(x)$ property. This process goes on until the indiscernibility cardinality of the subset function is the same as the universe's cardinality [20].

US, the Universal Set;
CA, the set of all conditional attributes;
(1) *Reduct* ← { }
(2) *Do*
(3) M ← *Redcut*
(4) *Compute SM (US)* = |*IND(US)*|
(5) ∀a ∈ (*CA* − *Reduct*)   *Find Maximum* |*IND(a)*)))|
(6)   *If two or more attributes have the same value,*
      Max ((|*IND(a)*)|)
      Find SIGNI(a) (Definition 2)
      Find a = *COR(a)*
    End if
(7) M ← *Redcut* ∪ {*a*}
(8) *Reduct* ← M
(9) *Until* |*Reduct*| == *SM (US)*
(10) *Return Reduct*

Algorithm 1. Soft set-based Unsupervised QR Algorithm

Table 10.2 Selected features and Time_Taken

Dataset	Total features	Selected features			Time_Taken (s)		
		USQR	USRR	SSUSQR	USQR	USRR	SSUSQR
CAD	279	234	205	184	12.2619	11.3342	10.1032
HDD	76	68	52	48	2.2777	1.7562	0.7562

10.5 Experimental Results

10.5.1 Dataset

Datasets are compiled from the Database of the UCI Repository [21, 22]. There are two cardiac arrhythmic datasets, namely Cardiac Arrhythmia Database (CAD and Heart Disease Database (HDD) taken in this experimental study. In CAD, there are 452 instances and 279 attributes in this database, 206 of which are linear values and the rest are nominal. It is one of the 16 classes to identify the presence and absence of cardiac arrhythmia. In the case of HDD dataset, 76 attributes and 920 instances are included. A comparison of the methods of Unsupervised QR, RR, and soft set-based QR is based on the subset, time_taken to discover subsets, and accuracy of classification. Two elements are part of this dataset: training and testing. 80% of the dataset is used for training and 20% for testing in this experiment.

10.5.2 Feature Selection

Four different classifiers are used by Naive Bayes (NB), IBk, J48, and LMT to test the accuracy of the various unregulated selection approaches. The collection of features is accomplished using the USQR, USRR, and SSUSQR algorithms. The selected feature set and time_taken are shown in Table 10.2.

Figure 10.1 illustrates how the time for all three unsupervised algorithms has passed. It is implied that the elapsed time should always be decreased by the SSUSQR algorithm.

10.5.3 Performance Assessment for Unsupervised Algorithms

In this research, the classification precision value is related to all three unsupervised feature selection algorithms. These classifications are employed first in the unreduced dataset and subsequently in the decreased dataset derived from USQR, USRR, and SSUSQR algorithms. Table 10.3 shows the overall classification accuracy.

Fig. 10.1 Time taken (in seconds) for feature selection

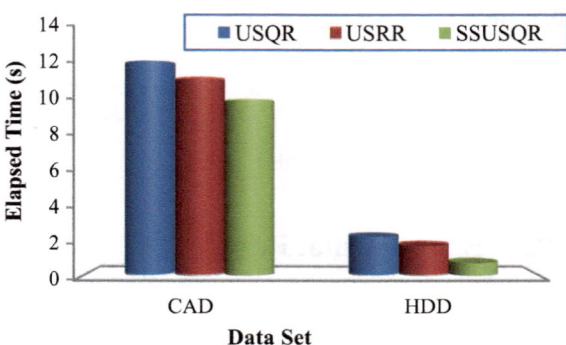

Table 10.3 Overall classification accuracy

Algorithm	Classification approach	Dataset	
		CAD	HDD
Un-reduct dataset (before feature selection)	Naive Bayes	90.4	56.6
	IBK	84.8	67.9
	J48	76.6	64.2
	LMT	89.8	63.2
USQR	Naive Bayes	63.5	57.5
	IBK	63.5	67.0
	J48	62.5	63.2
	LMT	61.8	62.3
USRR	Naive Bayes	80.4	48.5
	IBK	74.8	62.9
	J48	66.6	54.2
	LMT	79.8	54.2
SSUSQR	Naive Bayes	92.4	88.5
	IBK	86.8	77.9
	J48	78.6	74.2
	LMT	91.8	74.2

In the two cardiac arrhythmic datasets, the results of Naïve Bayes (NB), IBk, J48, and LMT classification approaches are shown in Figs. 10.2 and 10.3 based on the features selected using the soft computing approach. These figures demonstrate the effectiveness of the functional selection algorithms known as SSUSQR, USRR, and USQR. The high accuracy value in the SSUSQR algorithm.

Fig. 10.2 CAD dataset

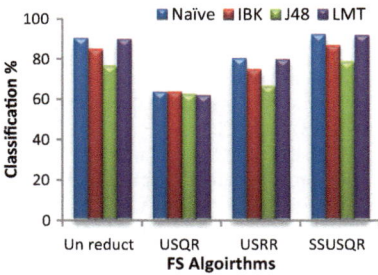

Fig. 10.3 HDD Dataset

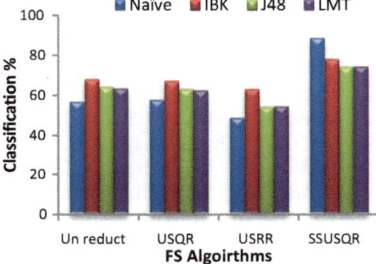

10.6   Conclusion

In the medical world, it is very helpful to identify the important risk factors and to prevent the diseases. In this study, the soft computing-based feature selection algorithms SSUSQR, USQR, and USRR are employed to evaluate the performance of heart disease datasets. It is also noted that the time of computation is evaluated by this comparative analysis. The empirical results show that the SSUSQR algorithm has a higher classification accuracy value of 92.4% and 88.5% for CAD and HDD datasets, respectively. It also decreases computational time as related to the existing feature reduction approaches. The optimization approaches functionality can be implemented in the future to enhance performance.

Declaration The competent authorities allowed us to use the images/data as provided in the study. We shall be entirely responsible for any dispute in the future.

References

1. Jabbar, M.A., Deekshatulu, B.L., Chandra, P.: Classification of heart disease using artificial neural network and feature subset selection. Glob. J. Comput. Sci. Technol. Neural Artif. Intell. **13**(3), 4–8 (2013)
2. Mustaqeem, A., Anwar, S.M., Majid, M.: Multiclass classification of cardiac arrhythmia using improved feature selection and SVM invariants. Comput. Math. Meth. Med. **2018**, 1–11 (2018)
3. Sahoo, S., Subudhi, A., Dash, M., Sabut, S.: Automatic classification of cardiac arrhythmias based on hybrid features and decision tree algorithm. Int. J. Autom. Comput. 551–561 (2020)
4. Molodtsov, D.: Soft set theory—first results. Comput. Math. Appl. **37**(4–5), 19–31 (1999)
5. Subanya, B., Rajalaxmi, R.R.: Feature selection using artificial bee colony for cardiovascular disease classification. In: 2014 International Conference on Electronics and Communication Systems (ICECS), pp. 1–6 (2014)
6. Acır, N.: Classification of ECG beats by using a fast least square support vector machines with a dynamic programming feature selection algorithm. Neural Comput. Appl. **14**(4), 299–309 (2005)
7. Kirtania, R., Mali, K.: Cardiac arrhythmia classification using optimal feature selection and k-nearest neighbour classifier. Int. J. Adv. Res. Comput. Sci. Softw. Eng. **5**(1), 54–58 (2015)
8. Lin, L.C., Yeh, Y.C., Chu, T.Y.: Feature selection algorithm for ECG signals and its application on heartbeat case determining. Int. J. Fuzzy Syst. **16**(4), 483–496 (2014)
9. Kumar, S.U., Inbarani, H.H.: Neighborhood rough set based ECG signal classification for diagnosis of cardiac diseases. Soft. Comput. **21**(16), 4721–4733 (2017)
10. Das, A.K., Sengupta, S., Bhattacharyya, S.: A group incremental feature selection for classification using rough set theory-based genetic algorithm. Appl. Soft Comput. **65**, 400–411 (2018)
11. Sang, B., Chen, H., Yang, L., Li, T., Xu, W.: Incremental feature selection using a conditional entropy-based on fuzzy dominance neighborhood rough sets. IEEE Trans. Fuzzy Syst. 1–14 (2021)
12. Kumar, S.S., Hannah Inbarani, H.: Modified soft rough set based ECG signal classification for cardiac arrhythmias. In: Big Data in Complex Systems, pp. 445–470. Springer International Publishing (2015)
13. Ganesan, J., Inbarani, H.H., Azar, A.T., Polat, K.: Tolerance rough set firefly-based quick reduct. Neural Comput. Appl. **28**(10), 2995–3008 (2017)
14. Zhao, J., Liang, J.M., Dong, Z.N., Tang, D.Y., Liu, Z.: Accelerating information entropy-based feature selection using rough set theory with classified nested equivalence classes. Pattern Recogn. **107**, 107–117 (2020)
15. Luz, E.J.D.S., Schwartz, W.R., Cámara, G., Menotti, D.: ECG-based heartbeat classification for arrhythmia detection: a survey. Comput. Meth. Programs Biomed. **127**, 144–164 (2016)
16. Gupta, V., Mittal, M.: Arrhythmia detection in ECG signal using fractional wavelet transform with principal component analysis. J. Inst. Eng. (India): Ser. B **101**(5), 451–461 (2020)
17. Herawan, T., Deris, M.M.: On multi-soft sets construction in information systems. In: International Conference on Intelligent Computing, pp. 101–110 (2009)
18. Herawan, T., Rose, A.N.M., Deris, M.M.: Soft set-theoretic approach for dimensionality reduction. In: International Conference on Database Theory and Application, pp. 171–178 (2009)
19. Velayutham, C., Thangavel, K.: Unsupervised quick reduct algorithm using rough set theory. J. Electr. Sci. Technol. **9**(3), 193–201 (2011)
20. Jothi, G., Inbarani, H.H.: Soft set-based quick reduct approach for unsupervised feature selection. In: 2012 IEEE International Conference on Advanced Communication Control and Computing Technologies (ICACCCT), pp. 277–281 (2012)
21. Blake, C.: UCI Repository of Machine Learning Databases (1998)
22. Rajinikanth, V., Satapathy, S.C., Dey, N., Fernandes, S.L., Manic, K.S.: Skin melanoma assessment using Kapur's entropy and level set—a study with bat algorithm. In: Smart Intelligent Computing and Applications, 193–202 (2019)

Chapter 11
Fuzzy-Based Methods for the Selection and Prioritization of Software Requirements: A Systematic Literature Review

Mohd. Nazim, Chaudhary Wali Mohammad, and Mohd. Sadiq

Abstract Software requirements (SRs) selection is one of the major research issues in the field of software engineering which is used to select the requirements from the list of the elicited SRs on the basis of their ranks. There are different fuzzy-based techniques for computing the ranking values of SRs. Based on our review, we found different studies which focus on the literature review of the SRs selection and prioritization. There is no study which synthesizes the SRs selection methods under a fuzzy environment. Therefore, to address this issue, this paper presents a systematic literature review (SLR) in the area of fuzzy-based methods for the selection of SRs. Four research questions (RQs) have been formulated to identify the research gaps in the literature: RQ1: What are the different fuzzy-based methods for the selection of the SRs? RQ2: Which fuzzy number is mostly used during the computation process? RQ3: What kinds of datasets have been used in the SRs selection methods? RQ4: Which system/application has been used to select the SRs? Search items were extracted from the Journals, Conferences, and Workshops based on the RQs. Our SLR has identified 54 different studies. Selected studies were accessed based on the RQs thus identifying the research issues in the literature. After the analysis of the 54 studies, it was identified that the area of SRs selection has lots of research issues which can be addressed by applying the other soft computing techniques to find out the ranking order of SRs.

Mohd. Nazim (✉) · C. W. Mohammad
Department of Applied Sciences, and Humanities, Computer Science and Technology Research Group, Jamia Millia Islamia (A Central University), New Delhi 110025, India

C. W. Mohammad
e-mail: cmohammad@jmi.ac.in

Mohd. Sadiq
Department of Computer Science and Automation, Indian Institute of Science, Bangalore 560012, India
e-mail: msadiq@jmi.ac.in

© The Author(s), under exclusive license to Springer Nature Singapore Pte Ltd. 2022   115
V. Bhateja et al. (eds.), *Evolution in Computational Intelligence*,
Smart Innovation, Systems and Technologies 267,
https://doi.org/10.1007/978-981-16-6616-2_11

11.1   Introduction

The notion of fuzzy logic was developed by Lotfi A. Zadeh in 1965 to deal with vagueness and imprecision. It is a multivalued logic and is used to handle the notion of partial truth [1]. There are different applications of fuzzy-based methods in science and engineering like management science, software engineering, electrical engineering, brain computing, etc. Different methods have been developed in software engineering to deal with vagueness and imprecision during the decision-making process of the software requirements (SRs) selection. The SRs selection is a vital research area of software engineering in which SRs are selected from the list of the identified SRs based on their ranking values [2]. The SRs selection is a multicriteria decision-making process (MCDM) in which different decision makers are involved to compute the ranking values of the SRs so that SRs can be selected during the different releases of software after the consensus of the stakeholders [3].

Different methods have been developed to select the SRs centered on the following concepts, i.e., ant colony optimization, integer linear programming, genetic algorithm, and fuzzy logic [4]. Based on the critical study of the SRs prioritization and selection methods, we found that fuzzy-based approaches have been employed to deal with imprecision and vagueness during the decision-making process. In these methods, different types of fuzzy numbers are used to prioritize and select the SRs from different types of datasets which are based on the requirements of some information systems. Based on this investigation, we formulated the RQs so that existing fuzzy-based methods can be synthesized and research gaps can be identified for further research. To achieve this objective, in this paper we have adopted the guidelines of Kitchenham [5] to perform the systematic literature review (SLR) because it is a very effective methodology to identify the research gaps in the literature.

The remaining part of this paper is organized as follows: The research method to perform the SLR in the area of fuzzy-based methods for SRs selection is discussed in Sect. 11.2. Section 11.3 describes the threats to validity. The results and discussion of our SLR are given in Sect. 11.4. Finally, conclusions and future work are given in Sect. 11.5.

11.2   Research Method

Kitchenham's guidelines [5] have been employed as a research method to determine the research issues in the area of fuzzy-based methods for the selection of SRs which includes the following steps: research questions (RQs), search strategy, study selection, and data synthesis.

A. Research Questions

The objective of this SLR is to identify the research gaps in the area of SRs selection and prioritization. To achieve this objective, the following research questions have been formulated as a part of the SLR:

- RQ-1: What are the different fuzzy-based methods for the selection of the SRs?
- RQ-2: Which fuzzy number is mostly used during the computation process?
- RQ-3: What kinds of datasets have been used in the SRs selection methods?
- RQ4: What are the different types of systems which have been used as a part of the case study in SRs selection and prioritization research?

B. Search Strategy

Based on the key terms available in the RQs, the following search string has been formulated for the extraction of primary studies from nine different electronic database resources, i.e., *"ACM Digital Library, IEEE XPlore, Google Scholar, ScienceDirect, Taylor and Francis, Semantic Scholar, Research Gate, Springer, and arXiv"*.

Search string: *((fuzzy logic OR fuzzy approaches OR fuzzy techniques) AND (software requirements selection OR software requirements prioritization OR selection of software requirements OR prioritization of software requirements))*.

C. Study Selection

In our SLR, we have extracted the data from the last 12 years. So we have selected those studies which have been published from 2009 to 2020. Figure 11.1 exhibits the complete selection process of the studies. In the first search stage, 139 studies were identified from the selected nine electronic database resources. For example, five studies were selected from ACM digital library.

To remove the duplicate and immaterial studies from the selected studies, the process of scrutinization was performed on the basis of the title of the studies. Consequently, 60 studies have been identified as the primary studies. 51 primary studies were identified in the second search process on the basis of their abstracts and conclusions. As a result, 14 studies were selected from the references of these 51 primary studies. After that, 65 studies were selected for the quality assessment (QA) process. After the process of QA, 54 studies were selected to give the answers to the RQs. The following questions were used for the QA of our selected studies to perform the evaluation of the quality and the relevance of our selected studies: **(QA-1)** *"Is the proposed method for the SRs selection clearly discussed?"* **(QA-2)** *"Is the SRs selection method applied on some dataset?"* **(QA-3)** *"Is the fuzzy-based method applied clearly?"* **(QA-4)** *"Does the research add value to the academia or software industry?"*. The evaluation of the selected 65 studies was performed on the basis of the following scores: (a) No = 0, (b) Partly = 0.5, and (c) Yes = 1. For example, the method presented in study S1 is clearly discussed so QA-1 is rated as 1.0. In S1, the method was explained by considering the requirements of an electronic healthcare system (EHCS), and a fuzzy-based method was used to deal with

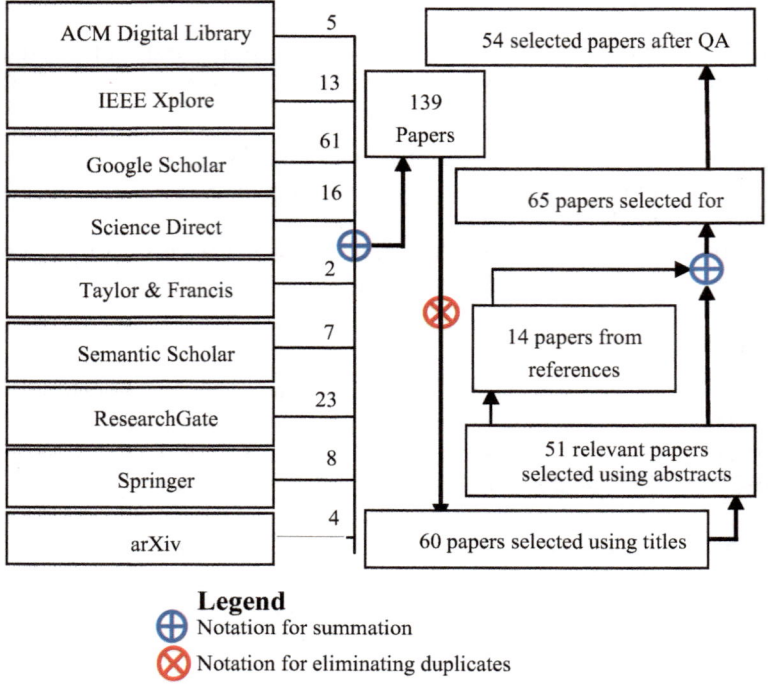

Fig. 11.1 Selection of studies

the vagueness and uncertainty in EHCS. Therefore, QA-2 and QA-3 are rated as 0.5 and 1.0, respectively. The study S1 was analyzed in light of the QA-4 and as a result, it was found that it is useful in academia or industry so it scores 1.0. This procedure was adopted to compute the values of the QA for the remaining studies. We have selected only those studies for a further analysis whose scores values are greater than or equal to 2.0. As a result, we have selected 54 studies. The list of 54 studies and their quality scores are exhibited in Appendix A and Appendix B, respectively.

D. Data Synthesis

In our work, the data of the selected 54 studies have been synthesized using bar charts and text. The bar charts are used to illustrate the data related to RQ-1 for showing the evolution of the fuzzy-based methods for the SRs selection and prioritization. The results of RQ-2, RQ-3, and RQ-4 are presented in the form of text.

11.3 Threats to Validity

In this study, we have generated a search string to identify the primary studies from different libraries. Practically, it's impossible to extract all studies from the terms which appear in the formulated RQs. In order to overcome this threat, we performed manual scrutiny to determine those primary studies that might be missed out during the initial search process of the SLR. The independent valuation of the QA was carried out to reduce the inaccuracy of the extracted data.

11.4 Results and Discussion

The objective of this section is to discuss the formulated RQs based on the selected 54 studies. The year-wise distribution of the selected studies is exhibited in Fig. 11.2.
The analysis of the RQs is given as follows [S1–S54].

RQ-1: What are the different fuzzy-based methods for the selection of the SRs?

Both crisp and fuzzy-based methods have been employed for the selection of the SRs. In this paper, we are only concerned with fuzzy-based methods. These methods have been developed to deal with imprecision and vagueness during the decision-making process. Based on our analysis, we have identified different methods for the selection of SRs using fuzzy logic. For example, Sen and Baracli [S13] focused on the NFRs in the domain of fuzzy "*quality function deployment*" (**Fuzzy QFD**). In their work, the NFRs were analyzed as a case study in Audio Electronics of Turkey's electronic industry under a fuzzy environment. Ramzan et al. [S11] applied the fuzzy

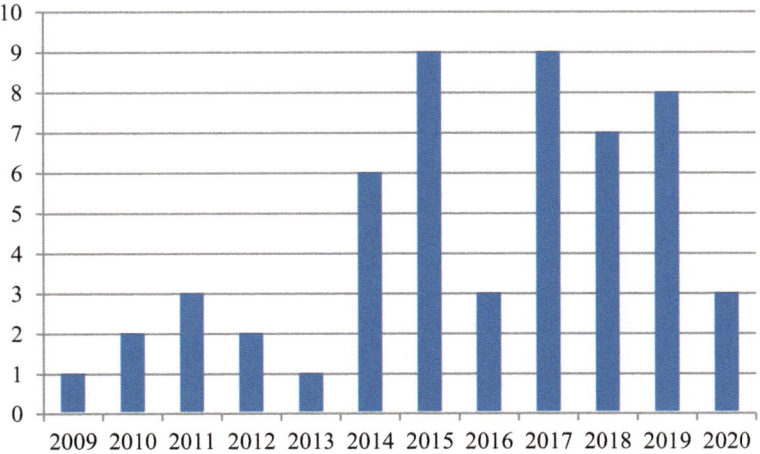

Fig. 11.2 Year-wise distribution of the selected studies

logic for the development of the multi-level "*values-based intelligent requirements prioritization*" (**VIRP**) for the prioritization of the SRs. In 2014, Achimugu et al. [S44] presented an "*Adaptive Fuzzy Decision Matrix Model*" (**AFDMM**) to prioritize the SRs on the basis of the weights assigned by the relevant stakeholders.

Sharif et al. [S29] presented a method named as Fuzzy "*Hierarchical Cumulative Voting*" (**Fuzzy HCV**). Momeni et al. [S14] proposed an approach which is based on "*Neuro-fuzzy system*" (**NFS**) to prioritize the software quality requirements. Sadiq et al. [S37] strengthen the goal-oriented requirements elicitation process (GOREP) using fuzzy set theory and developed a method for "prioritization of requirements using fuzzy-based approach in GORE" (**PRFGORE**) process. In 2015, an expert system, i.e., "*Priority Handler*" (**PHandler**), was proposed by Babar et al. [S3] for the prioritization of requirements by considering the VIRP and an "*analytical hierarchy process*" (AHP). Alrashoud and Abhari [S4] proposed a framework based on "*Fuzzy Inference System*" (**FIS**) for the selection of the most important requirements for the next release of software.

Jawale et al. [S23] proposed a new method for the requirements prioritization, i.e., "*Adaptive Fuzzy Hierarchical Cumulative Voting*" (**AFHCV**). The extent **fuzzy AHP** was used in [S21] for the prioritization of the stakeholders on the basis of the importance of the SRs. In another study, the fuzzy AHP (FAHP) and fuzzy TOPSIS (FTOPSIS) method were used to develop an integrated method for the selection of SRs [S18]. In Fig. 11.3, it is represented as "**integrated FAHP and FTOPSIS method**". Ahmad et al. [S36] developed the MoSCoW method using fuzzy logic, i.e., "**fuzzy-based MoSCoW**"; and applied it for the selection of the SRs. The fuzzy-based method was applied in the software testing area in which the testing requirements were generated from the selected set of the SRs [S39]. The fuzzy 2-tuple linguistic (**F2TL**) model was applied in [S50] for the prioritization of the SRs by considering the opinions of different stakeholders. An "*Adaptive network-based fuzzy inference system*" (**ANFIS**) was used in [S53] to deal with the problem of uncertainty in the process of releases planning. An artificial neural network fuzzy AHP (**ANN fuzzy AHP**) model was proposed by Singh et al. [S46] for the prioritization of the requirements.

To strengthen one of the processes of goal-oriented requirements engineering, i.e., requirements analysis, Mohammad et al. [S48] proposed a method called "*fuzzy attributed goal-oriented software requirements analysis*" (**FAGOSRA**) when different stakeholders participated during the decision-making process. Singh et al. [S40] proposed a method for SRs prioritization using "*logarithmic fuzzy trapezoidal AHP*" (**LFTAHP**). Fuzzy logic-driven security requirements (**FLDSR**) elicitation method was developed by Sadiq et al. [S17] in which the main emphasis was on the identification of those FRs which need more security during the development phase. An "*Intuitionistic Fuzzy Approach*" (**IFS**) was developed by Gupta et al. [S20] to support the stakeholder's view. Mougouei et al. [S9] proposed a framework based on fuzzy logic, i.e., "*Prioritization and Partial Selection*" (**PAPS**) to overcome the problem of ignoring the security requirements during the requirements selection process. In addition to these methods, the following methods are also represented in Fig. 11.3: "*User requirements prioritization*" (**URPCalc**) [S16], "*Graph-oriented*

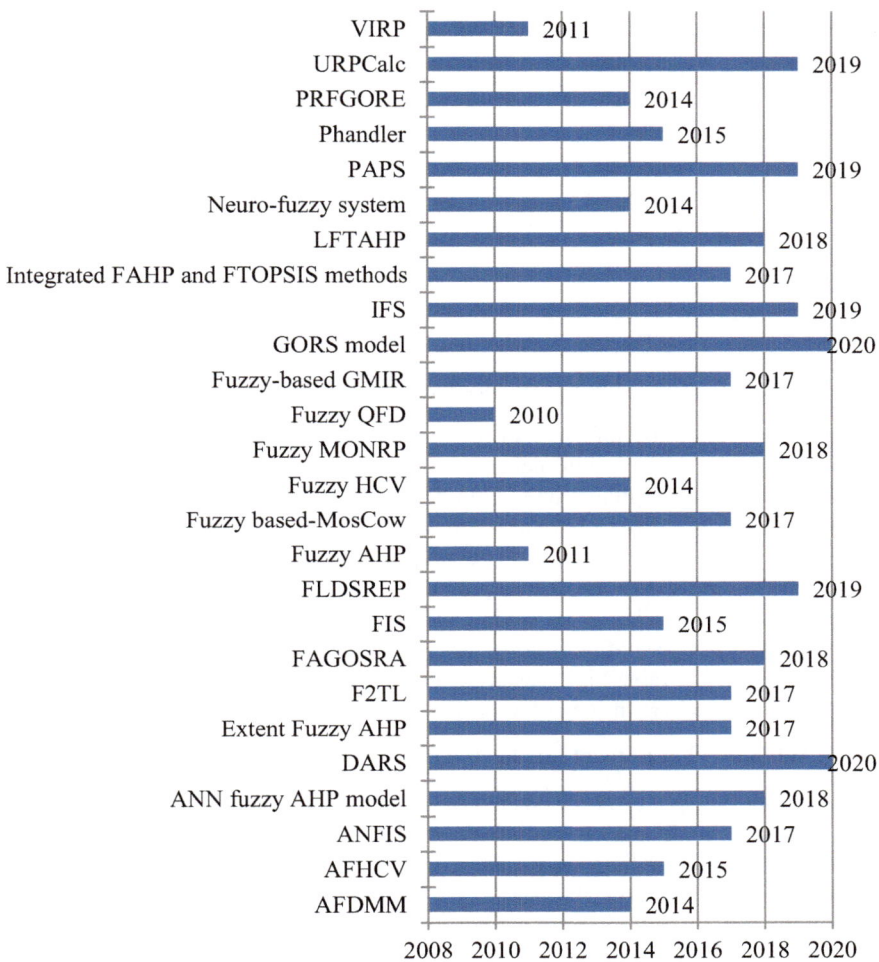

Fig. 11.3 Fuzzy-based SRs selection and prioritization methods

requirement selection" (**GORS**) model [S49], Fuzzy "*multi-objective NRP*" (**Fuzzy MONRP**) [S54], fuzzy analytic hierarchy process (**Fuzzy AHP**) [S5], "*Dependency-Aware Requirement Selection*" (**DARS**) method [S33], and "Adaptive Network-based Fuzzy Inference System" (**ANFIS**) [S53]. After the critical analysis of RQ-1, we have identified 26 different methods which are used for the prioritization and selection of the SRs under a fuzzy environment. The year-wise development of these methods is depicted in Fig. 11.3. From Fig. 11.3, it can be observed how different types of the fuzzy-based MCDM like fuzzy AHP, extent fuzzy AHP, fuzzy TOPSIS, fuzzy 2-tuple, fuzzy-based GMIR, etc. have been used to develop the SRs prioritization

Table 11.1 Types of fuzzy numbers used in SRs selection process

S. no.	Types of fuzzy numbers	No. of studies	References (see Appendix A)
1	Triangular fuzzy number	16	S1, S13, S17, S18, S21, S26, S27, S29, S36, S39, S46, S10, S25, S31, S37, and S41
2	Trapezoidal fuzzy number	2	S6 and S40
3	Fuzzy graph	2	S2 and S33

and selection methods. It is also observed that little attention is given to the neural networks during the SRs prioritization and selection process.

RQ-2: Which fuzzy number is mostly used during the computation process?

In fuzzy set theory, different types of fuzzy numbers are used to model the linguistic variables used by the decision makers during the evaluation of the FRs and NFRs of software, i.e., Triangular Fuzzy Numbers (TFN), Trapezoidal Fuzzy Numbers (TrapFNs), Gaussian Fuzzy Number, Pentagonal Fuzzy Numbers, and Bell-Shaped Fuzzy Numbers [S21, S37]. Based on our review, we found that the following types of fuzzy numbers are mostly used during the SRs selection and prioritization process, i.e., TFNs and TrapFNs. In 16 studies, TFNs have been used. Two studies have used TrapFNs. In two studies, a fuzzy graph has been used during the SRs selection process, as shown in Table 11.1. Based on our analysis, we found that TFNs are mostly used by the researchers in fuzzy-based SRs selection and prioritization methods.

RQ-3: What kind of dataset has been used in the SRs selection methods?

Based on the number of requirements of software, the datasets are classified into three parts, i.e., small dataset, medium dataset, and large dataset [6, 7]. Small dataset contains 1–14 SRs, medium dataset includes 15–50 SRs, and large dataset includes more than 50 requirements. The same classification of the dataset having a number of requirements has been used by Hujainah et al. [S5] for performing the SLR based on significance, stakeholders, techniques, and challenges in the area of SRs prioritization. In this study, we have also used the same classification of the SRs in small, medium, and large datasets. After a critical analysis, we found that in 16 studies, small dataset has been used during the experimental work. Seven studies have used medium dataset, and three studies have used large dataset for the analysis of the SRs; see Table 11.2.

Table 11.2 Datasets used in SRs selection and prioritization

S. no.	Types of dataset	No. of studies	References (see Appendix A)
1	Small dataset	16	S1, S9, S14, S17, S18, S20, S21, S25, S31, S36, S37, S39, S41, S44, S45, S46
2	Medium dataset	7	S4, S6, S12, S29, S33, S52, S49
3	Large dataset	3	S3, S13, S54

RQ4: What are the different types of systems which have been used as a part of the case study in SRs selection and prioritization research?

Based on our review, we have identified the following types of systems which are used as a part of the case studies, i.e., Academic Library Service, Audio Electronics of Turkey's electronic industry, Electronic Healthcare System, Institute Examination System (IES), Library Management System, Online Banking System, and Precious Messaging System (PMS). Among these systems, IES has been used in most of the studies during the experimental work.

11.5 Conclusions and Future Work

This paper presents a systematic literature review in the area of fuzzy-based SRs selection and prioritization. On the basis of the selected studies, i.e., S1–S54, we found 26 different methods in which fuzzy set theory has been used to deal with vagueness and imprecisions. To represent the linguistic variables mathematically, TFN is mostly used during the decision-making process. Datasets play an important role in SRs selection and prioritization research. Based on our critical analysis, we have identified that small datasets are mostly used to validate the steps of the SRs selection and prioritization methods, and the requirements of an IES are mostly used during the decision-making process. In future, we shall try to work on the following:

1. To extend the SLR by considering the other soft computing techniques like rough set theory, genetic algorithms, etc.
2. To develop the hybrid methods for SRs selection and prioritization by considering different soft computing techniques like rough set approach and fuzzy-based approach, fuzzy-based approach and genetic algorithm, fuzzy-based approach and neural networks, etc.

Appendix

Appendix A: Included Studies

Paper ID	Complete reference
S1	Gulzar, K., Sang, J., Ramzan, M., Kashif, M.: Fuzzy approach to prioritize usability requirements conflicts: an experimental evaluation. IEEE Access **5**, 13,570–13,577 (2017)
S2	Mougouei, D., Powers, D.: A fuzzy-based optimization method for integrating value dependencies into software requirement selection. arXiv: 2003.04806v1, pp. 1–15 (2020)

(continued)

(continued)

Paper ID	Complete reference
S3	Babar, M.I., Ghazali, M., Jawawi, D.N.A., Shamsuddin, S.M., Ibrahim, N.: PHandler: An expert system for a scalable software requirements prioritization process. Knowl. Based Syst. **84**, 179–202 (2015)
S4	Alrashoud, M., Abhari, A.: Perception-based software release planning. Intell. Autom. Soft Comput. **21**(2), 175–195 (2015)
S5	Hujainah, F., Bakar, R.B.A., Abdulgabber, M.A., Zamli, K.Z.: Software requirements prioritisation: a systematic literature review on significance, stakeholders, techniques and challenges. IEEE Access **6**, 71,497–71,523 (2018)
S6	Quamar, M.B., Gazi, Y.: On fuzzy qualitative and quantitative softgoal interdependency graph. Int. J. Comput. Appl. **122**(5), 30–35 (2015)
S7	Achimugu, P., Selamat, A., Ibrahim, R., Mahrin, M. N.: A systematic literature review of software requirements prioritization research. Inf. Softw. Technol. **56**(6), 568–585 (2014)
S8	Mishra, N., Khanum, A., Agarwal, K.: A review on approaches to optimize the requirement elicitation process. Int. J. Inf. Res. Rev. **2**(10), 1293–1298 (2015)
S9	Mougouei, D., Powers, D. M. W., Mougouei, E.: A fuzzy framework for prioritization and partial selection of security requirements in software projects. J. Intell. Fuzzy Syst. **37**(2), 1–17 (2019)
S10	Dabbagh, M., Lee, S.P.: An approach for prioritizing NFRs according to their relationship with FRs. Lect. Notes Softw. Eng. **3**(1), 1–5 (2015)
S11	Ramzan, M., Jaffar, A., Shahid, A.: Value based intelligent requirement prioritization (VIRP): expert driven fuzzy logic based prioritization technique. Int. J. Innov. Comput. Inf. Control **7**(3), 1017–1038 (2011)
S12	Achimugu, P., Selamat, A., Ibrahim, R.: Using the fuzzy multi-criteria decision making approach for software requirements prioritization. Jurnal Teknologi **77**(13), 21–28 (2015)
S13	Şen, C.G., Baraçlı, H.: Fuzzy quality function deployment based methodology for acquiring enterprise software selection requirements. Expert Syst. Appl. **37**, 3415–3426 (2010)
S14	Momeni, H., Motameni, H., Larimi, M.: A neuro-fuzzy based approach to software quality requirements prioritization. Int. J. Appl. Inf. Syst. (IJAIS) **7**(7), 15–20 (2014)
S15	Veena, N., D'Souza, R.: A survey on multi-criteria decision making methods in software engineering. Int. J. Innov. Sci. Res. Technol. **3**(7), 1–9 (2018)
S16	Bukhsh, F.A, Bukhsh, Z.A., Daneva, M.: A systematic literature review on requirement prioritization techniques and their empirical evaluation. Comput. Stand. Interf. **69**, 1–39 (2019)
S17	Ahmad J., Mohammad C. W., Sadiq M.: On software security requirements elicitation and analysis methods. Inf. Technol. Ind. **9**(1), 1–13 (2021)
S18	Sadiq, M., Afrin, A.: An integrated approach for the selection of software requirements using fuzzy AHP and fuzzy TOPSIS method. Int. J. Adv. Res. Dev. **2**(6), 170–183 (2017)
S19	Sadiq, M., Mohammad, C.W., Khan, S.: Methods for the selection of software requirements: a literature review. J. Eng. Technol. **8**, 108–128 (2019)

(continued)

(continued)

Paper ID	Complete reference
S20	Gupta, A., Gupta, C.: A novel collaborative requirement prioritization approach to handle priority vagueness and inter-relationships. J. King Saud Univ. Comput. Inf. Sci. 1–25 (2019)
S21	Sadiq, M.: A fuzzy set-based approach for the prioritization of stakeholders on the basis of the importance of software requirements. IETE J. Res. **63**(5), 1–14 (2017)
S22	Ejnioui, A., Otero, C.,Otero L.: A simulation-based fuzzy multi-attribute decision making for prioritizing software requirements. In: Proceedings of the 1st Annual Conference on Research in Information Technology, pp. 37–42. RIIT'12, New York, NY, USA: ACM (2012)
S23	Jawale, B., Bhole, A.T.: Adaptive fuzzy hierarchical cumulative voting: a novel approach toward requirement prioritization. Int. J. Res. Eng. Technol. **4**(5), 365–370 (2015)
S24	Bakhtiar, A., Hannan, A., Basit, A., Ahmad,J.: Prioritization of value based services of software by using AHP and fuzzy kano model. In: Proceedings of 3rd International Conference on Computational and Social Sciences, pp. 48–56, Johor Bahru, Malaysia (2015)
S25	Sadiq, M., Jain, S.K.: A fuzzy based approach for requirements prioritization in goal oriented requirements elicitation process. In: Proceeding of 2013 International Conference on Software Engineering and Knowledge Engineering, June 27–29, pp. 1–5 (2013)
S26	Jawale, B.B., Patnaik, G.K., Bhole, A.T.: Requirement prioritization using adaptive fuzzy hierarchical cumulative voting. In: IEEE 7th International Advance Computing Conference, pp. 95–102, Hyderabad (2017)
S27	Gambo, I., Ikono, R., Achimugu, P., Soriyan, A.: An integrated framework for prioritizing software specifications in requirements engineering. Int. J. Softw. Eng. Appl. **12**(1), 33–46 (2018)
S28	Babar, M. I., Ramzan, M., Ghayyur, S. A. K.: Challenges and future trendsin software requirements prioritization. In: International Conference on Computer Networks and Information Technology, pp. 319 –324 (2011)
S29	Sharif, N., Zafar, K., Zyad, W.: Optimization of requirement prioritization using computational Intelligence technique. In: International Conference on Robotics and Emerging Allied Technologies in Engineering, pp. 228–234 (2014)
S30	Malhotra, M., Bedi, R. P. S.: Analysis of software requirements prioritization techniques. In: Proceedings of 2nd International Conference on Computer Science Networks and Information Technology, pp. 195–200, Montreal, Canada (2016)
S31	Ashfaque, F., Kumar, R.: Elicitation of preference matrix and contribution values in goal models using fuzzy based approach. Adv. Comput. Sci. Inf. Technol. **2**(9), 38–42 (2015)
S32	Dhingra, S., Savithri, G., Madan, M., Manjula, R.: Selection of prioritization technique for software requirement using fuzzy logic and decision tree. In: Online International Conference on Green Engineering and Technologies, pp. 1–11 (2016)
S33	Mougouei, D., Powers, D. M. W.: Dependency-aware software requirements selection using fuzzy graphs and integer programming. arXiv:2003.05785v1, pp. 1–45 (2020)

(continued)

(continued)

Paper ID	Complete reference
S34	Ejnioui, A., Otero, C.E., Qureshi, A.A.: Software requirement prioritization using fuzzy multi-attribute decision making. IEEE Conference on Open Systems, pp. 1–6 (2012)
S35	Nayak, V., D'Souza, R.G.L.: Comparison of multi-criteria decision making methods used in requirement engineering. CiiT Int. J. Artif. Intell. Syst. Mach. Learn. **11**(5), 92–96 (2019)
S36	Ahmad, K.S., Ahmad, N., Tahir, H., Khan, S.: Fuzzy_MoSCoW: a fuzzy based MoSCoW method for the prioritization of software requirements. In: International Conference on Intelligent Computing, Instrumentation and Control Technologies, pp. 433–437 (2017)
S37	Sadiq, M., Jain, S.: Applying fuzzy preference relation for requirements prioritization in goal oriented requirements elicitation process. Int. J. Syst. Assur. Eng. Manage. **5**, 711–723 (2014)
S38	Ramzan, M., Jaffar, M.A., Iqbal, M.A., Anwar, S., Shahid, A.A.: Value based fuzzy requirement prioritization and its evaluation framework. In: Fourth International Conference on Innovative Computing, Information and Control, ICICIC, pp. 1464–1468 (2009)
S39	Sadiq, M., Neha.: Elicitation of testing requirements from the selected set of software's functional requirements using fuzzy based approach. In: International Conference on Computational Intelligence in Data Mining, Springer, pp. 429–437, Springer, India (2017)
S40	Singh, Y.V., Kumar, B., Chand, S.: A novel approach of requirements prioritization using logarithmic fuzzy trapezoidal AHP for enhancing academic library service. In: 2018 International Conference on Advances in Computing, Communication Control and Networking (ICACCCN), pp. 1164–1172, Greater Noida, India (2018)
S41	Sadiq, M., Jain, S. K.: A fuzzy based approach for the selection of goals in goal oriented requirements elicitation process. Int. J. Syst. Assur. Eng. Manage. **6**(2), 157–164 (2015)
S42	Devadas, R., Srinivasan, G.N.: Review of different fuzzy logic approaches for prioritizing software requirements. Int. J. Sci. Technol. Rese. **8**(9), 296–298 (2019)
S43	Alrezaamiri, H., Ebrahimnejad, A., Motameni, H.: Solving the next release problem by means of the fuzzy logic inference system with respect to the competitive market. J. Exp. Theoret. Artif. Intell. **32**(6), 959–976 (2019)
S44	Achimugu, P., Selamat, A., Ibrahim, R., Mahrin, M.N.: An adaptive fuzzy decision matrix model for software requirements prioritization. In: Nguyen, N.T., Kowalczyk, N.T. (Eds.) Advanced Approaches to Intelligent Information and Database Systems, pp. 129–138. Springer, Cham, Switzerland (2014)
S45	Lima, D., Freitas, F., Campos, G., Souza, J.: A fuzzy approach to requirements Prioritization. In: International Symposium on Search Based Software Engineering, pp. 64–69, Springer, Berlin Heidelberg (2011)
S46	Singh, Y.V., Kumar, B., Chand, S., Kumar, J.: A comparative analysis and proposing 'ANN fuzzy AHP model' for requirements prioritization. Int. J. Inf. Technol. Comput. Sci. **4**, 55–65 (2018)
S47	Mishra, N., Khanum, M.A., Agrawal, K.: Approach to prioritize the requirements using fuzzy logic. In: ACEIT Conference Proceeding, pp. 42–47 (2016)

(continued)

(continued)

Paper ID	Complete reference
S48	Mohammad, C.W., Shahid, M., Hussain, S.Z.: Fuzzy attributed goal oriented software requirements analysis with multiple stakeholders. Int. J. Inf. Technol. 1–9 (2018)
S49	Mougouei, D., Powers, D.M.W.: Modeling and selection of interdependent software requirements using fuzzy graphs. Int. J. Fuzzy Syst. **19**(6), 1812–1828 (2017)
S50	Gerogiannis, V. C.,Tzikas, G.: Using fuzzy linguistic 2-tuples to collectively prioritize software requirements based on stakeholders' evaluations. In: Proceedings of the 21st Pan-Hellenic Conference on Informatics 48, pp. 1–6 (2017)
S51	Ramzan, M.: Intelligent requirement prioritization using fuzzy logic. National University of Computer and Emerging Sciences, Islamabad, Pakistan, pp. 1–155 (2010)
S52	Sadiq, M., Khan, S., Mohammad, C. W.: Selection of software requirements using TOPSIS under fuzzy environment. Int. J. Comput. Appl. (2020)
S53	Alrashoud, M., Abhari, A.: Planning for the next software release using adaptive network-based fuzzy inference system. Intell. Decis. Technol. **11**(2), 153–165 (2017)
S54	Alrezaamiri, H., Ebrahimnejad, A., Motameni, H.: Software requirement optimization using a fuzzy artificial chemical reaction optimization algorithm. Soft Comput. **23**, 9979–9994 (2019)

Appendix B: Results of the Quality Scores of the Selected 54 Studies

Paper ID	QA-1	QA-2	QA-3	QA-4	Score
S1	1.0	0.5	1.0	1.0	3.5
S2	1.0	1.0	1.0	0	3.0
S3	0.5	1.0	1.0	0	2.5
S4	1.0	0.5	1.0	0	2.5
S5	1.0	0.5	0.5	0	2.0
S6	1.0	1.0	1.0	0	3.0
S7	1.0	0	1.0	0	2.0
S8	1.0	0	1.0	0	2.0
S9	1.0	1.0	1.0	0.5	3.5
S10	0.5	0.5	1.0	0.5	2.5
S11	1.0	0.5	0.5	0	2.0
S12	1.0	1.0	0.5	1.0	3.5
S13	1.0	1.0	1.0	0.5	3.5
S14	1.0	1.0	0.5	0	2.5
S15	0.5	0.5	1.0	0	2.0
S16	1.0	0	1.0	0	2.0
S17	1.0	1.0	1.0	0	3.0
S18	1.0	1.0	1.0	0.5	3.5
S19	1.0	0.5	0.5	0	2.0

(continued)

(continued)

Paper ID	QA-1	QA-2	QA-3	QA-4	Score
S20	1.0	1.0	0.5	0	2.0
S21	1.0	1.0	1.0	0	3.0
S22	1.0	0.5	1.0	0	2.5
S23	1.0	0.5	1.0	0	2.5
S24	0.5	1.0	0.5	0	2.0
S25	1.0	1.0	1.0	0	3.0
S26	1.0	0.5	0.5	0	2.0
S27	0.5	0.5	1.0	0.5	2.5
S28	1.0	0.5	1.0	0	2.5
S29	0.5	1.0	1.0	0	2.5
S30	0	0.5	1.0	0.5	2.0
S31	1.0	1.0	1.0	0.5	3.5
S32	1.0	1.0	1.0	0	3.0
S33	1.0	1.0	1.0	0	3.0
S34	1.0	0	1.0	0	2.0
S35	1.0	0	1.0	0	2.0
S36	1.0	1.0	1.0	0.5	3.5
S37	1.0	1.0	1.0	0	3.0
S38	1.0	0	1.0	0	2.0
S39	1.0	1.0	1.0	0	3.0
S40	1.0	1.0	1.0	1.0	4.0
S41	0.5	1.0	1.0	0	2.5
S42	1.0	0	1.0	0	2.0
S43	1.0	1.0	1.0	0	3.0
S44	0.5	1.0	0.5	0	2.0
S45	1.0	0.5	0.5	0	2.0
S46	0.5	1.0	1.0	0	2.5
S47	1.0	1.0	1.0	0	3.0
S48	1.0	1.0	1.0	0	3.0
S49	1.0	1.0	1.0	0	3.0
S50	1.0	0.5	1.0	0	2.5
S51	1.0	1.0	1.0	0	3.0
S52	1.0	1.0	1.0	0	3.0
S53	0.5	1.0	0.5	0	2.0
S54	0.5	1.0	1.0	0	2.5

References

1. Zadeh, L.A.: Fuzzy logic. Computer **21**(4), 83–93 (1988)
2. Garg, N., Sadiq, M., Agarwal, P.: GOASREP: Goal oriented approach for software requirements elicitation and prioritization using analytic hierarchy process. In: Proceedings of the 5th International Conference on Frontiers in Intelligent Computing Theory and Applications (FICTA), AISC, Springer, pp. 281–287, KIIT University, Bhubaneswar, Odisha, India (2016)
3. Sadiq, M., Khan, S., Mohammad, C.W.: Software requirements selection using consistent pairwise comparison matrices of AHP. Int. J. Comput. Sci. Eng. **6**(9), 168–175 (2018)
4. Pitangueira, A.M., Maciel, R.S.P., Barros, M.O.: Software requirements selection and prioritization using SBSE approaches: a systematic review and mapping of the literature. J. Syst. Softw. **103**, 267–280 (2015)
5. Kitchenham, B.: Guidelines for performing systematic literature reviews in software engineering, Version 2.3, EBSE Technical Report EBSE-2007–01, Keele University and University of Durham, pp. 1–57 (2007)
6. Ma, Q.: The effectiveness of requirements prioritization techniques for a medium to large number of requirements: a systematic literature review. Master of Computer and Information Sciences Dissertation, School of Computing and Mathematical Sciences, Auckland University of Technology, pp. 1–92 (2009)
7. Nazim, M., Mohammad, C. W., Sadiq, M.: Generating datasets for software requirements prioritization research. In: 2020 IEEE International Conference on Computing, Power and Communication Technologies, pp. 344–349, Greater Noida, India (2020)

Chapter 12
Investigation and Validation of Flow Characteristics Through Emergent Vegetation Patch Using Machine Learning Technique

Soumen Maji, **Apurbalal Senapati**, and **Arunendu Mondal**

Abstract Investigation on change in flow characteristics in an open channel flow is important to understand the flowing water ecosystem, which is important in various aspects like sediment deposition and water quality. A significant number of field and laboratory experiments have been carried out to characterize different types of flows in an open channel. Machine learning technique learns automatically from the given experimental data set and also improves from given data without explicitly being programmed. For years, various laboratory and analytical experiments have been carried out to establish the turbulent flow characteristics change. To validate the laboratory experimental data of turbulent flow characteristics with machine learning techniques is scanty in the literature. Here in the laboratory experiment, a wide rectangular channel with an emergent rigid vegetation patch has been used, and the velocity of water has been measured in different locations and directions. Out of three sets of laboratory experimental data, two sets of data have been considered as training data, and by using "Polynomial Regression Techniques" we validated stream-wise velocity of the third set of data. Interestingly, it has been found that both experimental and theoretical data closely matched with each other. This new approach using the machine learning technique may become important for future research in all types of vegetated flow.

S. Maji · A. Senapati (✉) · A. Mondal
Central Institute of Technology Kokrajhar, Kokrajhar 783370, Assam, India
e-mail: a.senapati@cit.ac.in

S. Maji
e-mail: s.maji@cit.ac.in

A. Mondal
e-mail: a.mondoal@cit.ac.in

© The Author(s), under exclusive license to Springer Nature Singapore Pte Ltd. 2022   131
V. Bhateja et al. (eds.), *Evolution in Computational Intelligence*,
Smart Innovation, Systems and Technologies 267,
https://doi.org/10.1007/978-981-16-6616-2_12

12.1 Introduction

In the aquatic system, vegetation is an essential part. Flow characteristics may change significantly while passing through any vegetation patch. Vegetation in the river ecosystem helps in sediment deposition and sometimes improves water quality as well. Hence, an impressive number of laboratory as well as field investigations were performed regarding the change in various flow characteristics in open-channel flows. Tracy et al. [1] used learning algorithms for turbulence modeling closure, which was trained on computational data to learn the characteristics of turbulence. Zhang and Duraisamy [2] focused on newly developed machine learning techniques (multiscale Gaussian process regression method) for use in turbulence modeling. Again Zhu et al. [3] compared with the machine learning models for the low Reynolds (Re) number flows based on direct numerical simulation data. It is worth mentioning that machine learning techniques can be applicable in any fluid medium with sufficient laboratory data, if available. Fukami et al. [4] used machine learning to reconstruct unsteady laminar and turbulent flows from spatially low-resolution data. Pandey et al. [5] stated Recurrent networks, which are found to be a promising tool to model dynamical processes in flows without the knowledge of the underlying Navier-Stokes or Boussinesq equations and with smaller training effort than other recurrent networks such as LSTM. To date, no attempt has been made to validate the flow characteristics through emergent vegetation using machine learning techniques. In this study, the authors have attempted to validate the machine learning technique with laboratory experimental data, which was performed with emergent vegetation in an open hydraulic channel.

12.2 Methodology

In this study, the authors have tried to investigate and validate the change in turbulent characteristics in a stream-wise direction through an emergent vegetation patch. In this regard, two types of setup were required like laboratory setup for measuring physical parameters of the flow and another one was the machine learning approach, which was used for validation purposes.

12.2.1 Laboratory Investigation

Laboratory experiments were performed in a wide rectangular channel with an emergent rigid vegetation patch. Figure 12.1 shows a schematic diagram of the top view of the experimental setup. Dimensions of the hydraulic flume were 1200 cm long, 91 cm wide, and 61 cm deep. Rails were provided at the top of the walls on both sides of the flume, and two carriages were arranged for depth and velocity measurements

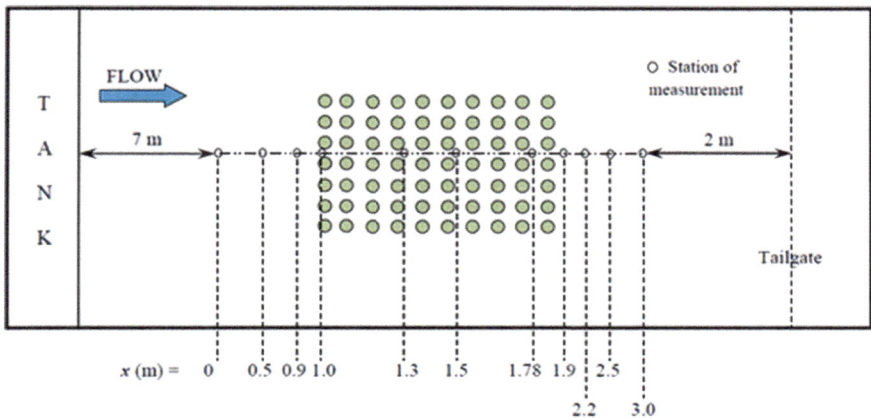

Fig. 12.1 Schematic diagram of the top view of laboratory experimental setup

(Fig. 12.2). To control the flow depth, a tailgate was put downstream (end of the hydraulic flume). Two centrifugal pumps were used to recirculate the flow between the flume and underground sump. Constant head of water was maintained throughout the experiment. A calibrated V-Notch was used for the measurement of discharge. A honeycomb baffle wall was deployed at the entrance of the flume to reduce the turbulence and non-uniformity in the incoming flow. Depth measurements were done by using a point gauge with the attachment of a graduated millimeter scale. The bed surface of the flume was made rough by cladding sand particles.

The test section was prepared as 300 cm long which starts at 700 cm from the entrance of the flume. The vegetation patch was located at the middle region of the test section (central area of the cross-sectional area of the flume). The vegetation characteristics of the patch were emergent, rigid, and sparse. It was made by seventy uniform acrylic cylindrical rods with dimensions 0.64 cm diameter and 30 cm length. The rods were adjusted perpendicularly as an array of 7 × 10 on a Perspex sheet which was fixed to the channel bed. In the patch, uniform spacing of 9 cm and 4 cm were maintained between centers of two consecutive cylinders along stream-wise

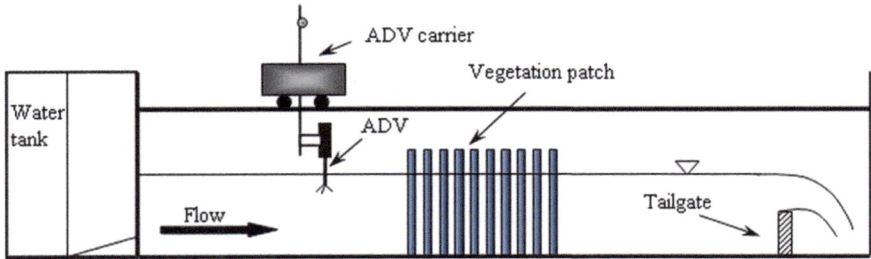

Fig. 12.2 Experimental model in cross-sectional view

Table 12.1 Details of flow conditions used in the laboratory experiments

Exp	h (m)	$\bar{u}$(m/s)	F_r	A_r
Re1Ar1	0.152	0.29	0.235	5.87
Re1Ar2	0.200	0.22	0.157	4.55
Re2Ar2	0.200	0.31	0.221	4.55

Where h: water depth; $\bar{u}$: Averaged velocity; F_r: Froude number; A_r: Aspect ratio.

and lateral directions, respectively. The dimensions of the patch were 81 cm long and 24 cm wide. Figures 12.1 and 12.2 clarify the setup of the vegetation patch in the flume.

A three-dimensional (3-D) Acoustic Doppler Velocimeter (ADV) with four down-looking probes was used to measure the point velocities across the channel section. To allow the stabilization of the flow field, the velocity measurement was taken 30 min after the flow was started. Along the centerline of the open channel through its cross-sectional area, the three-dimensional velocities were measured. The uniform water depth h before the flow entering into the vegetation was maintained as 15 cm for Re1Ar1 and 20 cm for both Re1Ar2 and Re2Ar2 by using a tailgate which was fixed at the downstream end of the flume. The detailed experimental conditions are given in Table 12.1. Velocity was measured at 15 points along the vertical direction which were 0.3, 0.5, 0.7, 0.9, 1.5, 2, 2.5, 3, 4, 5, 6, 7, 8, 9, and 10 cm above from the channel bed (12 cm and 15 cm when water depth was 20 cm). The stream-wise locations were situated in the exterior (x = 0 cm, 50 cm, 90 cm, and 100 cm), interior (x = 130 cm, 150 cm, and 178 cm), and downstream (x = 190 cm, 220 cm, 250 cm, and 300 cm) of the vegetation patch. The sampling rate was 100 Hz and the data was collected for 120 s duration at each vertical location, and this implies that $100 \times 60 \times 2 = 12,000$ instantaneous velocity data points were captured at each measuring point.

12.3 Machine Learning Approach

Machine learning is a technique that provides the ability to learn automatically and improve from given data without being explicitly programmed [6]. The main focusing point here is the data, and hence this technique can be adopted anywhere where suitable tagged data is available. Low-level representation in machine learning, the machine implies a mathematical model and the model need to train is known as the learning [7]. The training is done from experience, i.e. from past data. For the supervised learning technique, there are two phases, training and testing. In the training phase, the values of the model parameters are set to a specific value that can be interpreted as acquiring knowledge or skills from the experience.

Fig. 12.3 Example of a classification

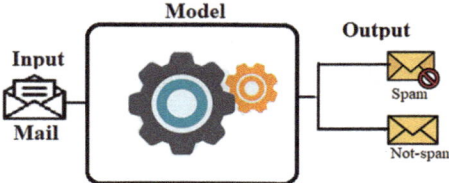

Fig. 12.4 Example of a regression (linear)

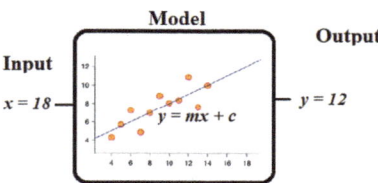

Broadly, machine learning is used for the regression or classification problem. Classification implies predicting a label of an input object (Fig. 12.3), whereas regression implies predicting a value of a given input (Fig. 12.4). Our problem is in the second category, i.e. a regression problem.

Model selection is the most crucial part to adopt the machine learning approach for a specific task [8]. So far from the above discussion, our problem is a regression problem, and there are numerous types of regression models like linear regression, polynomial regression, logistic regression, piecewise linear regression [9], etc. To choose the correct model must have to study the characteristic of the data points. Consider the Re1Ar1 data point of the experiment that is done at a distance of 300 with 15 different vertical depths as shown in Fig. 12.5 in the stream-wise direction, i.e. along the x-component.

This particular example clearly shows that the data points are not following the linearity and for the other data set along with the y-component and z-component, it is not following the linearity. Hence from the graphical observation, the polynomial

Fig. 12.5 Velocity in x-direction

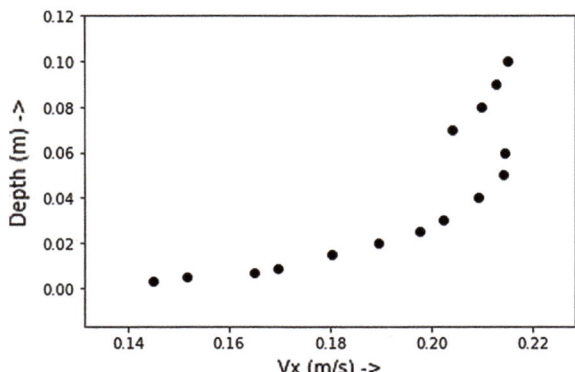

regression of degree two is best suited to fit the data. Before choosing the polynomial regression of degree two, we have trialed with the linear and cubic polynomial and the result shows that quadratic polynomial is dominating in terms of accuracy.

12.3.1 Polynomial Regression

In many real-life cases, the data points do not fit a linear regression, i.e. they cannot fit a straight line through all data points [8]. Rather, sometimes a curve is the best way to draw a line through the data points is the polynomial regression. The degree of the polynomial depends on the complex relationship of the data points. If the degree of the polynomial increases, then the complexity of the model also increases [10]. Therefore, the value of the degree must be chosen precisely. The low value faces the problem of under-fitting; in that case, the model won't be able to fit all the data points properly and if it is high, the model will go for the over-fitting problem. In our experiment, we have considered a polynomial of degree two. Mathematically, the model of a polynomial of degree two can be described by Eq. (12.1).

$$y = c_0 + c_1 x + c_2 x^2 \tag{12.1}$$

where c_0, c_1, and c_2 are the constants that have to be estimated in the training process.

Finding the value of the parameters c_0, c_1, and c_2 is known as the learning or training process. In order to do this, loss function is needed. There are various ways to define the loss function but in general, it is defined by the function of an error; it may be the mean square error. The loss function can be defined by Eq. (12.2).

$$J = f\left(y_{actl} - y_{pred}\right) \tag{12.2}$$

where y_{actl} is the actual value, i.e. the value from the data, and y_{pred} is the calculated or predicted value from the model. The model can be improved by decreasing the error iteratively. There is an algorithm to do this which is the gradient descent algorithm [11]. In this algorithm, it is done by changing the parameters iteratively and is known as training or learning.

12.3.2 Data Set

The detailed description of data is described above in Table 12.1; here, the format and volume of the data are focused and a sample data format is shown in Table 12.2. There are three sets of data to measure the stream velocity, where in sets contains at 15 different depths measured in meters (0.003, 0.005, 0.007, 0.009, 0.015, 0.02, 0.025, 0.03, 0.04, 0.05, 0.06, 0.07, 0.08, 0.09, 0.1) shown in second column in Table

Table 12.2 Sample data instances

Distance (cm)	Depth (m)	AvgVx (m/s)	AvgVy (m/s)	AvgVz (m/s)
…	…	…	…	…
…	…	…	…	…
0	0.04	0.308023139	−0.01983701	6.08067E-05
0	0.05	0.317947458	−0.021150604	−9.15614E-05
0	0.06	0.318096688	−0.022387368	−0.001125159
0	0.07	0.310325991	−0.021071419	0.000725015
0	0.08	0.319089329	−0.021184599	0.001856475
0	0.09	0.32099224	−0.02502235	0.000361428
0	0.1	0.326212355	−0.023846815	0.003926097
50	0.003	0.174241514	−0.012277089	−0.001272189
50	0.005	0.185135536	−0.010435847	0.000140378
50	0.007	0.193017014	−0.010657991	−0.000178517
50	0.009	0.204055493	−0.01030362	−0.000463234
50	0.015	0.222245545	−0.014612206	−0.00123144
…	…	…	…	…

12.2. These data are measured at distances (in cm) of 0, 50, 90, 100, 130, 150, 178, 190, 220, 250, 285, and 300 shown in the first column in Table 12.2. Hence, the total data point is taken as ($= 3 \times 15 \times 12$) 540. The velocity of the data having three components, i.e. in the x-direction, y-direction, and z-direction, is in third, fourth, and fifth columns, respectively. In the computational view, the training data is represented by the format (X, Y), where X is a pair of a component of distance and depth, i.e. X = [dist, depth], and y is the velocity components, i.e. Y = [Vx, Vy, Vz]. Two sets of data, i.e. ($2 \times 15 \times 12$) 360 data points are used for the training purpose, and some random data selected from the third set is considered for test purposes.

12.4 Result and Discussion

The system is trained by the data as discussed above. For the test and validation purposes, Re2Ar2 data set has been considered randomly. Figures 12.6 and 12.7 show the graphical presentation of experimental and system-generated data at the depth of 0.005 m and 0.05 m, respectively. From both Figs. 12.6 and 12.7, it can be expressed that system-generated data are closely similar to experimental data. Here, only stream-wise velocity data are validated. However, it can be extended that validation would be performed with lateral and transverse velocity also including other turbulent parameters in open-channel flows.

Fig. 12.6 Graphical presentation between experimented and system-generated data at the depth 0.005 m

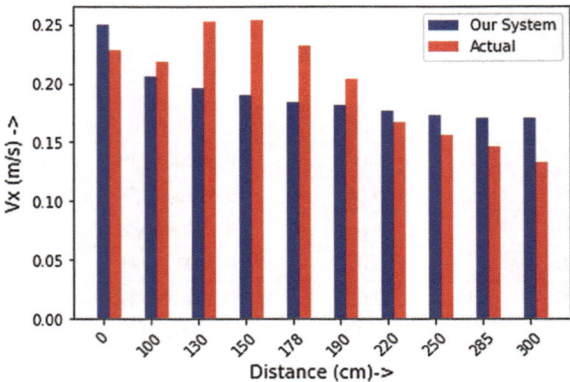

Fig. 12.7 Graphical presentation between experimented and system-generated data at the depth 0.05 m

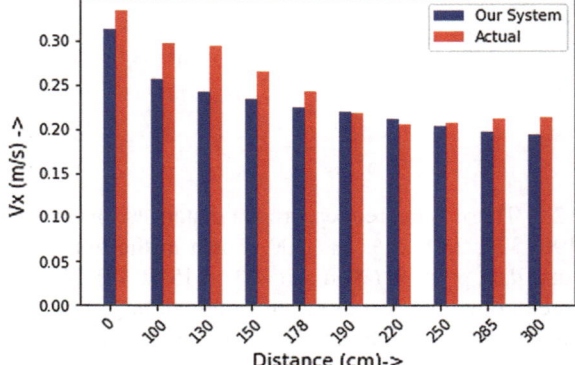

12.5 Conclusion

Turbulent flow characteristics change in due course of time is well established, especially for flows through vegetation patches. Various laboratory experimental and analytical investigations as well have been conducted for years. In this study, the authors have attempted validation of laboratory experimental data with machine learning techniques. Three sets of laboratory experiments have been performed (details are given in Table 12.1). For training purposes, the first two sets of data (Re1Ar1 and Re2Ar1) have been considered. Polynomial regression technique has been used for training the data. For validation, stream-wise velocity of the third set of data, i.e. Re2Ar2, has been taken. After comparing the system-generated data with experimental data, it is found that both data are very close to similar. This validation may be extended with other turbulent parameters also.

Declaration   We have taken permission from competent authorities to use the images/data as given in the paper. In case of any dispute in the future, we shall be wholly responsible.

References

1. Tracey, B.D., Duraisamy, K., Alonso, J.J.: A machine learning strategy to assist turbulence model development. In: 53rd AIAA Aerospace Sciences Meeting, p. 1287 (2015)
2. Zhang, Z.J., Duraisamy, K.: Machine learning methods for data-driven turbulence modeling. In: 22nd AIAA Computational Fluid Dynamics Conference, p. 2460 (2015)
3. Zhu, L., Zhang, W., Kou, J., Liu, Y.: Machine learning methods for turbulence modeling in subsonic flows around airfoils. Phys. Fluids **31**(1), 015105 (2019)
4. Fukami, K., Fukagata, K., Taira, K.: Super-resolution reconstruction of turbulent flows with machine learning. J. Fluid Mech. **870**, 106–120 (2019)
5. Pandey, S., Schumacher, J., Sreenivasan, K.R.: A perspective on machine learning in turbulent flows. J. Turbul. **21**(9–10), 567–584 (2020)
6. Chollet, F.: Deep Learning with Python, Manning Publications, ISBN: 9781617294433, pp. 4–6 (2017)
7. Senapati, A., Maji, S., Mondal, A.: Limitations and implications of doubling time approach in COVID-19 infection spreading study: a gradient smoothing technique, data preprocessing, active learning, and cost perceptive approaches for resolving data imbalance. IGI Glob. (2021). https://doi.org/10.4018/978-1-7998-7371-6
8. Senapati, A., Maji, S., Mondal, A.: Piece-wise linear regression: a new approach to predict COVID-19 spreading. In: IOP Conference Series: Materials Science and Engineering, Vol. 1020, No. 1, p. 012017. IOP Publishing (2021)
9. Senapati, A., Nag, A., Mondal, A., Maji, S.: A novel framework for COVID-19 case prediction through piecewise regression in India. Int. J. Inf. Technol. **13**, 41–48 (2021). https://doi.org/10.1007/s41870-020-00552-3
10. Deisenroth, M.P., Faisal, A.A., Ong, C.S.: Mathematics for Machine Learning, pp. 295–305. Cambridge University Press (2020)
11. Léon, B.: Online Algorithms and Stochastic Approximations. Online Learning and Neural Networks. Cambridge University Press. ISBN 978-0-52-1-65263-6

Chapter 13
Analysis of the Standard Objective Functions of RPL

Sneha Kamble and **B. R. Chandavarkar**

Abstract Low power and Lossy Network (LLN) is a network widely used to connect objects nowadays, and routing in such type of network is a new research area to work. ROLL-WG developed a routing protocol for LLN called Routing Protocol for low power and lossy network (RPL). Here, the Objective Function (OF) defines the way the node should select the parent node. The two standard OFs are OF-0, and MRHOF developed in 2010 and 2012. The paper explains these OFs and presents the experimental results of the two OFs in Contiki-3.0. The experiments were carried out by varying the number of nodes and varying data rates. The experimental results show that due to low hop count and fewer nodes getting involved in OF-0, it performs better than MRHOF, as MRHOF requires more stable links to transmit the data. Thus, OF-0 is useful in the scenarios where there are fewer nodes, and the destination is closer.

13.1 Introduction

In the new technical world, the need to connect everything to the internet has become a desire, which resulted in the new category of a network called Low Power and Lossy Network (LLN). The LLN nodes have low memory, low processing power, low data reliability, and limited power resources [2]. In any network, routing is an essential process before initiating the data transfer. A new IETF working group, Routing Over Low power and Lossy Network—Working Group (ROLL-WG), was developed for Routing in LLN. The group first studied the requirements of LLN and then analyzed the existing routing protocol, but none satisfied the criteria [7]. ROLL-WG then designed a routing protocol named Routing Protocol for Low power and Lossy network (RPL), which is a distance-vector routing protocol [11].

RPL forms a multi-hop connection between the nodes, and such a network is known as Destination Oriented Directed Acyclic Graph (DODAG). The DODAG is

S. Kamble (✉) · B. R. Chandavarkar
Department of Computer Science and Engineering, National Institute of Technology Karnataka, Surathkal, Karnataka, India

© The Author(s), under exclusive license to Springer Nature Singapore Pte Ltd. 2022   141
V. Bhateja et al. (eds.), *Evolution in Computational Intelligence*,
Smart Innovation, Systems and Technologies 267,
https://doi.org/10.1007/978-981-16-6616-2_13

formed by exchanging the control messages. RPL evaluates the shortest path and computes the rank based on Objective Function (OF) and routing metrics [11]. Once the DODAG is formed, the data transfer begins.

Modern applications have various ranges of requirements or have to focus on specified constraints. Therefore, the performance of the network has to be accordingly. For example, an application requiring high reliability should have a high packet delivery ratio. Improving the parent selection technique can help in such a situation. In RPL, the parent selection procedure depends on the Objective Function, which computes the rank and the routing metrics considered during the rank computation. The paper presents the experimental results and analysis of the two IETF standard OFs, i.e., Objective Function (OF-0) and Minimum Rank with Hysteresis Objective Function (MRHOF).

The rest of the paper is organized as follows: Sect. 13.2 briefs RPL and its working, Sect. 13.3 discusses the objective function, OF-0 and MRHOF, Sect. 13.4 presents the experimental results obtained for the two standard objective function and Sect. 13.5 concludes the paper.

13.2 RPL

ROLL-WG a working group design a new routing protocol for LLNs after studying the requirements of LLNs and analyzing the existing protocols. RPL is a distance vector routing protocol that forms a multi-hop ranked topology for nodes known as Destination Oriented Directed Acyclic Graph (DODAG). In DODAG, each node in the network forms a route to the destination node, usually the sink or low power and lossy network border router (LBR). The RPL control messages are used in the DODAG building process, which uses the Internet Control Message Protocol (ICMPv6) for the Internet Protocol version 6 messages. The three important RPL control messages are:

- DODAG Information Object (DIO): A message triggered by the root node to broadcast the DODAG details to the nodes present in the node's communication range.
- DODAG Information solicitation (DIS): The node uses DIS message to request network information, when it fails to receive the DIO message in the specified time interval.
- Destination Advertisement Object (DAO): The node uses DAO messages to reply to the DIO messages. DAO message is unicasted to the parent node or root node.

Once a node receives a DIO message, it checks if the DIO is received for the first time. If yes, then the node joins the DODAG directly. If a node receives the DIO message after joining the DODAG, it checks for its rank value through the new DIO received. If the rank value is less than the earlier one by some threshold value, then the node switches its parent; otherwise, not. Rank describes the position of the node corresponding to the root node. The rank value increases in the direction away from the root node. The rank is computed based on the Objective Function (OF). The two IETF standard OFs are Objective Function Zero (OF-0) and Minimum Rank with Hysteresis Objective Function (MRHOF).

13.3 Objective Function

The DODAG framing mechanism is dependent on the Objective Function. OF describes how a node should select and improve the route. Objective Code Point (OCP) identifies the OF stated in the DIO message. OF specifies a way the routing metrics are used to convert it to a Rank. Based on this rank value, a node decides whether to join the DODAG through that parent node. DODAG creation process affects the network performance; thus, route selection has to be efficient. The two standard objective functions specified by IETF are: Objective function Zero (OF-0) [9] and Minimum Rank with Hysteresis Objective Function (MRHOF) [5].

13.3.1 Objective Function (OF-0)

OF-0 connects to the other node and forwards its data packet towards the root using minimum hop count. OF-0 considers only the hop count factor while computing the rank; the node will select a parent nearer to the root, even if it has too many child nodes connected. The node will change its parent only when the node is unable to reach the root node. The node computes rank as shown in Eq. 13.1, using rank increment factor, given in Eq. 13.2.

$$Rank = Parent\_Rank + Increment\_Factor \qquad (13.1)$$

$$Increment\_Factor = (R_f * S_p + S_r) * MinHopRankIncrease \qquad (13.2)$$

In Eq. 13.2, R_f is rank factor, used to distinguish between the links. S_p is step of rank, which is the link property of a particular neighbor, and S_r is stretch of rank, used to pick another appropriate parent. The values of rank factor, step of rank, stretch of rank, and MinHopRankIncrease are 1, 1, 0, and 256, as defined in RFC 6552 [9]. Thus, the increment factor value results in 256.

13.3.2 Minimum Rank with Hysteresis Objective Function (MRHOF)

The Minimum Rank with Hysteresis Objective Function is intended to identify the path with the smallest path cost using the details mentioned in the DIO Metric Container. Currently, MRHOF works with two types of routing metrics, Expected

Transmission Count (ETX) [3] and energy. Here the ETX is the average number of re-transmissions of a packet at the CSMA level [1]. The first step in MRHOF is to find the path with a minimum path cost, that is, identify the parent node with minimum rank using the specified routing metric. Next, switching to the new parent node. The node will change its parent node only if the parent node's rank is lower than the current parent node by a given threshold whose numeric value is 192 [5]. The rank calculation in MRHOF is explained in [8]; where equation ?? is used to compute the rank increase factor using ETX, which is then added to the parent rank [6].

$$new\_etx = (recorded\_etx * ETX\_ALPHA + packet\_etx * (ETX\_SCALE -$$
$$ETX\_ALPHA))/ETX\_SCALE$$

As the devices in LLNs are memory-constrained, they can not track the ETX of many recent packets. Thus the information of the ETX of many recent packets is aggregated into a single value with the help of the Exponentially Weighted Moving Average (EWMA) filter. The $recorded\_etx$ is the old ETX, $packet\_etx$ is the current ETX, ETX_ALPHA and ETX_SCALE are the constants whose value is set to 90 and 100 respectively [1]. The initial value of $recorded\_etx$ is $RPL\_INIT\_LINK\_METRIC * RPL\_DAG\_MC\_ETX\_DIVISOR$, which are constants assigned with 2 and 256 values respectively, resulting the $recorded\_etx$ value in 512. The $packet\_etx$ is the number of transmissions required for the successful delivery of a packet, which is computed at CSMA level multiplied by the constant $RPL\_DAG\_MC\_ETX\_DIVISOR$, which is 256 [4].

13.4 Results and Analysis

13.4.1 Simulation Environment

The performance of the OFs mainly depends on the simulation environment, thus it is more likely that an objective function might perform better in certain environment and poor in other. For example, if the data rate is high, it is more likely that the ETX value will be high as there is chance of packet loss. Simulation environment includes the arrangement of nodes, the data rates, number of nodes, and node type (skymote, z1 mote, cooja mote, and so on). The assessment of the experiments carried out of the objective function is done using Contiki OS and Cooja simulator. The other experimental setup details are presented in Table 13.1.

Table 13.1 Analysis of various OFs based on ETX variants

Network Parameter	Value
Deployment area	100 m × 100 m
Deployment type	Random
Number of nodes	15, 30, 45, 60
Data rate	1 packet/min, 10 packet/min, 30 packet/min
Simulation time	30 min
Simulation iterations	15

13.4.2 Analysis

The Objective function OF-0 and MRHOF are analyzed on the performance metrics Packet Delivery Ratio (PDR), Latency, network setup time, and traffic overhead. The computation of these metric is discussed below:

Packet Delivery Ratio is computed as a ratio between the number of packets successfully received by the receiver to number of packets sent by the sender. The performance of OF-0 over MRHOF for packet delivery ratio is better but with small margin. As the number of nodes or the data rate increases, performance of both OF's starts degrading as shown in Figs. 13.1 and 13.2. Figure 13.1 shows three plots for three data rates for steady node, whereas Fig. 13.2 are the plots for mobile nodes.

$$Packet\,Delivery\,Ratio = (Total\,Packets\,Received\,/\,Total\,Packets\,Sent) * 100 \tag{13.3}$$

Latency is the time taken by the packet from the time it is sent by the sender to the time it is received by the receiver. The performance for latency of OF-0 is slightly better over MRHOF for data rate 1 packet/min. For data rate 10 packet/min and 30 packet/min the latency value decreases for both OF-0 and MRHOF but as the nodes increase MRHOF needs more time than OF-0. Same is depicted in the Figs. 13.3

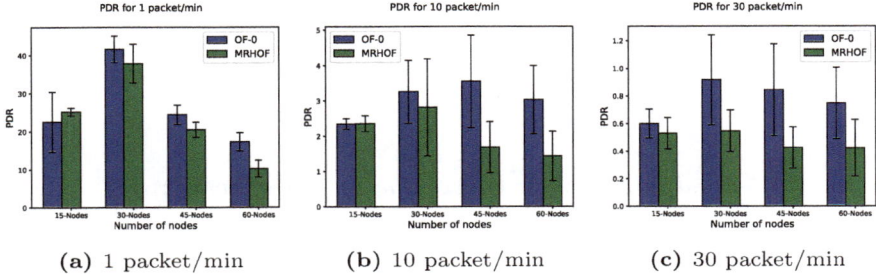

(a) 1 packet/min   (b) 10 packet/min   (c) 30 packet/min

Fig. 13.1 Packet delivery ratio for steady node

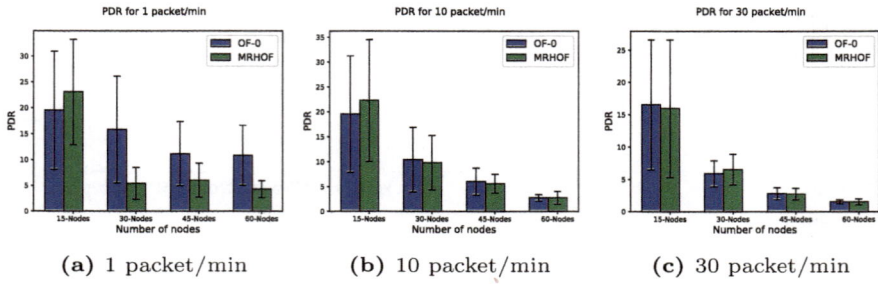

(a) 1 packet/min   **(b)** 10 packet/min   **(c)** 30 packet/min

Fig. 13.2 Packet delivery ratio for mobile node

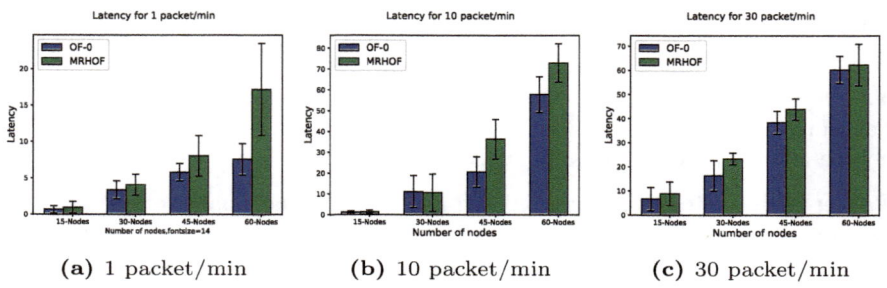

(a) 1 packet/min   **(b)** 10 packet/min   **(c)** 30 packet/min

Fig. 13.3 Latency for steady node

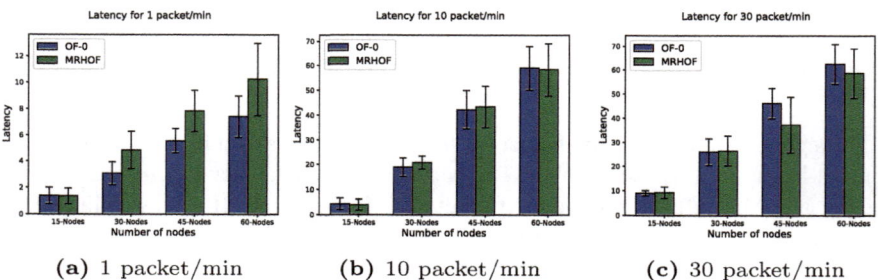

(a) 1 packet/min   **(b)** 10 packet/min   **(c)** 30 packet/min

Fig. 13.4 Latency for mobile node

and 13.4. Figure 13.3 shows three plots for three data rates for steady node, whereas Fig. 13.4 are the plots for mobile nodes.

$$TotalLatency = \sum_{k=1}^{n}(RecvTime(k) - SentTime(k)) \qquad (13.4)$$

Network Setup Time is the time taken by the network to complete the DODAG formation. As observed from Figs. 13.5 and 13.6 the setup time for MRHOF is more as compared to OF-0 when the data rate is low. As the data rate increases the time

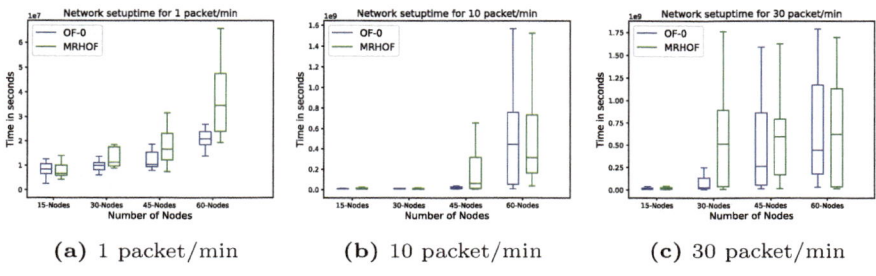

(a) 1 packet/min   **(b)** 10 packet/min   **(c)** 30 packet/min

Fig. 13.5 Network setup time for steady node

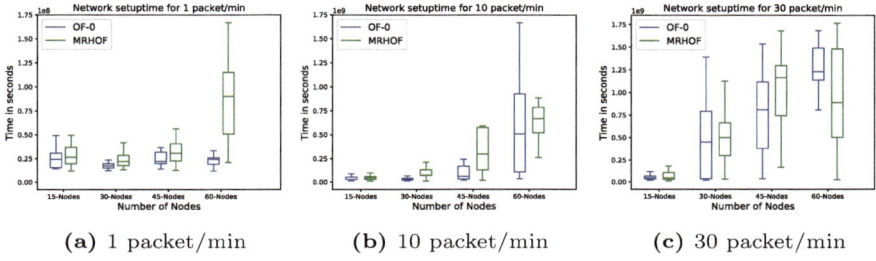

(a) 1 packet/min   **(b)** 10 packet/min   **(c)** 30 packet/min

Fig. 13.6 Network setup time for mobile node

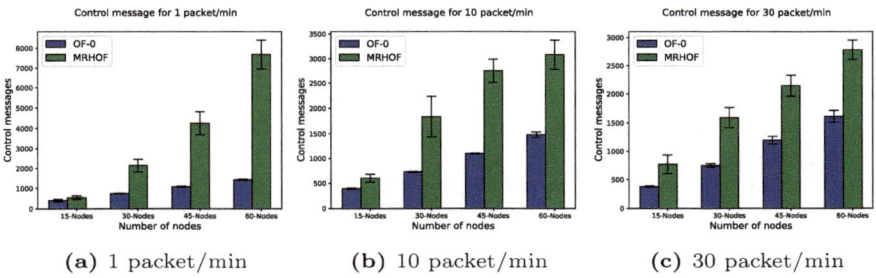

(a) 1 packet/min   **(b)** 10 packet/min   **(c)** 30 packet/min

Fig. 13.7 Control messages overhead for steady node

consumed for both the OFs. Figure 13.5 shows three plots for three data rates for steady node, where Fig. 13.6 are the plots for mobile nodes.

$$ConvergenceTime = LastDIOjoinedDAG - FirstDIOsent \qquad (13.5)$$

Control Overhead is the count of the total control messages used to form and manage the connection of the DODAG. The control messages used to build the DODAG are less for OF-0 objective function when compared to MRHOF. The difference between the OF-0 and MRHOF decreases as the data rate increases from 1 packet per min to 30 packet per min. Figures 13.7 and 13.8 depict the plots for the same. Figure 13.7

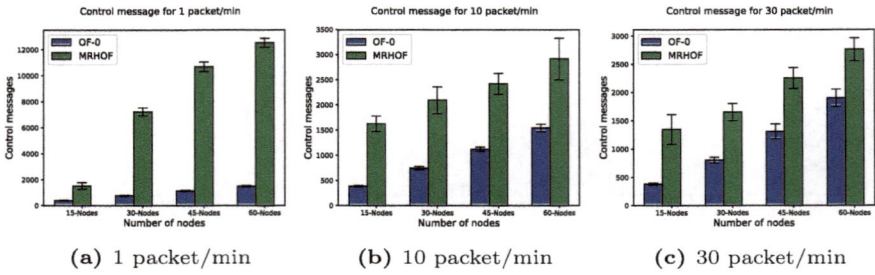

(a) 1 packet/min   **(b)** 10 packet/min   **(c)** 30 packet/min

Fig. 13.8 Control messages overhead for mobile node

shows three plots for three data rates for steady node, whereas Fig. 13.8 are the plots for mobile nodes.

$$ControlTrafficOverhead = \sum_{k=1}^{n} DIO(k) + \sum_{k=1}^{n} DIS(k) + \sum_{k=1}^{n} DAO(k)$$

$$(13.6)$$

The experimental results show that OF-0 performs better than MRHOF. In MRHOF there occurs a parent switch when a node finds a better node, as parent, which causes network imbalance thus reduces the performance of the network. In the case of OF-0 the node is connected to one node full time or till its lifetime. MRHOF searches for more stable link qualities than short paths having poor link qualities which leads to more nodes getting involved to transfer the data and more energy consumption. Whereas in OF-0, the retransmission is less as nodes use less hops to reach the root.

13.5   Conclusion

Internet of Things is a modern technology used for monitoring and automating applications like industry, home, building, and urban. Routing is one of the crucial step in such real time applications and the routing protocol designed for IoT is RPL. In RPL, objective function is responsible for parent selection. The routing metrics and objective function is selected based on the application requirement. The paper provides experiments of the two standard OF, OF-0 and MRHOF. The experimental results show that the OF-0 performs better than MRHOF in terms of the performance metrics such as PDR, network setup time, latency, and control message overhead. Further the experiments can be performed for real testbeds, and also check its performance in more denser networks.

References

1. Bezunartea, M., Bahn, M., Gamallo, M., Tiberghien, J.: Impact of cross-layer interactions between radio duty cycling and routing on the efficiency of a wireless sensor network - a testbed study involving contikiMAC and RPL (03 2017). https://doi.org/10.1145/3018896.3018935
2. Chen, S., Xu, H., Liu, D., Hu, B., Wang, H.: A vision of IoT: applications, challenges, and opportunities with china perspective. IEEE Internet Things J. 1(4), 349–359 (2014). https://doi.org/10.1109/JIOT.2014.2337336
3. De Couto, D.S.J., Aguayo, D., Bicket, J., Morris, R.: A high-throughput path metric for multi-hop wireless routing. MobiCom '03, Association for Computing Machinery, New York, NY, USA, pp. 134–146 (2003). 10.1145/938985.939000, https://doi.org/10.1145/938985.939000
4. Dunkels, A., Gronvall, B., Voigt, T.: Contiki-a lightweight and flexible operating system for tiny networked sensors. In: 29th Annual IEEE International Conference on Local Computer Networks, pp. 455–462. IEEE (2004)
5. Gnawali, O.: The minimum rank with hysteresis objective function (2012)
6. Ko, J., Eriksson, J., Tsiftes, N., Dawson-Haggerty, S., Terzis, A., Dunkels, A., Culler, D.: Contikirpl and tinyrpl: Happy together. In: Workshop on Extending the Internet to Low Power and Lossy Networks (IP+ SN), vol. 570 (2011)
7. Levis, P., Tavakoli, A., Dawson-Haggerty, S.: Overview of existing routing protocols for low power and lossy networks. Internet Engineering Task Force, Internet-Draft draftietf-roll-protocols-survey-07 (2009)
8. "rpl objective function modification and simulation in cooja" (2014). https://anrg.usc.edu/contiki/index.php/RPL_objective_function_modification_and_simulation_in_cooja
9. Thubert, P.: Rpl objective function 0. Draft-ietf-roll-of0-03 (work in progress) (2010)
10. Vasseur, J., Kim, M., Pister, K., Dejean, N., Barthel, D.: Routing metrics used for path calculation in low-power and lossy networks. In: RFC 6551, pp. 1–30. IETF (2012)
11. Winter, T.: Rpl: Ipv6 routing protocol for low-power and lossy networks (2012)

Chapter 14
Comparative Performance Analysis of Recent Evolutionary Algorithms

Debojyoti Sarkar and Anupam Biswas

Abstract The Evolutionary Algorithms (EAs) are employed to solve complex, multimodal, and multidimensional real-world optimization problems. The non-deterministic nature and randomness of EAs ascertain a larger diversity of the search heuristic, and hence, promises tremendous potential in reaching a better solution. Genetic Algorithm (GA) is a prime example of an EA. In the present time, with the rapid increase of EAs in numbers, a need for algorithmic evaluation has become prominent. In this paper, we have summarized the algorithmic framework of EAs and presented the computational performances of four recent EAs, tested upon a set of fifteen CEC-2015 benchmark functions.

14.1 Introduction

Optimization problems are those that demand a minimization or maximization of a quantity, an entity or an attribute. Such problems are inherent to nature and are open to human perception. The problem of optimization can be seen in almost every sector, such as production industries, where profit is to be maximized and man-power is to be minimized; telecommunication sectors, where optimal path is to be found for data routing; machine learning and artificial intelligence, where suitable neural network architectures are sought, etc. Optimization algorithms have come a long way in addressing numerous such real-world optimization problems. In the last decade, the world has seen an extensive research on, and application of optimization algorithms [1]. Evolutionary strategies such as GA, PSO, ACO, etc., are preferred over the conventional approaches for their ability to deal with strongly non-linear, multidimensional and ill-behaved dynamic engineering problems.

D. Sarkar (✉) · A. Biswas
Department of Computer Science and Engineering, National Institute of Technology Silchar, Silchar 788010, Assam, India
e-mail: debojyoti_rs@cse.nits.ac.in

A. Biswas
e-mail: anupam@cse.nits.ac.in

V. Bhateja et al. (eds.), *Evolution in Computational Intelligence*,
Smart Innovation, Systems and Technologies 267,
https://doi.org/10.1007/978-981-16-6616-2_14

The problem of optimization is not a modern one. Any operation can or should be optimized in terms of expense, length, and benefit. Optimization problems can be tackled through experimentation, simulation or algorithms. While the former two methods are expensive, specific, and comply with human intervention, the optimization algorithms provide a generality of problem-solving in the sense that a single algorithm can solve an entire class of problems.

14.2 Evolutionary Algorithms

14.2.1 The Generic Framework for Evolutionary Algorithms

The EAs are inspired by Darwin's 'survival of the fittest' theory. The methodology of EAs can be summarized in the following steps:

(a) Selection of Individuals- Each individual in an EA is nothing but a potential solution in the search space. The individuals are selected such that their positions lie within the predetermined bounds.
(b) Population- A certain number of individuals are collectively called a population. The population size, too, is predefined.
(c) Iterations- It signifies evolution stage. Iterations will continue for a certain number of times or may prematurely stop if the best solution is already obtained.

   i. Fitness evaluation- The population is fed to a function and the corresponding function values (fitness values) are noted.
   ii. Selection- Individuals in the population having the best fitness are selected for further regeneration.
   iii. Exploitation- The process of *digging deeper* for a solution. It occurs once all the individuals converge to a particular point and with iterations, betters itself. Exploration is achieved through communication between the individuals. A communication enables the individuals to have an idea of their global best and generate a tendency to rush toward the intermediate global best leaving its own search space. For example, in GA, the *crossover* operator is responsible for exploitation. In PSO, tuning the constant attached to the *social component* to a higher value leads to greater exploitation.
   iv. Exploration- This refers to the search about the entire landscape. A greater exploration raises the probability of finding new and better solutions. An individual, when prioritizes its local best over global best, refrains itself from rushing toward the global solution, and hence increases exploration. For example, in GA, exploration is achieved by the *mutation* operator, whereas in PSO, an increased weight on the *cognitive component* serves the purpose.
   v. New population- After the individuals undergo all the operations pertaining to exploration and exploitation, they transform into new individuals and form the new population for the particular iteration (or generation).

(d) The final population obtained, consists of the best evolved individuals and the best individual from the final population is the best solution.

14.2.2   Evaluation of Evolutionary Algorithms

In order to evaluate an algorithm, a knowledge about its fundamental structure and heuristics, is of paramount importance. Modern day optimization algorithms, such as Evolutionary Algorithms (EAs), are largely stochastic and non-deterministic. The random trait of the stochastic algorithms, at times, provides a faster convergence and accurate solutions. The evolution in EAs is seen through a series of well balanced exploration and exploitation. This also leaves a substantial gap for criticism about the balance of the processes which are directly accountable for the algorithm's accuracy and efficacy. Certain evaluation metrics can be set for determining the efficiency of algorithms. A *convergence graph* tells us whether an algorithm is able to find the optimal solution. *Type of convergence* is about finding a global solution or a local optima. *Speed of convergence* is a metric to identify how fast the global best solution is obtained. The significant change in performance when the starting points are randomly changed or shifted, determines an algorithm's *robustness*.

14.3   Recent Evolutionary Algorithms

Adaptive Cuckoo Search Using Levy's Flight (CS) The CS algorithm was inspired by certain cuckoo species' obligate brood parasitism, in which they lay their eggs in the nests of other species' host birds [2]. The intruding cuckoos may create direct confrontation with certain host birds. The 3 major stages are: Generation of a new candidate solution (eggs) by perturbing the current solution using Levy's Flight Distribution (LFD), replacing some nest with new solutions where the host bird throws the egg away or abandon the nest with a certain probability and greedy (elitist) selection strategy.

Gradient-Based Optimizer The GBO, which is based on the gradient-based Newton's method, examines the search space using two major operators: the gradient search rule (GSR) and the local escaping operator (LEO), as well as a set of vectors [3]. The GSR uses a gradient-based approach to increase the exploration inclination and speed up the convergence rate in order to obtain better search space positions. The proposed GBO is able to perform exploitation through LEO. The Newton's approach is an efficient way to solve equations numerically [4]. This approach is a root-finding algorithm that uses the Taylor series' initial terms. The GBO is formulated using a new version of Newton's method developed by Weerakoon and Fernando [5].

Improved Moth-Flame Optimization The Moth-Flame Optimization (MFO) algorithm is a search algorithm that uses a transverse orientation mechanism. The moths appear to hold a fixed angle with respect to the moon in this mechanism. The global search capability and convergence speed of MFO are declining. An Improved Moth-Flame Optimization (IMFO) fixes these shortcomings [6].

Spotted Hyena Optimizer The action of spotted hyenas inspired Spotted Hyena Optimizer (SHO). The social interaction between spotted hyenas and their collaborative actions is the core idea behind this algorithm. Searching for prey, encircling prey, and assaulting prey are the three fundamental stages of SHO. The methodology can be summarized into four processes: Encircling prey, Hunting, Attacking prey—exploitation and Search for prey—exploration

14.4 Algorithmic Analysis

14.4.1 Experimental Setup

All the algorithms are implemented by using MATLAB R2019a and executed on a Workstation with Intel(R) Xeon(R) processor and 128 GB of RAM. The operating system is 64-bit Microsoft Windows 10 Pro. The four algorithms run 51 times over the 15 benchmark functions defined in the CEC-15 test suite. All other tunings are according to the protocols provided by 'CEC 2015 Competition on Single Objective Multi-Niche Optimization' [7].

14.4.2 Benchmark Functions

Benchmark functions are a set of functions that are used as standards for evaluating or measuring performances of optimization algorithms. These functions are complex in nature, have multiple optimal points, providing a fair and unbiased idea about working of an algorithm. In CEC-15 test suite, out of 15 functions, 8 are extended simple functions (f1-f8) while the rest are hybrid functions (f9-f15). These functions are highly scalable minimization problems, incorporated with rotation and shift operations, thus, helping create a relation among various dimensions and also placing the optima at different positions in a search space.

14.4.3 Performance Analysis

The best value, worst value, mean, median, and standard deviation (SD) are calculated on the data gathered from the 51 runs shown in Tables 14.1 and 14.2. The success rate (SR) is shown in Table 14.3.

$$SR = \frac{\text{No. of runs with an optimum sol.}}{\text{Total No. of runs}}$$

The optimum can either be the global optimum or any of the local optima.

Table 14.1 Performance analysis of CS and GBO

		CS					GBO				
	D	Best	Worst	Mean	Median	SD	Best	Worst	Mean	Median	SD
f1	5	100	220	153.33	100	38.08	100	220	141.56	140	30.94
	10	180	398.3	313.02	140	53.72	100	441.35	261.21	260	64.38
	20	420	1371.59	833.24	260	238.71	260	620	440.88	420	75.91
f2	2	200	200	200	200	0	200	220	200	200	0
	5	200	200	200	200	0	200	200	200	200	0
	8	200	200	200	200	0	200	300.01	223.13	200	30.2998
f3	2	300	300	300	300	0	300	300	300	300	0
	3	300	300	300	300	0	300	300	300	300	0
	4	300	300	300	300	0	300	300	300	300	0
f4	5	400	400.08	400.01	400	0.03	400	401.2	400.58	400	0.3145
	10	400.17	401.58	400.83	400.25	0.3304	400.16	402.87	401.16	400.87	0.6129
	20	401.39	404.34	401.97	402	1.3306	401.49	406.28	403.47	402	1.1268
f5	2	500	500	500	500	0	500	500	500	500	0
	3	500	500	500	500	0	500	500	500	500	0
	4	500	500	500	500	0	500	500	500	500	0
f6	4	500	600	600	600	0	600	600	600	600	0
	6	600	600	600	600	0	600	601.48	600	600	0.2905
	8	600	600	600	600	0	600	600	600	600	0
f7	6	700	700	700	700	0	700	703.26	700.44	700	1.1346
	10	700	700	700	700	0	700	709.79	702.52	700	2.9761
	16	700	700.19	700	700	0.0265	700	725.6	709.46	705.6	7.2242
f8	2	800	800	800	800	0	800	800	800	800	0
	3	800	800	800	800	0	800	800	800	800	0
	4	800	800	800	800	0	800	800	800	800	0
f9	10	900	900	900	900	0.0004	905.64	968.84	923.76	931.15	20.1675
	20	900	903.2342	900.076	900	0.4543	906.96	931.11	922.24	931.11	10.585
	30	900	914.84	910.95	914.01	5.6540	913.46	930.35	925.89	930.31	6.8229
f10	10	1070	1070	1070	1070	0	1070	1080	1070.78	1070	2.7152
	20	1030	1080	1070.42	1080	16.7707	1080	1080	1080	1080	0
	30	1040	1080	1069.256	1080	17.6937	1080	1080	1080	1080	0
f11	10	1100	1100.89	1100	1100	0.1989	1100	1103.38	1101.69	1100	0.7265
	20	1100	1100.42	1100.05	1100	0.0870	1101.26	1106.86	1103.28	1102	1.053
	30	1100	1103.76	1101.584	1100	1.3344	1102.38	1105.99	1103.86	1103.69	0.8835
f12	10	1201.72	1255.09	1209.57	1208.6	7.4416	1200.43	1739.48	1266.03	1201.49	144.23
	20	1261.12	2780.6	2094.43	1967.11	380.6075	1630.09	2914.04	2395.22	2494.12	312.6352
	30	2113.35	4041.92	3355.48	3372.04	233.9468	2299.84	3728.59	3173.41	3228.25	273.25
f13	10	1303.62	1495.341	1438.29	1440.6	31.7847	1420.9	1673.17	1530.51	1533.25	55.95
	20	1523.29	1802.06	1729.16	1739.49	45.0713	1656.42	2135.78	1892.38	1902.04	118.59
	30	1872.9	2206.56	2095.33	2102.46	66.7506	2078.88	2816.42	2396.74	2418.86	167.0024
f14	10	1455.03	2405.05	2129.06	2162.44	173.3251	1520	2748.82	2309.09	2339.87	264.78
	20	3330.24	4339.4	3799.69	3811.09	237.5543	2867.29	4523.644	3717.94	3589.46	416.24
	30	4721.65	6684.95	5641.52	5612.2	437.6957	3655.55	6204.13	4898.79	5000.33	689.1167
f15	10	1640	1640	1640	1640	0	1640	1891.32	1700.99	1640	82.37
	20	1644.31	2202.21	1942.45	2038.94	209.0722	2082.72	2578.64	2290.1	2253.08	128.96
	30	1950.66	2493.29	2290.2	2289.86	107.3376	2436.96	3648.72	2782.42	2764.51	239.53

Table 14.2 Performance analysis of IMFO and SHO

		IMFO					SHO				
	D	Best	Worst	Mean	Median	SD	Best	Worst	Mean	Median	SD
f1	5	100	220	153.33	140	38.0876	326.69	2838.96	1116.95	1120.89	552.1875
	10	180	398.3	313.02	300.09	53.72	2962.08	19161.17	9443.47	8793.59	3682.347
	20	420	1371.59	833.24	801.24	238.71	26477.13	56575.66	39565.47	39413	8217.67
f2	2	200	200	200	200	0	200.71	291.55	237.67	232.1	28.1689
	5	200	260.09	207.45	200	17.89	328.46	1946.19	847.26	732.85	323.3778
	8	200	313.42	234.54	215.42	39.38	771.56	6553.72	3247.04	3230.04	1193.626
f3	2	300	300	300	300	0	300	300.3	300.05	300	0.0541
	3	300	300	300	300	0	300.11	303.18	300.88	300	0.6313
	4	300	300	300	300	0	300.02	305.28	302.76	300	1.2544
f4	5	400	401.58	400.71	400	0.4082	403.64	416.1	407.68	404.44	2.5563
	10	400	405	402.88	400	0.9613	420.49	473.41	441.34	440.07	11.3419
	20	405.16	428.03	413.04	406.56	5.2027	469.3	544.59	514.4	514	13.7497
f5	2	500	500	500	500	0	500	501.38	500.12	500	0.2536
	3	500	500	500	500	0	500.07	502.79	500.99	501.23	0.5915
	4	500	500	500	500	0	500.86	507.16	502.83	502.45	1.2868
f6	4	600	600	600	600	0	641.78	2540.19	1057.36	829.6	476.9507
	6	600	612.95	600.64	600	2.2836	733.67	26279.79	4742.5	3908.76	4168.4
	8	600	600	600	600	0	2971.46	32641.85	14331.09	12582.42	9212.341
f7	6	700	703.26	700.12	700	0.64	716.19	1120.66	799.52	757.06	100.6443
	10	700	713.16	702.8	700	3.0326	771.91	8040.73	2650.98	2401.63	1415.72
	16	700	741.08	709.77	702.22	8.5744	1937.97	33005.97	10678.06	9190.03	6708.325
f8	2	800	800	800	800	0	800	800	800	800	0.0006
	3	800	800	800	800	0	800.02	801.99	800.25	800.21	0.2825
	4	800	800	800	800	0.0006	800.03	802.06	800.45	800.34	0.4305
f9	10	907.33	968.84	937.09	931.15	11.0341	969.6	2497.64	1486.63	1345.7	400.4581
	20	907.75	932.9	930.81	931.12	3.3127	1408.3	10600.38	2977.28	2493.71	1597.16
	30	930.31	933.81	930.49	930.4	0.4883	2292.31	14303.55	6474.87	5830.26	2797.879
f10	10	1070	14249.3	1458.35	1070	2036.2	13400.72	32562.58	19925.87	19440.21	4343.25
	20	1030	27821.82	6730.38	1080	9293.481	23878.12	42333.03	27981.93	26234.86	4863.4
	30	1040	51554	11492.19	1080	15383.39	36509.87	40254.87	38157.83	38092.99	902.7601
f11	10	1100	1106.08	1102.53	1102.36	1.3953	1110.163	1140.311	1128.22	1129.13	6.6873
	20	1102.33	1115.91	1105.7	1105.49	2.2336	1137.37	1167.52	1151.7	1151.2	7.2075
	30	1104.88	1132.16	1108.76	1107.79	4.51	1148.66	1197.34	1171.4	1169.03	10.806
f12	10	1200.88	1678.92	1330.02	1352.4	99.15	1768.88	2619.36	2241.67	2237.93	189.1528
	20	1636.39	2715.34	2516.84	2536.86	145.9658	3432.83	4488.03	4007.27	3978.18	235.642
	30	3041.44	3387.96	3191.66	3187.51	83.874	4767.27	6078.56	5542.79	5609.25	287.5467
f13	10	1300	1701.69	1511.69	1502.05	69.9168	2332.18	50096.05	14141.07	10448.17	10984.72
	20	1653.67	8586.82	2098.02	1915.04	980.78	11383.26	108315.8	40478.71	37648.15	18585.09
	30	2230.99	34124.77	4977.53	3100.12	5839.449	194480.1	1191951	445051.6	393120	194342.5
f14	10	2101.77	2885.7	2517.21	2518.45	162.536	2978.72	3575.28	3382.65	3389.77	121.2827
	20	3035.53	4017.47	3556.73	3546.55	206.7934	5198.71	6113.45	5577.98	5512.62	244.5
	30	3975.67	5351.94	4654.29	4674.23	301.4416	7419.63	9063.82	8482.5	8496.04	358.61
f15	10	1640	1967.81	1718.27	1640	97.0031	2949.98	47190.43	14228.71	12902.86	9161.248
	20	2095.28	6073.67	2571.91	2387.85	740.8211	164400	2272282	824012.4	611048	513140
	30	2169.52	15613.08	4084.39	2936.86	2645.45	3211089	2201135	878362.2	780395.9	495496.3

Table 14.3 Success rate of the 4 algorithms for 13 functions

	f1			f2			f3			f4			f5			f6			f7		
D	5	10	20	2	5	8	2	3	4	5	10	20	2	3	4	4	6	8	6	10	16
CS	98.04	98.04	7.84	100	100	58.82	100	100	100	100	1.97	0	100	100	100	100	100	100	100	100	98.04
GBO	98.04	13.72	0	100	100	58.82	100	100	100	19.6	1.96	0	100	100	100	100	96.07	100	86.27	47.05	11.76
IMFO	88.24	3.92	0	100	100	58.82	100	100	100	9.8	0	0	100	100	100	100	84.31	100	96.08	27.45	1.96
SHO	0	0	0	0	0	0	29.42	0	1.96	0	0	0	19.61	0	0	0	0	0	0	0	0

	f8			f9			f10			f11			f12			f13		
D	2	3	4	10	20	30	2	3	4	10	20	30	10	20	30	10	20	30
CS	100	100	100	20	88.23	5.89	10	100	100	10	100	100	10	70.59	100	10	20	30
GBO	100	100	100	0	0	0	100	100	100	100	0	0	68.62	0	0	3.9216	0	0
IMFO	100	100	92.16	0	0	0	96.08	70.59	64.7	1.96	0	0	21.57	0	0	1.96	0	0
SHO	52.94	0	0	0	0	0	0	0	0	0	0	0	0	0	0	0	0	0

14.4.4 Convergence Analysis

The convergence is obtained by plotting the best fitness value of the population against the iteration number. Figure 14.1 shows the convergence curves of the respective algorithms separately for the 15 functions. It is to be noted that convergence evaluation of only one dimension from each function is represented in the graphs.

14.5 Discussions

All the Algorithms perform well for functions f5 and f8. Performances of CS, GBO, and IMFO are fair for f2-4, f6-7, and f11. No algorithms converge for functions f12-15. SHO's performance is unsatisfactory, except for f3, f5 and f8, where it performs fairly, and it can be safely stated that SHO is the worst performing algorithm among all four. Observation of the convergence graphs would reveal that the algorithms are done with all the exploration–exploitation at a very early stage in case of functions f1-3 and f5-7, at a mid-stage for functions f4, f10 and f11, and at a later stage for f12 and f14. The measure of mean or median, however, does not speak much

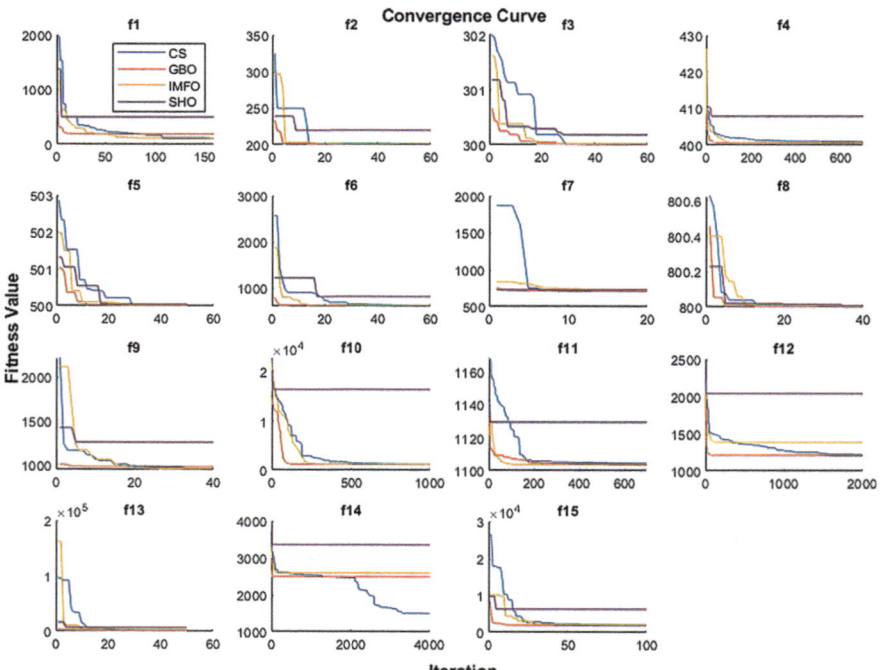

Fig. 14.1 Convergence curves: iteration versus fitness value

about exploration and exploitation. A future prospect can be an independent analysis of exploration and exploitation, studied by keeping track of the diversification and intensification of the population with each run [8], by a distance measure.

14.6 Conclusion

Optimization algorithms are an integral part of any domain we can think of. With the rapid increase in algorithm designs, it becomes an utmost necessity to compare, group, or rank these algorithms in context of the problem in hand. With an abundance of statistical tools, it is possible to perform parametric or non-parametric analysis on the experimental data, which will unfold a lot of information about the meta-heuristics employed in the algorithm design, hence providing deeper insights in evaluation. The experimental data represented in this paper gives us a gist about the algorithm and the benchmark functions, and can be regarded as a first step into a much more accurate and elaborated research.

Acknowledgements This work is supported by the Science and Engineering Board (SERB), Department of Science and Technology (DST) of the Government of India under Grant No. ECR/2018/000204 and Grant No. EEQ/2019/000657.

References

1. Parsopoulos, K.E., Vrahatis, M.N.: Recent approaches to global optimization problems through particle swarm optimization. Nat. Comput. **1**(2), 235–306 (2002)
2. Yang, X.-S., Deb, S.: Cuckoo search via lévy flights. In: World Congress on Nature & Biologically Inspired computing (NaBIC). IEEE, pp. 210–214 (2009)
3. Ahmadianfar, I., Bozorg-Haddad, O., Chu, X.: Gradient-based optimizer: a new metaheuristic optimization algorithm. Inf. Sci. **540**, 131–159 (2020)
4. Bazaraa, M.S., Sherali, H.D., Shetty, C.M.: Nonlinear Programming: Theory and Algorithms. Wiley, New York (2013)
5. Weerakoon, S., Fernando, T.: A variant of newton's method with accelerated third-order convergence. Appl. Math. Lett. **13**, 87–93 (2000)
6. Pelusi, D., Mascella, R., Tallini, L., Nayak, J., Naik, B., Deng, Y.: An improved moth-flame optimization algorithm with hybrid search phases. Knowl.-Based Syst. **191**, 105–277 (2020)
7. Liang, J., Qu, B., Suganthan, P., Chen, Q.: "Problem definitions and evaluation criteria for the CEC 2015 competition on learning-based real- parameter single objective optimization. Technical Report201411A. Com- putational Intelligence Laboratory, Zhengzhou University, Zhengzhou China and Technical Report, Nanyang Technological University, Singapore, vol. 29, pp. 625–640 (2014)
8. Olorunda, O., Engelbrecht, A.P.: "Measuring exploration/exploitation in particle swarms using swarm diversity. In: IEEE Congress on Evolutionary Computation (IEEE World Congress on Computational Intelligence), pp. 1128–1134 (2008)

Chapter 15
Medical Image Protection Using Blockchain for E-healthcare System

Punam Prabha, Yash Janoria, Harsh Raj, Uday Patidar, and Kakali Chatterjee

Abstract Basically, the modern E-healthcare system is a Cloud-based system where all the details of Patient Health Information (PHI) on disease diagnosis and scan reports are stored and shared among the stockholders of the system. However, the security of these sensitive records against unauthorized access and misuse is of major concern. In this context, a Blockchain-based record-sharing system could play an important role for secure transactions of medical information. In this paper, a smart contracts and Inter Planetary File System (IPFS)-based secure medical image protection system has been proposed to encrypt the session data. Moreover, it records every transaction of patient data in the Blockchain for better traceability. The performance of the proposed framework is evaluated in simulation to show its effectiveness.

15.1 Introduction

Nowadays, the E-healthcare system is serving as the backbone of the world healthcare management system for solving COVID-19 pandemics. The valuable information of healthcare stakeholders such as text, image, video of x-ray, and lab reports could be conveyed electronically for remote access by healthcare professionals and other stockholders [1]. However, the remote accessibility may invite threats such as unauthorized access, data modification, data interference and misuse of data. In this context, Blockchain technology can be used as a tamper-proof framework to perform encapsulation on confidential patient's information including medical images. It is based on a decentralized collision-resistance mechanism which uses a distributed public ledger to avoid tampering of the created block [2, 3]. In addition, the security concern of medical images can be increased by applying the multiple watermark

P. Prabha (✉) · Y. Janoria · H. Raj · U. Patidar · K. Chatterjee (✉)
Department of Computer Science and Engineering, National Institute of Technology Patna, Patna 800005, India
e-mail: punamp.phd18.cs@nitp.ac.in

K. Chatterjee
e-mail: kakali@nitp.ac.in

© The Author(s), under exclusive license to Springer Nature Singapore Pte Ltd. 2022
V. Bhateja et al. (eds.), *Evolution in Computational Intelligence*,
Smart Innovation, Systems and Technologies 267,
https://doi.org/10.1007/978-981-16-6616-2_15

(copyrights) protection technique. The technique provides the actual copyright information on the basis of a timestamp and order of access of the medical image [4]. Subsequently, traditional Hash algorithms such as MD5, SHA-256, and SHA-512 can be implemented on it [5] to find out the structural changes at ease [6]. However, perceptual Hash function is more effective than the traditional Hash function in terms of image quality and clarity because neither it alters the image size nor it removes watermark details.

The implementation of robust protection to the medical images is very challenging. In this context, the generation of a quick response (QR) code using perceptual Hash function is very relevant to collect information regarding the number of attacks performed on the watermarked images. Moreover, the defected QR code can be easily detected by any electronics or mobile devices [7, 8].

The proposed paper intends to establish a secure medical image protection framework using Blockchain technology for the E-healthcare system. First of all, an Ethereum-based smart contract is created which keeps all details of the user. Subsequently, the fingerprint of the watermarked medical image is created and stored in Inter Planetary File System (IPFS). The extensive simulation results using MetaMask are shown to validate the proposed watermarking technique in association with the Blockchain technology.

The present paper is organized in seven sections. Section 15.2 overviews the relevant research work done in the field of Blockchain network-based E-healthcare system. In Sect. 15.3, the security issues and their challenges are highlighted. The proposed framework is explained in detail in Sect. 15.4. Moreover, the simulation study and results using the MetaMask software platform are discussed in Sect. 15.5. In Sect. 15.6, the performance of the proposed framework is evaluated. At last in Sect. 15.7, the findings of the proposed research work are concluded.

15.2   Related Work

Medical imaging devices used in IoT-based systems are the prime target for hackers due to the lack of suitable security measures [9]. The emerging Blockchain techniques offer privacy preservation in E-healthcare sectors for securing patient records efficiently against different cyber-attacks [10, 11]. Recent experimentations in [12] show that MRI and CT scans can easily be tampered without any trace. To overcome such serious issues, the Blockchain-based copyright management technique is implemented in [5]. The work was not absolute due to leakage of privacy and lack of supporting simulation results. In [13], a copyright technique is implemented in a Blockchain network with the help of the ElGamal cryptographic encryption method. However, it encountered difficulty in building one time password due to huge number of modulo arithmetic operations. Moreover, it implements centralized server databases. Hence, a gap is formed between the actual copyright user and other involved members. To remove the shortcomings, an improved version of ElGamal is used to generate one-time passwords efficiently [1]. A robust watermarking technique on medical image is proposed in [14] to handle two or more than two simultaneous attacks. However, the centralized storage place suffers from drawbacks such as large storage spaces, slow processing, and loss of content of information due to server failure and server hacking. To resolve the issues of centralized server, Inter Planetary File System (IPFS) is used in the Blockchain network [15]. IPFS focuses only on the content of information apart from its path of location and the name of the folder where it is stored. All the stored information in IPFS is kept in a tabular form and secured by cryptographic hash functions. Thus, all the saved information will be found out with the help of an index of a distributed hash table. Based on the above literature survey, the main objective of the present work is to implement copyright image protection using watermarking. The proposed method is based on the generation of QR code which utilizes perceptual hash function in a Blockchain network that stores digital images on IPFS.

15.3 Security Issues and Challenges

Some security concerns on medical images and the appropriate solutions to avoid the leakage or tampering of information using Blockchain technology are listed as follows:

1. *Safety from unwanted swapping of documents*—Lab reports of different patients may be mismatched when a number of files are available at the same time. Such issues can be resolved using watermarking on medical images.
2. *Tamper proof*—It is practically impossible to manipulate the encrypted watermarked medical images. The distributed public ledger is responsible for unauthorized handling.
3. *Storage security*—IPFS is used for storing the watermarked medical image and providing access to information by implementing secure distributed hash index.
4. *Multiple copyright security*—Multiple copyrights on medical images can be easily traced by observing spend timestamp and embedding the order of the watermarks.
5. *Hiding credential detail*—The easy and efficient sharing of medical image is done by scanning its QR code using any electronic device instead of mandating the credential details such as user-id and password.

15.4 Proposed Framework

The proposed E-healthcare multimedia data protection framework is shown in Fig. 15.1. The working of the framework is based on two subsystems: Patient Registration System and PHI Transaction System.

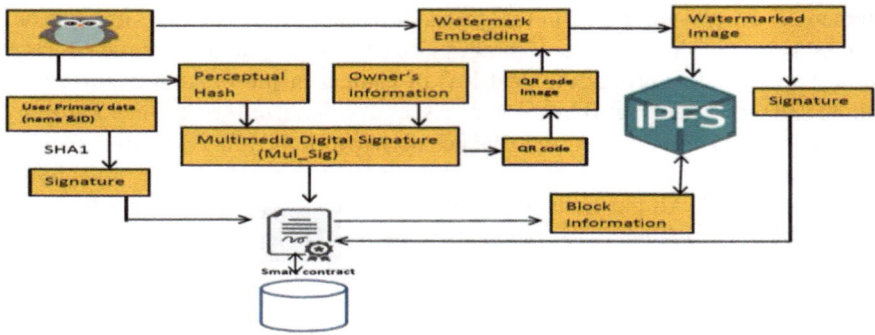

Fig. 15.1 Watermarked image in the proposed E-healthcare system

15.4.1 Patient Registration System

During the registration process, the personal information of patients such as names, ids, and medical images are entered. This information is passed through the perceptual hash function to generate a hash value followed by the watermark embedding process. A QR code is generated using the owner's personal information and the perceptual hash of the image. IPFS is used to share and allocate text and images using a temporary shared key. The folder is uploaded in IPFS with a hash index for further tracing the file.

15.4.2 Patient Health Information Transaction System

Patient Health Information Transaction System (PHITS) is a database which is used to store all the confidential information and secure transactions performed by the patient and healthcare stakeholders. The patient and the authorized stockholders use a temporary shared key to encrypt multimedia data such as prescription (text) and medical images. The patient (owner) deploys a smart registration contract where the digital fingerprint of the medical image and his personal information is stored. The permission to write data is only allowed to the owner and the doctor (partial). Other registered users (lab technician, nurses, etc.) can only view the data. Any transaction done by the user is verified as per the rules offered in smart contract. The user has to enter the encrypted IPFS hash in the smart contract to complete verification process. If transaction is exact, the user sends an approval to the contract and the transaction is recorded in the contract through the chain.

15.5   Implementation and Result

The proposed work is implemented in a MetaMask platform, which is a deployment process model given in Fig. 15.2. First of all, the details of the owner, i.e., name, email, medical image, and title, are recorded in the Blockchain-based database of the E-healthcare system. By clicking on the submit button, this information passes to the Blockchain network for checking of duplication and privacy using the perceptual hashing technique. The authenticity of this information is verified and registered in the Blockchain. Next, this information is transferred to the Flask server (localhost: 5000) for watermarking using Discrete Cosine Transform (DCT) and Discrete Wavelet Transform (DWT). A QR code is generated using the owner's personal information and the perceptual hash of the medical image. Subsequently, the generated QR code and the watermarked image are displayed on the screen. Meanwhile, it requests for uploading generated watermarked images in IPFS. IPFS stores the image in a folder and gives a unique hash address for the uploaded image. Now, the watermarked image is ready for the deployment on the Blockchain. By clicking on deploy to Blockchain button, MetaMask asks for the deduction of ethers from linked account. As soon as the payment process is completed, the medical image is successfully deployed to the Blockchain.

The step-by-step deployment process of the proposed framework is discussed as follows:

1. First of all, the owner information containing medical image is provided in Ethereum platform to get copyrights.
2. On submission, the generated perceptual hash of the image is compared with the perceptual hash of all the images present in the Blockchain. The perceptual hashing algorithm is designed to point out the slight variations in it, if any.
3. Now, the watermarked image and its corresponding QR code are uploaded to IPFS which generates a conforming unique hash key to retrieve it in future.
4. After clicking on "Deploy to Blockchain" button, MetaMask asks for the owner's permission for the deduction of Gas from the current Ethereum account.

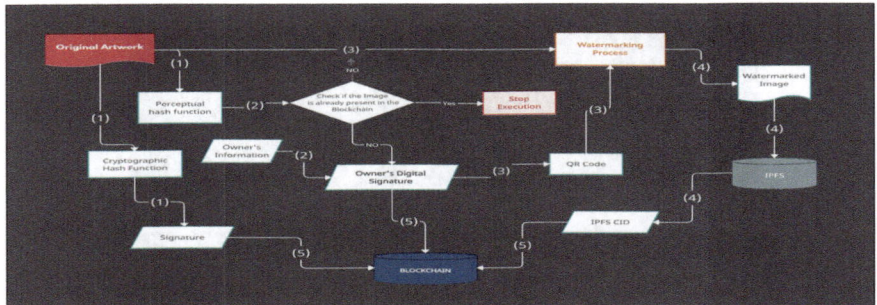

Fig. 15.2 Schematic diagram of proposed watermarked scheme

5. On confirmation, some amount of ethers is deducted from the owner's account and a new block is created in the Blockchain.
6. The new block has owner details such as title, unique id, address in IPFS, and perceptual hash of the image.
7. The QR code corresponding to the new block contains all the relevant information which approves the authenticity and ownership of the image. Finally, this QR code can be scanned using any mobile device.

15.6 Performance Evaluation

The proposed framework is performed with a system configured with Windows 10 operating system having 8 GB hard-disk, 4 GB RAM, 512 GB crucial SSD, and 2 GB NVIDIA 1650 graphic card. The aforementioned configuration is available on Acer laptop which is capable to perform well for the initial transactions. However, it may cause some problems on huge data storage. The proposed framework requires 19% CPU on 2.62 GHz, 68% memory, at 16 Kbps Internet speed as shown in Fig. 15.3. The proposed system has been implemented on Ethereum smart contract using Ganache, MetaMask. Web Ether Wallet is used to complete the transaction. IPFS is used for storing the distributed hash address of medical images which are uploaded on Blockchain. A Python lightweight backend framework, named as Flask, is used for the generation of the watermarked QR code of the uploaded medical image. The simulation results show that the proposed system performs satisfactorily at assumed expectations and is able to detect the copied image even with some damages, and modifications occurred on QR code successfully (Figs. 15.4, 15.5, 15.6, 15.7, and 15.8).

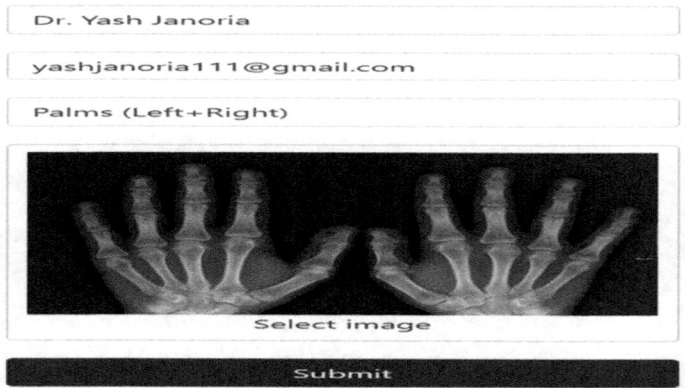

Fig. 15.3 Deployment of medical image in blockchain network of E-healthcare system—step 1

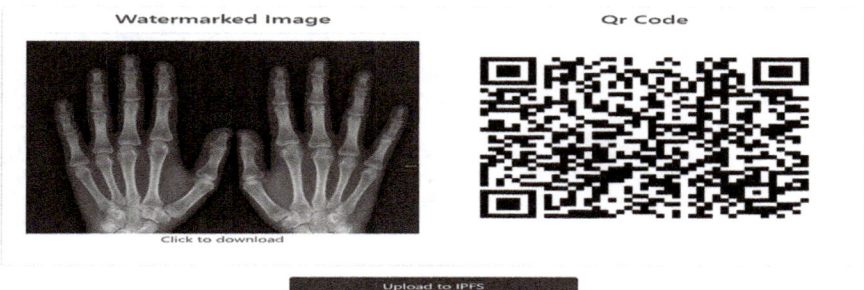

Fig. 15.4 Deployment of medical image in blockchain network of E-healthcare system—step 2

Fig. 15.5 Deployment of medical image in blockchain network of E-healthcare system—step 4

```
0: BN {negative: 0, words: Array(11), length: 10, red: null}
1: "10010101011111110110101001111111101011111101101111010110111111011"
2: "QmToD1pdW5Tr9TVsBQci2Ry9XmLfNUZahUE7wDVq1882mR"
3: "Dr. Yash Janoria"
4: "yashjanoria111@gmail.com"
5: "Palms (Left+Right)"
imageID: BN {negative: 0, words: Array(11), length: 10, red: null}
image_title: "Palms (Left+Right)"
ipfsAddress: "QmToD1pdW5Tr9TVsBQci2Ry9XmLfNUZahUE7wDVq1882mR"
owner_email: "yashjanoria111@gmail.com"
owner_name: "Dr. Yash Janoria"
pHash: "10010101011111110110101001111111101011111101101111010110111111011"
```

Fig. 15.6 Deployment of medical image in blockchain network of E-healthcare system—step 5

15.7 Conclusion

In the present paper, a watermarked medical image is deployed in a Blockchain-based E-healthcare system. The aim of the paper is to solve the security problem of lab reports consisting of confidential images in such a scenario when multiple types of similar images are kept in IPFS. The problem of detecting the valid medical image and authorized owner is solved using watermarking and perceptual hash technique so

Fig. 15.7 Deployment of medical image in blockchain network of E-healthcare system—step 7

that copyright sequence of medical images can be predicted effortlessly. Moreover, Blockchain technology is involved in E-healthcare system as well to restrict any kind of misbehavior performed by unauthorized users. The distributed public ledger takes care of all transactions so that malicious actions would be captured. The proposed framework is extensively tested on Ethereum smart contract, Ganache, MetaMask. The simulation results validate the effectiveness of the proposed framework for securing medical image using a Blockchain-based E-healthcare system.

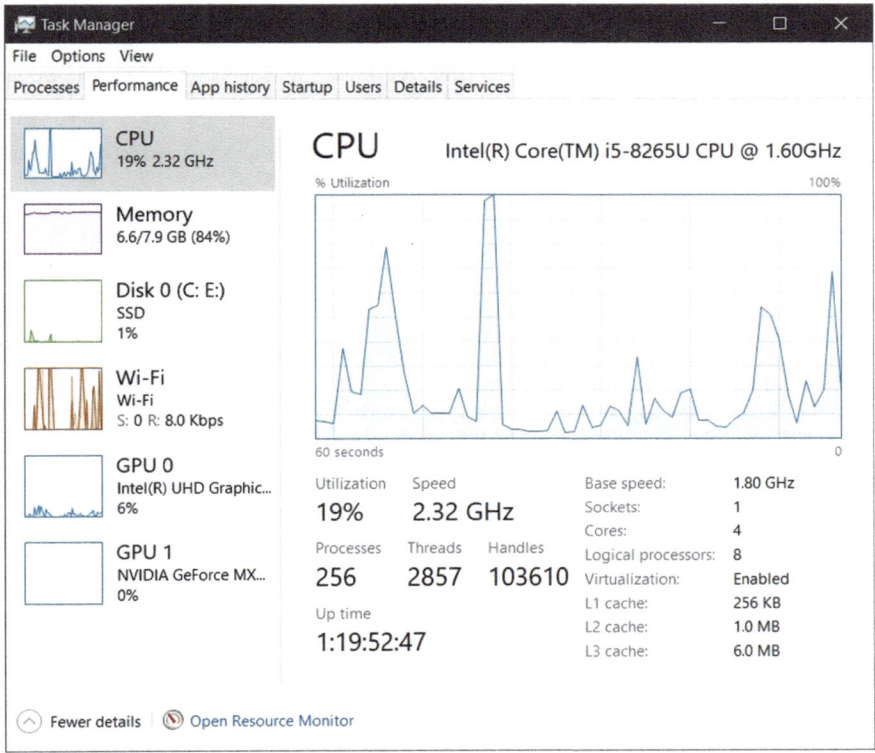

Fig. 15.8 System utilization during deployment of medical images in Ethereum-based Blockchain technology

References

1. Peng, W., Yi, L., Fang, L., XinHua, D., Ping, C.: Secure and traceable copyright management system based on blockchain. In: IEEE 5th International Conference on Computer and Communications (ICCC) (2019)
2. Jose, D.M., Karuppathal, R., Kumar, A.V.A.: Copyright protection using digital watermarking. In: National Conference on Advances in Computer Science and Applications with International Journal of Computer Applications (NCACSA), International Journal of Computer Applications (IJCA) (2012)
3. Savelyev, A.: Copyright in the blockchain era: promises and challenges. Comput. Law Secur. Rev. **34**(3), 550–561 (2018)
4. Fujimura, S., et al.: BRIGHT: A concept for a decentralized rights management system based on blockchain. In: 2015 IEEE 5th International Conference on Consumer Electronics-Berlin (ICCE-Berlin), Berlin (2015)
5. Meng, Z., Morizumi, T., Miyata, S., Kinoshita, H.: Design scheme of copyright management system based on digital watermarking and blockchain. In: 2018 IEEE 42nd Annual Computer Software and Applications Conference (COMPSAC) (2018)
6. Zauner, C.: Implementation and benchmarking of perceptual image hash functions (2010)

7. Nishane, S., Umale, V.M.: Digital image watermarking based on DWT using QR code. Int. J. Curr. Eng. Technol. **5**(3), 1530–1532 (2015)
8. Chow, Y.W., Susilo, W., Tonien, J., Zong, W.: A QR code watermarking approach based on the DWT-DCT technique. In: Australasian Conference on Information Security and Privacy, Cham, 2017, pp. 314–331
9. Mahler, T., et al.: Know your enemy: characteristics of cyber-attacks on medical imaging devices. Comput. Res. Repos. (CoRR) (2018). arXiv:1801.05583
10. Zhang, A., Lin, X.: Towards secure and privacy-preserving data sharing in e-health systems via consortium blockchain. J. Med. Syst. **42**(8), 1–18 (2018)
11. Azim, A., Islam, M.N., Spranger, P.E.: Blockchain and novel coronavirus: towards preventing COVID-19 and future pandemics. Iberoamerican J. Med. **2**(3), 215–218 (2020)
12. Solomon, S.: Israeli researchers show medical scans vulnerable to fake tumors (2019). https://www.timesofisrael.com/israeli-researchers-show-medical-scans-vulnerable-to-fake-tumors/
13. Ma, Z., Huang, W., Gao, H.: Secure DRM scheme based on blockchain with high credibility. Chin. J. Electron. **27**(5), 1025–1036 (2018)
14. Parah, S.A., Sheikh, J.A., Ahad, F., Loan, N.A., Bhat, G.M.: Information hiding in medical images: a robust medical image watermarking system for E-healthcare. Multimedia Tools Appl. **76**(8), 10599–10633 (2017)
15. Benet, J.: IPFS-content addressed, versioned, p2p file system (2014)

Chapter 16
Synthetic Data Augmentation of MRI using Generative Variational Autoencoder for Parkinson's Disease Detection

Yamini Madan, Iswarya Kannoth Veetil, Sowmya V, Gopalakrishnan EA, and Soman KP

Abstract Machine learning models are being increasingly proposed for the automated classification of Parkinson's disease from brain imaging data such as magnetic resonance imaging (MRI). However, the problem of class imbalance is a major setback in deriving the maximum benefit from using these techniques. We propose in overcoming the class imbalance between Parkinson's disease (PD) and normal cohorts (NC) subjects by using variational autoencoders (VAEs) as the generative model that follows a probabilistic regeneration and aim to experiment over the effect of latent variables in generating new MR images of the subjects to improve the detection of PD. The efficiency of the proposed method with and without data augmentation is compared and evaluated using a deep learning classifier model at the subject level. The results obtained using a model trained with data augmentation show a significant increase of 6% in the performance of the classification model.

16.1 Introduction

Parkinson's disease (PD), being the second most common neurodegenerative disorder [1], is a progressive disorder where the condition of the subjects deteriorates with time. It occurs due to the accumulation of iron resulting in loss of dopamine leading to de-pigmentation in the substantia nigra region of the midbrain [3, 4]. Symptoms in the later stage include tremors and stiffness in limbs resulting in locomotive problems, speech disorder and impaired posture [2]. Deep learning (DL) is evolving rapidly and gaining a lot of popularity in the medical field. Early symptoms of PD can be

Y. Madan (✉) · I. K. Veetil · S. V · G. EA · S. KP
Center for Computational Engineering and Networking (CEN) Amrita School of Engineering,
Coimbatore Amrita Vishwa Vidyapeetham, Coimbatore, India
e-mail: ea_gopalakrishnan@cb.amrita.edu

S. KP
e-mail: kp_soman@amrita.edu

© The Author(s), under exclusive license to Springer Nature Singapore Pte Ltd. 2022   171
V. Bhateja et al. (eds.), *Evolution in Computational Intelligence*,
Smart Innovation, Systems and Technologies 267,
https://doi.org/10.1007/978-981-16-6616-2_16

identified by leveraging the abilities of deep learning to extract important features and predict the possibility of disease in a subject, thereby slowing or preventing the offset of the disease to critical stage. However, lack of availability of sufficient images in each of the classes under study leads to overfitting. This class imbalance issue can be overcome using data augmentation techniques. Simple data augmentation techniques do not produce new variations in features which can be overcome by using generative modelling that produces realistic images with diverse variations while maintaining the original morphological structure, which is explored in this study.

Variational autoencoder (VAE) is a generative model that uses latent representational learning to reconstruct data. The encoder portion of the VAE learns a low-dimensional representation of the input data by building a probabilistic model which is the latent vector. The decoder generates new images by sampling the statistical measures to form a latent vector [5]. The difference between autoencoders and VAE is that the latent vector values are forwarded to the decoder directly in autoencoders, while in VAE the model learns the statistical measure—mean and standard deviation of the input data distribution during the training process and the input to the decoder is sampled from the corresponding normal distribution, which improves the model's ability to generate new images. In this paper, we have used the VAE architecture to generate NC images to overcome the class imbalance between the NC class and its counterpart PD.

16.2   Related Work

The application of DL is widespread in the medical field as follows. Different CNN architectures have been used to identify the presence of musculoskeletal abnormalities in radiographic images [6], bio-metric recognition using ear morphology in [7] and tuberculosis detection from microscopic specimen images [8]. The most prominent in identifying PD was by [9] who uses the AlexNet architecture to identify PD from T2 MRI. As an extension of the above work, an ensemble DL architecture is used by [10] to identify PD from T2 MRI. The exact slice range used in training has not been described clearly in both the works. Simple data augmentation techniques have been used to overcome class imbalance, and the model performance with and without data augmentation is compared. The results have been discussed at the slice level. As subject-level classification aids the medical field to analyse the results better, [11] made use of a transfer-learned DL model and verified for generalisability to detect PD. A clear demarcation on slices of interest is described. For Identifying the region of interest [12] improves the model performance as features distinct to the application lie here. To classify diabetic retinopathy with a class imbalance problem, [13] have used GANs in generating synthetic images of the proliferative class to overcome class imbalance. The classifier model performance before and after data augmentation is compared which shows the strong potential of generative models in the medical field. Speech impairment, which is one of the characteristic signs of PD, has been explored by [2] using an autoencoder neural network for detecting PD from speech signals on two different datasets. The performance of the proposed DNN is better when compared to that of other standard ML classifiers for both datasets. Variational autoencoders have been used by [5] to determine the effect of latent variable size for the reconstruction of new MNIST images. A latent vector of size 10 captures all the possible variation in styles of the digits which is not so in the case of smaller latent vector size. Deep variational autoencoders have also been used as a feature extractor by [14] to prevent overfitting due to high dimension of fMRI and lack of sufficient data in the identification of ADHD. The latent variables generated are used in estimating functional brain networks. From the literature, using VAE as a generative model seems promising in generating new images from the sampled latent vector which is a compact and efficient way of representing the features of data in an unsupervised manner.

In this work, we propose to investigate the feasibility and generative capability of VAE with respect to PD classification. To the best of our knowledge, this is the first work to address the issue of class imbalance in MRI of PD for classification using VAE. Rest of the work is structured as follows: Section 16.3 gives an insight into the details of the dataset used in this study. Section 16.4 describes the overall methodology used, experiments and analysis of the results obtained.

16.3   Dataset Description

Parkinson's Progressive Marker's Initiative (PPMI) is a widely used public repository for research on PD diagnosis [?]. In this study we have used the T2-weighted MRI images from the PPMI dataset. T2w MRI captures the iron load deposition in the brain better than the T1 modality [3]. The chosen sample is split in the 80:20 ratio and consists of 116 PD and 84 NC subjects used for training and 19 NC and 19 PD subjects for testing. The test data is split and maintained separately before adding the training data as input to the generative model so as to avoid data leakage. The scans are acquired along the axial plane under a field strength of 3T using TrioTim scanner. Each subject under study has 48 slices, however, the most significant 20 slices are used [11].

16.4   Methodology

In the first subsection we have described the preprocessing techniques applied on the MRI images. The experimental design is explained in the forthcoming subsection, followed by an analysis of the results obtained. The DL model is trained on only real images, an augmentation of synthetic images is generated using VAE to the real images, and the performance of both the models is compared on the test data at subject level which is an area of novelty in our work.

16.4.1 Data Preprocessing

The raw data obtained in NifTi format is preprocessed to normalise the images in two steps: bias field correction (BFC) and skull stripping. BFC is done using Slicer 3D software [15] which corrects the non-uniformity in intensity of images which removes the bias in regions of the brain as seen in Fig. 16.1. The bias generally occurs due to imperfections in magnetic field while scanning. Pixels of the cranium are stripped from the image using FSL BET [16] which reduces the unnecessary features to be learned by the deep learning model (see Fig. 16.2). The slices of each subject are stacked to form a three-channel image for architectural compatibility with the DL model.

16.4.2 Experiments and Results

The real training dataset consists of 2320 PD images and 1680 NC images belonging to 116 PD and 84 NC subjects, 20 slices of which are considered for each subject. This inequality in number of PD and NC training images results in a case of class imbalance. The improved dataset has 600 synthetic images generated by VAE of which 400 images used in training give optimal results (Fig. 16.3).

The metrics used to evaluate the generative performance of VAE are Structural Similarity Index Measurement (SSIM) and peak signal-to-noise ratio (PSNR). SSIM

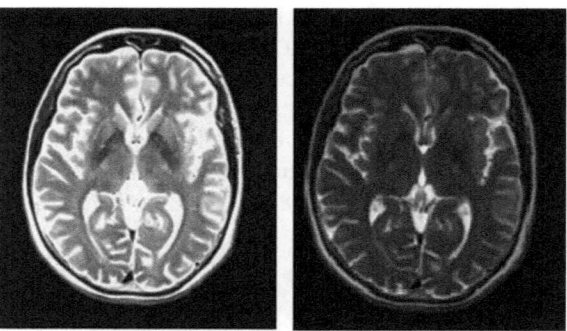

Fig. 16.1 Bias field correction

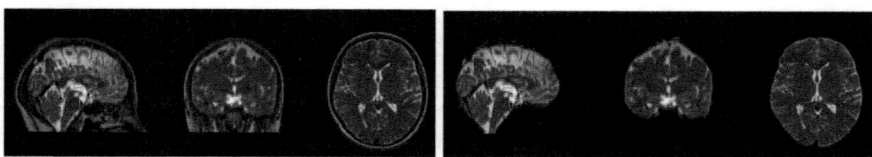

Fig. 16.2 Skull stripping

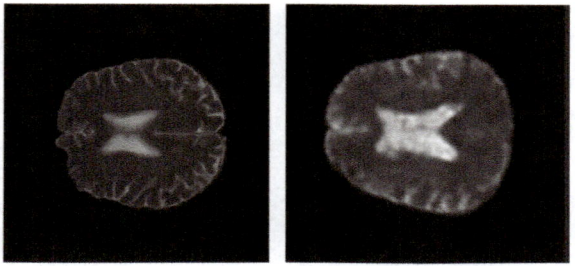

Fig. 16.3 **a** Real image of a NC. **b** Synthetic image generated by VAE

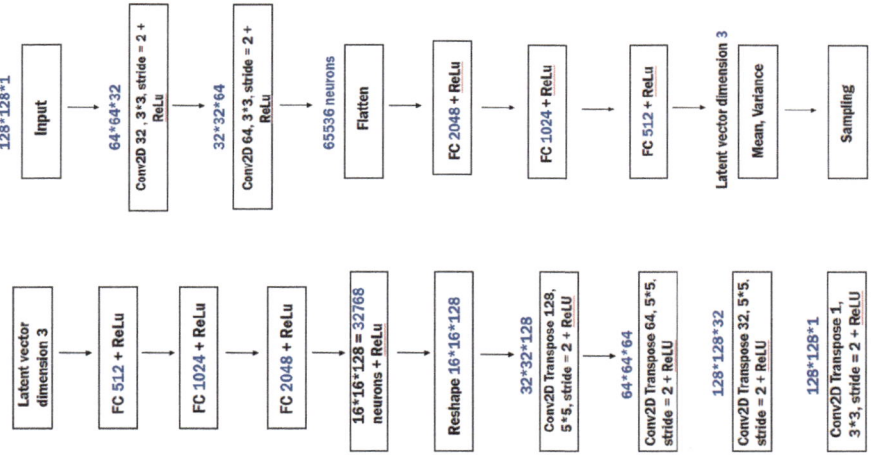

Fig. 16.4 Encoder and decoder architecture of the proposed variational autoencoder model

captures the information about the similarity between the original and generated images. PSNR measures the ratio of highest intensity value with mean squared error of real and generated image (metrics are scaled to 100) [17]. Metrics used to evaluate test data are accuracy, specificity, sensitivity, precision and F1 score. Specificity determines the number of PD subjects correctly classified. Higher specificity indicates a lesser number of mispredictions of a normal cohort to be diseased. Sensitivity or recall determines the number of NC subjects correctly classified. Higher sensitivity indicates lesser chances of missing out on identifying to detect the disease when it is present. Precision determines the number of PD predictions that are actually true.

The architecture diagram for the VAE with the number and size of hidden layers is shown in Fig. 16.4. The model is trained for 200 epochs with a batch size of 32. Adam optimiser with a default learning rate of 0.001 is used. We have experimented with different latent vector dimension to identify the best suitable vector size for the reconstruction of images such that the diversity in the pattern of images is maintained. A sample of the real image with synthetic image generated is shown in Fig. 16.3. It

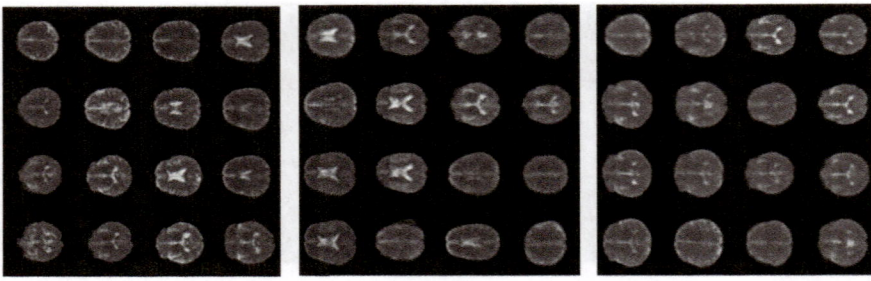

Fig. 16.5 Generated images for different latent sizes. **a** 5, **b** 35, **c** 500

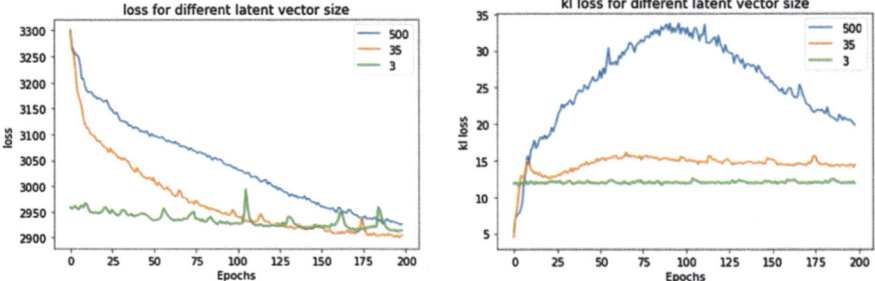

Fig. 16.6 Total loss and Kullback Leiber loss over different latent dimension vector size

Table 16.1 Metrics assessing quality of generated images by VAE

Latent space dim	Mean SSIM	Mean PSNR
3	**68.2**	**68.62**
35	66.3	67.67
500	66.8	67.31

is observed that the reconstruction of sampled images is blurry in lower dimensions, while in the midrange latent space pattern diversity is captured which is not so in a higher latent space (see Fig. 16.5). This is reflected in the image quality metrics as well.

Kullback Leiber (kl) loss signifies the error between the original and reconstructed image distribution, based on which the probability distribution learnt by the model is altered during training. From Fig. 16.6 we can see that the kl loss is higher for a bigger latent vector dimension and the total loss is least for latent vector dimension 3. As seen in Table 16.1, the SSIM and PSNR are maximum for dimension value 3, signifying that the structure of reconstructed images are similar to that of the original images, and the noise ratio is also low.

Densenet201 architecture is found to be a classifier model that works efficiently in identifying PD [11]. The synthetic images are incremented starting from 200 to

Table 16.2 Comparing classifier model performance with and without augmentation

	Specificity	Sensitivity	Precision		F1 score		Accuracy
			PD	NC	PD	NC	
Real	0.58	0.63	0.6	0.61	0.61	0.59	0.60
VAE	0.68	0.63	0.66	0.65	0.64	0.66	0.66

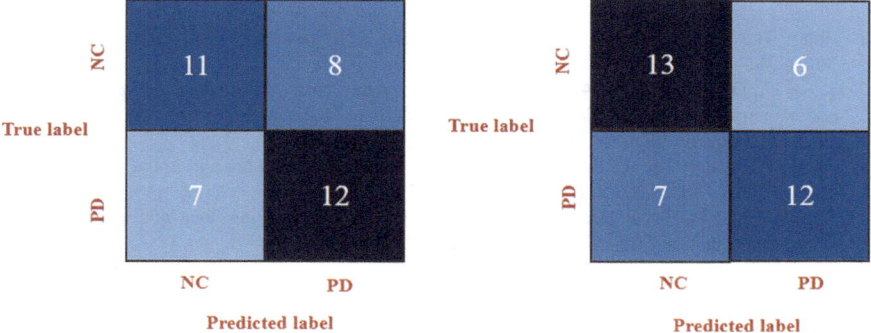

Fig. 16.7 Confusion matrix for test data at subject level. **a** Model trained only with real data. **b** Model trained with real and synthetic data

determine the performance of the deep learning model. It is noted that the metrics do not improve beyond 400 images, which eventually has been chosen as the number of synthetic images given as input along with the real images. The performance of DenseNet201 classifier is shown in Table 16.2. Early stopping is used to prevent overfitting of the model. The confusion matrix for the test data is shown in Fig. 16.7. There is a significant improvement of 6% in accuracy and a superlative performance in identifying both classes under study by the model trained with augmentation when compared to just the real images.

16.5 Conclusion

The class imbalance between NC and PD is overcome by generating images of slices of interest of normal cohorts using VAE. The data is split into train and test before data augmentation to avoid data leakage. On experimenting with the latent vector dimension, a lower vector represents features of NC images the best. The images generated by VAE maintain the distribution and diversity in variants of input NC images. The classifier performance has significantly improved when testing on a model trained with real and synthetic data as seen from the metrics. The latent vector size is an important component in tuning the VAE model. The synthetic images generated are blurry and repetitive patterns are found in the features for higher latent vector size. The time taken to train the model is a little expensive. As future work, the image generating ability of other generative models in terms of divergence of images produced and quality can be compared and assessed.

References

1. Pinter, B., et al.: Mortality in Parkinson's disease: a 38-year follow-up study. Mov. Disord. **30**(2), 266–269 (2015)
2. Kose, U., Deperlioglu, O., Alzubi, J., Patrut, B.: Diagnosing parkinson by using deep autoencoder neural network. In: Deep Learning for Medical Decision Support Systems. Studies in Computational Intelligence, vol 909. Springer, Singapore (2021)
3. Pyatigorskaya, N., et al.: A review of the use of magnetic resonance imaging in Parkinson's disease. Ther. Adv. Neurol. Disord. **7**(4), 206–220 (2014)
4. Shinde, S., et al.: Predictive markers for Parkinson's disease using deep neural nets on neuromelanin sensitive MRI. NeuroImage: Clin. **22**, 101748 (2019)
5. Xu, Q., et al.: Different latent variables learning in variational autoencoder. In: 2017 4th International Conference on Information, Cybernetics and Computational Social Systems (ICCSS). IEEE (2017)
6. Harini, N., Ramji, B., Sriram, S., Sowmya, V., Soman, K.: Musculoskeletal radiographs classification using deep learning. In: Deep Learning for Data Analytics, pp. 79-98. Elsevier (2020)
7. Gopika, P., Sowmya, V., et al.: Transferable approach for cardiac disease classification using deep learning. In: Deep Learning Techniques for Biomedical and Health Informatics, pp. 285–303. Academic Press (2020)
8. Simon, A., Ravi, V., Vishvanathan, S., Sowmya, K.P.: Shallow CNN with LSTM layer for tuberculosis detection in microscopic image. Int. J. Rec. Technol. Eng. **7**, 56–60 (2019)
9. Sivaranjini, S., Sujatha, C.M.: Deep learning based diagnosis of Parkinson's disease using convolutional neural network. Multimed. Tools Appl. **79**(21), 15467–15479 (2020)
10. Rajanbabu, K., Veetil, I.K., Sowmya, V., Gopalakrishnan, E.A., Soman, K.P.: Ensemble of deep transfer learning models for parkinson's disease classification. In: Accepted in Fourth International Conference Soft Computing and Signal Processing (ICSCSP) (2020)
11. Madan, Y., Veetil, I.K., et al.: Deep learning based approach for Parkinson's Disease Detection using Region of Interest. In: Accepted in 4th International Conference on Intelligent Sustainable Systems (ICISS) (2021)
12. Bhateja, V., et al.: Haralick features-based classification of mammograms using SVM. In: Information Systems Design and Intelligent Applications, pp. 787–795. Springer, Singapore (2018)
13. Balasubramanian, R., et al.: Analysis of adversarial based augmentation for diabetic retinopathy disease grading. In: 2020 11th International Conference on Computing, Communication and Networking Technologies (ICCCNT). IEEE (2020)
14. Qiang, N., Dong, Q., Sun, Y., Ge, B., Liu, T.: deep variational autoencoder for modeling functional brain networks and ADHD identification. In: 2020 IEEE 17th International Symposium on Biomedical Imaging (ISBI), pp. 554–557 (2020)
15. Fedorov, A., et al.: 3D slicer as an image computing platform for the quantitative imaging network. Magn. Reson. Imag. **30**(9), 1323–1341 (2012)
16. McCarthy, P.: FSLeyes (2020)
17. Islam, J., Zhang, Y.: GAN-based synthetic brain PET image generation. Brain Inform. **7**, 1–12 (2020)

Chapter 17
Recursive Visual Cryptography Scheme with PRWP and Additional Basis Matrix

T. E. Jisha and Thomas Monoth

Abstract Recursive visual cryptography scheme (RVCS) is a high-security image secret sharing scheme in which the shares are recursively encoded into sub-shares at numerous levels. Here we presented a new scheme for RVCS with a perfect reconstruction of white pixels (PRWP) and an additional basis matrix (ABM) which boosts contrast. The method has been illustrated on the basis of investigational examinations. We made an evaluation between the conventional schemes and the proposed scheme through investigational analysis. The results revealed that our scheme improves the visual quality and ensures security and reliability.

17.1 Introduction

Visual cryptography scheme (VCS) is an image secret sharing model to encrypt a furtive image into n transparencies (*or* shares) and the overlying of k transparencies can decipher the clandestine image which was initiated in 1994 by Naor and Shamir. The intruders do not gain any intimation about the secret if the shares are less than k and also the secret cannot be obtained by stacking these $k − 1$ shares. This feature of VCS ensures security [1, 2]. The decryption can be done by overlapping transparencies with no computation. RVCS is constructed based on conventional VCS where the image is encoded into n shares and the shares can be encoded again into p sub-shares recursively. The recursion occurs at different levels hierarchically [3, 4].

The reliability and security of the system are remarkably increased in the RVCS due to encryption at different levels but the contrast of the decoded image is notably decreased. The contrast, pixel expansion, security and reliability are the crucial problems in the research vicinity of the RVCS and it necessitates more attention. The RVCS can be used to construct many authentication systems with passwords and

T. E. Jisha (✉)
Department of Information Technology, Kannur University, Kannur 670 567, Kerala, India

T. Monoth
Department of Computer Science, Mary Matha Arts and Science College, Kannur University, Mananthavady, Wayanad 670 645, Kerala, India

biometrics because the encoding can be done at each level and ensures the secrecy of the image. It also guarantees the authenticity of the system [4–9].

The fundamental intention of the recommended technique is to develop an RVCS that progresses the visual quality of the reconstructed image and preserves reliability, security and authenticity. We introduced a new RVCS with PRWP in place of black pixels and with ABM [3, 10, 11]. The forthcoming segments of the paper are: segment 2 exemplifies RVCSs. Segment 3 illustrates the proposed scheme, the investigational outcomes and the scrutiny of different models. The finishing segment sketches the conclusion.

17.2 Recursive Visual Cryptography Scheme

Encryption in traditional VCS is at one level but in RVCS is at many levels. At the initial level, the image can be encoded into shares and the encoded shares can be encoded into sub-shares recursively at the subsequent levels [3, 4, 11]. The two-level RVCS can be demonstrated by a (2, 2) scheme in Fig. 17.1 with a tree structure. The first level consists of two shares, the second level consists of four shares and so on. Here the original and decrypted images and the shares at all levels are of the same size.

In Fig. 17.1 SI represents a secret image. $Share_1$ and $Share_2$ are the shares at the initial level and $share_{11}$, $share_{12}$, $share_{21}$ and $share_{22}$ are the second-level shares. The image can be decoded in different ways as stated below:

$$SI = Share_1 + Share_2$$
$$SI = Share_1 + Share_{21} + Share_{22}$$
$$SI = Share_2 + Share_{11} + Share_{12}$$
$$SI = Share_{11} + Share_{12} + Share_{21} + Share_{22}$$

There are different combinations to decode the image since it enhances both reliability and security. RVCS is performed by perfect reconstruction of both black pixels and white pixels [10, 12].

Fig. 17.1 Two-level (2, 2) RVCS

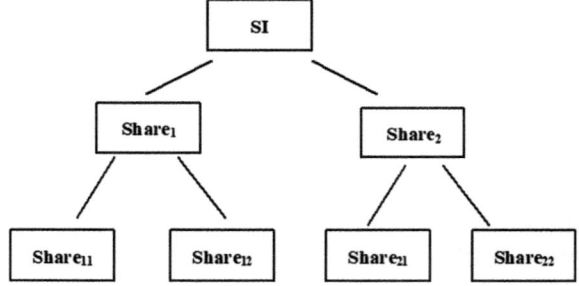

17.2.1 RVCS with PRWP

The main shortcoming of RVCS is that there is a reduction in contrast to the reconstructed image as the encryption level increases. The application of PRWP overcomes the contrast reduction up to a limit because most of the natural image contains more white pixels compared to black pixels. In PRWP the reconstruction of pixels (white) is full but half of the pixels are lost in the case of black. The basis matrices are

$$BW^{rb} = \begin{bmatrix} 1010 \\ 1010 \end{bmatrix}, BW^{rw} = \begin{bmatrix} 1010 \\ 0101 \end{bmatrix}$$

For obtaining black pixels select one of the columns from BW^{rb} arbitrarily and for white from BW^{rw}. For instance, the columns selected from BW^{rb} and BW^{rw} are

$$V^{rb} = \begin{bmatrix} 1 \\ 1 \end{bmatrix}$$

$$V^{rw} = \begin{bmatrix} 1 \\ 0 \end{bmatrix}$$

An illustration of two-level RVCS (size invariant) with random basis column pixel expansion is portrayed in Fig. 17.2.

Here we recognized that the contrast degradation of the recovered image decreases as the encryption level increases [12–18].

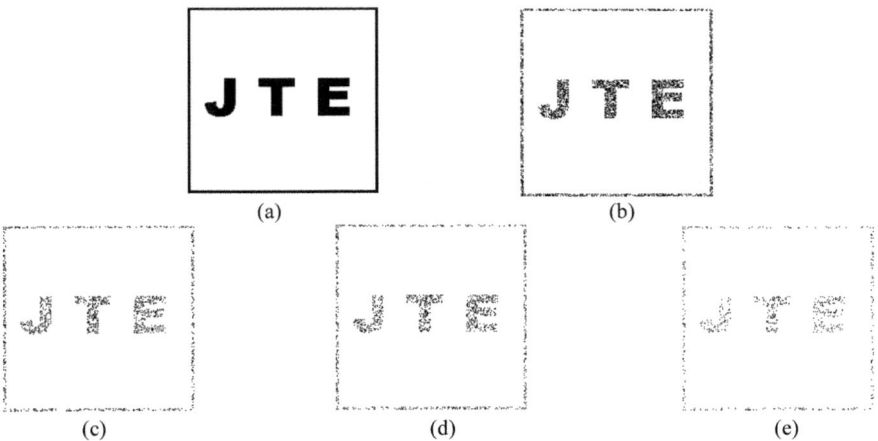

Fig. 17.2 **a** SI. **b** Share$_1$ + Share$_2$. **c** Share$_1$ + Share$_{21}$ + Share$_{22}$. **d** Share$_2$ + Share$_{11}$ + Share$_{12}$. **e** Share$_{11}$ + Share$_{12}$ + Share$_{21}$ + Share$_{22}$

17.2.2 RVCS with PRBP and ABM

RVCS with PRBP and ABM was introduced in [4]. This scheme increases the contrast of the reconstructed image at various levels. Here the black pixels are reconstructed in full but the white pixels are half lost. To conquer this loss, add ABM to the white pixels side in order to increase the number of white pixels to some extent.

The basis matrices are

$$BW^{rb} = \begin{bmatrix} 1010 \\ 0101 \end{bmatrix}, BW^{rw} = \left\{ \begin{bmatrix} 1010 \\ 1010 \end{bmatrix}, \begin{bmatrix} 1000 \\ 1000 \end{bmatrix} \right\}$$

Select arbitrarily one column vector from BW^{rb} for black and for white from BW^{rw}. An illustration of two-level RVCS with PRBP and ABM is portrayed in Fig. 17.3.

Here we detected that RVCS with ABM tremendously increased the contrast of the recovered images at different levels. The usage of ABM increases the number of white pixels. Hence the loss of white pixels in the reconstruction is reduced compared to RVCS without ABM [5, 10, 19].

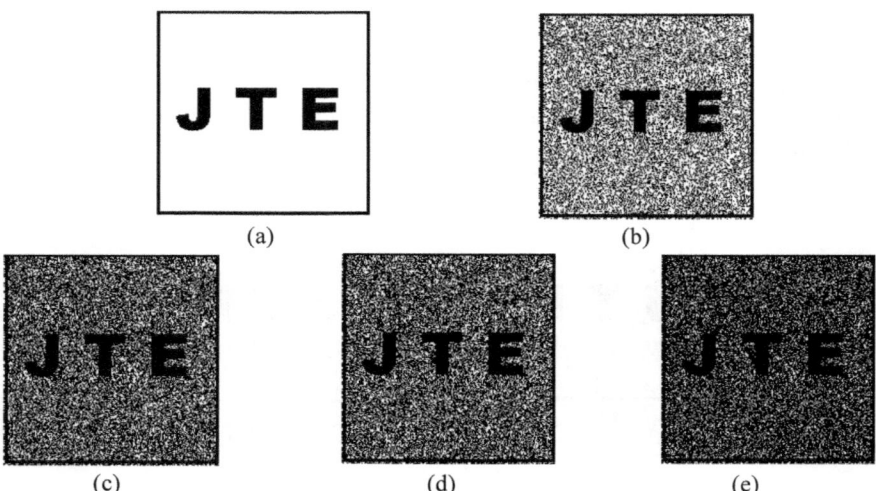

Fig. 17.3 **a** SI. **b** $Share_1 + Share_2$. **c** $Share_1 + Share_{21} + Share_{22}$. **d** $Share_2 + Share_{11} + Share_{12}$. **e** $Share_{11} + Share_{12} + Share_{21} + Share_{22}$

17.3 The Proposed Scheme: RVCS Based on PRWP and ABM

The negative aspect of the RVCS with PRWP is the half loss in the reconstruction of black pixels. To overcome the decrease in the number of black pixels we can apply ABM. The proposed system preserves reliability and security and enriches the visual quality of the decoded image as the level increases. Here we employed the enhanced contrast, reliability and security attributes of the RVCS with PRWP and additional contrast enhancement attribute of RVCS with ABM.

The encryption of secret image (*SI*) is performed at multiple levels. Initially, *SI* is encrypted into *n* shares, which can be then recursively encrypted into *p* sub-shares at subsequent levels.

$$SI = Share_1, Share_2 \ldots Share_n$$
$$Share_n = Shrae_{n1}, Share_{n2}, \ldots, Share_{np}$$

Image can be decoded with any *k* shares based on Eq. (17.1).

$$SI = \sum_{i=1}^{k} Share_i \tag{17.1}$$

Here we implemented a two-level (2, 2) RVCS with PRWP and ABM. In this scheme, *SI* can be represented with binary *n* x *m* basis matrices

$$BW^{rw} = \begin{bmatrix} 1010 \\ 0101 \end{bmatrix}, BW^{rb} = \left\{ \begin{bmatrix} 1010 \\ 1010 \end{bmatrix}, \begin{bmatrix} 1000 \\ 1000 \end{bmatrix} \right\}$$

In this scheme 1 stands for white and 0 stands for black. For encryption of white pixels, one of the column vectors is arbitrarily selected from the BW^{rw} and for black pixels from BW^{rb}. For instance, the vectors are used as the following:

$$V^{rw} = \begin{bmatrix} 1 \\ 0 \end{bmatrix} \text{ used for white and}$$

$$V^{rb} = \begin{bmatrix} 1 \\ 1 \end{bmatrix} \text{ used for black}$$

The selection of pixels in RVCS based on PRWP is exactly reverse to the RVCS based on PRBP [11–14, 17, 19]. In this scheme, we are adding ABM in the side of basis matrices of black pixels which increases the number of black pixels when the decryption occurs.

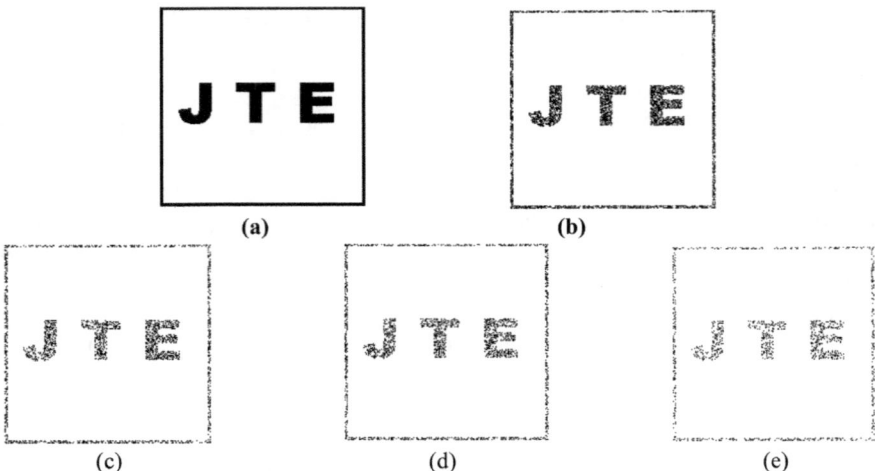

Fig. 17.4 **a** SI. **b** Share$_1$ + Share$_2$. **c** Share$_1$ + Share$_{21}$ + Share$_{22}$. **d** Share$_2$ + Share$_{11}$ + Share$_{12}$. **e** Share$_{11}$ + Share$_{12}$ + Share$_{21}$ + Share$_{22}$

17.3.1 Experimental Results

The experimentations were illustrated with (2, 2) RVCS with PRWP and ABM for various echelons of encoding. Decrypted images of various encryption levels are portrayed in Fig. 17.4.

From Figs. 17.2, 17.3 and 17.4, we realized that the visual quality of the reconstructed images in the suggested scheme is superior to the RVCS with PRWP and RVCS with PRBP and ABM (conventional RVCS with ABM) which are illustrated in Sects. 17.2.1 and 17.2.2 [3–5, 11, 18, 20–24].

The following are the main benefits of the suggested model:

- The method is non-expansible RVCS.
- It guarantees security and reliability.
- It boosts the visual perception of overlapped images.
- The model is exploited with PRWP and ABM.

17.3.2 Experimental Result Analysis

Here we investigated four RVCS schemes: Conventional RVCS, conventional RVCS with ABM, RVCS with PRWP and the proposed model (RVCS with PRWP and ABM). We analyzed the security, reliability and visual quality of the collected images. The investigational study of existing and the proposed models are portrayed in Tables 17.1 and 17.2 with respect to the mean square error (MSE), peak signal-to-noise ratio (PSNR) and Structural Similarity Index Measure (SSIM) ratio. PSNR and

Table 17.1 Analysis of conventional RVCS and conventional RVCS with ABM

Overlapping of shares and sub-shares	Conventional RVCS					Conventional RVCS with ABM				
	White	Black	MSE	PSNR	SSIM	White pixels	Black pixels	MSE	PSNR value	SSIM
$Share_1 + Share_2$	17,842	22,158	0.44	3.57	0.10	18,144	21,856	0.33	4.86	0.13
$Share_1 + Share_{21} + Share_{22}$	8967	31,033	0.66	1.79	0.05	14,024	25,976	0.54	2.71	0.07
$Share_2 + Share_{11} + Share_{12}$	9069	30,931	0.66	1.81	0.05	13,918	26,199	0.54	2.69	0.07
$Share_{11} + Share_{12} + Share_{21} + Share_{22}$	4591	35,409	0.77	1.13	0.03	8820	31,180	0.67	1.77	0.05

Table 17.2 Analysis of RVCS with PRWP and the proposed model

Overlapping of shares and sub-shares	RVCS with PRWP					Proposed model				
	White pixels	Black pixels	MSE	PSNR value	SSIM	White pixels	Black pixels	MSE	PSNR value	SSIM
$Share_1$ + $Share_2$	37,655	2345	0.06	12.56	0.86	37,079	2921	0.04	13.87	0.88
$Share_1$ + $Share_{21}$ + $Share_{22}$	38,832	1168	0.08	10.71	0.80	38,211	1789	0.07	11.60	0.83
$Share_2$ + $Share_{11}$ + $Share_{12}$	38,816	1184	0.08	10.73	0.80	38,161	1839	0.07	11.63	0.83
$Share_{11}$ + $Share_{12}$ + $Share_{21}$ + $Share_{22}$	39,415	585	0.10	10.03	0.77	38,886	1114	0.09	10.64	0.80

SSIM stand for visual quality measurement between the furtive and decoded image. High PSNR and SSIM ratio points to enhanced contrast of the decoded image. Low MSE value leads to higher PSNR and SSIM value. Considering Tables 17.1 and 17.2 we noticed that the proposed model's PSNR and SSIM ratio in the overlapped images is superior to the conventional RVCSs. We also discovered that our method met the security and reliability requirements.

From Tables 17.1 and 17.2 we noticed that the PSNR value of the proposed scheme is 13.87 at the first level of decryption against 12.56 in RVCS with PRWP, 4.86 in conventional RVCS with ABM and 3.57 in conventional RVCS. Similarly, the SSIM value is greater in the proposed scheme as against the previous schemes. Hence, we stated that our model outstands the existing models in the case of decrypted image's contrast. The reconstructed images from these four schemes are illustrated in Table 17.3. From the figures, we proved that our scheme is optimum in the case of contrast.

The graphical portrayal of PSNR and SSIM values of prior schemes and the proposed model are exemplified in Figs. 17.5 and 17.6. From the portrayal we avowed that the visual qualities of the decoded images are greatly advanced in the proposed scheme.

Table 17.3 Decrypted images of various RVCS

	Conventional RVCS	Conventional RVCS with ABM	RVCS with PRWP	Proposed model
Share$_1$ + Share$_2$				
Share$_1$ + Share$_{21}$ + Share$_{22}$				
Share$_2$ + Share$_{11}$ + Share$_{12}$				
Share$_{11}$ + Share$_{12}$ + Share$_{21}$ + Share$_{22}$				

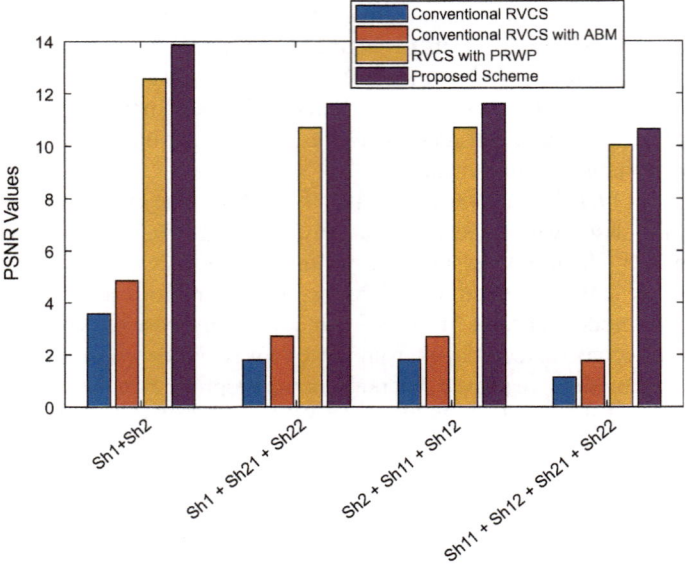

Fig. 17.5 Portrayal of PSNR values

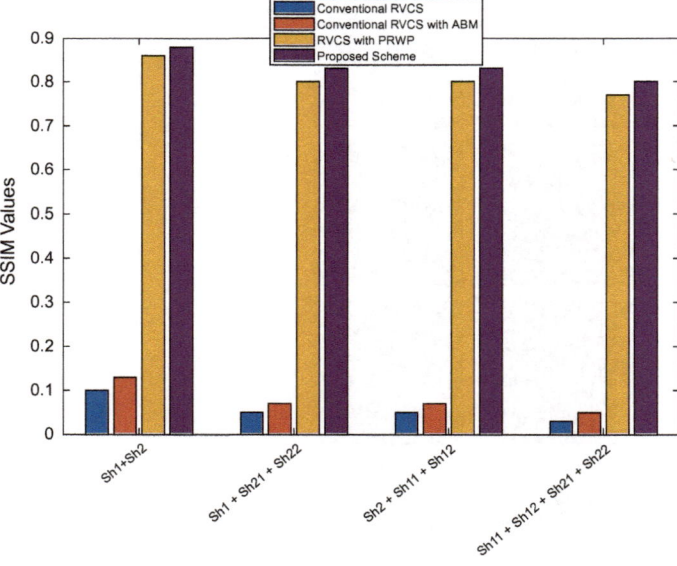

Fig. 17.6 Portrayal of SSIM values

17.4 Conclusion

Here we proposed a new recursive VCS which is size invariant, upholds reliability, security and also advances contrast of the decoded images at the various levels of decryption. We designed a recursive VCS with PRWP and ABM. Through the experimental results and graphical portrayals the prior and proposed schemes are illustrated and analyzed. Our further study will give attention to the proposal of color RVCS with enriched visual quality, reliability and security.

References

1. Naor, M., Shamir, A.: Visual cryptography. In: De Santis, A. (Ed.) Advances in Cryptology—EUROCRYPT 1994. LNCS, vol. 950, pp. 1–12. Springer, Heidelberg (1995). https://doi.org/10.1007/BFb0053419
2. Pandey, D., Kumar, A., Singh, Y.: Feature and future of visual cryptography based schemes. In: Singh, K., Awasthi, A.K. (Eds.) QShine 2013. LNICST, vol. 115, pp. 816–830. Springer, Heidelberg (2013). https://doi.org/10.1007/978-3-642-37949-9_71
3. Monoth, T., Babu, A.P.: Recursive visual cryptography using random basis column pixel expansion. In: 10th International Conference on Information Technology (ICIT 2007), Orissa, 2007, pp. 41–43. IEEE Xplore. https://doi.org/10.1109/ICIT.2007.32
4. Monoth, T.: Contrast-enhanced recursive visual cryptography scheme based on additional basis matrices. In: Smart Intelligent Computing and Applications, vol. 105, pp.179–187. Springer, Singapore (2019). https://doi.org/10.1007/978-981-13-1927-3_18
5. Monoth, T., Babu Anto, P.: Analysis and design of tamperproof and contrast-enhanced secret sharing based on visual cryptography schemes, Ph. D. thesis, Kannur University, Kerala, India (2012). http://shodhganga.inflibnet.ac.in
6. Ateniese G., Blundo C., De Santis A., Stinson D.R.: Constructions and bounds for visual cryptography. In: Meyer F., Monien B. (Eds.) Automata, Languages and Programming. ICALP 1996. Lecture Notes in Computer Science, vol. 1099, pp. 416–428. Springer, Berlin, Heidelberg (1996). https://doi.org/10.1007/3-540-61440-0_147
7. Weir, J., Yan, W.A.: Comprehensive study of visual cryptography. In: Shi Y.Q. (Eds.) Transactions on Data Hiding and Multimedia Security V. Lecture Notes in Computer Science, vol. 6010, pp. 70–105. Springer, Berlin, Heidelberg (2010). https://doi.org/10.1007/978-3-642-14298-7_5
8. Jisha, T.E., Monoth, T.: Recent research advances in black and white visual cryptography schemes, soft computing for problem solving. In: Advances in Intelligent Systems and Computing, vol. 1048, pp. 479–492. Springer, Singapore (2020). https://doi.org/10.1007/978-981-15-0035-0_38
9. Chavan, P.V., Atique, M.: Secured approach for authentication using threshold-based hierarchical visual cryptography. Int. J. Inf. Privacy Secur. Integr. 2(2), 159–175(2015). https://doi.org/10.1504/IJIPSI.2015.075440
10. Monoth, T., Babu Anto, P.: Contrast-enhanced visual cryptography schemes based on perfect reconstruction of white pixels and additional basis matrix. Computational intelligence, cyber security and computational models. In: Advances in Intelligent Systems and Computing, vol. 412. pp. 361–368. Springer, Singapore (2016). https://doi.org/10.1007/978-981-10-0251-9_34
11. Jisha, T.E., Monoth, T.: Optimal contrast and size-invariant recursive VCS using perfect reconstruction of white pixels. Evolution in Computational Intelligence. In: Advances in Intelligent Systems and Computing, vol. 1176, pp. 181–189. Springer, Singapore (2021). https://doi.org/10.1007/978-981-15-5788-0_17

12. Jisha, T.E., Monoth, T.: WiP: Security enhanced size invariant visual cryptography with perfect reconstruction of white pixels. In: ICISS 2019, LNCS 11952, pp. 279–289, Springer (2019). https://doi.org/10.1007/978-3-030-36945-3_15
13. Lin, T.H., Shiao, N.S., Chen, H.H., Tsai, C.S.: A new non-expansion visual cryptography scheme with high quality of recovered image. In: IET International Conference on Frontier Computing. Theory, Technologies and Applications. IEEE Xplore (2010) https://doi.org/10.1049/cp.2010.0571
14. Huang, Y.-J., Chang, J.-D.: Non-expanded visual cryptography scheme with authentication. In: IEEE 2nd International Symposium on Next-Generation Electronics (ISNE)—IEEE (2013). https://doi.org/10.1109/ISNE.2013.6512319
15. Chow, Y.W., Susilo, W., Wong, D.S.: Enhancing the perceived visual quality of a size invariant visual cryptography scheme. Information and Communications Security. Lecture Notes in Computer Science, vol. 7618. Springer, Berlin, Heidelberg (2012)
16. Ito, R., Kuwakado, H., Thanka, H.: Image size invariant visual cryptography. IEICE Trans. Fund. **E82-A**(10) (1999)
17. Liu, F., Guo, T., Wu, C.K., Qian, L.: Improving the visual quality of size invariant visual cryptography scheme. J. Vis. Commun. Image Represent **23**(2), 331–342 (2012). Elsevier. https://doi.org/10.1016/j.jvcir.2011.11.003
18. Chen, Y.-F., Chan, Y.-K., Huang, C.-C., Tsai, M.-H., Chu, Y.-P.: A multiple-level visual secret-sharing scheme without image size expansion. Inf. Sci. **177**(21), 4696–4710 (2007)
19. Mohan, A., Binu, V.P.: Quality improvement in color extended visual cryptography using ABM and PRWP. In: International Conference on Data Mining and Advanced Computing (SAPIENCE)ss. IEEE Xplore, (2016). https://doi.org/10.1109/SAPIENCE.2016.7684159
20. Yan, B., Wang, Y.F., Song, L.Y., et al.: Size-invariant extended visual cryptography with embedded watermark based on error diffusion. Multimed. Tools Appl. **75**, 11157 (2016). https://doi.org/10.1007/s11042-015-2838-4
21. Yan, B., Xiang, Y., Hua, G.: Improving the visual quality of size-invariant visual cryptography for grayscale images. An Analysis-by-Synthesis (AbS) approach. IEEE Trans. Image Process. **28**(2). https://doi.org/10.1109/TIP.2018.2874378
22. Yan, B., Wang, Y.-F., Song, L.-Y., Yang, H.-M.: Size-invariant extended visual cryptography with embedded watermark based on error diffusion. Multimed. Tools Appl. **75**(18), 11157–11180 (2016). https://doi.org/10.1007/s11042-015-2838-4
23. Ou., Duanhao, Sun, W., Wu., Xiaotian: Non-expansible XOR-based visual cryptography scheme with meaningful shares. Signal Process. **108**, 604–621 (2015). https://doi.org/10.1016/j.sigpro.2014.10.011(Elsevier)
24. Sharma, R., Agrawal, N.K., Khare, A., Pal, A.K.: An improved size invariant n, n extended visual cryptography scheme. Int. J. Bus. Data Commun. Netw. **12**(2) (2016)

Chapter 18
Mobility-Aware Application Placement for Obstacle Detection in UAM Using Fog Computing

D. Malarvizhi and S. Padmavathi

Abstract Urban air mobility (UAM), defined as safe and efficient air traffic operations in a metropolitan area for manned aircraft and unmanned aircraft systems, is one of the major research areas for enabling seamless transport and communication. UAM is one of the emerging transportation technologies considering the traffic congestion faced in most of the cities. Internet of Things, being a platform for connected things, can aid in addressing the challenges faced by the UAM industry. The scope of this work is to provide high-level details on the challenges posed to UAM. This paper proposes an infrastructure for UAM vehicles (UAV) to use the fog computing layer, allowing the UAM vehicles to have a smoother flight in the air with automated obstacle detection, thereby reducing the risks of accidents.

18.1 Introduction

Aviation technologies and concepts have reached a maturity level to enable transport with urban air mobility (UAM) using efficient manned and unmanned vehicles to perform on-demand and scheduled operations [1]. Types of operations could be of several purposes, including emergency medical evacuations, search and rescue operations, humanitarian missions, newsgathering, ground traffic flow assessment, weather monitoring, package delivery, inspecting infrastructure, taking photos and passenger transport. By 2050, the urban population is predicted to increase to 66%. With the adoption of UAM, the environmental impact can be reduced, due to the use of electric propulsion.

A number of technical challenges must be overcome to achieve mature UAM operations at a higher density that can be accommodated by the current air traffic control (ATC) system [2].

D. Malarvizhi (✉) · S. Padmavathi
Thiagarajar College of Engineering, Department of Information Technology, Madurai, India

S. Padmavathi
e-mail: spmcse@tce.edu

© The Author(s), under exclusive license to Springer Nature Singapore Pte Ltd. 2022
V. Bhateja et al. (eds.), *Evolution in Computational Intelligence*,
Smart Innovation, Systems and Technologies 267,
https://doi.org/10.1007/978-981-16-6616-2_18

The major challenge in UAM is to have a system in place to avoid collisions and crashes within the UAM vehicles [3]. The UAM vehicles are usually targeted to fly at a lower altitude, which is an airspace that is not currently managed and controlled by the air traffic control (ATC) system. This poses serious challenges in the operability of UAM vehicles.

With the advent of the Internet of Things (IoT), devices, services and people are ubiquitously connected almost all the time, also generating a lot of data. The backbone of IoT communication is machine to machine (M2M). In M2M, machines can communicate with each other without human intervention [4]. Internet of Things offers a variety of solutions enabling machines/devices to talk with each other and make intelligent decisions. An IoT application acts on the large flurry of data generated by the Edge nodes, which in turn are analysed and processed further to obtain estimates and make decisions based on the inferences. With IoT, the major processing happens in the Cloud due to the scalable and elastic nature of the infrastructure with a huge amount of data storage capabilities. The UAM vehicles can be analogous to the IoT Edge nodes talking with each other and can be modelled in an IoT architecture with the data processing happening in the Cloud. UAM vehicles operating in an unmanaged airspace will require faster responses with the real-time data generated. Also, considering the dynamic nature of UAM vehicles, support for mobility should be considered in processing the data.

Fog computing is an additional layer that sits in between the Cloud and the Edge device that enables faster processing and quicker decisions with real-time data [5]. Fog computing offers a variety of benefits compared to Cloud computing in terms of real-time data processing in addition to conserving network bandwidth. Fog computing is suitable to be deployed for latency-sensitive applications such as video conferencing, intelligent surveillance, smart vehicle parking system and smart waste management system.

The paper is organized as follows: Section 18.2 discusses the related work. Section 18.3 presents an overview of UAM and its challenges. Section 18.4 discusses the background of IoT and fog computing. Section 18.5 briefs the proposed architecture and the simulation environment used. Section 18.6 presents the results and analysis based on metrics such as latency, network usage and energy consumption. Section 18.7 concludes the paper.

18.2 Related Work

Urban air mobility has evolved as the integration of the third dimension to urban transport networks, including on-demand sharing mobility. Various literatures [6–8] have explored the challenges in adopting the UAM and have come up with recurring factors along with exploratory factor analyses. Factors such as the number of hyper commuters, those who travel more number hours in a single day can contribute more to UAM's adoption, however, other factors such as high incomes and good weather also play an important role [9]. Understanding the service demand and predicting consumer adoption are critical, and researchers still try to investigate the factors for users to adopt urban air mobility [10]. Vascik and Hansman et al. [11] emphasized the significance of noise and altitude levels as areas of concern for community acceptance of urban air mobility systems. Infrastructure is another main challenge in UAM, Fadhil et al. [12] proposed a system to select ground infrastructure based on the GIS-based analysis.

In [13], the authors proposed an obstacle detection system using multiple sensors and cameras. The collisions are avoided by position controlling based on the position captured from the cameras. However, the system fails to cover 360° and also distance control is not handled. A distance-controlled-based collision avoidance is focused in [14] using quadrotors with a low computational burden. It also results in reducing the simultaneous localization and mapping, which is a time- and memory-consuming process. Marcin et al. [3] proposed a detection algorithm, which is capable of tracking an unspecified number of targets based on an RGB-D camera and a Bin-Occupancy filter measurement. In [15], laser scanners and other range finders are used to avoid obstacles based on the measurements obtained.

Other limitations in building robust obstacle detection and collision avoidance systems lie in capturing and processing the data. Deep learning-based architectures and solutions have been discussed in [16] for solving the data processing requirements for highly autonomous drones. The aircrafts in UAM are often constrained in terms of size, weight and power; hence the usage of lightweight sensors could help. Abdulla et al. [17] have also proposed a monocular camera-based system to detect the collision state of the approaching obstacle.

Also, a number of aircrafts will be increased as a result of UAM; hence it is necessary to have proper flight plans to deal with dynamic obstacles. In [18], a guidance algorithm using the Markov decision process (MDP) is proposed to compute flight plans for resource-constrained aircrafts. Geometric programming-based optimization tools are presented in [19] to ensure vehicle design and thereby optimize the cost rates compared to other mobility operations. The utilization of Edge nodes for such resource-constrained aircrafts could result in increased energy and power consumption and this paper aims at implementing an obstacle detection system in the fog computing layer. Having all the processing near the data generation helps in attaining the solutions in real time with reduced latency.

18.3 UAM Challenges and Overview

UAM includes aircraft operations for tasks in metropolitan areas that include public safety, medical evacuations and rescue, newsgathering, ground traffic assessment, weather monitoring and package delivery in addition to passenger-carrying flights [20]. The initial phase of UAM is expected to be driven by human pilots on board for flight operations and complying with current airspace rules and regulations. The future of UAM is targeted to be autonomous systems that may require to be piloted remotely with minimal or no human intervention.

As UAM is marching towards a broader picture for the aerospace industry, it also faces a lot of serious challenges in addition to technological challenges. This section briefs about the challenges faced by the UAM sector.

18.3.1 Air Traffic Control and Infrastructure

There are a lot of practical challenges towards creating a regulatory management system for the UAM vehicles: small drones susceptible to changing weather, need for a dynamic real-time mapping system for flight, connectivity for the tracing and interoperability of the drones. Developed cities also face issues with building the infrastructure (Vertiports) for the take-offs and landing due to limited space availability.

18.3.2 Cost and Accessibility

Flying costs are always expected to be higher than driving on roads and this applies to UAM vehicles used for passenger transport. Considering the infrastructure needs of UAM, cost and accessibility will be a major challenge.

18.3.3 Safety

Safety is always the topmost challenge whether it comes to delivering goods or transporting people. Consideration of safety is the biggest obstacle that delays the development and deployment of UAM.

18.4 IoT and Fog Computing

The Internet of Things (IoT) describes the network of connected physical objects over the Internet. The objects could be embedded with sensors, software and other technologies for the purpose of connecting and exchanging data with other devices and systems over the Internet. The devices used to generate the source of data are the Edge nodes and the data exchanged from multiple Edge nodes are processed at the Cloud. Cloud computing can be used to embed intelligence in the processing or perform high computation tasks which are not possible at the data collection nodes (Edge). Fog computing is a newly introduced paradigm, which extends the standard Cloud computing to the Edge. Therefore, it is also called edge computing. It is a Micro

Fig. 18.1 IoT architecture
with Fog layer

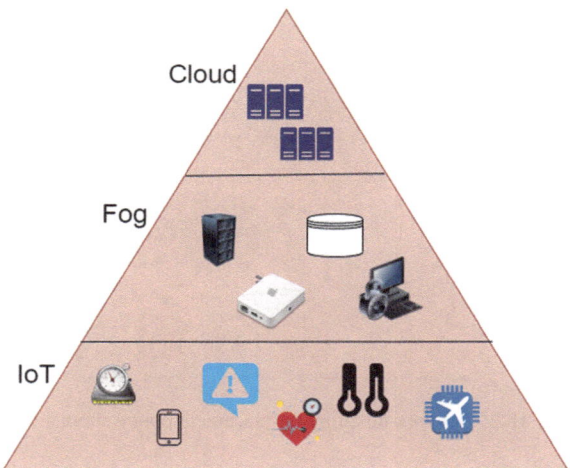

Datacenter (MDC) responsible for providing computation, storage and networking services between the end nodes in IoT and traditional clouds.

Figure 18.1 shows the three-layer IoT architectural model involving Fog as an intermediate processing node between the Edge and the Cloud. Extending the concept of IoT to UAM, the drones or UAM vehicles can be depicted as Edge nodes which are the source for air traffic data. The data could include the speed, altitude and position of the vehicle, direction in which it is moving, weather at the current location, terrain information, and other Edge nodes in the same airspace zone.

To manage the air traffic for UAM, the UAM vehicles or the drones should be capable of interacting with each other and making decisions. Fog nodes can be installed at different ground locations and can be mapped to service at different airspace zones. Drones flying in the airspace will dynamically connect to the Fog node at the flying zone and exchange data with other vehicles. The dynamic nature of the UAV calls for a need for addressing mobility issues in connecting to the Fog devices. Fog nodes are to be designed to process the data received from the Edge UAM nodes and arrive at intelligence to identify obstacles in the flight path based on the weather and traffic data received from other nodes in the same airspace zone. This enables the UAM nodes to talk to each other through the Fog and achieve seamless flight avoiding collisions and accidents to ensure safety.

18.5 Proposed Architectural Framework

Unmanned aerial vehicles or UAM vehicles are equipped with a large number of sensors and actuators to perceive information from an external environment. This leads to the generation of a huge volume of data that needs to be processed at the Cloud or Fog devices for real-time interactions. Fog nodes are not like conventional

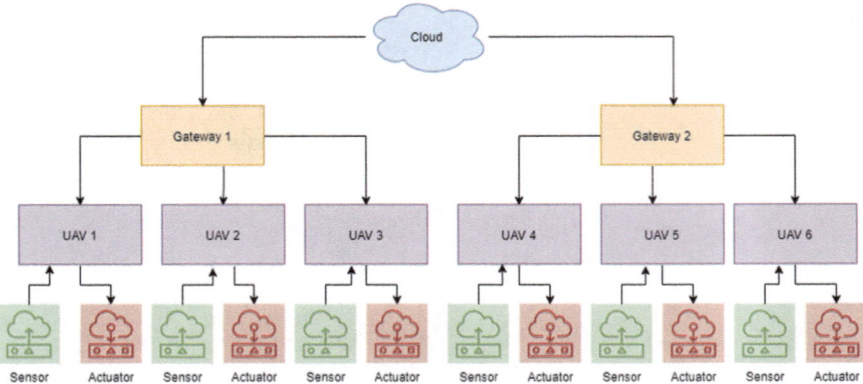

Fig. 18.2 Network topology for placement policy simulation

cloud datacentres and the development of large-scale applications in such resource-constrained nodes is difficult. Processing of the data becomes even more complex, considering the mobile nature of the UAM vehicles. With the proposed architecture, data from the UAVs are processed at the associated Fog nodes considering the QoS deadline and the resource availability at the Fog device. The framework also takes into accountability, the mobility of the Edge UAM nodes and applies the application management policy for processing and the results are simulated using iFogSim [21].

The proposed solution considers the IoT devices to be homogeneous at the same layer (Fog and Edge). Figure 18.2 shows the topology of the three-layer IoT architectural model used for simulation. The application model is assumed to receive data from the IoT sensor to the Client module running on the UAV Edge device and forwarded to the Storage module at the Cloud for storage. The Main module sits in between the Client module and the Storage module to process the received data. Figure 18.3 shows the application model of the computational processing occuring between the Edge and the Cloud.

For the proposed application management policy, the Main module is requiring a certain amount of resources to perform computation. To cater to the needs of the end devices within their QoS deadline, computation offloading happens at the Fog nodes to request additional resources for the processing. Figure 18.4 depicts the flowchart used in application placement at the Fog devices.

The proposed application management assumes that the UAV Edge devices are bound to a specific Fog node. In real-time scenarios, UAM vehicles move around the airspace at a faster rate and it is practically not possible to be bound to a specific processing node. Considering the mobility of the nodes into account, Edge nodes are dynamically associated with the nearest Fog nodes for real-time processing and faster responses. As the nodes keep changing position continuously, association to a Fog node happens based on the minimum distance criteria (Fig. 18.5).

At any point in time, the minimum distance between the Edge node and the Fog node is calculated using Eq. (18.1).

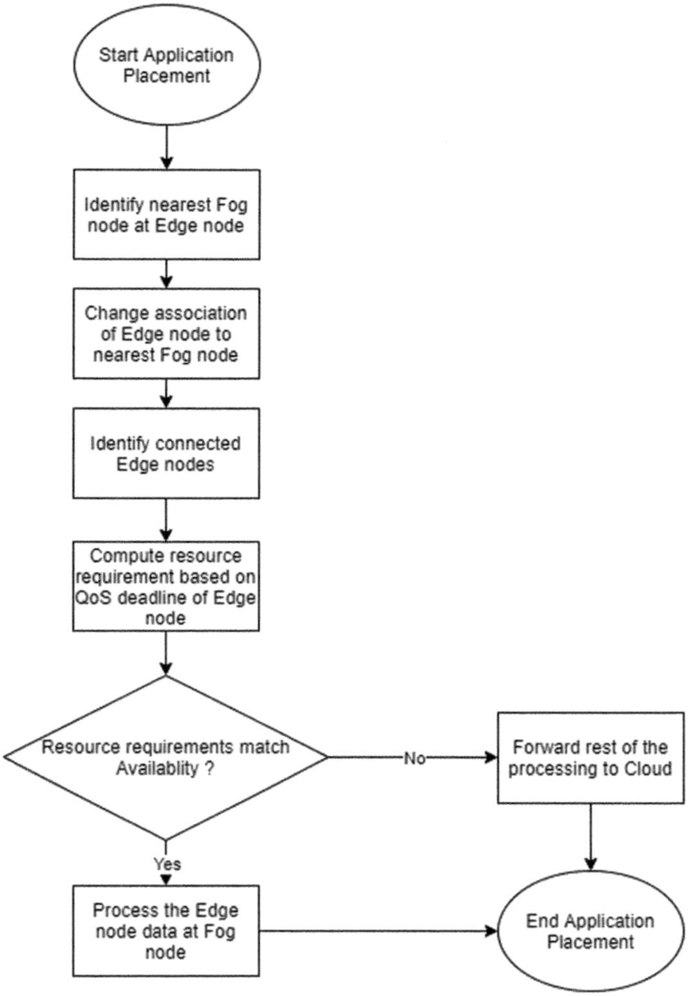

Start Application Placement
Identify nearest Fog node at Edge node
Change association of Edge node to nearest Fog node
Identify connected Edge nodes
Compute resource requirement based on QoS deadline of Edge node
Resource requirements match Availablity ?
No
Forward rest of the processing to Cloud
Yes
Process the Edge node data at Fog node
End Application Placement

Fig. 18.3 Flowchart of application placement policy

IoT Sensor
Sensor
Raw Data
Client Module
Main Module
Store Data
Storage Module
Result Data
Actuator
Response

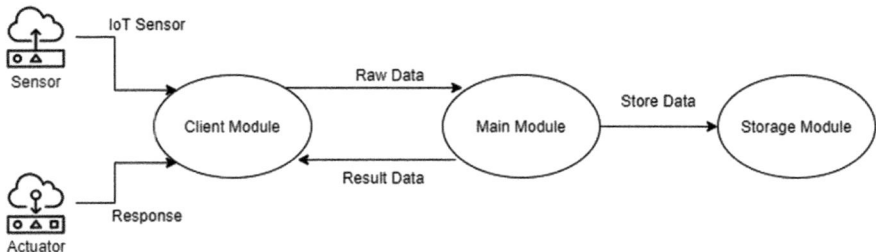

Fig. 18.4 Application model used for placement policy

Fig. 18.5 Energy
consumption without
mobility

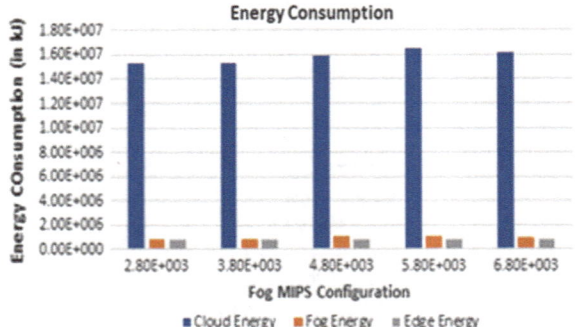

$$Dist_{min} = \sqrt{(Fog_X - Edge_X)^2 + (Fog_Y - Edge_Y)^2}. \qquad (18.1)$$

where (FogX, FogY) are the position coordinates of the Fog node and (EdgeX, EdgeY) are the position coordinates of the UAV node. The mapping of Fog and Edge nodes are redone and the application placement happens with the new topology created.

18.6 Results and Discussions

The proposed algorithm for addressing mobility and application placement is simulated using iFogSim. UAM vehicles are modelled as Edge nodes in the iFogSim simulation environment and a number of Fog nodes are predefined in the simulation. Fog nodes are designed with different processing capabilities to show the application placement based on resource availability. The processing capabilities used for the simulation are 2800 MIPS, 3800 MIPS, 4800 MIPS, 5800 MIPS and 6800 MIPS for the Fog nodes. The Main module for processing the UAV data is designed to be running with a constant resource requirement.

Edge nodes are designed to receive and send data from a single IoT sensor. Edge node is also designed to contain an IoT actuator to function according to the response received from the processing. UAV nodes can be extended to send data from multiple sensors such as speed, position, altitude, direction and other parameters that aid in the flight path to the destination. Table 18.1 describes the association of UAV to Fog nodes before and after addressing mobility. The association of UAV to Fog node is computed and updated based on Eq. (18.1). The simulation results are analysed based on the metrics like network bandwidth usage, energy consumption of the devices and latency in the data processing.

Figure 18.6 shows the performance of the simulation based on energy consumption of the devices at each layer of IoT architecture with a static association of the Edge devices. Figure 18.7 shows the performance results based on energy consumption with addressing mobility in the UAV nodes. It is evident from the results that the

Table 18.1 Mobility changes in UAV

Fog MIPS	Config	Edge devices					
		e-0–0	e-0–1	e-0–2	e-1–0	e-1–1	e-1–2
2800	MP	0	0	0	1	1	1
	M + MP	1	1	0	1	1	1
3800	MP	0	0	0	1	1	1
	M + MP	1	1	1	1	1	0
4800	MP	0	0	0	1	1	1
	M + MP	1	0	0	0	0	0
5800	MP	0	0	0	1	1	1
	M + MP	1	1	1	1	0	0
6800	MP	0	0	0	1	1	1
	M + MP	0	1	0	1	1	0

MP—Module Placement
M + MP—Module Placement with Mobility

Fig. 18.6 Energy
consumption with mobility

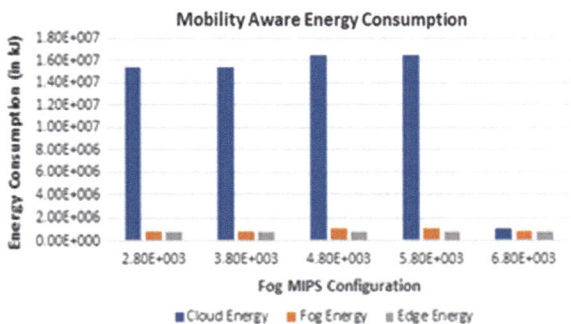

energy consumption at the Fog devices are minimal as compared to Cloud processing and is not affected by the mobility of UAV nodes. Fig. 18.8 shows the performance of the simulation based on latency. The results show that a tradeoff happens in increasing the computational power of the fog nodes against the latency involved in the communication of devices. Fig. 18.9. Energy consumption without mobilityFig. 18.6Energy consumption with mobility shows the performance results based on network bandwidth usage. Network bandwidth usage is observed to remain the same when the edge nodes are static or mobile.

Figure 18.9 shows the effectiveness of the algorithm in addressing mobility with a linear decrease in the response time from the Main module. As evident from the linear decrease in response time, UAV nodes can detect obstacles in the flight path in near real time and avoid collisions and accidents. Response time is critical to this application, as this could lead to fatal accidents in the case of manned UAM vehicles used for passenger transport or material damage in case of unmanned vehicles.

Fig. 18.7 System latency
with and without mobility

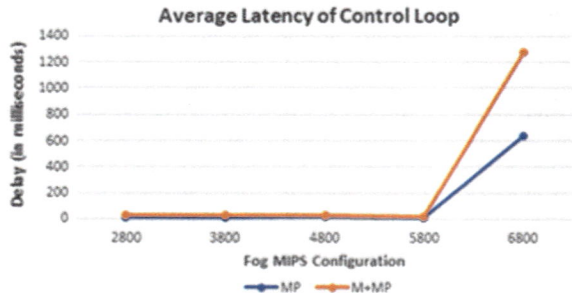

Fig. 18.8 Network
bandwidth usage with and
without mobility

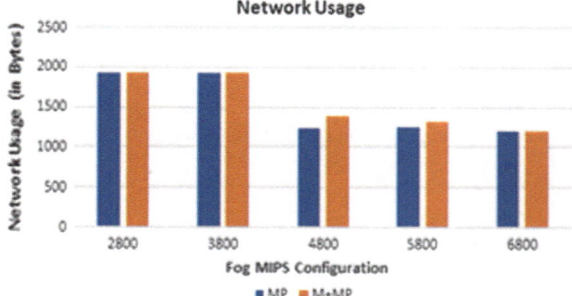

Fig. 18.9 Response time
from the Main module with
and without mobility

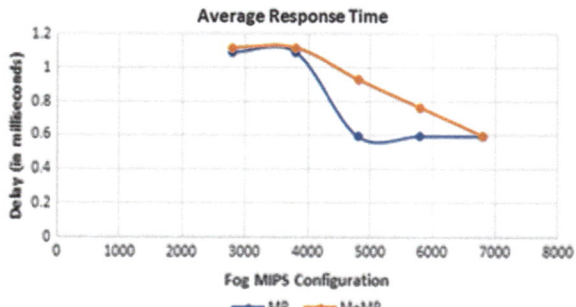

18.7 Conclusion

Fog computing offers several advantages over Cloud computing with respect to
latency and improved performance. However, Fog computing cannot completely
replace Cloud processing, which is known for its high-computational capabilities.
This paper proposes a methodology for integrating UAM into the IoT world to
overcome the challenges currently faced in enabling UAM. The paper also proposes
an intelligent framework for Fog computing to address mobility issues with UAM and
adapt based on the requirement of application's computational resources to provide

improved performance in terms of latency, usage of the network traffic and the energy consumption of the devices. This could also aid in the deployment of more unmanned vehicles and enable them to operate with minimal or no human intervention. The proposed approach can be expanded in the future to incorporate more intelligence based on the purpose for which the UAM vehicle is supposed to be used.

References

1. Straubinger, Anna, Raoul Rothfeld, Michael Shamiyeh, Kai-Daniel Büchter, Jochen Kaiser, and Kay Olaf Plötner. An overview of current research and developments in urban air mobility–Setting the scene for UAM introduction." Journal of Air Transport Management 87 (2020): 101852
2. Metzger, U., Parasuraman, R.: The role of the air traffic controller in future air traffic management: An empirical study of active control versus passive monitoring. Hum. Factors **43**(4), 519–528 (2001)
3. Odelga, Marcin, Paolo Stegagno, and Heinrich H. Bülthoff. "Obstacle detection, tracking and avoidance for a teleoperated UAV." In 2016 IEEE international conference on robotics and automation (ICRA), pp. 2984–2990. IEEE, 2016
4. Vaidian, Iulia, Muhammad Azmat, and Sebastian Kummer. "Impact of Internet of Things on Urban Mobility." (2019): 4–17
5. Faraci, G., Grasso, C., Schembra, G.: Fog in the Clouds: UAVs to Provide Edge Computing to IoT Devices. ACM Transactions on Internet Technology (TOIT) **20**(3), 1–26 (2020)
6. Al Haddad, Christelle, EmmanouilChaniotakis, Anna Straubinger, Kay Plötner, and Constantinos Antoniou. "Factors affecting the adoption and use of urban air mobility." Transportation research part A: policy and practice 132 (2020): 696–712
7. Rothfeld, Raoul, Milos Balac, and Constantinos Antoniou. "Modelling and evaluating urban air mobility–an early research approach." Transportation Research Procedia 41 (2019)
8. Sharma, Prashin, Cosme A. Ochoa, and Ella M. Atkins. "Sensor constrained flight envelope for urban air mobility." In AIAA Scitech 2019 Forum, p. 0949. 2019
9. Antcliff, Kevin R., Mark D. Moore, and Kenneth H. Goodrich. "Silicon valley as an early adopter for on-demand civil VTOL operations." In 16th AIAA Aviation Technology, Integration, and Operations Conference, p. 3466. 2016
10. Binder, Robert, Laurie A. Garrow, Brian German, Patricia Mokhtarian, Matthew Daskilewicz, and Thomas H. Douthat. "If You Fly It, Will Commuters Come? A Survey to Model Demand for eVTOL Urban Air Trips." In 2018 Aviation Technology, Integration, and Operations Conference, p. 2882. 2018
11. Vascik, Parker D., and R. John Hansman. "Scaling constraints for urban air mobility operations: air traffic control, ground infrastructure, and noise." In 2018 Aviation Technology, Integration, and Operations Conference, p. 3849. 2018
12. Fadhil, Dimas Numan. "A GIS-based analysis for selecting ground infrastructure locations for urban air mobility." inlangen]. Master's Thesis, Technical University of Munich (2018)
13. Becker, Marcelo, Rafael Coronel B. Sampaio, Samir Bouabdallah, Vincent de Perrot, and Roland Siegwart. "In-flight collision avoidance controller based only on OS4 embedded sensors." Journal of the Brazilian Society of Mechanical Sciences and Engineering 34, no. 3 (2012): 294–307
14. Gageik, N., Benz, P., Montenegro, S.: Obstacle detection and collision avoidance for a UAV with complementary low-cost sensors. IEEE Access **3**, 599–609 (2015)
15. Kise, M., Qiang Zhang, and N. Noguchi. "An obstacle identification algorithm for a laser range finder-based obstacle detector." Transactions of the ASAE 48, no. 3 (2005): 1269–1278

16. Fraga-Lamas, P., Ramos, L., Mondéjar-Guerra, V., Fernández-Caramés, T.M.: A review on IoT deep learning UAV systems for autonomous obstacle detection and collision avoidance. Remote Sensing **11**(18), 2144 (2019)
17. Al-Kaff, Abdulla, Fernando García, David Martín, Arturo De La Escalera, and José María Armingol. "Obstacle detection and avoidance system based on monocular camera and size expansion algorithm for UAVs." Sensors 17, no. 5 (2017): 1061
18. Bertram, Josh, Xuxi Yang, Marc W. Brittain, and Peng Wei. "Online flight planner with dynamic obstacles for urban air mobility." In AIAA Aviation 2019 Forum, p. 3625. 2019
19. Brown, A., Harris, W.L.: Vehicle Design and Optimization Model for Urban Air Mobility. J. Aircr. **57**(6), 1003–1013 (2020)
20. Thipphavong, David P., Rafael Apaza, Bryan Barmore, VernolBattiste, Barbara Burian, Quang Dao, Michael Feary et al. "Urban air mobility airspace integration concepts and considerations." In 2018 Aviation Technology, Integration, and Operations Conference, p. 3676. 2018
21. Mahmud, R., Rajkumar, B.: Modelling and simulation of fog and edge computing environments using iFogSim toolkit. Fog and edge computing: Principles and paradigms (2019): 1–35

Chapter 19
Identify Twitter Data from Humans or Bots Using Machine Learning Algorithms with Kendalls Correlation

R. Sangeethapriya and J. Akilandeswari

Abstract This study introduces bot detection techniques for identifying Twitter bot profiles and determining their susceptibility in today's online debate. Bots are common in social media. Bot accounts are problematic even though they can detect intrusions, disseminate misinformation, and disseminate unverified content, all of which can harm public opinion on a variety of issues, including product sales and election organizations. This paper proposes a framework for detecting bots using a machine learning method based on Kendall's tau-b (b) and a variety of data, including the number of followers, friends, listed counts, favorites count, and confirmed statuses counts, default profile, and default profile image.

19.1 Introduction

Twitter has been one of the fastest-growing social networking platforms. Users can share news, express their opinions, and discuss current events. Users can follow others who share their interests or viewpoints. People send tweets to their followers in real time. If the details are retweeted, they will reach a larger audience [18]. The number of tweets increases at random during special events such as sports or award shows. Users are bombarded with tweets, some of which are sent by bots. Bot detection is critical for identifying false users and safeguarding legitimate users against misinformation and criminal intent. A Twitter bot is the technology that automatically generates tweets for users. Bots are created with the intent of spamming. The dominance of human users as the primary drivers of Internet traffic is coming to an end (Phillip George Efthimion et.al 2016). Bots generated more Internet traffic than people. A bot is a piece of software that performs automated tasks online. Bots are a common occurrence on social media. According to some estimates, about 48 million Twitter

R. Sangeethapriya (✉) · J. Akilandeswari
Department of IT, Sona College of Technology (Autonomous), Salem, Tamil Nadu, India

J. Akilandeswari
e-mail: akilandeswari@sonatech.ac.in

© The Author(s), under exclusive license to Springer Nature Singapore Pte Ltd. 2022   203
V. Bhateja et al. (eds.), *Evolution in Computational Intelligence*,
Smart Innovation, Systems and Technologies 267,
https://doi.org/10.1007/978-981-16-6616-2_19

accounts are automated. There are a variety of bots, including false follower bots. Bots that imitate human behavior and aim to spread information while appearing as human users are more difficult to identify. Bots can be used for a wide range of tasks, with many of them providing services to users. Bots are classified as "good" or "bad" according to how openly they reveal their true identities. These social spambots can be used for a variety of purposes, but they can be difficult to detect even by humans [19]. 'Social trolls and bots' can serve a variety of purposes, but they can be difficult to detect even by human observers. Successful bots declare and mark themselves, whereas bots that are unsuccessful do not associate with the web services they access. Approximately 44% of Internet bot traffic is classified as good, while the remaining 56% is classified as bad [13]. The ability to detect bot accounts on social media sites like Twitter is critical for maintaining a healthy information exchange environment.

(1) Spreading rumors and fake news is the bots' harmful intent. (2) To bring others' reputations into disrepute. (3) Misleading communication to steal credentials. (4) The user is taken to a fake website. (5) To influence the popularity of a person or a group by changing their opinions. Datasets are gathered via Kaggle. The number of followers, friends, location, name of screen, verified, Favorite, URL, id, description, and listed count are all included. The coefficient of Kendall's tau-b (b) correlation is used to extract features. The coefficient of the Spearman correlation is used to derive attributes. The data collection system is being trained to detect bots. Decision Tree, Multinomial Nave Bayes, Random Forest, and logistic regression have all been used. In real time, the greatest precision algorithm is employed to evaluate the data.

19.2   Literature Review

Bots are automated software applications capable of carrying out a wide range of tasks and doing so regularly. Bots accounted for 50% of all web traffic in 2016 [11]. They can be used for input fetchers, product crawlers, monitoring, and web search bots. Eiman [7] suggested (Server Message Block) SMB detection methods only on Twitter and omitted any other malicious behavior. Authors gave a thorough description of and model and compared the 13 approaches proposed for detection, which followed their methodology, mostly used in 2016 and 2017.

Clayton A. [5] proposed Bot or Not, which used an RF technique to focus on over 1000 variables. When bots begin sending tweets through the social network, Bot's average user score drops precipitously. This is because Bot's capabilities provide the number of tweets and retweets from a particular address. Swedish researchers found bots twittering well about the Swedish electoral process using a (Random Forest) RF method. The most important element is multiplying the number of likes received [12] by the group of contacts the profile has. They discovered that the time between retweets is linked to many of the most relevant attributes. Oentaryo et al. [15] used four classifiers to significant variance accounts from three kinds of bots: webcast, phishing, and consumption: Logistic Regression (LR), (Naive Bayes) NB, (Random Forest) RF, and (Support Vector Machine) SVM. The major elements used to accurately distinguish the behavioral characteristics of different bot types are profile and follow characteristics, post, retweet, hashtag, and URL.

Erdem Beğenilmiş et al. [16] used supervised ML algorithms based on (Random Forest) RF, (Support Vector Machine) SVM, and (Logistic Regression) LR to detect organized behaviors. They distinguished three distinct groups using user- and temporal-based elements. Their methodology relied on social behavior features in hashtag-based comment collections obtained by searching for relevant hashtags.

Muhammad [15] detected irregular behaviors in OSNs. They found that malicious users engage in a variety of activities directed at different subjects. This is compared to normal users, who engage in range of events directed at a number of subjects. The Leave-One-Botnet-Out model would enable the training of supervised ML algorithms on statistics using several kinds [6] of social bots. This process is based on cross-validation, with such a subset of existing evidence retained and used for N number of folds testing. Naif Radi Aljohani et al. [13] proposed a deep learning model and a graph of convolutional networks to distinguish for both person and bot on Twitter accounts. The dataset contains 4540 annotated users and is deployed to build an independent language system.

19.3   Implementation

19.3.1   Data Collection Process

The entire dataset was gathered from Twitter, Kaggle, UCI, and Data.gov. The dataset contains various attributes such as the number of followers, friends, identified, likes, confirmed status, and the default profile. Data collection is among the most important aspects of any machine learning project. We feed data to the algorithms because it is the input. Therefore, the efficiency and accuracy of the algorithms depend on the correctness and consistency of the collected data. So, the output would be like the data itself.

19.3.2   Data Cleaning

Data pre-processing is the process of collecting data from a task and making it usable for other information. Data collected may be in an unorganized format and several null values, invalid data values, and unnecessary data will be available. The basic steps are to clean and substitute all the data with the estimated data. But our dataset should be full to construct a successful machine learning model. That is why we use some imputation techniques to substitute some probable values for the NaN values. In this work, Heatmap is used to display the missing association between every two columns. A value near -1 means that if one variable exists, then it is very likely

Table 19.1 A detailed systematic review conducted in the bot detection domain

Reference	Approach	Feature	Strength	Short comings
Phillip G. et al.	SVM, logistic regression	Id, profile picture, screen name, and friends followers	Merge the strength of conventional l spambots with social spambots	Bots can avoid detection by raising the number of human-like tweets they send
Mohammad Alauthman et al.	Reinforcement Learning and, Control Channel and J48, Naive Bayes, and Nearest-Neighbor	Retweeting number of followers	The technology is being used to improve the performance of its detection bot, according to the company	The method does not use a false positive rate, so it does not detect botnetsat the same time as a human detection rate.
Shamsul Haq et al.	Trees j48 classifier, Naïve Bayes classifier and Rule decision table Classifier	Network Traffic	The accuracy of bot detection was improved by clustering, grouping, and hybrid method	The cost of computation is high
M. Sarreddy et al.	Graph-based method	Network structure content and behavior feature	To quickly classify the bot, graph visualization is used	A Large network would be a cost of measurement
Abdulrahman Alarifia,	Entropy and tree-based classification	Profile name, screen name and description	Accuracy is high	Requires high computational costs and more time to classify the bot
Amous. Azaria et al. (2016)	Crowdsourcing	Account, Profile and content	Accuracy is high	High computational cost

that the other variable is missing. The decision tree, multinomial naive base, random forest, and logistic regression algorithms are used.

19.4 Algorithms

19.4.1 Decision Tree Algorithm

The Decision Tree algorithm is used to detect bots in social networks. The algorithm uses a series of algorithms to analyze the number of viewers, mates, statuses count, standard profile, default profile image, and other features. Algorithm 1 requires: detects bots depending on the total number of people watching.

Each subtree is built successively just on a training sample that would have been marked further off the tree route. If no features are left on the node, the final classification mark is decided to leave on that node. The entire training set will be considered the root and the information gain used to determine which attribute to classify with each node.

19.4.2 Multinomial Naive Bayes

Algorithm 2 Displays Application Characterizations for the Multinomial Classifier for Naive Bayes.

The probability of data d is P(d) if the hypothesis is true, and P(c) if it is not true. P(a) is the probability that inference d is correct, while P(b) is possible to predict whether it is true. The odds of being right in the case of a given hypothesis are P(e) P(f).

$$P(c|d) = (P(d|c) * P(c)/P(d) = (P(c/d)) \qquad (19.1)$$

19.4.3 Random Forest

Algorithm 3 Details the Implementation of the Random Forest Algorithm for Detecting Bots

In this interactive, we look into how a tree is built from "k" to "l" over time. We then predict the final outcome for each of these trees and measure votes for each one. The results are shown by adding up the number of nodes per second.

19.4.4 Logistic Regression

Algorithm 4 Shows the Specifics of the Implementation to Detect Bots Using the Algorithm Logistic Regression.

The logistic function is used to predict the relationship between different features in a graph. It can be used to measure map and predict the relationships between real and fake values. In this example, A is the number of points that have been added to the top of the graph.

19.5   Kendall's Correlation and Compare the Model of Machine Learning

Kendall's tau coefficient of determination is designed to represent the association between two ordinal variables (not necessarily interval). Its closest approximation (referred to as) could be as follows.

$$\tau = \frac{nc - nd}{nc + nd} = \frac{nc - nd}{n(n-2)/2} \tag{19.2}$$

where, *nc* number of concordant pair, A concordant pair is a pair of observations (s1, t2) and (s2,t2) that have the property s1 > t2 and s1 > t2 or s1 < t2 and s1 < t2.

nd number of discordant pairs. Discordant pairs: A pair of observations (s1, t2) and (s2, t2) that follows the property s1 > t2 and s1 < t2 or s1 < t2 and s1 > t2.

The number of pairs n, the tau scale has a range of 1.00 to + 1.00. The meaning is based on the sign and the significance. A higher value indicates a more positive relationship. A higher value indicates a strong relationship, while a lower correlation showed a negative relationship. Twitter datasets are checked against the Spearmen correlation coefficient. Kendall's correlation outperforms Spearman correlation in terms of performance and F1 score. The use of a heatmap in this application resulted in balanced data in Twitter datasets in this application. This problem arises not only in binary data but also with multiclass data. Kendall correlation is preferred over others because it has a lower gross error sensitivity (GES) and a lower asymptotic variance (AV) This section provides a high-level overview of the algorithmic aspects of system implementation. It uses four algorithms: Decision Tree, Multinomial Nave Bayes, Random Forest, and logistic regression.

19.6   Experimental Result and Discussion

The data are divided into 70% train data and 30% test data. The positive is the one that correctly defines the actual value, while the false positive occurs when true numbers are incorrectly defined as false values. To differentiate between classes, the best accuracy is plotted in ROC curves against the false positives.

Figure 19.1 Bots and not bot friend's versus followers counts are displayed and Fig. 19.2 Number of bots and not bots counts are displayed.

In Fig. 19.3, the correctness of the training samples using the Decision Tree algorithm is 95.85%, while the accuracy of the test data is 93.95% and in Fig. 19.4, the correctness of the training samples using the multinomial Naive Bayes algorithm is 95.85%, while the accuracy of the testing data is 93.95%.

As shown in Fig. 19.5, the reliability of the training samples using the Random Forest algorithm is 94.93% and 93.63% for testing dataset, and Figure 19.6 depicts the train and test data of logistic regression ROC.

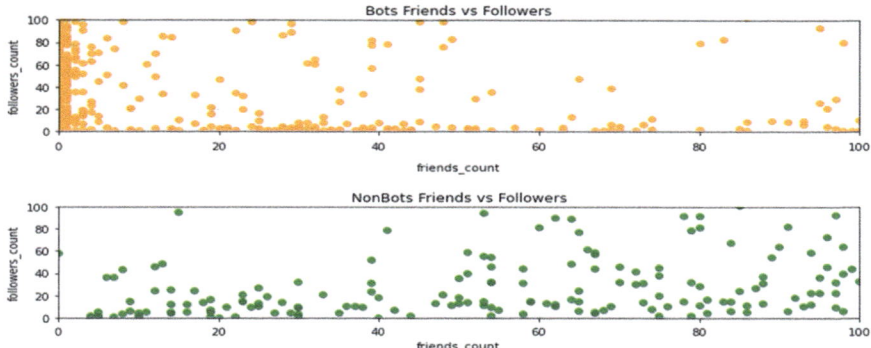

Fig. 19.1   Bots and not bots friends versus follower bots

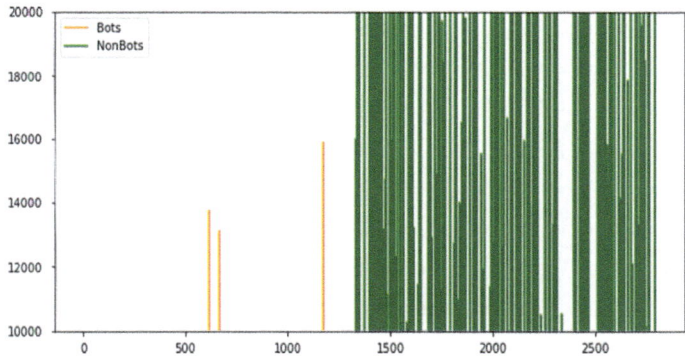

Fig. 19.2   Bots and not

Fig. 19.3   Decision tree
ROC curve using Kendall's

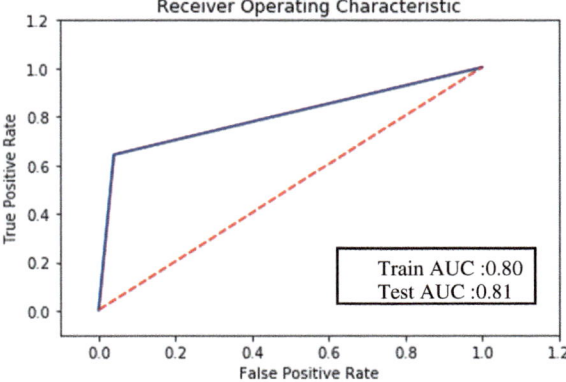

Fig. 19.4 Multinomial
Naive Bayes ROC

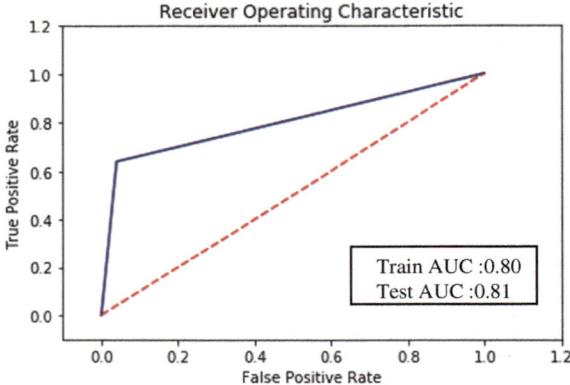

Fig. 19.5 Random forest
ROC curve

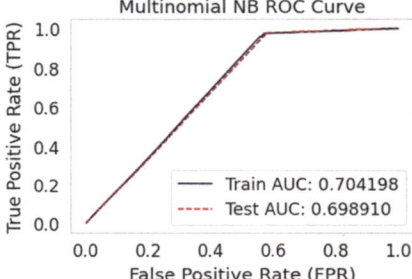

Fig. 19.6 Logistic
regression

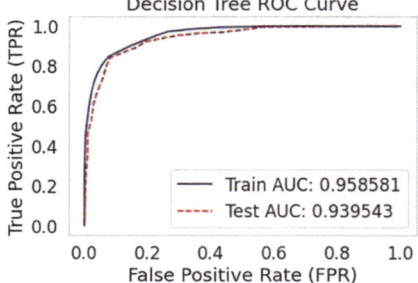

19.7 Conclusion, Future Scope, and Limitations

Researchers propose an algorithm for detecting bots on Twitter. The decision tree algorithm has 98.01% accuracy and 93.9% data for test data. It can be improved even further by incorporating more variables about user interaction regularities and tweet quality, they say. In the future, one can test other deep learning classifiers or a broader range of datasets. The current structure performs admirably on the

Table 19.2 Performance comparison of all algorithms implemented using Spearman correlation and Kendall's correlation

S. No	Algorithm name	Accuracy (%) using Spearman Correlation	Accuracy (%) using Kendall's Correlation
1	Decision Tree	Train data Accuracy: 95.66% Test data Accuracy: 93 0.71%	Train data Accuracy: 9 5.85% Test data Accuracy: 93.9%
2	Multinomial Naïve Bayes	Train data Accuracy: 5 5.6% Test data Accuracy: 55.5%	Train data Accuracy: 7 0.41% Test data Accuracy: 69.89%
3	Random Forest	Train data Accuracy: 91.86% Test data Accuracy 90.02%	Train data Accuracy: 94.93% Test data Accuracy 93.63%
4	Logistic Regression	Train data Accuracy: 79% Test data Accuracy 80.1%	Train data Accuracy: 80% Test data Accuracy: 81.1%

dataset. However, further experiments in collaboration with behavioral experts and to confirm the similarity among emotional expressions observed in the document as well as potential insider tasks are required to validate the application's viability in the actual world. Our actual implementation has the drawback of not being suitable for multi-level categorical issues because it uses a specific feature and the ability to interact between the attributes is not possible. When a huge set of features are used, the model has become time-consuming.

References

1. Abdulrahman, A., Mansour, A., AbdulMalik Al-Salmanb: Twitter turing test: Identifying social machines, pp. 332–346 (2016)
2. Amous Azaria,Skylur,Vadim Kagan,Aram Galstyan Kristina: The Darpa Twitter Bot Challenge, Vol 49, pp 38–46 (2016)
3. Ashish Mehrotra, Mallidi Sarreddy and Sanjay Singh (2016) Detection of Fake Twitter Followers using Graph Centrality Measures in Contemporary Computing and Informatics (IC3I),2nd International Conference on. IEEE, pp. 499–504.
5. Clayton A. Davis, Onur Varol, Emilio Ferrara ,Alessandro Flammini, Filippo Menczer(2016) Bot Or Not: A System to Evaluate Social Bots, Proceedings of **the 25th** international conference companion on world wide web. International World Wide Web Conferences Steering Committee pp 273–274. https://doi.org/10.1145/2872518.2889302.
6. Christoph Besel, Juan Echeverria, Shi Zhou (2018) Full Cycle Analysis of a Large-scale Botnet Attack on Twitter In IEEE/ACM international conference on advances in social networks analysis and mining (ASONAM) IEEE, pp 170–177.

7. Eiman Alothali , Nazar Zaki , Elfadil A. Mohamed , and Hany Alashwal (2018) Detecting Social Bots on Twitter: A Literature Review, International conference on innovations in information technology (IIT), IEEE, pp-175–180
8. Erdem Beğenilmiş, Suzan Uskudarli (2018) Organized Behavior Classification of Tweet Sets using Supervised Learning Methods, In Proceedings of the 8th international conference on web intelligence, mining, and semantics. Association for Computing Ma- chinery. WIMS' 1https:// doi.org/10.1145/3227609.3227665
10. Johan Fernquist and Lisa Kaati, Ralph Schroeder (2020), Political Bots and the Swedish General Election. https://doi.org/10.1109/ISI.2018.8587347
11. Juan Echeverr, Emiliano De Cristofaro, Nicolas Kourtellis , Ilias Leontiadis , Gianluca Stringhini , and Shi Zhou (2018) LOBO – Evaluation of Generalization Deficiencies in Twitter Bot Classifiers. In Proceedings of the 34th annual computer security applications conference pp. 137–146.
12. Majd Latah, Levent Toker (2019) Artificial intelligence enabled software-defined networking: A comprehensive Overview, Vol. 8 Issue no. 2, pp.79–99.
13. Mariam Orabi, Djedjiga Mouheb, Zaher Al Aghbari, Ibrahim Kamel (2020) Detection of Bots in social media: A Systematic Review, https://doi.org/10.1016/j.ipm.2020.102250
14. Mohammad Alauthman, Nauman Aslam, Mouhammd Alkasassbeh, Suleman Khan, Ahmad AL-qerem, Kim-Kwang Raymond Choo (2019) An Efficient Reinforcement Learning- Based Botnet Detection approach, https://doi.org/10.1016/j.jnca.2019.102479
15. Muhammad Al-Qurishi, Student Member, IEEE, M. Shamim Hossain (2018), Leveraging Analysis of User Behavior to Identify Malicious Activities in Large-Scale Social Networks, Vol. 14, no. 2, IEEE Transactions on Industrial Informatics, pp 799–813, https://doi.org/ https://doi.org/10.1109/TII.2017.2753202
16. Naif Radi Aljohani, Ayman Fayoumi,Saeed-Ul Hassan (2020) Bot prediction on social networks of twitter in altmetrics using deep graph convolutional networks
17. Phillip George Efthimion, Scott Payne, Nicholas Proferes (2018), Supervised machine learning bot detection techniques to identify social twitter bots, vol. 1, No. 2, Art. 5, pp1–70, https:// scholar.smu.edu/datasciencereview/vol1/iss2/5
18. Battur, R., Yaligar, N.: Twitter bot detection using machine learning algorithms. Vol. 8, Issue 7, pp. 304–307 (2018). https://doi.org/10.17148/IJARCCE.2021.10417
19. Oentaryo, R.J., Murdopo, A., Prasetyo, P.K., Lim, E.P.: On Profiling Bots in social media. In: International conference on social informatics, pp 92–109 (2016), https://doi.org/10.1007/978-3-319-47880-7_6
20. Shamsul, H., Yashwant, S.: Botnet detection using machine learning. In: 5th IEEE International Conference on Parallel, Distributed and Grid Computing, pp.240–245 (2019)
21. Hoang, X.D., Nguyen, Q.C.: Botnet Detection Based on Machine Learning Techniques Using DNS Query Data, pp 1–11 (2018). https://doi.org/10.3390/fi10050043.

Chapter 20
Neologism Related to COVID-19 Pandemic: A Corpus-Based Study for the Bengali Language

Apurbalal Senapati and Amitava Nag

Abstract Language is changing over time, and it is a common phenomenon for all languages. Generally, it is a slow but continuous process. The new words come or adopt in the languages, and some existing words become less frequent use or becoming obsolete in written or verbal communication. The term neologism is implying a newly coined word or expression or a phrase that is entering for common use. But sometimes or some special events like War, New disease, Computer, Internet, etc. make the change rapidly, and the COVID-19 pandemic is one such latest event. Note that the infectious disease caused by a newly discovered coronavirus is termed COVID-19, and it is now the official name of coronavirus disease. It has led to an explosion of neologism in the context of disease and several other social contexts. During this period, many new words were coined in the languages and many of these terminologies are rapidly becoming a part of our daily life. For example, some established terms like "lockdown", "quarantine", "isolation", "pandemic", etc. increased quickly the use in our daily terminology. From the linguistic point of view, the study of such change or adaptation and its quantization is very much important. This study attempted a corpus-based computational approach to explore the adaptation or creation of new words during the outbreak of COVID-19 in the Bengali language. The main components of this work are the creation of the corpus related to the COVID-19 and an algorithm to find out the neologism. For this study, a news corpus has been used. The corpus is created from the news article related to the COVID-19 from January 2020 to February 2021.

A. Senapati (✉) · A. Nag
Department of Computer Science and Engineering, Central Institute of Technology
Kokrajhar, Kokrajhar 783370, Assam, India
e-mail: a.senapati@cit.ac.in

A. Nag
e-mail: amitava.nag@cit.ac.in

20.1   Introduction

The COVID-19 pandemic has affected the people's health system, socio-economic, cultural activity across the globe [1]. It also affected the linguistic change specially coined and adoption of new phrases almost in all the languages. History shows that during epidemic and pandemic, there is a significant contribution in neologism [2]. During the pandemic period, the main focusing research area was related to the COVID-19 in various aspects among the people of the entire world. Every day people come across new terminologies like COVID-19, social distancing, lockdown, quarantine, isolation, etc. and that affects neologism in languages rapidly. The language is not static but dynamic, i.e. the lexicons of a language are changing continuously, and generally it is a slow process. But the COVID-19 situation flips this scenario and rapidly neologism influences the lexicons of languages. The term neologism itself was adopted from French in the early nineteenth century [3], however, the root "neos" comes from Latin, the meaning is *new* and in Greek logos means *word*. According to Jansen et al. [4], the term neologism implies toward linguistic category that signifies the coined of new terms and their modification. Whereas, according to Stenetorp [5], "a neologism is a lexeme that is not described in dictionaries". Neologism is an important aspect in the linguistic point of view and language aspects. Though neologism is a common phenomenon in the language, it is a slow process. But sometimes some events influence it rapidly in the languages. Like the incident of the Word War, SARS (Severe Acute Respiratory Syndrome), AIDS (Acquired immunodeficiency syndrome), Computer, Internet, etc. coined so many new terms and vocabulary in the language rapidly.

From history, it shows that, with the spreading of the diseases, vocabulary and new words also emerged. It has also happened in 1738, yellow fever appeared and Spanish influenza spread it was in 1890. In 1878 Poliomyelitis spread, and Polio spread in 1911. In recent decades, the SARS (in the year 2003), AIDA (in the year 1982), etc. are spread. Since early 2020, the COVID-19 pandemic has caused an outburst in neology, Oxford English Dictionary presented a newly coined word till April 2020 [6]. During the pandemic, numerous works can be found related to the COVID-19. Most of the works are related to mathematical modeling, COVID-19 prediction, clinical investigation, socio-economic situation, and related to neologism. It is found that the COVID-19 pandemic is one such event that coined many vocabularies rapidly in all languages. Many researchers of different disciplines have already done several investigations of COVID-19 on different aspects [7–13] and many of them tried to verify the modern classical model [14] and theories [15] in the present context.

20.2 Related Work

During the pandemic outbreak, there are several works conducted on neologism and studies its influence on socio-cultural activity. Still, people are working on that from a different perspective. In this survey, some earlier studies on neologism related to their various types of etymological and next to the recent study related to the COVID-19 are outlined here.

Krishnamurthy [16] tried to explain the word-formation through a model as the neologism model. He described the model into three components. The first component of the model describes the word-formation mechanism. It is based on morphological properties like compounding, blending, affixation, coinage, and acronym. The second component describes the borrowing of words from other languages. The end-users used the terms from other languages because of the unavailability of the equivalent terms in their targeted languages. Gradually, these borrowing words are adopted in the targeted language vocabulary. The lexical derivation is described in the third component. It is the derivation process of neologism by applying the existing word formation rules.

John studied the neologism over time in a chronological fashion [17]. He tried to identify the etymological sources for a new word are creating, borrowing, combining, shortening, blending, shifting, and from an unknown source. He described creating means the creation of a new word from nothing or scratch or at least not from the existing words. From the theoretical point of view, it is possible but in the reality, there must be some context. Borrowing is the adaptation of a new word from another language. Combining is the process of creating new words through a combination of existing words or word parts by the prefixes, suffixes, or compounded. Shortening is the process of creating new words by omitting some part of an existing word. Blending is the process of creating new words by combining and shortening simultaneously. Shifting is the derivation process of creating a new lexeme from the existing one and does not follow a regular morphological process. Source unknown, sometimes the origins are unknown in the whole or partial.

Alyeksyeyeva et al. [7, 9] describe the cultural changes and study the language–culture correlation because of the neologisms, where they described the process of neologisms and investigated the newly coined terms that are found and coined a new term as coronaculture. They also elaborated the different theories of neologism mainly three ways of neologism, namely, stylistic, etymological, and denotational.

Chen et al. [8] have done a corpus-based analysis to find the insight from the corpus. They use an automated approach for the analysis and use statistical parameters. The results of the frequency-based analysis of the token give valuable information on neologism.

Akut has done [10] the morphological investigation of the neologisms during the COVID-19 outbreak. The results show that the newly coined terms are the content words, namely, nouns and verbs.

Khalfan et al. [11] had done this study from a linguistic point of view. The study tried to explore the neologisms related to COVID-19 and shoe language–mind relationship, in the discipline of linguistic relativity.

Katermina et al. [12] studied the reflection of COVID-19 in different social aspects and especially in the neologism in the medical domain too. They have analyzed a hundred neologisms related to the coronavirus pandemic including the name of the disease and social phenomenon.

Estabraq et al. [13] tried to focus neologism in social media and especially in teeter data. For that purpose, they have considered a volume of three million tweets collected from the interval January 2020 to May 2020. They have also listed a set of COVID-19-related expressions and their usage along with the expressions through lexical deviation.

Thorne [18], the British linguist and lexicographer, studied the impact of the outbreak on the English language and concluded that some of the languages are newly created, some of them are re-purposed old language. He tried to study the formation and adoption of the new terms in linguistic aspects with the different phases of the pandemic.

20.3   Corpus Creation

This neologism study is based on descriptive statistics in terms of frequency analysis and word ranking in a text corpus. To study the coronavirus-related neologism, the text corpus must be related to the coronavirus or at least it should contain all the relevant terms related to the COVID-19. But so far no such dedicated COVID-19 corpus is available rather enormous articles related to the COVID-19 are available in a scattered way. In that context, it motivated us to develop a corpus dedicated to the COVID-19. Since our target is to study the neologism of the Bengali language during the coronavirus pandemic, hence we need a Bengali corpus related to the coronavirus. This section describes the corpus creation-related issue.

A corpus can be defined as a collection of machine-readable authentic texts (including transcripts of spoken data) that are sampled to be representative of a particular natural language or language variety [19]. Here, we are only dealing with the text corpora. The texts should be selected as much as diverse varieties of texts with proper ratios. The corpora are one of the important resources for language processing. The development of a written text corpus involves various factors like size of the corpus, representativeness of texts, determination of target users, quality of the text, selection of time-span, selection of documents source, etc. The formatting is also an important issue for the corpus for maintenance and further development purpose [20, 21].

In the study of neologism, we need a COVID-19 related corpus that should contain all the information related to the COVID-19. Since our study is only dedicated to identifying the neologism for the Bengali language hence we set the specific criteria on the COVID-19 corpus that it should contain all the terms related to the coronavirus pandemic. Since the beginning of the year 2020, the topic COVID-19 was the most cultivated topic and most of the newspaper covered the issues related to COVID-19 with the highest priorities and continuing till today. Hence, the collection of news articles related to the COVID-19 from the beginning till date may be considered as COVID-19 corpus.

For this purpose, we have collected all the news articles related to COVID-19 from a leading Bengali newspaper Anandabazar Patrika [22]. This corpus contains all the news articles related to COVID-19 from 23 January 2020 to 22 February 2021. Since the first article related to corona, we have found it on 23 January 2020, hence we have considered the news article from this date. This is done by crawling all the webpage on the time intervals 23rd January 2020–22th February 2021 to and after filter all the COVID-19 related articles. We termed this corpus the "COVID-19 corpus", and there are approximately 12,000 news articles.

The news corpus creation of the web is a critical job. There are various tools and techniques to download the text from the web but the raw web text contains other irrelevant. Before processing that text, it must need to clean the unwanted garbage and must be organized in a specific format, otherwise, it may lead to a wrong result. In the corpus creation, we have used two steps sequentially, the download of the raw text from the wen and the cleaning. Figure 20.1 gives the block diagram of corpus creation from the web source of Anandabazar Patrika [22].

We have used the Python library to download the text from the given Uniform Resource Locator (URL). Since we are only interested in the COVID-19 corpus, hence filtering is done in the URL level. In our study, we have seen that most of the URL contains some key terms related to the text. For example, the news article of Anandabazar Patrika, dated 17 June, 2020 (in Fig. 20.2), and its URL contains the key term *corona*.

For the URL filtering, we have considered a pre-defined list of corona related keywords {"covid", "corona", "mask", "quarantine", "sanitize", "sanitizer", "sanitizing", "sanitization", "pandemic", "virus", …."}. But, we have seen that there is some news article related to corona but its URL does not contain any corona-related keywords, which are listed in the keyword list (in Fig. 20.3). In that case, search the text content and if found any corona-related key term then also considered text related to the corona.

Fig. 20.1 Block diagram of the corpus creation process

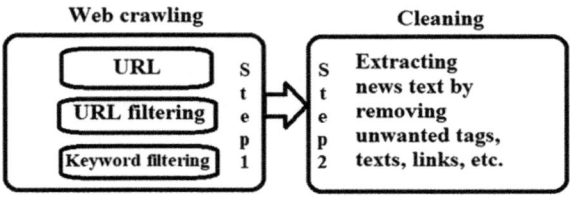

Fig. 20.2 URL contains key term *corona* (https://www.anandabazar.com/west-bengal/24-parganas/the-number-of-corona-positives-is-increasing-without-any-awareness-1.1163785)

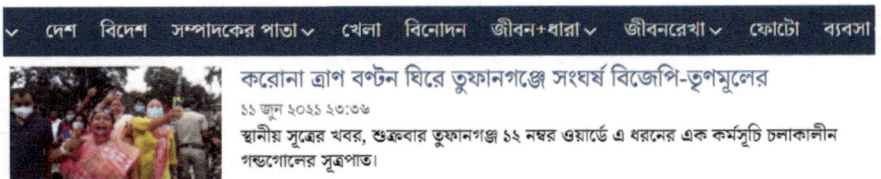

Fig. 20.3 URL does not contain key term-related *corona* (https://www.anandabazar.com/west-bengal/north-bengal/tmc-bjp-clash-in-cooch-behar-over-relief-material-distribution-dgtld/cid/1286406)

20.4 Methodology to Retrieve COVID-19 RelatedTerms

Since the target is to find out all the newly coined terms related to the COVID-19 or the words popularized during the pandemic period. For this purpose, we have considered the set-different strategies. For this operation first, the COVID-19-related news articles have been separated. This is done by the keyword matching strategies. A set of COVID-19-related keywords have been collected manually, and all the articles considered the COVID-19-related article if it contains any one of these keywords. For computational efficiency, first, it checks the news headline, and it headline contains any keywords then it considered as COVID-19-related article otherwise goes to test the body of the article. This process partitions the news articles into two categories, the COVID-19 related and other categories.

In the experiment, consider *set A* that contains all the tokens (words) from the COVID-19-related article, and consider *set B* that contains all the terms from the other articles (tokens or words from a non-COVID-19 related news article). Note that, the set has an inherited property that it contains all unique terms, i.e. not duplicated. Now in the ideal case, the result is *set difference operation (A - B)* contains all the terms related to the COVID-19 (shown in Fig. 20.4). But, the result set contains all the terms related to the COVID-19 and their morphological forms, some name entities, some invalid tokens, etc. The interesting fact is that the *result set (A - B)* is a relatively very small set where almost all the stop words are removed, most of the named entities, action words, etc. are removed. Finally, all the COVID-19-related terms are extracted by manual checking. Next, the frequency of each term is calculated in the COVID-19-related article to find how frequently the terms are using.

20.5 Result

Next, the frequency of the COVID-19-related terms is calculated from the COVID-19 corpus. In that corpus, there are 1,51,673 different tokens (including the inflectional forms) and after removing the stop and action words rank them with their frequency. The rank list shows that the most frequent word is করোনা (corona) and its inflections. We have got 105 such newly adopted terms in Bengali considered as the neologism for the Bengali language because of the pandemic. Some of the newly coined most frequent COVID-19 terms are listed in Table 20.1.

20.6 Conclusion

The proposed approach of the corpus-based study presents significant optimizing results for corpus analysis in processing the significant volume of COVID-19-related text. The notable contributions of this work are summarized as: (1) proposed set difference method for identifying neologism in computationally efficient, (2) experimentally it establishing the fact that like the previous pandemics COVID-19 also added to new words and vocabulary, (3) the frequency study (Table 20.1 shows the rank list or the high-frequency terms) gives quantitative information of the, (4) created a COVID-19 related corpus, which will help the further research.

The experiments show that the Bengali language also adopted several newly coined terms. Also, it is noted that most of the newly adopted terms are also

Fig. 20.4 Set difference to retrieve the COVID-19-related terms

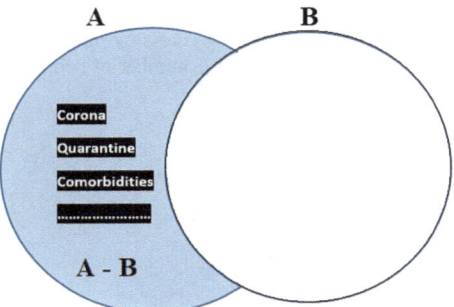

Table 20.1 Frequency-wise rank list of Bengali COVID-19-related newly added terms

Sl No	Bengali terms	Meaning
1	করোনা, কোভিড (and its inflections)	COVID-19 stands for novel coronavirus disease 2019, which refers to the year of its initial detection. COVID-19 is an illness related to the current pandemic; the illness is caused by the virus SARS-CoV-2
2	লকডাউন (and its inflections)	An emergency measure in which people are restricted from certain areas in an attempt to control exposure or transmission of disease
3	কোয়রান্টিন (and its inflections)	The act of refraining from any contact with other individuals for some time prescribed by the medical/government agency
4	কন্টেনমেন্ট জোন (and its inflections)	A geographical area designated by medical/administrative people with limited access in or out to contain an outbreak
5	আইসোলেশন (and its inflections)	The act of separating oneself from others
6	কো-মর্বিডিটি (and its inflections)	It implies that more than one illness or disease occurring in one person at the same time
7	স্যানিটাইজেশন (and its inflections)	It is the scientific process of making something completely clean and free from bacteria
8	গৃহ-নিভৃতবাস (and its inflections)	The act of separating oneself from others
9	অনলাইন (and its inflections)	It is a virtual environment that is controlled and connected through the network with electronics devices like computers, smartphones, etc
10	ওয়েবিনার (and its inflections)	It is an online event that is hosted by an organization/company and broadcast to a select group of individuals through their computers via the Internet
11	সুপার স্প্রেডার (and its inflections)	A person who transmits an infectious disease or agent to an unexpected or unusually large number of other people
12	অতিমারির (and its inflections)	an epidemic occurring worldwide, or over a very wide area, crossing international boundaries and usually affecting a large number of people

adopted in Bengali languages in the original form of English but not in translation form.

Declaration *We have* taken permission from competent authorities to use the images/data as given in the paper. In case of *any dispute* in the future, we shall be wholly responsible.

References

1. Still, S. COVID-19 Health System Response, Quarterly of the European Observatory on Health Systems and Policies, Volume 26, Number 2, 2020, ISSN 1356 – 1030.
2. Muhammad, A., Zhiyong, D., Iram, A. and Nisar, M. Linguistic Analysis of Neologism Related to Coronavirus (COVID-19). Available at SSRN: https://ssrn.com/abstract=3608585 or http://dx.doi.org/https://doi.org/10.2139/ssrn.3608585
3. Jesenská, P. Analysis of Neological Anglicisms used in Slovak from Aspects of Orthography and Frequency in the Slovak National Corpus, European Journal of Social and Human Sciences, 2016, Vol. (12), Is. 4.
4. Janssen, M. `Between Inflection and Derivation: ParadigmaticLexical Functions in Morphological Databases". Em: East WestEncounter: *second international conference on Meaning & Text Theory.*Moscovo: Rússia (2005).
5. Stenetorp, P. Automated Extraction of Swedish Neologisms using a Temporally Annotated Corpus. TRITA-CSC-E 2010:017 ISRN-KTH/CSC/E–10/017—SE, (2010).
6. McPherson, F., Stewart, P. and Wild, K. Presentation on "The language of Covid-19: special OED update", OED, 2020, https://public.oed.com/wp-content/uploads/The-Language-of-Covid-19-webinar_10-09-20_presentations.pdf, Accessed on 12–06–2021.
7. Alyeksyeyeva1, I. O., Chaiuk, T. A., and Galitska, E., A. Coronaspeak as Key to Coronaculture: Studying New Cultural Practices Through Neologisms. International Journal of English Linguistics; Vol. 10, No. 6; ISSN 1923–869X E-ISSN 1923–8703, (2020).
8. Chen, L.C., Chang, K.H., Chung, H.Y.: A Novel Statistic-Based Corpus Machine Processing Approach to Refine a Big Textual Data: An ESP Case of COVID-19 News Reports. Appl. Sci. **10**, 5505 (2020). https://doi.org/10.3390/app10165505
9. Alyeksyeyeva, I.O., Chaiuk, T.A., Galitska, E.A.: Coronaspeak as Key to Coronaculture: Studying New Cultural Practices Through Neologisms. International Journal of English Linguistics **10**(6), 202–212 (2020). https://doi.org/10.5539/ijel.v10n6p202
10. Katherine B. A. Morphological Analysis of the Neologisms during the COVID-19 Pandemic, International Journal of English Language Studies (IJELS) (Aug. 2020), ISSN: 2707–7578, doi:https://doi.org/10.32996/ijels.2020.2.3.11
11. Khalfan, M., Batool, H. and Shehzad, W. Covid-19 Neologisms and their Social Use: An Analysis from the Perspective of Linguistic Relativism, Linguistics and Literature Review 6 (2): 117- 129 (2020), doi: https://doi.org/10.32350/llr.62.11
12. Katermina, V. V. and Lipiridi, S. C. Reflection on the Social and Psychological Consequences of the Coronavirus Pandemic in the New Vocabulary of the Non-professional English Language Medical Discourse, Proceedings of the Research Technologies of Pandemic Coronavirus Impact (RTCOV 2020), (44–49), ISSN: 2352–5398, ISBN: 978–94–6239–268–7, doi: https://doi.org/10.2991/assehr.k.201105.009.
13. Estabraq, R. I., Suzanne, A. K., Hussain H. M., Haneen A. H. A Sociolinguistic Approach to Linguistic Changes since the COVID-19 Pandemic Outbreak. Multicultural Education Volume 6, Issue 4, (2020). doi: https://doi.org/10.5281/zenodo.4262696.
14. Krishnamurthy, S. The chutnification of English: An examination of the lexis of Salman Rushie's Midnight's Children. *Journal of social and cultural studies.* (2010) http://hdl.handle.net/10628/230

15. Štekauer, P. On the Theory of Neologisms and Nonce-formations. Australian Journal of Linguistics 22:1 (2002) pp. 97 ff.
16. Krishnamurthy, S. The chutnification of English: An examination of the lexis of Salman Rushie's Midnight's Children. Journal of social and cultural studies. http://hdl.handle.net/10628/230
17. Algeo, J. Fifty Years Among the New Words: A Dictionary of Neologisms, 1941–1991, Cambridge University Press 978-0-521–44971-7
18. Thorne, T. (2020, June 1). Spotlight on COVID: Pandemic language and the role of linguistics. King's College London. News Centre. Retrieved on 28–02–2021, https://www.kcl.ac.uk/news/spotlight-on-covid-pandemic-language-and-the-role-of-linguists-1
19. Indurkhya, N., and Damerau, F. J. Handbook of Natural Language Processing, Second Edition (Chapman & Hall/CRC: Machine Learning & Pattern Recognition), pp. 163–177.
20. Koeva, S., Stoyanova, I., Leseva, S., Dimitrova, I., Dekova, R., and Tarpomanova, E,. The Bulgarian National Corpus: Theory and Practice in Corpus Design. Journal of Language Modelling Vol 0, No 1 (2012), pp. 65–110.
21. Ali, R., Khan, M. A., Ahmad, I., Ahmad, Z., and Amir, M. A State-of-the-art a review of corpus linguistic Journal, (January 2011).
22. ABP Pvt. Ltd. 6, Prafulla Sarkar St. Kolkata-700001, West Bengal, India, www.anandabazar.com.

Chapter 21
Recovering ROI of Medical Image Through Curvelet Transform-Based Watermarking Method

Rayachoti Eswaraiah and Tirumalasetty Sudhir

Abstract Telemedicine is a forward-thinking research area that is gaining traction. A remote specialist relies heavily on scan images or therapeutic images and data of the patient for diagnostic purposes. While sending a scan image to a remote specialist, intruders may change the Region of Interest (ROI). If the ROI in a clinical picture has been modified, the remote specialist is responsible for restoring it. This paper uses a novel robust watermarking method based on Curvelet Transform to retrieve the ROI in a medical picture. The proposed method hides ROI information in a diagnostically unimportant part of the medical picture (Region of Non-Interest) using Curvelet Transform. Experiments show that using this novel method, the ROI in a medical image can be restored to its original state.

21.1 Introduction

As a result of the healthcare industry's growth, most hospitals' information management systems now provide a range of medical imaging services. Computer Tomography, X-Rays, Magnetic Resonance Imaging, and Ultrasound are the medical imaging methods used to draw diagnostic conclusions [1, 2]. Users can access, create, and distribute digital content thanks to rapidly emerging communication technologies and the availability of low-cost, high-bandwidth internet connections. As a consequence, E-Health services are being created. Medical professionals and patients may use e-health systems to submit and interpret medical images for diagnosis resolution.

Due to the rapid advancement of diseases, obtaining the most appropriate medical diagnosis becomes a challenging decision that necessitates collaboration among a number of medical organizations. The word "telemedicine" was defined by the American Telemedicine Association. Telemedicine is the transfer of medical data from one location to another through electronic infrastructures in order to improve

R. Eswaraiah (✉) · T. Sudhir
Department of Computer Science & Engineering, Vasireddy Venkatadri Institute of Technology, Guntur, Andhra Pradesh 522508, India

© The Author(s), under exclusive license to Springer Nature Singapore Pte Ltd. 2022   223
V. Bhateja et al. (eds.), *Evolution in Computational Intelligence*,
Smart Innovation, Systems and Technologies 267,
https://doi.org/10.1007/978-981-16-6616-2_21

a patient's health. Via collaborative video, mobile phones, email, and wireless resources, telemedicine also offers a range of applications and facilities [3].

Furthermore, telemedicine can be used for a number of purposes, including remote consultation, service, and monitoring [4]. The rapid developments in the fields of telemedicine and communication technologies have resulted in a slew of issues, including the simple copying, manipulation, and redistributing of digital material without compromising its credibility. When medical data is shared across vulnerable networks for accurate medical diagnosis, it can be extorted. During their lifespan, medical images can be extorted in a number of ways as they move through different image processing systems.

Digital watermarking schemes are used in health data supervision systems to monitor data access and retrieval, protect patient data privacy, and preserve data integrity. In digital watermarking, there are two main phases. The first is embedding, and the second is extraction. During the embedding process, data is concealed within the image. The extraction process extracts secret data and uses it to vouch for the image's trustworthiness and authenticity [5].

There are two distinct regions in medical images: ROI and RONI. ROI refers to areas that should be considered critical in clinical analysis, while RONI refers to the rest. The value of enhancing ROI's prominence during the concealing phase cannot be overstated. Since this region is used by doctors for analytic purposes, even a small mutilation within this space may have significant repercussions, as it may lead to misdiagnosis. Furthermore, any modifications made to ROI during transmission must be highlighted, and ROI must be recouped without incident. This recovery prevents picture retransmission and incorrect diagnosis. The primary data for recovering ROI information is hidden in RONI [6, 7].

Section 21.2 of the paper is arranged in such a way that it shows the work of previous scholars, Sect. 21.3 elaborates the proposed plan, Sect. 21.4 investigates the investigational findings, and Sect. 21.5 supports the conclusion.

21.2   Related Work

A watermarking scheme was developed by A Al-Haj and A Amer [6]. To provide secrecy and authenticity, robust watermarks are inserted into the RONI of an image using a blind scheme based on DWT and SVD transformation techniques. To ensure image integrity, fragile watermarks are inserted in the ROI of the image using a reversible scheme. Integrity implementation at the block level allows for localized identification of tampered regions. This approach can be used on medical images of different modalities. LSBs of pixels in ROI are accumulated to establish ROI recovery data.

A watermarking scheme was suggested by Khor et al. [7]. This approach is ROI-RONI based and only applies to ultrasound medical pictures. The ROI knowledge is hidden in RONI. A hash value created by the SHA-256 technique is used to detect tampering within ROI. The PSNR value produced was greater than 40 dB, and the imperceptibility results were fine. This approach provided a high degree of robustness by withstanding a variety of attacks. The ROI that has been tampered with is restored in its original condition.

A fragile watermarking system was introduced by Surekha and Rohit [8]. To address credibility validation, tamper identification, and secure authentication, patients' biometrics are inserted into their medical images. DCT and CS theory were used to enhance privacy. On CT, X-Ray, and MRI images, this approach was tested. This method allows for the embedding of a large payload, as well as high imperceptibility and superior tamper detection.

A novel authentication approach was proposed by Thabit and Khoo [9]. This approach is ROI-RONI based and can be used on medical images of different modalities. For hiding information within ROI and RONI, two separate watermarking techniques, both based on SLT, are used. The ROI recovery information is provided using the ROI's IWT coefficients and then hidden in the image's RONI. The tampered blocks in ROI are detected and the ROI is recovered using this process. This scheme is immune to a number of accidental assaults.

A watermarking scheme was developed by Arda and Guzin [10]. The image is divided into Border Area (BA) and Center Image Area (CIA) using this technique. Later, the CIA is divided into non-overlapping 4×4 blocks. Blocks are split into two groups: expandable and non-expandable. Using the Modified Difference Expansion and LSB techniques, the EPR and payload are embedded into expandable blocks. Using the LSB process, the compressed position map, its hash value, and the last two bits of pixels are all inserted into BA. The areas of the medical picture that have been tampered with are correctly described. This approach can be used on medical images with larger ROIs and various modalities.

Al-Haj et al. [11] suggested a crypto-watermarking scheme for securely sharing medical images between healthcare organizations. This ROI-RONI approach is applicable to medical images of different modalities and protects the ROI from distortion. The patient's data and the hospital logo are both hidden in the RONI of the image using DWT and SVD techniques to ensure authenticity. A cryptographic hash watermark is hidden in the ROI to ensure its integrity. Until contact, the ROI is encrypted to help locate tampered regions and preserve confidentiality. This algorithm respects the ROI's strict and content-based honesty. Via symmetric encryption, it also facilitates tamper localization and confidentiality for the ROI.

Badshah et al. [12] suggested a watermarking scheme based on ROI-RONI that is only applicable to ultrasound pictures. The hash value of ROI is used to authenticate ROI. The ROI recovery data is provided by first converting the pixels within the ROI to binary and then concatenating all binary strings together. The ROI recovery data and hash value are combined to create a watermark. The watermark is then compressed lossless using the LZW compression method and hidden in the image's RONI.

A ROI-RONI-based watermarking technique was developed by Liu et al. [13]. The details about ROI are hidden within the image's border. ROI hides the hash value and patient information. In the RONI of the image, the ROI recovery data and position map are hidden.

21.3 Proposed Work

21.3.1 Discrete Curvelet Transform (DCuT)

Candes and Donoho [14] invented DCuT. For representing an image, the Fast Fourier Transform (FFT) and Discrete Wavelet Transform (DWT) both require a large number of frequency coefficients and wavelet base functions. DCuT is a multiresolution transform that overcomes FFT and DWT's limitations. Uneven-Spaced Fast Fourier Transform (USFFT) based DCuT and Frequency Wrapping (FW) based DCuT are the two types of DCuT used in image processing. In contrast to the USFFT-based DCuT [14], the FW-based DCuT requires less computation and is simpler to implement. As a consequence, DCuT based on FW is used in a number of image processing applications, including watermarking and compression.

21.3.2 Embedding Algorithm

The physician will apply the succeeding embedding algorithm on the medical image afore communicating it to the far-flung physician. The detailed steps of embedding algorithm are as follows:

Step 1: Convert pixels within the ROI to binary form and concatenate them to create ROI recovery data. R is used to denote the recovery data.
Step 2: Using the SHA-1 hashing method, calculate the ROI hash value. H1 stands for the measured hash value.
Step 3: Concatenate R and H1 to make the watermark. W stands for the created watermark.
Step 4: As shown in Fig. 21.1, divide the RONI of the medical image into four sections.
Step 5: In RONI, for each component P,

      Step 5.1: Apply Frequency Wrapping based DCuT on P.
      Step 5.2: In the last layer of the Detail scale band, for each matrix M.
      Step 5.2.1: For each coefficient $C(i,j)$ in M, perform the following steps.

      Step 5.2.1.1: Embed one bit of watermark data using (21.1)

$$C'(i, j) = \begin{cases} C(i, j) * (-\alpha) & \text{if } C(i, j) > 0 \text{ and } w(i) = 0 \\ C(i, j) * \alpha & \text{if } C(i, j) \leq 0 \text{ and } w(i) = 0 \\ C(i, j) * \alpha & \text{if } C(i, j) > 0 \text{ and } w(i) = 1 \\ C(i, j) * (-\alpha) & \text{if } C(i, j) \leq 0 \text{ and } w(i) = 1 \end{cases} \qquad (21.1)$$

Fig. 21.1 Segmentation of
RONI into four parts

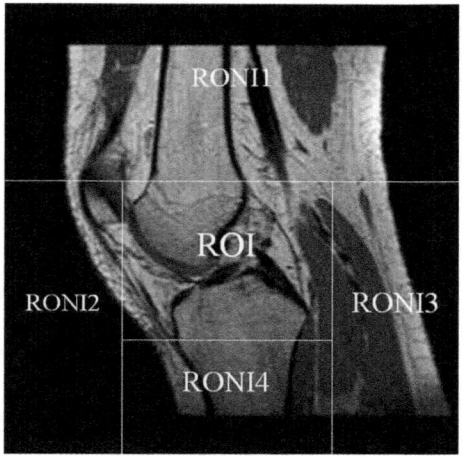

where $w(i)$ denotes a watermark bit, α is an embedding parameter, $C(i, j)$ denotes a curvelet coefficient, and $C'(i, j)$ denotes a modified curvelet coefficient. For better robustness, the value of α is chosen to be greater than 10 [15].

Step 5.3:   Apply inverse DCuT on P.

21.3.3 Embedding Algorithm

Before making diagnostic decisions, the remote physician will use the following extraction algorithm on the received medical image. The following are the detailed steps of the extraction algorithm:

Step 1:   Using the SHA-1 hashing method, calculate the ROI hash value. H2 stands for the calculated hash value.
Step 2:   Divide the RONI of the watermarked medical image into four parts.
Step 3:   In RONI, for each component P,

    Step 3.1:   Apply Frequency Wrapping based DCuT on P.
    Step 3.2:   In the last layer of the Detail scale band, for each matrix M.
        Step 3.2.1: For each coefficient $C(i, j)$ in M.
        Step 3.2.1.2: Extract one bit of watermark data using (21.2)

$$w = \begin{cases} 0 & \text{if } C(i, j) \leq 0 \\ 1 & \text{if } C(i, j) > 0 \end{cases}$$

Step 3.3:   Apply inverse DCuT on *P*.

Step 4:   Concatenate all of the extracted bits to get watermark data.
Step 5:   Divide the extracted watermark into R (recovery data) and H1 (hash value).
Step 6:   If H1 and H2 match, the algorithm will be terminated. In either case,
Step 7:   Divide *R* into 8-bit blocks to obtain ROI pixels.
Step 8:   Using the pixels obtained in the previous step, recover the ROI.

21.4 Experimental Results

Various gray scale medical images are used to examine the recitation of the proposed system. In the tests, medical photos of various sizes and forms were used. Images from MRI scans, PET scans, ultrasound scans, and CT scans were used.

21.4.1 Imperceptibility

Experiments on medical images of different sizes and modalities were carried out in order to determine the recital of the proposed technique. Figure 21.2 illustrates a couple of the medical images used in the experiments. All of the images are in an

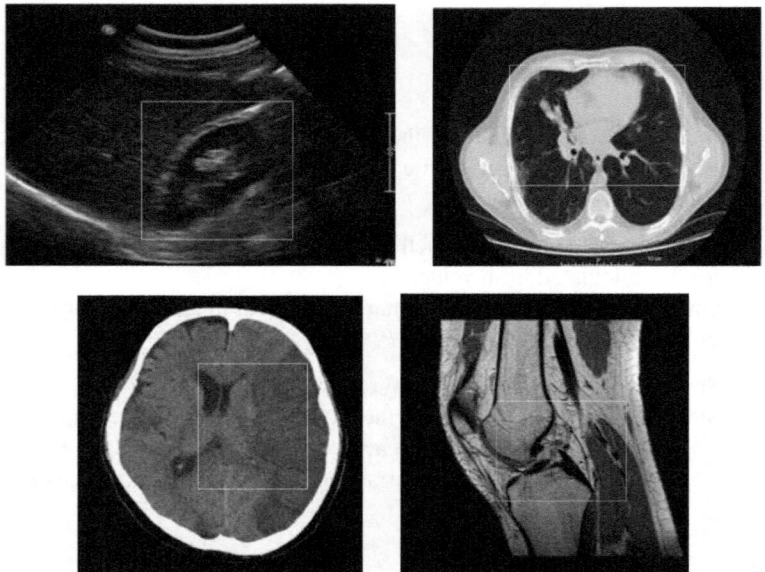

Fig. 21.2 Cover images: ultrasound of liver, CT scan of chest, PET scan of brain, and MRI scan of knee

8-bit greyscale format. These images include a liver ultrasound, a CT scan of the chest, a PET scan of the brain, and an MRI scan of the knee.

The RONI of each image is sliced into four parts, as defined in the embedding algorithm, and then the ROI hash value and recovery data are embedded as a watermark in the RONI. Figure 21.3 displays the watermarked photos that resulted. Table 21.1 displays the specifics of the cover medical photos as well as the PSNR value. The PSNR value for all medical images is greater than 50 dB. As a consequence, the suggested scheme meets the imperceptibility criterion.

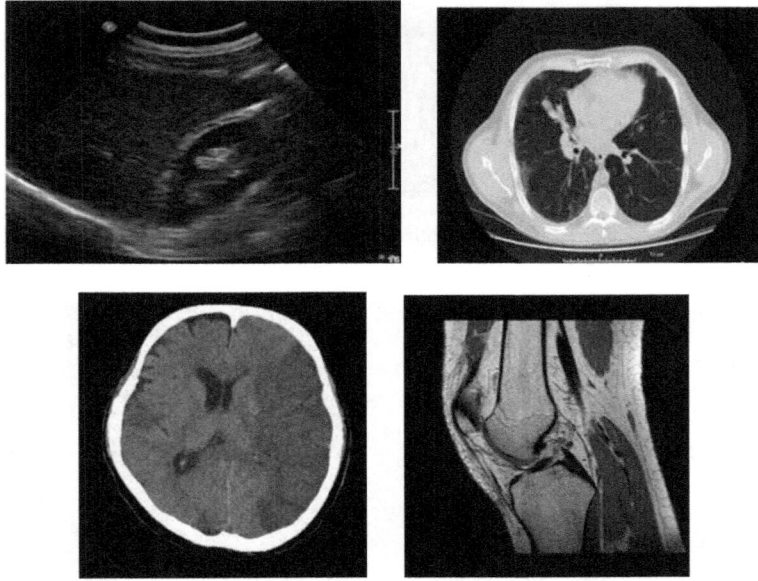

Fig. 21.3 Watermarked images: ultrasound of liver, CT scan of chest, PET scan of brain, and MRI scan of knee

Table 21.1 Particulars of images used in the experiments

Modality of image	Size of image	Size of ROI	Size of watermark (recovery data + hash value)
Ultrasound	270 × 210	110 × 116	102,336
CT scan	270 × 230	130 × 100	104,256
PET scan	200 × 210	74 × 110	65,376
MRI scan	230 × 250	108 × 80	69,376

21.4.2 Tamper Detection and Recovery

The proposed method defined the changes made to the ROI and recovered it. Figures 21.4, 21.5, 21.6, and21.7 show the proposed scheme's tamper detection and recovery capabilities. The results of the research show that the proposed technique detects and restores changes in the ROI of an image made by an intruder during transmission.

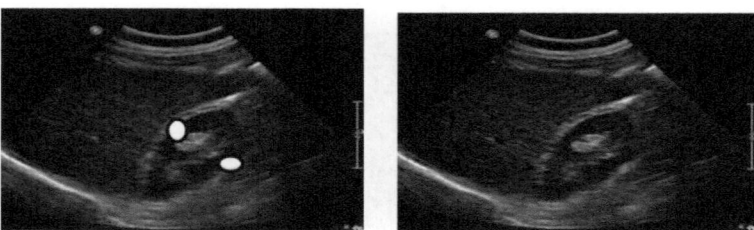

Fig. 21.4 Watermarked ultrasound image with tampers inside ROI and recovered ultrasound image

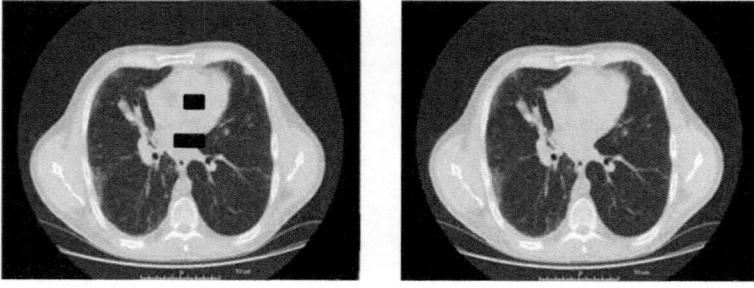

Fig. 21.5 Watermarked CT scan image with tampers inside ROI and recovered CT scan image

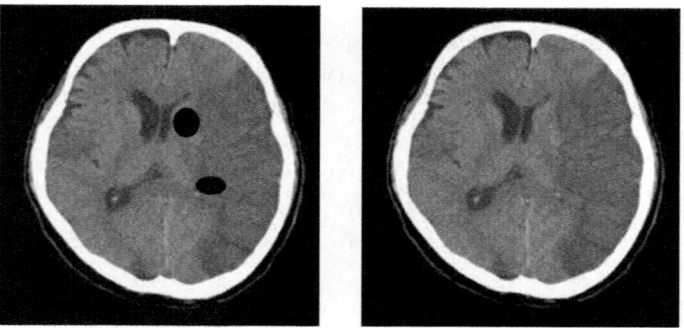

Fig. 21.6 Watermarked PET scan image with tampers inside ROI and recovered PET scan image

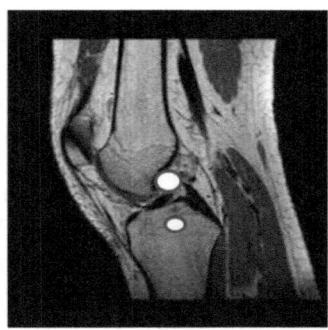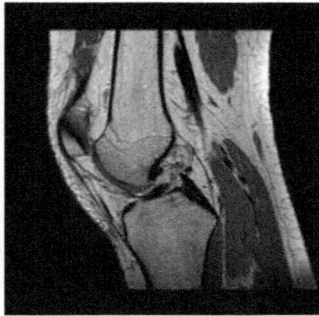

Fig. 21.7 Watermarked MRI scan image with tampers inside ROI and recovered MRI scan image

21.4.3 Comparison with the State-of-the-Art Methods

Table 21.2 shows a close correlation between the proposed approach and best-in-class techniques. The following factors are considered in the relative investigation: hiding falsification inside ROI, recovering ROI when it is changed, applicability of the proposed technique for medical images of different sizes and modalities, and providing robustness to information hidden inside RONI. Table 21.2 lists the additional functionality that the proposed technique supports.

Table 21.2 Proposed scheme versus state-of-the-art methods

Method	Concealing falsification inside ROI	Recuperation of ROI when is altered	Pertinent to medical images of diverse size and modality	Robustness to data concealed inside RONI	Utilization of compression method
Al-Haj [6]	Present	Not possible	Yes	Afford	No
Khor [7]	Not present	Not possible	No	Does not afford	Yes
Surekha [8]	Present	Not possible	No	Afford	No
Thabit [9]	Present	Not possible	Yes	Afford	No
Arda [10]	Present	Not possible	Yes	Does not afford	Yes
Al-Haj [11]	Not present	Not possible	Yes	Afford	No
Badshah [12]	Not present	Possible	No	Does not afford	Yes
Liu [13]	Present	Possible	Yes	Does not afford	Yes
Proposed method	Not present	Possible	Yes	Afford	No

21.5 Conclusion

In telemedicine, remote specialist reckons intemperately on scan images or medicinal images and data of patient for devising diagnostic conclusion. While imparting a scan image (clinical picture) to remote authority, the significant part (ROI) in it may be altered by interlopers. We devised a new method of watermarking that is both accurate and safe. The proposed method prevents changes within ROI. Proposed method accurately recognizes tampered areas in ROI, and restores ROI. The proposed scheme also ensures that the data hidden within RONI is safe, and it can be applied to medical images of different sizes and modalities.

References

1. Coatrieux, G., Lecornu, L., Sankur, B., Roux, Ch.: A Review of image watermarking applications in healthcare. In: International Conference of the IEEE EMBS, pp. 4691–4694 (2006)
2. Chao, H.M., Hsu, C.M., Miaou, A.G.: A data hiding technique with authentication, integration and confidentiality for electronic patient records. In: IEEE Transactions on Information Technology in Biomedicine, pp. 46–53 (2002)
3. American Hospital Association, The Promise of Tele-health for Hospitals, Health Systems and their Communities, TrendWatch (2015)
4. Dinya, E., Toth, T.: Health Informatics: e-Health and Telemedicine. Semmelweis University, Institute of Health Informatics (2013)
5. Pan, W., Coatrieux, G., Cuppens-Boulahia, N., Cuppens, F., Roux, C.: Medical image integrity control combining digital signature and lossless watermarking. In: International Conference on Data Privacy Management and Autonomous Spontaneous Security, pp. 153–162. Springer, Berlin (2010)
6. Al-Haj, A., Amer, A.: Secured telemedicine using region-based watermarking with tamper localization. J. Dig. Imag. 737–750 (2014)
7. Khor, H.L., Liew, S.C., Zain, J.M.: Region of interest-based tamper detection and lossless recovery watermarking scheme (ROI-DR) on ultrasound medical images. J. Dig. Imag. 328–349 (2017)
8. Surekha, B., Rohit, T.: Crypto-watermarking scheme for tamper detection of medical images. Comput. Methods Biomech. Biomed. Eng. Imag. Vis. 345–355 (2019)
9. Thabit, R., Khoo, B.E.: Medical image authentication using SLT and IWT schemes. Multimed. Tools Appl. 309–332 (2017)
10. Arda, U., Guzin, U.: A new medical image watermarking technique with finer tamper localization. J. Dig. Imag. 665–680 (2017)
11. Al-Haj, A., Mohammad, A., Amer, A.: Crypto-watermarking of transmitted medical images. J. Dig. Imag. 26–38 (2016)
12. Badshah, G., Liew, S. C., Zain, J. M., Ali, M.: Watermark compression in medical image watermarking using lempel-ziv-welch (LZW) lossless compression technique. J. Dig. Imag. 216–225 (2016)
13. Liu, Y., Qu, X., Xin, G., Liu, P.: ROI-based reversible data hiding scheme for medical images with tamper detection. IEICE Trans. Inf. Syst. 769–774 (2015)
14. Candes, E.J., Donoho, D.L.: New tight frames of curvelets and optimal representations of objects with piecewise C2 singularities. Commun. Pure Appl. Math. 219–266 (2004)
15. Leung, H.Y., Cheng, L.M., Cheng, L.L.: A robust watermarking scheme using selective curvelet coefficients. Int. J. Wavelets Multi Resolut. Inf. Process. 163–181 (2009)

Chapter 22
Water Table Depth Forecasting Based on Hybrid Wavelet Neural Network Model

Niharika Patel, Arun Kumar Bhoi, Dilip Kumar Paika, Abinash Sahoo, Nihar Ranjan Mohanta, and Sandeep Samantaray

Abstract Knowledge about water table depth fluctuation is vastly necessary for appropriate water resources planning and sustainable development. In this paper, wavelet based artificial neural network (W-ANN) is developed for water table depth (WTD) forecasting. Relative performance of proposed W-ANN was compared with regular ANN models considering different performance indicators. The WI and NSE values for W-ANN model were 0.9924 and 0.9901 and for ANN model was 0.9538 and 0.9519 respectively. It was revealed that application of wavelet analysis can increase forecasting accuracy of conventional ANN with more model accuracy and reliability. Level of disintegration in wavelet analysis should be found based on seasonality and periodicity of data series.

22.1 Introduction

Recent decades have seen an increase in water demand globally along with impact of climate changes, which has led to augmented inconsistency in depletion of groundwater table. In several catchments, groundwater (GW) is a main water source for drinking, household, industrial and agricultural users specifically in arid and semi-arid areas where rainwater is limited and fresh water is limited [17]. Yet, in many parts of the world groundwater provisions for industrial, municipal and agricultural purposes have been exploited heavily [10]. Correct estimation of WTD is extremely essential for water resources modelling, artificial recharge, design and

N. Patel · A. K. Bhoi · D. K. Paika
Department of Civil Engineering, GIET University Gunpur, Gunpur, Odisha, India

A. Sahoo (✉) · S. Samantaray
Department of Civil Engineering, NIT Silchar, Assam, India

S. Samantaray
e-mail: sandeep1139_rs@civil.nits.ac.in

N. R. Mohanta
Department of Civil Engineering, NIT Raipur, Raipur, Chhattisgarh, India

© The Author(s), under exclusive license to Springer Nature Singapore Pte Ltd. 2022
V. Bhateja et al. (eds.), *Evolution in Computational Intelligence*,
Smart Innovation, Systems and Technologies 267,
https://doi.org/10.1007/978-981-16-6616-2_22

management of irrigation systems and hydrologic water balance study. ANN tools have been extensively used for understanding mechanisms groundwater and for anticipating aquifer reaction to external variations like extraction and recharge [5, 15]. Soft computing techniques such as ANN, adaptive neuro-fuzzy inference system (ANFIS), and support vector machine (SVM) have been successfully employed in several engineering fields like rainfall–runoff modelling [7, 21], estimation of sediment concentration [12, 19], flood modelling [18, 20], stream flow forecasting [13, 16], and fluctuation in GW level [22, 24].

Adamowski and Chan [1] projected a new hybrid technique namely W-ANN to forecast GW level at two stations in Chateauguay catchment of Quebec, Canada, and assessed its performance with ANN and ARIMA models. Findings from their study revealed that W-ANN model provided more precise forecasting of ground water level than other applied models. Yoon et al. [26] developed ANN and SVM models for predicting variations in ground water level of two wells in Korea. Obtained result demonstrated that performance of SVM was superior to simple ANN for GW level forecasting. Kisi and Shiri [9] explored potential of a wavelet based ANFIS (W-ANFIS) model for forecasting ground water depth of two wells in Illinois, U.S. Comparison of results obtained from hybrid model with standalone ANFIS model showed that W-ANFIS model performed better. Developing and investigating the accuracy of ANN, ANFIS, W-ANN and W-ANFIS models for GW level forecasting [14, 23]. Results indicated that using WT as a preprocessing technique improves the forecasting ability of conventional artificial intelligence models. Ebrahimi and Rajaee [2] investigated potential of ANN, multi linear regression (MLR), SVM, W-ANN, W-MLR and W-SVM approaches for simulating GW level fluctuations of two wells in Qom plain of Iran. Their study indicated that application of WT improved performance of simple models with W-SVM model showing best results. Jeihouni et al. [6] studied the impact of climatic parameters w.r.t ground water level in Shabestar Plain, Iran, using ANN and fuzzy based models. The objective to this study is to predict WTD utilising hybrid W-ANN approach which is the first application of proposed method in selected study area.

22.2   Study Area

The district Balangir is located in western part of Odisha which is bounded in the east by Subarnapur, in west by Nuapada, in south by Kalahandi and in north by Bargarh (Fig. 22.1). It falls between $20°11'40$ and $21°05'08$ North latitude and $82°41'15$–$83°40'22$ East longitude. It covers an area of 6575 km^2. Here maximum temperature measured is 48.7 °C and minimum is 16.6 °C. Balangir experiences an average rainfall of 1215.6 mm.

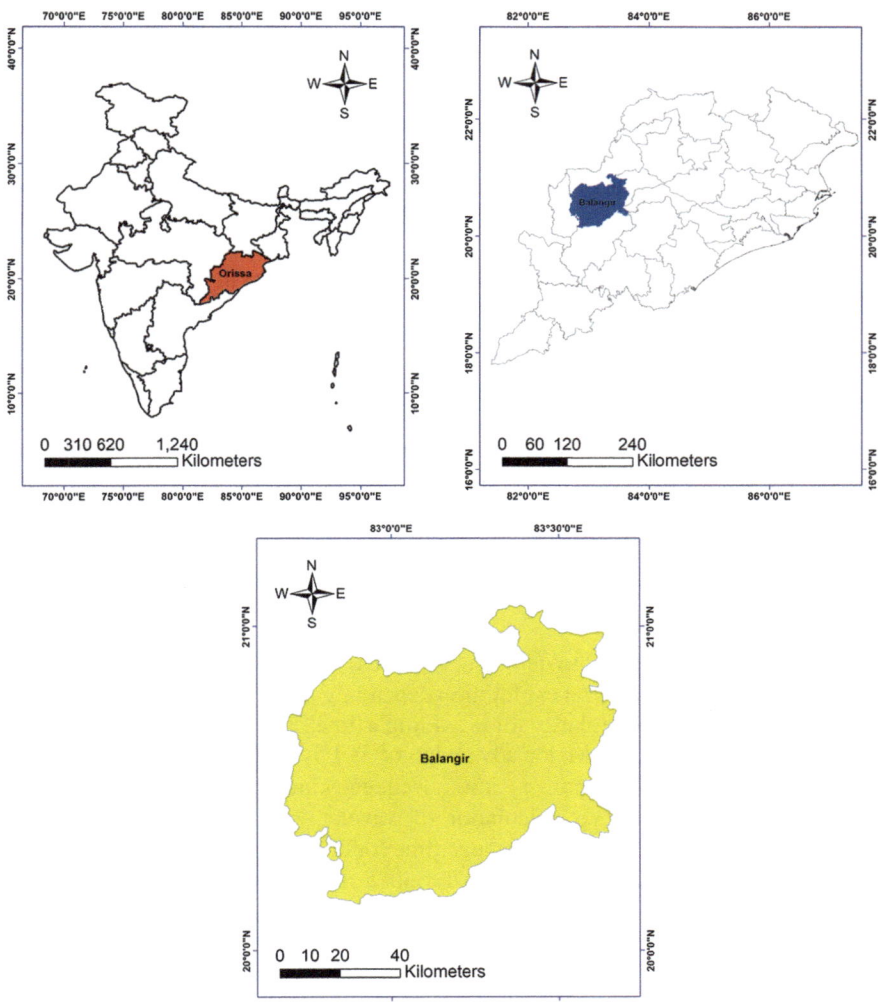

Fig. 22.1 Selected study area

22.3   Methods

22.3.1   ANN

ANN is a powerful non-linear modelling approach governed by working of a human brain. Identification and learning interconnected patterns amid input datasets and output values can be done using ANN. They can be defined as a structure of simple processing neurons or nodes, interrelated to each other in a particular manner, carrying out simple numerical operations. A usual three-layer NN comprises of a number of components called nodes. The three layers are input layer which represents different input parameters, hidden layer comprising of several hidden neurons and single output layer comprising of target values [4]. Most extensively utilised NN to model hydrological process is multilayer perceptron (MLP) that has the potential of memory association and non-linear pattern recognition [3]. In MLP, neurons are organised in layers, and every neuron is allied simply with neurons in adjoining layers.

Objective of ANN models are to generalise a relationship in order of

$$Y^m = f(X^n), \tag{22.1}$$

where X^n input vector of n-dimension comprising of variables $x_1...x_i, ..., x_n$; whereas Y^m—output vector of m-dimension comprising of resultant variables of interest $y_1...y_i, ..., y_n$.

22.3.2 Wavelet Transform

Wavelet transform is utilised to de-noise, compress and decompose data of different types (Fig. 22.2). It is a spectral exploration depending on time which untangles time-series in a space of time regularity for providing a time scale depiction of procedures and their interactions [25]. Major advantage of WT is their capability to instantaneously attain info about signal's position, frequency and time [8]. Continuous WT can function at all scales, yet calculation of wavelet coefficients at each potential scale is a difficult task which needs a huge time for computation, as well as produces enormous quantity of data. In discrete WT, only a subsection of positions and scales are selected for making calculations. DWT executes disintegration of a signal into a set of functions [11],

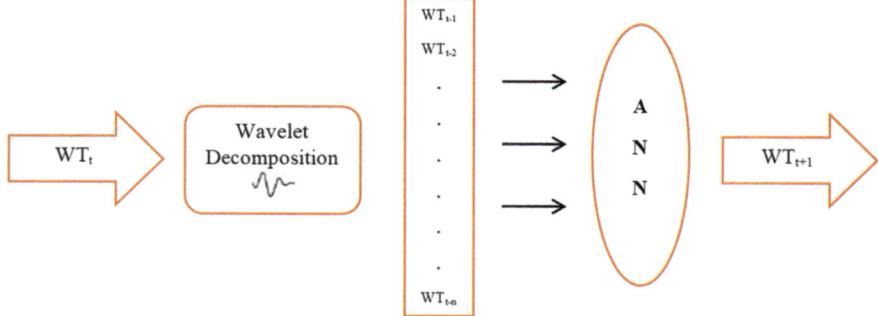

Fig. 22.2 Schematic diagram of W-ANN

$$\psi_{j,k}(x) = 2^{\frac{j}{2}}\psi_{j,k}(2^j x - k),$$ (22.2)

where $\psi_{j,k}(x)$ is generated from main wavelet function $\psi(x)$, amplified by j, interpreted by k. A condition has to be satisfied by mother wavelet

$$\int \psi(x)dx = 0.$$ (22.3)

For signal $f(x)$, DWT is found in following manner:

$$c_{j,k} = \int\limits_{-\infty}^{+\infty} f(x)\psi_{j,k}^*(x)dx,$$

$$f(x) = \sum_{j,k} c_{j,k}\psi_{j,k}(x),$$

where $c_{j,k}$—approximate coefficient of a signal.

22.3.3 Model Preparation

Monthly water table data for five month lag time is collected from IMD, Pune for a period of 1990–2019. Data values from 1990 to 2013 are utilised for training and from 2014 to 2019 for testing model efficacy. WI, NSE and R^2 are applied for performance evaluation of models.

$$R^2 = \left(\frac{\sum_{k=1}^{N}\left(A_i - \overline{A}_1\right)\left(B_i - \overline{B}_1\right)}{\sqrt{\sum_{k=1}^{N}\left(A_i - \overline{A}_1\right)^2 \sum_{k=1}^{N}\left(B_i - \overline{B}_1\right)^2}}\right)^2$$

$$\mathrm{WI} = 1 - \left[\frac{\sum_{i=1}^{N}(A_i - B_i)^2}{\sum_{i=1}^{N}\left(\left|B_i - \overline{A}_1\right| + \left|A_i - \overline{A}_1\right|\right)^2}\right]$$

$$\mathrm{NSE} = \left[\frac{\sum_{i=1}^{N}(A_i - B_i)^2}{\sqrt{\sum_{i=1}^{N}\left(A_i - \overline{A}_1\right)^2}}\right]$$

where A_i and B_i are measured and predicted WTD; $\overline{A}_i$ and $\overline{B}_i$ are mean of measured and predicted WTD.

22.4   Results and Discussion

Assessment of various models is done by using statistical indexes. In present study, different structures are utilised for evaluating effectiveness of model performances. For each input combination NSE, WI and R^2 are used for performance evaluation in both training as well as testing phases. For ANN the preeminent value of NSE and WI are 0.9603 and 0.9637, 0.9519 and 0.9538 at training and testing stages respectively. On a similar way for W-ANN prominent value of WI and NSE are 0.9963 and 0.9939, 0.9924 and 0.9901 for training and testing stage. Results reveal that model V proves best value of performance for both the technique. This shows that proposed W-ANN model maintains its high effectiveness when decreasing (to a certain amount without affecting training set) or substituting training samples. In Table 22.1, results for various model performance of Balangir watershed are presented.

The testing results of best neural network models with observed water table depth series are presented in Fig. 22.4. Scatter plot diagrams of predicted verses observed data with linear regression trend are shown in Fig. 22.3. Plotted figures clearly show that W-ANN model on a whole performed superiorly than ANN model. Predicted values of W-ANN are closer to observed values compared to ANN models as illustrated in Fig. 22.4; scatter plot representation shows that prediction of W-ANN are more closer to linear curve with little outliers (Fig. 22.3).

Table 22.1   Model performance

Model	Technique	Input combination	Training period			Testing period		
			WI	NSE	R^2	WI	NSE	R^2
I	ANN	WT_{t-1}	0.9395	0.9354	0.9316	0.9301	0.9282	0.9259
II		WT_{t-1}, WT_{t-2}	0.9461	0.9438	0.9387	0.9386	0.9357	0.9321
III		$WT_{t-1}, WT_{t-2},$ WT_{t-3}	0.9528	0.9492	0.9461	0.9429	0.9401	0.9368
IV		$WT_{t-1}, WT_{t-2},$ WT_{t-3}, WT_{t-4}	0.9596	0.9565	0.9498	0.9498	0.9463	0.9405
V		$WT_{t-1}, WT_{t-2},$ $WT_{t-3}, WT_{t-4},$ WT_{t-5}	0.9637	0.9603	0.9552	0.9538	0.9519	0.9443
I	W-ANN	WT_{t-1}	0.9799	0.9775	0.9743	0.9715	0.9683	0.9654
II		WT_{t-1}, WT_{t-2}	0.9862	0.9832	0.9779	0.9792	0.9767	0.9716
III		$WT_{t-1}, WT_{t-2},$ WT_{t-3}	0.9895	0.9866	0.9817	0.9867	0.9835	0.9775
IV		$WT_{t-1}, WT_{t-2},$ WT_{t-3}, WT_{t-4}	0.9924	0.9893	0.9871	0.9898	0.9879	0.9813
V		$WT_{t-1}, WT_{t-2},$ $WT_{t-3}, WT_{t-4},$ WT_{t-5}	0.9963	0.9939	0.9905	0.9924	0.9901	0.9867

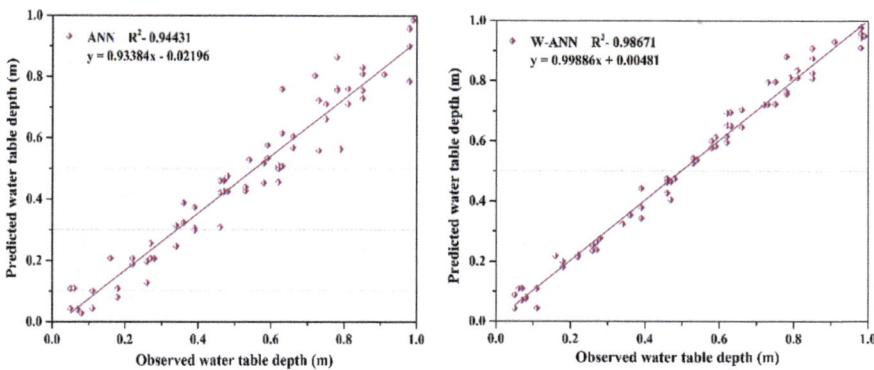

Fig. 22.3 Scatter plot of simulated data and observed data

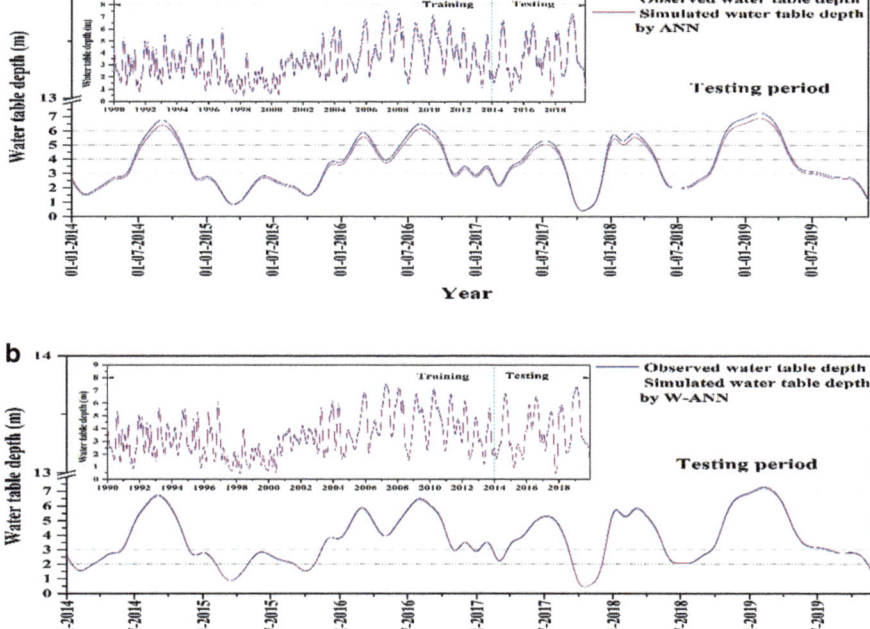

Fig. 22.4 Prediction outcomes for water table depth by **a** ANN and **b** WANN

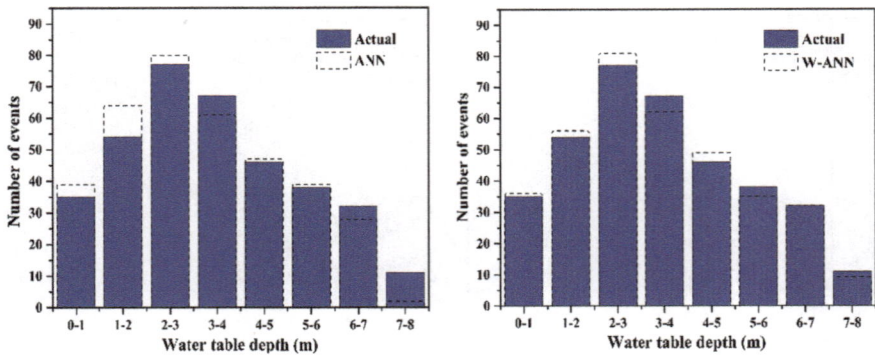

Fig. 22.5 Histogram plot for ANN and W-ANN approach

This is because of solid mathematical procedures in conjunction models: data pre-processing, parameter optimisation and mechanism of cross validation. These components give robust guarantee for adaptability and flexibility in apprehending inherit structures of non-stationary time series. Histogram plots of forecasting and actual water table depth for ANN and W-ANN considered as best models developed in present study is depicted in Fig. 22.5.

22.5   Conclusion

Accuracy of W-ANN technique in forecasting water table depth was investigated in present study for helping watershed managers on planning and managing groundwater supplies in a more sustainable and effective way. W-ANN models were compared to regular ANN model by applying them with different input combinations. Wavelet analysis is a preprocessing technique which can enhance forecasting capability of models by apprehending valuable information on different resolution stages. Comparison outcomes specified that W-ANN model performed superiorly compared to conventional ANN technique in groundwater depth forecasting with WI and NSE value are 0.9924 and 0.9901.

References

1. Adamowski, J., Chan, H.F.: A wavelet neural network conjunction model for groundwater level forecasting. J. Hydrol. **407**(1–4), 28–40 (2011)
2. Ebrahimi, H., Rajaee, T.: Simulation of groundwater level variations using wavelet combined with neural network, linear regression and support vector machine. Glob. Planet. Change **148**, 181–191 (2017)

3. Ghose, D.K., Samantaray, S.: Integrated sensor networking for estimating ground water potential in scanty rainfall region: challenges and evaluation. In: Computational Intelligence in Sensor Networks, pp. 335–352. Springer, Berlin, Heidelberg (2019)
4. Haykin, S.: Neural networks, a comprehensive foundation, 2nd edn, pp. 135–155. Prentice-Hall, Englewood Cliffs (1999)
5. Igboekwe, M.U., Amos-Uhegbu, C.: Fundamental approach in groundwater flow and solute transport modelling using the finite difference method. Earth Environ Sci **556**, 301–328 (2011)
6. Jeihouni, E., Eslamian, S., Mohammadi, M., Zareian, M.J.: Simulation of groundwater level fluctuations in response to main climate parameters using a wavelet-ANN hybrid technique for the Shabestar plain. Iran. Environ. Earth Sci. **78**(10), 1–9 (2019)
7. Jimmy, S.R., Sahoo, A., Samantaray, S., Ghose, D.K.: Prophecy of runoff in a river basin using various neural networks. In: Communication Software and Networks, pp. 709–718. Springer, Singapore (2021)
8. Khalil, B., Broda, S., Adamowski, J., Ozga-Zielinski, B., Donohoe, A.: Short-term forecasting of groundwater levels under conditions of mine-tailings recharge using wavelet ensemble neural network models. Hydrogeol. J. **23**(1), 121–141 (2015)
9. Kisi, O., Shiri, J.: Wavelet and neuro-fuzzy conjunction model for predicting water table depth fluctuations. Hydrol. Res. **43**(3), 286–300 (2012)
10. Konikow, L.F., Kendy, E.: Groundwater depletion: a global problem. Hydrogeol. J. **13**(1), 317–320 (2005)
11. Misiti, M., Misiti, Y., Oppenheim, G., Poggi, J.M.: Wavelet Toolbox for Use with Matlab. The Mathworks Inc., Natick, Massachusetts, USA (1996)
12. Mohanta, N.R., Patel, N., Beck, K., Samantaray, S., Sahoo, A.: Efficiency of River Flow Prediction in River Using Wavelet-CANFIS: a Case Study. In Intelligent Data Engineering and Analytics, pp. 435–443. Springer, Singapore (2021a)
13. Mohanta, N.R., Biswal, P., Kumari, S.S., Samantaray, S., Sahoo, A.: Estimation of sediment load using adaptive neuro-fuzzy inference system at Indus River Basin, India. In: Intelligent Data Engineering and Analytics, pp. 427–434. Springer, Singapore (2021b)
14. Moosavi, V., Vafakhah, M., Shirmohammadi, B., Behnia, N.: A wavelet-ANFIS hybrid model for groundwater level forecasting for different prediction periods. Water Resour. Manag. **27**(5), 1301–1321 (2013)
15. Praveena, S.M., Aris, A.Z.: Groundwater resources assessment using numerical model: A case study in low-lying coastal area. *International Journal of Environmental Science & Technology,* 7(1), 135–146 (2010)
16. Sahoo, A., Samantaray, S., Ghose, D.K.: Stream flow forecasting in mahanadi river basin using artificial neural networks. Procedia Comput. Sci. **157**, 168–174 (2019)
17. Samantaray, S., Rath, A., Swain, P.C.: Conjunctive use of groundwater and surface water in a part of Hirakud command area. Int. J. Eng. Technol. (IJET) **9**(4), 3002–3010 (2017)
18. Sahoo, A., Samantaray, S., Bankuru, S., Ghose, D.K.: Prediction of flood using adaptive neuro-fuzzy inference systems: a case study. In: Smart Intelligent Computing and Applications, pp. 733–739. Springer, Singapore (2020)
19. Sahoo, A., Samantaray, S., Ghose, D.K.: Prediction of flood in barak river using hybrid machine learning approaches: a case study. J. Geol. Soc. India **97**(2), 186–198 (2021a)
20. Sahoo, A., Barik, A., Samantaray, S., Ghose, D.K.: Prediction of sedimentation in a watershed using RNN and SVM. In: Communication Software and Networks, pp. 701–708. Springer, Singapore (2021b)
21. Samantaray, S., Sahoo, A.: Prediction of runoff using BPNN, FFBPNN, CFBPNN algorithm in arid watershed: a case study. Int. J. Knowl.-Based Intell. Eng. Syst. **24**(3), 243–251 (2020)
22. Samantaray, S., Sahoo, A., Ghose, D.K.: Assessment of groundwater potential using neural network: a case study. In: International Conference on Intelligent Computing and Communication, pp. 655–664. Springer, Singapore (2019)
23. Seo, Y., Kim, S., Kisi, O., Singh, V.P., X.: Daily water level forecasting using wavelet decomposition and artificial intelligence techniques. J. Hydrol. **520**, 224–243 (2010)

24. Sridharam, S., Sahoo, A., Samantaray, S., Ghose, D.K.: Estimation of water table depth using wavelet-ANFIS: a case study. In: Communication Software and Networks, pp. 747–754. Springer, Singapore (2021)
25. Tayyab, M., Zhou, J., Dong, X., Ahmad, I., Sun, N.: Rainfall-runoff modeling at Jinsha River basin by integrated neural network with discrete wavelet transform. Meteorol. Atmos. Phys. **131**(1), 115–125 (2019)
26. Yoon, H., Jun, S.C., Hyun, Y., Bae, G.O., Lee, K.K.: A comparative study of artificial neural networks and support vector machines for predicting groundwater levels in a coastal aquifer. J. Hydrol. **396**(1–2), 128–138 (2011)

Chapter 23
Application of RBFN and FFA for Flood Prediction: A Case Study

Abinash Sahoo, Mrutyunjay Nayak, Rayudu Samuel Raju, Sandeep Samantaray, and Nihar Ranjan Mohanta

Abstract Flood is one of the most chaotic natural disaster, caused due to many meteorological factors. Complex nature of flood and its response makes it a very important and challenging task of researchers. A method to forecast flood accurately and timely is important. The present paper deals with studying the potential of a hybrid artificial neural network model combining radial basis function network (RBFN) with firefly algorithm (FFA) for flood prediction at one gauging station of Mahanadi river basin, Odisha, India. To assess performance of hybrid model, coefficient of determination (R^2), Root Mean Square Error (RMSE) and Willmott Index (WI) were used. Obtained results specify that projected RBFN-FFA model gives more precise prediction outcomes than simple RBFN model.

23.1 Introduction

Flood hazards continue occurring in several countries across the globe and especially in India. Forecasting flood occurrences is a very significant flood control measure helping to reduce loss of life and economy, and also a non-structural measure that can be rationally used for water resources and water energy purposes [15]. Modelling hydrological events in their entirety is impossible because of difficulty in determining all related constraints and absence of statistical evidence; hence, usage of simulation techniques such as data driven models is vital [10, 14]. Artificial neural network

A. Sahoo (✉) · S. Samantaray
Department of Civil Engineering, NIT Silchar, Assam, India

S. Samantaray
e-mail: sandeep1139_rs@civil.nits.ac.in

M. Nayak · R. S. Raju
Department of Civil Engineering, GIET University, Gunupur, Odisha, India

N. R. Mohanta
Department of Civil Engineering, NIT Raipur, Raipur, Chhattisgarh, India

© The Author(s), under exclusive license to Springer Nature Singapore Pte Ltd. 2022   243
V. Bhateja et al. (eds.), *Evolution in Computational Intelligence*,
Smart Innovation, Systems and Technologies 267,
https://doi.org/10.1007/978-981-16-6616-2_23

approach is one such technique, and its appropriateness for hydrological research modelling is proved by findings from many studies.

Lin and Chen [13] used RBFN for constructing a rainfall-runoff model considering data from Fei–Tsui reservoir in north Taiwan. Chang et al. [4] investigated reliability and suitability of auto-configuring RBFN model for flood forecasting at River Lan-Young located in north-eastern Taiwan. Xie et al. [28] developed a flood forecasting model based on chaotic theory and RBFN. They concluded that application of chaotic theory and RBFN improved forecasting accuracy of developed flood model. Ruslan et al. [16] applied RBFN technique for predicting flood level at River Kelang situated in Malaysia. Results showed that RBFN can be considered as an appropriate method to predict flood level. Chen et al. [7] used random forest (RF) algorithm and RBFN model for evaluating flood risk level of Yangtze River Delta located in China.

Optimisation algorithms can be efficient to optimise training of data driven models [5, 6, 11]. FFA is a relatively novel optimization algorithm having certain advantages over other similar algorithms. Kavousi-Fard et al. [9] proposed hybrid model composed of SVM and FFA for providing electrical load forecast on short term basis. From their study it was concluded that application of FFA optimization algorithm increases the prediction accuracy of simple ANN models. Liu et al. (2017) projected a deep learning technique by incorporating stack auto-encoders and BP to predict discharge in a river that helps in flood forecasting. The performance of projected model was compared with SVM, BPNN, and RBFN models and found that the hybrid model performed better than all other techniques. Agarwal and Bhanot [1] used hybrid RBFN-FFA model for face recognition and found that applied hybrid model outperformed other existing methods in face recognition. Anjarsari et al. [2] applied hybrid RBFN-FFA algorithm for detecting if someone is liable to high cholesterol or not using iris images. Shang et al. [24] developed a new combined ANN-FFA model to predict ground vibration made by bench blasting and compared its efficiency with other neural network models. Results revealed that proposed ANN-FFA model performed best compared to other models. In present study, a hybrid RBFN-FFA technique is established to predict flood events which is a novel application in current study. Combination of RBFN and FFA has enhanced the performance simple RBFN model in flood prediction.

23.2 Study Area

Mahanadi is a key river flowing in east-central India originating from Pharsiya village in Raipur district of Chattisgarh at a height of about 442 m above MSL (Fig. 23.1). It lies amid $19°8'N–23°32'N$ latitude and $80°28'E–86°43'E$ longitude having an area of 1,41,589 km^2. Approximate geographical area is 4.3% of entire India, with its major portion flowing in Chhattisgarh and Orissa and rest in Madhya Pradesh, Maharashtra and Jharkhand. Total river length from its source to convergence at Bay of Bengal is around 851 kms of which 357 kms flows in Chhattisgarh and rest 494 kms in Odisha. Mean annual precipitation of Mahanadi basin is 1458 mm mostly occurring

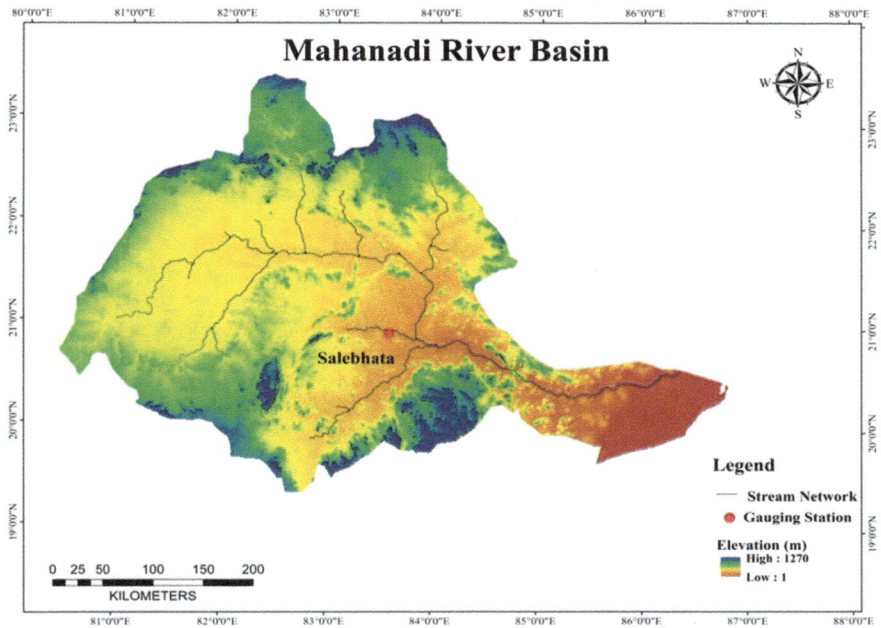

Fig. 23.1 Proposed study area

during June–September (monsoon season). In almost every year devastating floods have occurred in the basin.

23.3 Methodology

23.3.1 Radial Basis Function Network (RBFN)

BFN is a feed forward network having three layers. Because of very simple topological architecture and capability of performing suitable learning procedure in clear way, RBFN has been broadly utilised as universal function classifier or estimator for solving nonlinear problems. Architecture of RBFN presented in Fig. 23.2 comprises of input, hidden and output layers. Input is received by input layer, and information is passed to hidden layer [3]. Hidden layer comprises of RBF elements, also known as hidden neurons [17, 19, 20]. Each jth RBF unit has an associated center C_j, spread (σ_j) and base function $\varnothing_j$.

Nonlinear base function $\varnothing_j$ is function of input radial distance from midpoint of jth RBFN component. Most common applied base functions like multi-quadric, thin spline, Gaussian, and inverse multi-quadric are utilised in RBFN. Present study uses Gaussian base function which can be expressed as

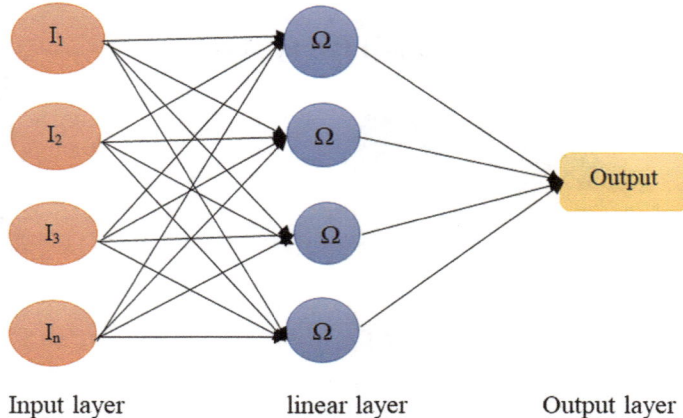

Input layer   linear layer   Output layer

Fig. 23.2 Architecture of RBFN

$$\varnothing_j = e^{\frac{\|x - C_j\|^2}{2\sigma_j^2}}, \tag{23.1}$$

where C_j signifies center of jth RBFN component. Constraint σ—spread of RBFN and represents its width.

23.3.2 Firefly Algorithm (FFA)

Among many meta-heuristic algorithms FFA is one which was motivated from flashing behaviour of fireflies Yang [29]. It has been extensively used in numerous field of studies for optimisation problems [21–23, 25]. Fireflies utilise flashing light as a specific mode of link amid individuals. Yet, between individuals connection signal can be fragile because of continuous motion of fireflies [30, 8]. In problem space for optimisation, FFA arbitrarily allocates fireflies. Light intensity $I(r)$ varies monotonically and exponentially with distance in the form of

$$I(r) = I_0 e^{-\gamma r^2}, \tag{23.2}$$

where γ—coefficient of light absorption, r—distance, and I_0—actual light intensity. If intensity is proportionate to attractiveness, then we have following equation:

$$\beta(r) = \beta_0 e^{-\gamma r^2}. \tag{23.3}$$

Distance amid two random fireflies i and j at x_i and x_j is specified using Cartesian expression

$$r_{i,j} = |x_i - x_j| = \sqrt{\sum_{k=1}^{d} (x_{i,k} - x_{j,k})^2}, \tag{23.4}$$

where $x_{i,k}$—kth component of coordinate x_i of the ith firefly.

Movement of firefly i engrossed to another more attractive (livelier) firefly j is determined by

$$x_i = x_i + \beta_0 e^{-\gamma r^2}(x_i - x_j) + \alpha\left(rand - \frac{1}{2}\right), \tag{23.5}$$

where second part defines the attractiveness, and third part describes arbitrariness from a Gaussian distribution. In present study FFA is used to tune RBFN parameters for predicting flood. Flowchart of RBF-FFA is illustrated in Fig. 23.3.

23.3.3 Evaluating Constraint

Precipitation (R_t) and flood (F_t) data of monsoon season (June–October) are collected from IMD, Bhubaneswar, for a period of 1990–2019. The data collected from 1990 to 2010 (70% of data) are used to train and from 2011 to 2019 (30%) are used to test the network. Three constraints WI, R^2, and RMSE are applied to get the paramount model

$$\text{RMSE} = \sqrt{\frac{1}{N} \sum_{k=1}^{N} (E_c - E_o)^2} \tag{23.6}$$

$$R^2 = \left(\frac{\sum_{k=1}^{N} (E_o - \overline{E}_o)(E_c - \overline{E}_c)}{\sqrt{\sum_{k=1}^{N} (E_o - \overline{E}_o)^2 + \sum_{k=1}^{N} (E_c - \overline{E}_c)^2}}\right)^2 \tag{23.7}$$

$$\text{WI} = 1 - \left[\frac{\sum_{k=1}^{N} (E_o - E_c)^2}{\sum_{k=1}^{N} (|E_c - \overline{E}_o| + |E_c - \overline{E}_c|)^2}\right] \tag{23.8}$$

23.4 Results and Discussions

Performance of ANN models based on RBFN and RBFN-FFA algorithms was evaluated using the Willmott Index, R^2 and RMSE. Performance assessment of different input scenarios utilizing RBFN algorithm are given in Table 23.1. Four types of input

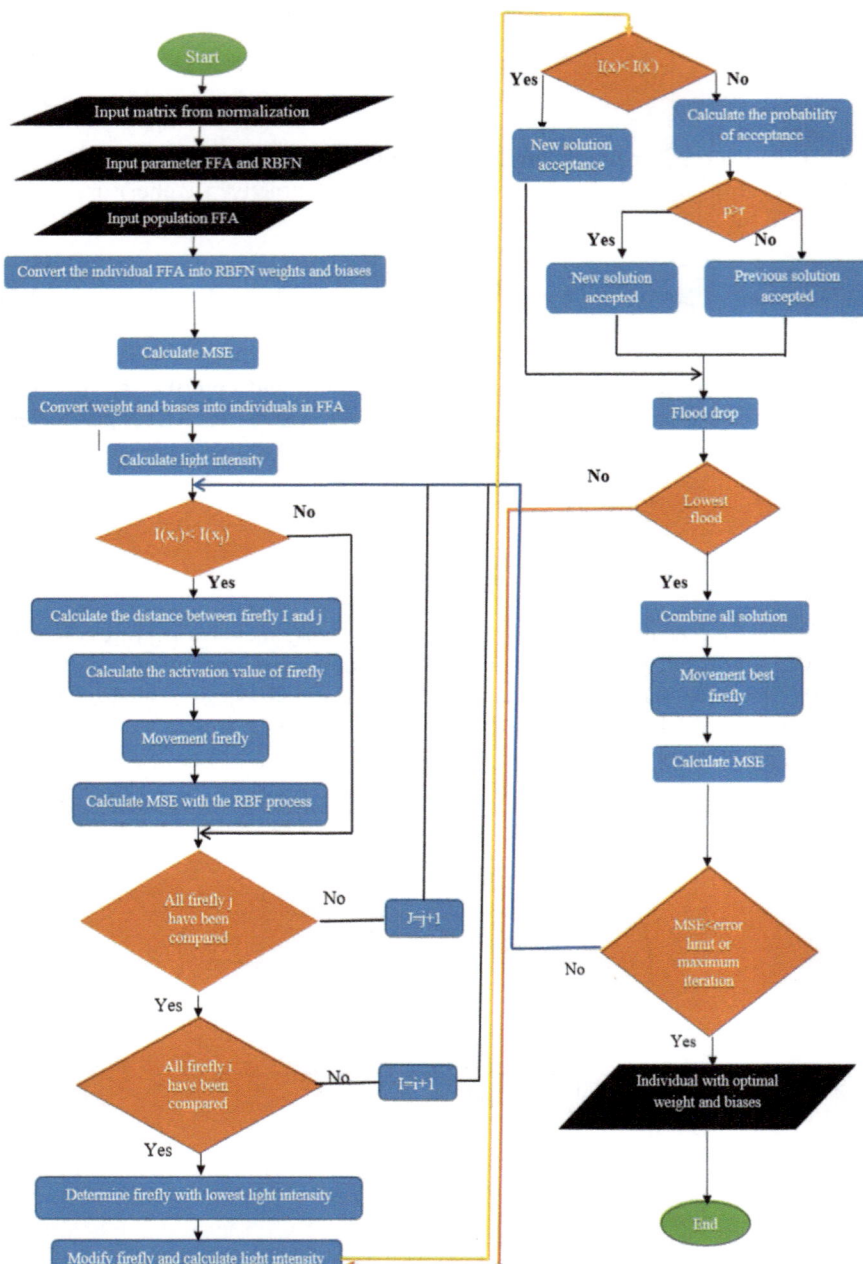

Fig. 23.3 Flow-chart of RBFN-FFA

Table 23.1 Performance of various models

Technique	Model	Input scenario	Training period			Testing period		
			R^2	RMSE	WI	R^2	RMSE	WI
RBF	Model-I	R_{t-1}	0.9254	0.0688	0.9316	0.9165	0.0825	0.9244
	Model-2	R_{t-1}, R_{t-2}	0.9297	0.0605	0.9388	0.9218	0.0778	0.9296
	Model-3	$R_{t-1}, R_{t-2}, R_{t-3}$	0.9323	0.0579	0.9421	0.9271	0.0735	0.9362
	Model-4	$R_{t-1}, R_{t-2}, R_{t-3}, R_{t-4}$	0.9368	0.0528	0.9468	0.9321	0.0669	0.9415
RBF-FFA	Model-I	R_{t-1}	0.9743	0.0463	0.9819	0.9638	0.0592	0.9723
	Model-2	R_{t-1}, R_{t-2}	0.9786	0.0397	0.9853	0.9665	0.0536	0.9774
	Model-3	$R_{t-1}, R_{t-2}, R_{t-3}$	0.9825	0.0331	0.9887	0.9717	0.0484	0.9802
	Model-4	$R_{t-1}, R_{t-2}, R_{t-3}, R_{t-4}$	0.9861	0.0275	0.9913	0.9776	0.0415	0.9842

combination are considered for assessing the model. Results indicate that correlation coefficient for RBFN-FFA is better than simple RBFN model considering model 1. RBFN-FFA predicted flood with a R^2 value of 0.9638, while RBF model exhibited R^2 value of 0.9165. The higher R^2 value specifies that RBFN-FFA model is more precise in flood prediction.

In the same way, for two month lag (model 2) flood prediction has been made for Salebhata using RBFN and RBFN-FFA techniques and their quantitative indices are summarized in Table 23.1. It can be observed that RBFN-FFA has lowest RMSE of 0.0415 compared to RBFN model in testing phase. Similarly, for RBFN-FFA R^2 value is 0.9776 which is much better than RBFN. Figure 23.4 presents scatter plot diagrams providing assessment between actual and simulated flood utilizing RBFN and RBFN-FFA during training and testing phases. Figure 23.4 illustrates that in

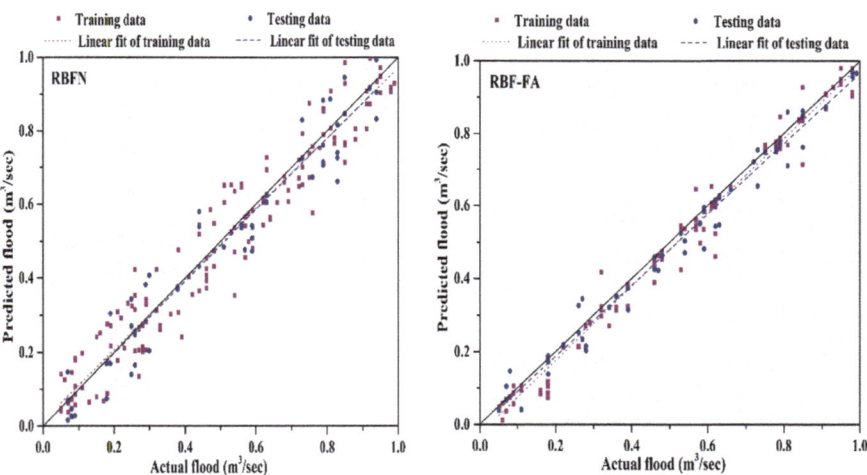

Fig. 23.4 Scatter plots of actual verses predicted flood

both cases estimations given by RBFN-FFA model are denser in neighbourhood of straight line having very fewer scattered estimations than that of RBFN. Finally a conclusion can be made that for present study RBFN-FFA has provided more precise results contrary to RBFN.

Figure 23.5 shows comparison of observed and simulated flood utilising testing data for RBFN-FFA and RBFN techniques. Results illustrate that estimated peak flood is 3895.84 m^3/S, 4146.786m^3/S for RBFN, RBFN-FFA against actual peak 4260.542 m^3/S for the station Salebhata. It clearly indicates simulated flood value

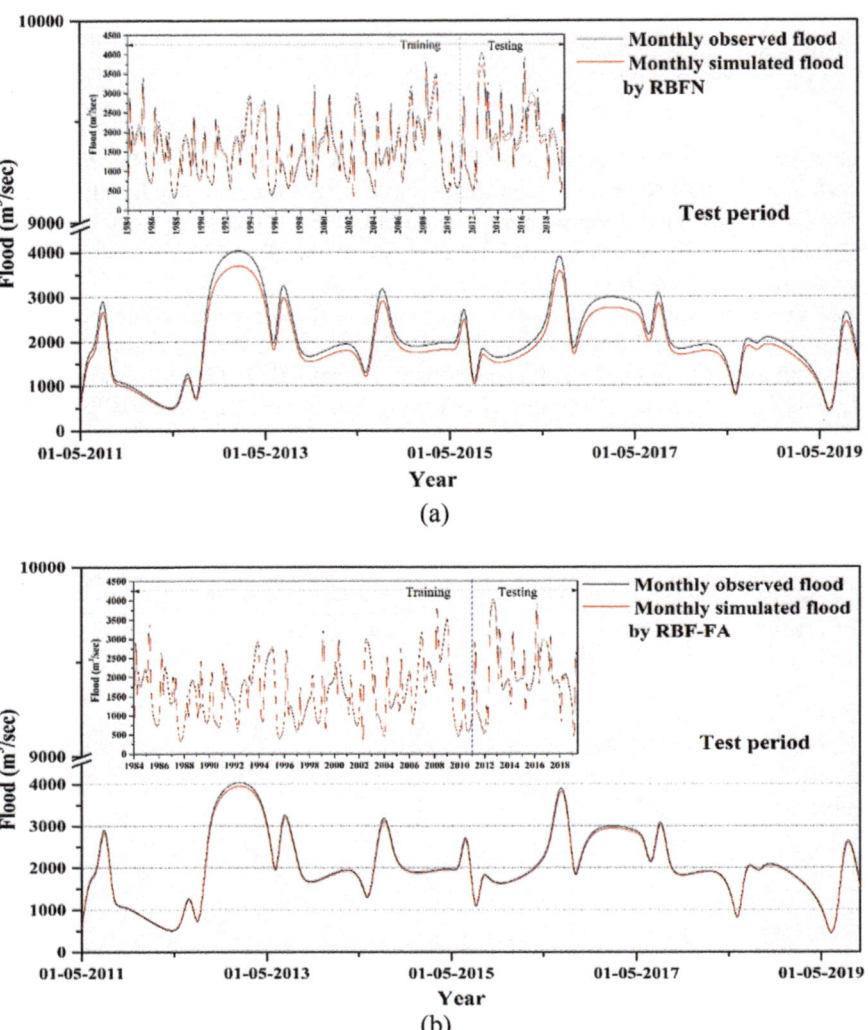

(a)

(b)

Fig. 23.5 Comparison between simulated and observed flood using **a** RBFN and **b** RBF-FFA

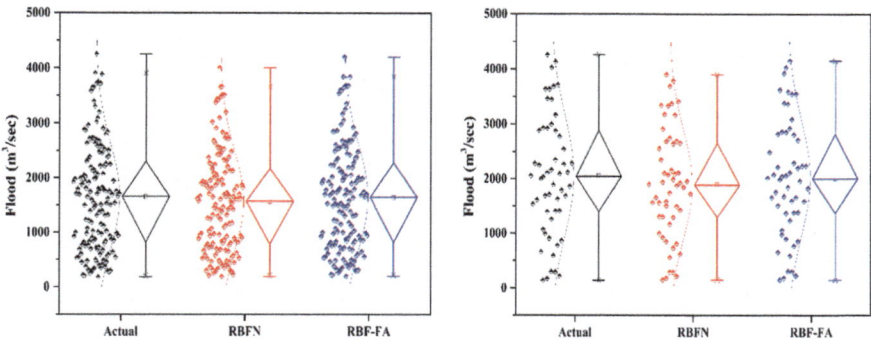

Fig. 23.6 Box plot of actual and predicted flood at **a** Bikaner and **b** Jodhpur

using RBFN-FFA approach is very closer to the actual data set than the RBFN algorithm.

The box plot of the model predictions and actual flood is illustrated in Fig. 23.6. Result indicated that high relative error is achieved when least number of inputs is utilised to construct prediction models. Here also flood estimation by RBFN-FFA model is more accurate than RBFN model. Assessment of various modelling criteria's (RBFN and RBF-FFA) with different combination of inputs (i.e., Model 1–Model 4) validates that most accurate predictions depend upon model parameters for any given input data which is the main reason for better performance by different machine learning tools with various input data sets. Therefore, RBFN-FFA algorithm might be more reliable for flood predictions in river gauge station.

23.5   Conclusion

Floods are the most devastating natural hazards universally. Therefore, prediction of flood water level is of great significance for incorporated watershed management in line with objectives for sustainable development. A new hybrid RBFN-FFA model was developed for flood prediction at one gauging station of Mahanadi river basin, Odisha. Performance of the developed hybrid model was assessed by applying different statistical evaluation parameters. Based on the applied indices, results indicated that RBFN-FFA model predicted flood events with maximum accuracy as compared to simple RBFN model for the same input data set. Proposed methodology of this study can also be applied for estimating different other hydrological parameters.

References

1. Agarwal, V., Bhanot, S.: Radial basis function neural network-based face recognition using firefly algorithm. Neural Comput. Appl. **30**(8), 2643–2660 (2018)
2. Anjarsari, A., Damayanti, A., Pratiwi, A.B., Winarko, E.: Hybrid radial basis function with firefly algorithm and simulated annealing for detection of high cholesterol through iris images. In: IOP Conference Series: materials Science and Engineering, vol. 546, No. 5, p. 052008. IOP Publishing (2019)
3. Bishop, C. M.:Neural networks for pattern recognition. Oxford University Press, Oxford (1995)
4. Chang, L.C., Chang, F.J., Wang, Y.P.: Auto-configuring radial basis function networks for chaotic time series and flood forecasting. Hydrol. Process.: Int. J. **23**(17), 2450–2459 (2009)
5. Chau, K.W.: A split-step particle swarm optimization algorithm in river stage forecasting. J. Hydrol. **346**(3–4), 131–135 (2007)
6. Chen, X.Y., Chau, K.W., Busari, A.O.: A comparative study of population-based optimization algorithms for downstream river flow forecasting by a hybrid neural network model. Eng. Appl. Artif. Intell. **46**, 258–268 (2015)
7. Chen, J., Li, Q., Wang, H., Deng, M.: A machine learning ensemble approach based on random forest and radial basis function neural network for risk evaluation of regional flood disaster: a case study of the Yangtze River Delta, China. Int. J. Environ. Res. Public Health **17**(1), 49 (2020)
8. Fister, I., Fister Jr, I., Yang, X.S, Brest, J.: A comprehensive review of firefly algorithms. Swarm Evol. Comput. **13**, 34–46 (2013)
9. Kavousi-Fard, A., Samet, H., Marzbani, F.: A new hybrid modified firefly algorithm and support vector regression model for accurate short term load forecasting. Expert Syst. Appl. **41**(13), 6047–6056 (2014)
10. Kisi, O., Genc, O., Dinc, S., Zounemat-Kermani, M.: Daily pan evaporation modeling using chi-squared automatic interaction detector, neural networks, classification and regression tree. Comput. Electron. Agric. **122**, 112–117 (2016)
11. Gholami, V.C.K.W., Chau, K.W., Fadaee, F., Torkaman, J., Ghaffari, A.: Modeling of groundwater level fluctuations using dendrochronology in alluvial aquifers. J. Hydrol. **529**, 1060–1069 (2015)
12. Guven, A., Kişi, Ö.: Daily pan evaporation modeling using linear genetic programming technique. Irrig. Sci. **29**(2), 135–145 (2011)
13. Lin, G.F., Chen, L.H.: A non-linear rainfall-runoff model using radial basis function network. J. Hydrol. **289**(1–4), 1–8 (2004)
14. Mosavi, A., Bathla, Y., Varkonyi-Koczy, A.: Predicting the future using web knowledge: state of the art survey. In: International Conference on Global Research and Education. pp. 341–349. Springer, Cham (2017)
15. Rath, A., Samantaray, S., Bhoi, K.S., Swain, P.C.: Flow forecasting of hirakud reservoir with ARIMA model. In: International Conference on Energy, Communication, Data Analytics and Soft Computing (ICECDS), pp. 2952–2960. IEEE (2017)
16. Ruslan, F.A., Zain, Z.M., Adnan, R.: Modelling flood prediction using radial basis function neural network (RBFNN) and inverse model: a comparative study. In: 2013 IEEE International Conference on Control System, Computing and Engineering, pp. 577–581. IEEE (2013)
17. Sahoo, A., Samantaray, S., Bankuru, S., Ghose, D.K.: Prediction of flood using adaptive neuro-fuzzy inference systems: a case study. In: Smart Intelligent Computing and Applications, pp. 733–739. Springer, Singapore (2020)
18. Sahoo, A., Singh, U.K., Kumar, M.H., Samantaray, S.: Estimation of flood in a river basin through neural networks: a case study. In: Communication Software and Networks, pp. 755–763. Springer, Singapore (2021)
19. Samantaray, S., Sahoo, A., Ghose, D.K.: Assessment of groundwater potential using neural network: a case study. In: International Conference on Intelligent Computing and Communication, pp. 655–664. Springer, Singapore (2019)

20. Samantaray, S., Sahoo, A.: Appraisal of runoff through BPNN, RNN, and RBFN in Tentulikhunti Watershed: a case study. In: Frontiers in Intelligent Computing: Theory and Applications, pp. 258–267. Springer, Singapore (2020a)

21. Samantaray, S., Sahoo, A.: Assessment of sediment concentration through RBNN and SVM-FFA in arid watershed, India. In: Smart Intelligent Computing and Applications, pp. 701–709. Springer, Singapore (2020b)

22. Samantaray, S., Sahoo, A., Ghose, D.K.: Assessment of sediment load concentration using SVM, SVM-FFA and PSR-SVM-FFA in arid Watershed, India: a case study. KSCE J. Civil Eng. 1–14 (2020c)

23. Sekhar, G.C., Sahu, R.K., Baliarsingh, A.K., Panda, S.: Load frequency control of power system under deregulated environment using optimal firefly algorithm. Int. J. Electr. Power Energy Syst. **74**, 195–211 (2016)

24. Shang, Y., Nguyen, H., Bui, X.N., Tran, Q.H., Moayedi, H.: A novel artificial intelligence approach to predict blast-induced ground vibration in open-pit mines based on the firefly algorithm and artificial neural network. Nat. Resour. Res. 1–15 (2019)

25. Singh, M., Patel, R.N., Neema, D.D.: Robust tuning of excitation controller for stability enhancement using multi-objective metaheuristic Firefly algorithm. Swarm Evol. Comput. **44**, 136–147 (2019)

26. Soleymani, S.A., Goudarzi, S., Anisi, M.H., Hassan, W.H., Idris, M.Y.I., Shamshirband, S., Noor, N.M., Ahmedy, I.: A novel method to water level prediction using RBF and FFA. Water Resour. Manag. **30**(9), 3265–3283 (2016)

27. Taormina, R., Chau, K.W., Sivakumar, B.: Neural network river forecasting through baseflow separation and binary-coded swarm optimization. J. Hydrol. **529**, 1788–1797 (2015)

28. Xie, J.C., Wang, T.P., Zhang, J.L., Shen, Y.: A method of flood forecasting of chaotic radial basis function neural network. In: 2010 2nd International Workshop on Intelligent Systems and Applications, pp. 1–5. IEEE (2010)

29. Yang, X.S.:Firefly algorithms for multimodal optimization. In: International symposium on stochastic algorithms, pp. 169–178. Springer, Berlin, Heidelberg (2009)

30. Zhou, J., Nekouie, A., Arslan, C.A., Pham, B.T, Hasanipanah, M.: Novel approach for forecasting the blast-induced AOp using a hybrid fuzzy system and firefly algorithm. Engineering with Computers **36**(2), 703–712 (2019)

Chapter 24
Improvements to Vanilla Implementation of Q-Learning Used in Path Planning of an Agent

Aritra Bhuiya and Suresh Chandra Satapathy

Abstract The Vanilla Reinforcement Learning algorithm for path planning in a 2-D world suffers from high mean path length and large iterations to optimize its path. An approach involving multi agents sharing common knowledge of the world has been simulated to improve the results. Multi agents increase the exploration area of the map without affecting the explore–exploit dilemma. Keeping in mind that in the real world agents may have vision, a simple yet effective vision-based technique has been proposed where an agent can locate its goal if it lies in its line of sight and not blocked by obstacles. An environment is designed in Python 3 that uses dynamically generated maps of different sizes to simulate the results.

24.1 Introduction

Many algorithms such as A*, h*, Dijkstra were used in the path for planning. But these algorithms have a huge limitation. All such heuristical path planning algorithms just take distance into account. Factors such as no of turns, path quality, congestion, etc., are completely beyond the scope of these algorithms. Moreover, these algorithms rely upon the knowledge of the world and fail to work in unknown environments. But these factors can easily be accounted for by training the robots using Reinforcement learning. But Reinforcement Learning takes some time in the initial iterations to find the path. This makes them slower than A* or any other heuristical algorithm for small maps. To improve this, multi agents sharing common knowledge of the world can be used as the rate of exploration increases. The major disadvantage revealed in the results is that in reinforcement learning, the agents take a blind approach which results in longer path finding time (initially) and longer path lengths. It is revealed in

A. Bhuiya (✉) · S. C. Satapathy
Kalinga Institute of Industrial Technology, Bhubaneswar 751024, India
e-mail: suresh.satapathyfcs@kiit.ac.in

© The Author(s), under exclusive license to Springer Nature Singapore Pte Ltd. 2022   255
V. Bhateja et al. (eds.), *Evolution in Computational Intelligence*,
Smart Innovation, Systems and Technologies 267,
https://doi.org/10.1007/978-981-16-6616-2_24

the simulations that a robot often moves away from the goal even when it is nearby in order to explore, because the robot relies only on the Q-maps and not on vision or any direction. This is unlike in the real world where an agent may be fitted with a camera and even a low range vision can provide huge improvement to the algorithm.

24.2 Related Work

Masehian and Sedighizadeh [1] have reviewed the most popular classical and heuristic approaches throughout a 35 year period in Robot Motion Planning. One of the early classical approaches was the skeleton roadmap approach, where set of feasible motions were reduced to a network of 1-D lines. This reduces the problem into a graph solving problem. But these methods are slow. The authors also reviewed the use of Neural Networks and agreed that they are computationally efficient and can produce on real time dynamic collision free path. Xianyi and Meng [2] have used Hopfield Neural Networks in a fuzzified environment. Li et al. [3] in their paper gave a solution to the path planning problem in a warehouse. Their algorithm is aimed at warehouse robots on a 2D grid trying to find an optimal path with the least computation time. An autonomous robot would be assigned a task to pick up goods from a location (co-ordinates) and start loading. Then it would go to the next goods' location and perform the same operation. Once the loading tasks were completed, the robot would get back to its initial parking position. The path calculation would be done using the robot's built-in CPU, but the task would be assigned by a central controller. The authors categorized this type of path planning algorithm as a TSP problem. Zheng et al. [4] has used A* algorithm to find an optimal path. They have added inclination of the path to the A* cost function but have not suggested any way to optimize the long path calculation time and this is exactly where Li et al. [3] have targeted their work. Numerous authors have used A* algorithm to find an optimal path, but none, except Li et al. [3] have tried to solve the long calculation time. The authors [3] found out that the computation time of heuristic search algorithms was significantly huge and to solve this issue, have introduced a novel algorithm named LCP. Bae et al. [5] has proposed a noble algorithm for using multi agents for path planning. They have used Deep Q learning for obtaining the Q-values and multiple agents to collaborate and divide the wide area into smaller areas for efficiency. However, their approach needs entire information of the environment and is not suited to unknown environments. Sichkar [6] has improved the learning rate of Q-learning by modifying the way the algorithm uses to select the best state. The author has used non-greedy policies, modifying the algorithm to 'SARSA' to improve the result in early episodes. Various approximation methods were also suggested to speed up the learning process. Butyrev et al. [7] use deep reinforcement learning to reach a state using continuous trajectory, taking kinematic and dynamic constraints into account. Jiang et al. [8] have modeled the environment as a graph and graph convolution adapts to make the agents learn to cooperate.

Fig. 24.1 The environment
built with Python 3 for
A-star

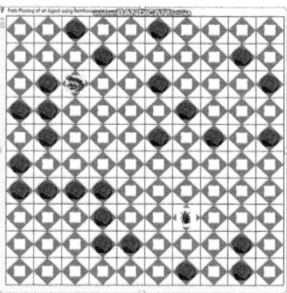

24.3   Reinforcement Learning

Reinforcement Learning is a branch of Machine Learning where an agent chooses an action from a state-action table, such that the action maximizes its reward. In other words, at every state, each action offers some reward or penalty and the agent is expected to perform rationally by maximizing its reward. The most important aspect of reinforcement learning is building the Q-Table or state-action table. Reinforcement learning can also be applied on a multi-agent scenario. Agents learn from individual action as well as co-operative actions. In this work, model free Q-learning is used, where an agent can move in four directions and have no prior knowledge of its environment.

24.4   Implementation

24.4.1   Environment

All the simulations shown in this work have been carried out on an Alienware M15 notebook having 16 GB of RAM, Intel i7 8th gen Processor and Nvidia RTX 2070 GPU with 8 GB of VRAM with Max Q design. All Programs were written in Python 3.7 and run on Windows 10. The environment supports one agent, one goal and multiple obstacles. The environment uses a picture of frog to represent the agent, an insect to represent the goal and rock to represent obstacle. The agent is allowed to move in only four directions: up, down, left and right. The environment is made in ways that it can also represent the Q-values for each legal action. The environment runs on one process thread while the agent computes on another thread of the same process. The environment can ask the user to design the map as well as use an existing one. Maps of various sizes were randomly generated with 30% obstacles (Figs. 24.1 and 24.2).

Fig. 24.2 Environment for
reinforcement learning

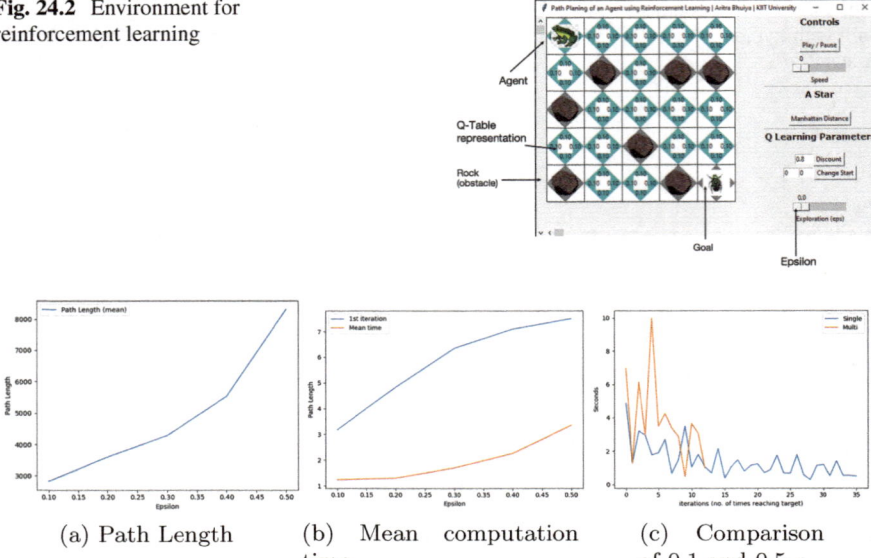

Fig. 24.3 Effect of Epsilon on reinforcement learning

(a) Path Length   (b) Mean computation time   (c) Comparison of 0.1 and 0.5 ϵ

24.4.2 Implementation of Reinforcement Learning (Q-Learning)

Q-Learning Q-learning (A model-free learning) technique is used for the simulation. A vanilla implementation was used and its details are beyond the scope of this work as it is a very popular technique.

Epsilon-Greedy Strategy Epsilon-Greedy strategy has been used and is the backbone of the Q-learning technique. A value of 0.1 ϵ represents the agent will be exploring 10% of the time and exploiting the best known action 90% of the time. As the results indicate, having a higher value of Epsilon will delay the agent as it will keep on searching for better path. $\epsilon = 0$ will also delay the agent as the agent will never try to explore (Fig. 24.3).

Running the algorithm on various map sizes yields an interesting result. A detailed study of computation time with map size has been given (Table 24.1).

The Map length indicates the length of the square grid world. The 1st iteration represents the time taken by the algorithm to find the goal for the first time. This time is mostly the worst case time.

Table 24.1 Reinforcement learning map length and computation time

Map length	Mean time	1st iteration time
5	0.0016	0.1906
15	0.0073	0.3975
25	0.1197	1.1502
35	1.1301	2.7474
45	2.3257	4.7422
55	6.6124	13.9841
65	8.5480	19.8340

Fig. 24.4 Comparison of reinforcement learning and A-star: computation time versus grid size

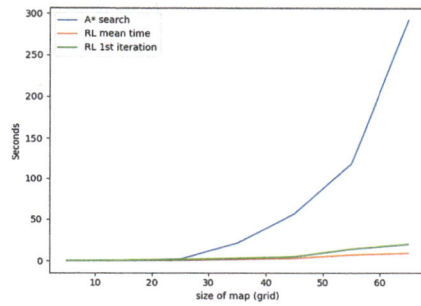

Fig. 24.5 A-star algorithm: computation time versus map size

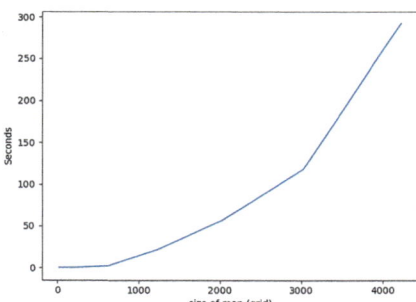

24.4.3 Comparison with A-Star Algorithm

As verified from Fig. 24.4, A-star performance degrades exponentially from 35 × 35 grid. Reinforcement Learning on the other hand maintains its linearity. But this comes at a cost. Reinforcement Learning does depend on exploration and suffers from high path cost (Fig. 24.5).

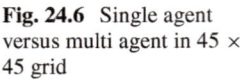
Fig. 24.6 Single agent versus multi agent in 45 × 45 grid

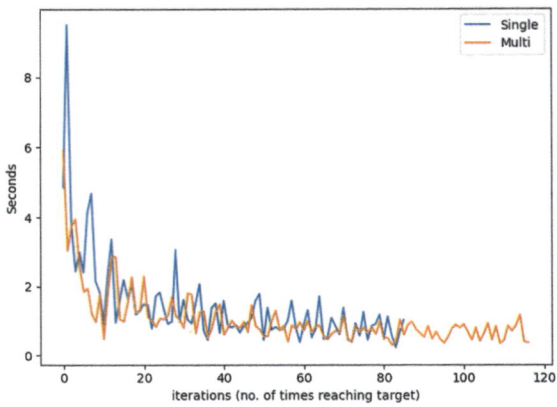

24.5 Proposed Improvements

The disadvantages of a vanilla implementation of Reinforcement Learning discourage one from using it for practical scenarios. However, small improvements can be made in order to improve the performance.

24.5.1 Multi-agent Reinforcement Learning

The agent is in an unknown environment. It is only when an agent explores that it comes to learn about the environment and accordingly update its Q-Value. One assumption can be if more than one agent is working toward a common goal, the searching may be faster.

- At the location of the main agent, an extra agent is created (Shadow agent).
- Collision among agents is ignored (like a shadow for simplicity).
- Computation time and path length of main agent is only logged.
- Shadow Agents may have different epsilon values.
- Shadow Agents and main agent share common knowledge of the world.

Multi-threading in Python 3 is used in order to implement this.

From Fig. 24.6, it is clear that an auxilary agent along with the main agent reduce the path finding time. In fact the speed up for this particular data is 1.38. The extra iterations performed by the multi agent is due to the shorter iteration of the multi agents. As the data has been recorded for both single and multi agents under the same time limit, the multi agent scenario can perform more iterations due to its faster search.

Upon running the algorithm for various map sizes, the data is shown (Table 24.2).

Figure 24.10 reveals a dip in the speed up at 75 × 75 grid. The estimated reason for this is the 75 × 75 grid differs on the smaller maps in obstacle percentages. Since, the maps are randomly generated, for the same obstacle percentage the 75

Table 24.2 Comparison of single and multi agent reinforcement learning

Length of side	Single	Multi	Speed up
35	0.56	0.44	1.27
45	1.42	1.03	1.37
50	0.92	0.66	1.39
65	6.95	4.14	1.67
75	9.99	6.21	1.61

Fig. 24.7 Single agent versus multi agent running time on different maps

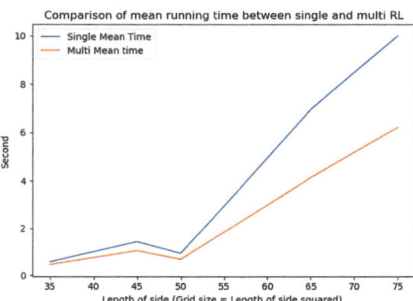

Fig. 24.8 Single agent versus multi agent mean path length comparison

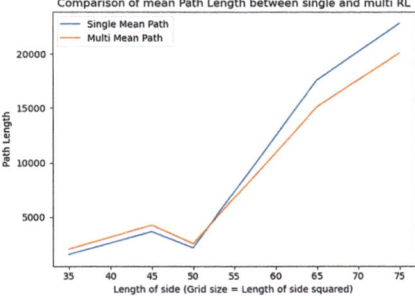

× 75 grid has no solution. Therefore that grid was generated with reduced obstacle percentage. Reducing the obstacles causes the single agent to perform better and thus a reduced speed up. However, the multi agent does perform better even in this situation (Figs. 24.7 and 24.8).

24.5.2 Vision to Agent

The vanilla implementation still possesses a high path length. This can be a problem if path length is of greater importance. Since, the algorithm performs in an unknown environment, it makes sense that it will explore. In real world, an agent may have

Fig. 24.9 Single agent
versus single agent with
vision (15^2) mean time
comparison

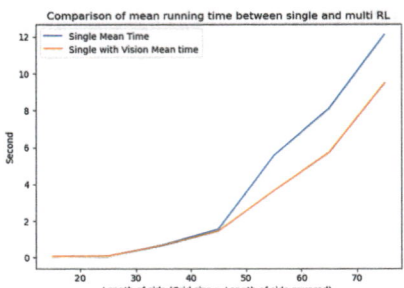

Fig. 24.10 Single agent
versus single agent with
vision mean path length
comparison

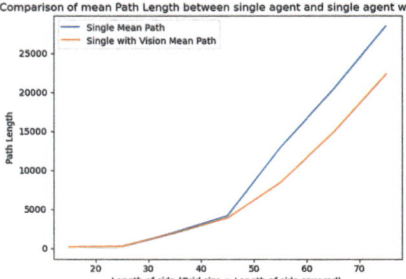

vision sensors such as a camera and this can greatly improve the path cost. A blind
agent may come near the goal but turn away and search for longer routes.

An algorithm for vision has been designed keeping in mind that:

– An agent can only see x^2 in top, left, right and bottom directions.
– An agent's vision can be obstructed by obstacles.
– If an agent sees its goal, it will directly reach for its goal and update Q-values
  accordingly (Figs. 24.9 and 24.10).

24.6 Conclusion

The claim made by Zheng and Zheng [4] is confirmed that A* algorithm suffers from
high computational time compared to reinforcement Learning. It is also true that Q-
learning in its vanilla implementation yields higher mean path length. The vanilla
algorithm can be improved by using multiple agents to co-ordinate and explore the
entire map and share Q-Values which guides an agent from its source to its goal.
The extra agents update the Q-values throughout the map significantly decreasing
the iterations before an optimum path is found. Bae et al. [5] in their paper has also
used multiple agents but their approach requires complete knowledge of the world,
whereas the proposed technique only needs partial knowledge of the world. The
multi-agent approach shown in this work requires no complete knowledge of the

world and can work in any unknown environments. This is ideal when robots need to navigate a large area like warehouses, maze-solving problems, etc. The path length is also improved significantly by giving the agent a sense of vision. Because of the Epsilon-Greedy Strategy, an agent might move away from the goal even if the goal is right next to it. Vision eliminates this shortcoming. The Q-values are used when the goal is not in vision.

'We have taken permission from competent authorities to use the images/data as given in the paper. In case of any dispute in the future, we shall be wholly responsible.'

References

1. Masehian, E., Sedighizadeh, D.: Classic and heuristic approaches in robot motion planning—A chronological review. Proc. World Acad. Sci. Eng. Technol. 101–106 (2007)
2. Xianyi, Y., Meng, M.: A neural network approach to real-time motion planning and control of robot manipulators. Proc. IEEE/SMC **4**, 674–679 (1999)
3. Yang, B., Li, W., Wang, J., Yang, J., Wang, T., Liu, X.: A novel path planning algorithm for warehouse robots based on a two-dimensional grid model. https://doi.org/10.1109/ACCESS.2020.2991076
4. Zheng, T., Xu, Y., Zheng, D.: AGV path planning based on improved A-star algorithm. In: Proceedings of the IEEE 3rd Advanced Information Management, Communicates, Electronic and Automation Control Conference (IMCEC), pp. 1534–1538 (2019)
5. Bae, H., Kim, G., Kim, J., Qian, D., Lee, S.: Multi-robot path planning method using reinforcement learning. Appl. Sci. **9**, 3057 (2019)
6. Sichkar, V.N.: Reinforcement learning algorithms in global path planning for mobile robot. In: 2019 International Conference on Industrial Engineering, Applications and Manufacturing (ICIEAM), pp. 1–5 (2019). https://doi.org/10.1109/ICIEAM.2019.8742915
7. Butyrev, L., Edelhäußer, T., Mutschler, C.: Deep reinforcement learning for motion planning of mobile robots (2019). arXiv:1912.09260
8. Jiang, J., Dun, C., Huang, T., Lu, Z.: ICLR 2020 (2020)

Chapter 25
Emotion Recognition from Speech Using Multiple Features and Clusters

A. Revathi, Bagathi Neharika, and Gayathri G

Abstract Speech Emotion Recognition (SER) using speech signals to detect our emotions is gaining popularity in the field of Human–Computer Interactions. The emotional state of a speaker is identified from a speech by using Mel Frequency Cepstral Coefficients (MFCC) feature and Gammatone Cepstral Coefficients (GTCC) as features with less dimensionality, and classification is done based on vector quantization (VQ) modelling and minimum distance classifier. The source used was the Berlin database which has the recorded utterances in various emotions like anger, boredom, sad and neutral spoken by actors. The speech signals are first digitized and pre-emphasized, after which it gets converted into frames. The frames are then multiplied with hamming window to reduce damping at higher frequencies. Then from the windowed speech signal, Mel Frequency Cepstral Coefficients (MFCC) feature and Gammatone Cepstral Coefficients (GTCC) are extracted. The extracted features are applied to the VQ models and based on minimum distance; the emotion is classified. The Unsupervised machine learning algorithm K-MEANS is used as a classifier and then the comparison is carried out between the accuracy of MFCC and GTCC features in distinguishing the emotions such as anger, sadness, boredom, and neutral.

25.1 Introduction

Emotion recognition aims to spontaneously recognize the present emotion of a person from their voice. It is emanated from the in-depth interpretation of the speech signal, extracting the features that have some specific characteristic specifications which consist of data of emotions derived from the speaker's speech signals. These specifications and relevant classification methods help to recognize states of emotion from the speaker's speech signal [1]. Recently, emotion recognition from speech is emerging in the fields of computer science and artificial psychology; furthermore, an

A. Revathi (✉) · B. Neharika · G. G (✉)
SASTRA Deemed To Be University, Tirumalaisamudram, Thanjavur, Tamil Nadu, India
e-mail: revathi@ece.sastra.edu

© The Author(s), under exclusive license to Springer Nature Singapore Pte Ltd. 2022   265
V. Bhateja et al. (eds.), *Evolution in Computational Intelligence*,
Smart Innovation, Systems and Technologies 267,
https://doi.org/10.1007/978-981-16-6616-2_25

eminent research topic of signal processing and pattern recognition [2]. Nowadays where online education is becoming pervasive, understanding the student's state of emotion and resolving appropriately can escalate the performance as teaching quality is improved [3]. It can be implemented in current trending bots like Alexa, Siri, chatbots, and many more. Generally, speech is an acoustic waveform that results from the vibration of the vocal cords [4]. The acoustic features like MFCC and GTCC can be extracted from the speech signal for classification [5, 6]. The classification can be done by employing the K-means clustering algorithm [7]. In this clustering, the feature vectors from the trained speaker's utterances are converted into emotion-specific clusters [8]. This process of feature extraction is carried out in two phases namely the training and the test phase [9]. Since Gammatone cepstral coefficients mimic the human auditory system more precisely than the Mel cepstral coefficients the accuracy of the overall system will increase by extracting GTCC [10]. This technique can also be used in the customer service division of any organization where it can be used to identify the customer's state of satisfaction or behaviour towards any particular situation. It can also be used for sales and marketing purposes where the customer's viewpoint is important. We can make use of it in the diagnosis of patients or with video game users. This is a great way of intensifying efficiency among people with their work and competently deciphers their problems. This can enhance the lives and quality of life. Thus, Speech Emotion Recognition is beneficial for applications that involve human–computer interactions such as speech synthesis, forensics, customer service, education, medical analysis, and music processing such as music genre classification and mood classification [11].

25.2 Speech an Overview

As we all know speech is the best fastest way of interacting amongst humans. Speech is the output of the vocal tract system excited by the vibration of vocal cords due to quasi-periodic pulses of air for voiced speech or noise excitation for unvoiced speech. In other words, speech is the output of the vocal tract system excited by the vibration of vocal cords due to acoustic air pressure from the lungs. This acoustic air pressure is a function of time. The speech signal can be normally segmented into voiced segment and unvoiced segment. The source of excitation for the voiced segment is a quasi-periodic pulse of air whereas for the unvoiced segment it is a white noise signal. So, depending on how you shape your vocal tract different sounds can be produced such as fricatives, semivowels, consonants, diphthongs, and vowels.

25.2.1 Formalising the Speech

The vocal tract frequency response convolved with the glottal pulse results in the speech signal. This vocal tract frequency response acts as an impulse response. Let the

Speech signal be $x(t)$, Glottal pulses represented as e(t), excitation signal produced by the vocal tract system, and $h(t)$ be the vocal tract frequency response, i.e., impulse response.

$$\text{Speech, } s(t) = e(t) * h(t).\tag{25.1}$$

Normally the log spectrum resembles the Speech, Spectral Envelope resembles the Vocal tract Frequency response and the Spectral details resemble the Glottal pulses. In the spectral envelope, the formants carry the identity of sound.

In terms of digital signal processing Vocal tract acts as a filter, by filtering the glottal pulses which consists of high-frequency noisy signals from the carrier information about the pitch. The speech signal $s(n)$ is given by

$$s(n) = g(n) * h(n).\tag{25.2}$$

And,

$g(n) =$ excitation signal.

$v(n) =$ impulse response of vocal tract system.

Because of the digital processing of speech signals, before the extraction of features from the audio signals, let us get to know how the speech signal should be used for the analysis. First and foremost, the analog speech signal has to be digitized and pre-emphasized before it is being used for analysis. Later, spectral analysis techniques are carried out to extract the features from the audio speech signal. It involves two steps. They are.

(1)   Analog speech signal from sound pressure wave is converted to digitized form.
(2)   Significant frequency components are emphasized i.e., Digital Filtering.

The role of digitization is to produce HIGH Signal to Noise Ratio (SNR) on the sampled data of audio speech. After digitization comes the amplification process which is done by a pre-emphasis filter that boosts the spectrum of the audio signal up to 20 dB/decade since the natural voice signal has a negative slope of 20 dB/decade. In pre-emphasis, voice signals at high frequencies are amplified due to the effect of damping at higher frequencies. Here

$$y(n) = x(n) - a * x(n - 1)\tag{25.3}$$

is our pre-emphasis filter, with filtering co-efficient $a = 0.95$. In general, there are two major concerns in any speech communication system.

(1)   Preservation of the Information contained in the Speech signal.
(2)   Representation of the Speech signal in an easier manner so that modifications can be done to the signal, without degrading the original information content present in the speech signal.

Hence these two major concerns have to be dealt with before further processing of speech signals in any communication system.

25.2.2 Framing and Windowing

As the speech signal is not stationary for an infinite length, framing is done so that for a short period it remains stationary since the glottal system cannot vary at once. So framing is the splitting of the speech signal into smaller chunks. Before framing, the speech signal is filtered first using a pre-emphasis filter. Framing is done as machines cannot do computations with infinite data points, as the signal will be cut off at either end leading to information loss. In framing, the speech signal is divided into frames of 8 to 16 ms length and shifted with overlapping of up to 50% so that the next frame contains information about the previous frames. The frames of the speech signal are then multiplied with the window. Usually, windows enhance the performance of an FFT to extract spectral data. The window which we preferred for multiplication with frames of the speech signal is the Hamming window. The Hamming window is used to reduce the ripple caused in the signal so that we can get a clear idea of the original signal's frequency spectrum. Hamming window $w(n)$ is given by

$$w(n) = 0.54 - 0.46 \cos\left(2\pi n / (N - 1)\right), \quad 0 \le n \le N - 1. \quad (25.4)$$

25.2.3 Understanding the Cepstrum

We know that the speech signal can be represented as follows:

$$\text{Speech, } s(t) = e(t) * h(t). \quad (25.5)$$

$$\text{Taking Fourier Transform, } s(w) = E(w) * H(w). \quad (25.6)$$

$$\text{Taking log on both sides, } \log(s(w)) = \log(E(w)) * \log(H(w)). \quad (25.7)$$

By using the log magnitude spectrum, we can separate the vocal tract information and the glottal pulse information which could not be done by normal spectrum. After which by taking IDFT of log magnitude spectrum, cepstrum is obtained. The physical separation of the information that is relative to the spectral envelope (vocal tract frequency response) is in the lower end of the quefrency domain and the information relative to spectral details (glottal pulse excitation) is in the higher end. Thus, the excitation $e(t)$ can be removed after passing it through a low pass lifter.

Thus, after analysis of speech signals, it can be used to train the speaker models by extracting the feature vectors of each specified emotion. Later, the emotion model is constructed and test feature vectors are given to the classifier after which the emotional state of the test feature is being tracked.

25.3 Feature Extraction

25.3.1 Mel-Frequency Cepstral Coefficients (MFCC) [12]

Mel Frequency Cepstrum is a short-time power spectrum representation of a speech signal obtained after computation of Discrete Cosine transformation. The word Cepstrum, Quefrency, Liftering, and Harmonic is just the wordplay of the word Spectrum, Frequency, Filtering and Harmonic respectively. The Former terms correspond to the frequency domain, and the later terms correspond to time-domain representation. MFCC is based on the characteristics of the hearing perception of humans It is a fact that human ears use a non-linear frequency unit to simulate the auditory system of humans. Usually, human ears perceive frequency logarithmically. This, in turn, necessitates an ideal audio feature that can be depicted in the time–frequency domain and has a perceptually relevant amplitude and frequency representation. One such audio feature is Mel Frequency Cepstral Coefficients.

Figure 25.1 describes the procedure for MFCC extraction.

The speech waveform after undergoing pre-emphasis, frame blocking, and windowing, Discrete Fourier Transform of the signal is computed after which the log amplitude spectrum is obtained. Then Mel-Scaling is performed. Mel-Scale is a logarithmic scale. It is a perceptually relevant or perceptually informed scale for pitch. It is a fact that equidistant on the scale has the same "perceptual" distance. In the Mel filter bank, the difference between the Mel points is the same resulting in null weight whereas the difference between frequency points is not the same on a scale of frequencies. After Mel-Scaling, the log amplitude spectrum undergoes Discrete Cosine Transform and the Cepstrum of the speech signal is obtained. The advantages of MFCC are as follows.

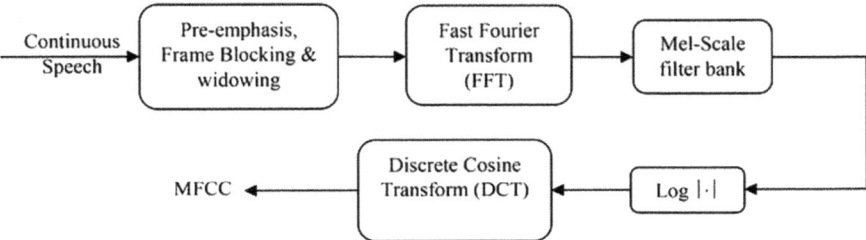

Fig. 25.1 Block diagram of MFCC extraction

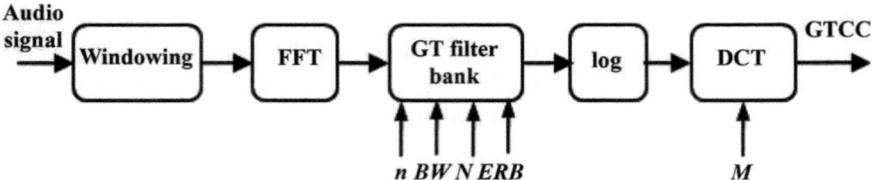

Fig. 25.2 Block diagram of GTCC extraction

1. MFCC describes the "LARGE" structures of the spectrum that is, and it focuses on the phonemes.
2. It ignores the fine spectral structures like pitch.
3. Works well in speech and music processing.

25.3.2 Gammatone Cepstral Coefficients (GTCC) [10–16]

During the hearing, the Gammatone cepstral coefficients capture the movement of the basilar membrane in the cochlea. GFCCs model the physical changes more closely and accurately that occur within the ear during the hearing and are therefore more representative of a human auditory system than Mel frequency cepstral coefficients. Unlike Mel filter bank, here Gammatone filter bank is used as it models the human auditory system and thereby uses ERB scale. This Gammatone filter bank is often used in cochlea simulation's front end thereby transforming complex sounds into multichannel activity as observed in the auditory nerve. The GFB is designed in such a way that the centre frequency of the filter is distributed in proportion to their bandwidth and is linearly spaced on the ERB scale. Generally, ERB (Equivalent Rectangular Bandwidth) scale gives an approximation to the filter bandwidths in human hearing. That is, it models the filter either as rectangular bandpass or band stop filters. The Gammatone filter is given by

$$g(t) = at^{n-1}e^{-2\pi bt}\cos(2\pi f_c t + \phi), \qquad (25.8)$$

where a is the amplitude factor, t is the time in seconds, n is the filter order, f_c is the centre frequency, b is the bandwidth, and φ is the phase factor. Figure 25.2 indicates the modules used for GTCC extraction.

25.3.3 How Many MFCC and GTCC Coefficients?

Traditionally, the first 12–14 coefficients are computed for the MFCC analysis and 42 coefficients for the GTCC analysis, since they retain the information about formants and the spectral envelope (vocal tract frequency response) which is the requirement of the analysis. Whereas, the higher-order coefficients retain information about the fast-changing spectral details (glottal pulse excitation).

25.3.4   Why Discrete Cosine Transform Instead of IDFT?

- As it is a simplified version of Fourier Transform
- By using Discrete Cosine Transform, we will be able to get only the real-valued coefficients thereby neglecting the imaginary values whereas in IDFT both real and imaginary valued coefficients will be computed.
- Discrete Cosine Transform decorrelates energy in different Mel bands.
- It reduces the dimension to represent the spectrum.

25.4   K-Means Clustering [6]

K-means clustering belongs to the group of exclusive clustering where data points belong exclusively to one group. K-means clustering is a vector quantization method, specifically from signal processing, that aims to partition k clusters to n observations in which each observation belongs to the cluster that has the nearest mean (cluster centres or cluster centroid), serving as the preliminary model of the cluster. We use k-means clustering rather than hierarchical clustering because k-means works on actual observations creating a single level of clusters rather than the variation between every pair of observations leading to a multilevel hierarchy of clusters. The variable K indicates the number of clusters. The algorithm then runs iteratively to assign each data points to one of the k groups based on the features provided. K-means works by evaluating the Euclidean distance between the data points and the centre of the clusters, thereby assigning the data point to the nearest cluster. Based on the extracted Mel frequency cepstral coefficients and Gammatone cepstral coefficients from the trained set of speakers, k-means clusters are formed. According to a k-means algorithm, it iterates over again and again, unless and until the data points within each cluster stop changing. At the end of each iteration, it keeps track of these clusters and their total variance and repeats the same steps from scratch but with a different starting point. The K-means algorithm can now compare the result and select the best variance out. For finding the value of K, you have to use the Hit and Trial method starting from $K = 1$. This $K = 1$ is the worst-case scenario as the variations among the dataset is large. Each time when you increase the number of clusters, the variation decreases. When the number of clusters is equal to the number of data points, then in that case variation will be zero. In this project, the optimal k-value is found to be 4 as the variation is reduced and a further increase in the number of clusters does not affect the variation much. That particular value of K is termed to be Elbow point. We have preferred k-means as a classifier rather than other mentioned classifiers since it is very simple to carry out and also applies to large sets of data like the Berlin database. It has the specialty of generalization to clusters of any shape and size. Each cluster is assumed to be an emotion feature, cluster centroid is formed using a k-means algorithm and finally, emotion data is classified.

25.5   Results and Discussions

A random speaker's utterances are tested against the trained cluster models to predict the emotional state of the test speaker. Such that the minimum distance obtained from the test phase is used to determine the recognition rate and further, we can develop a confusion matrix or average recognition rate table. The results obtained from both the features are tabulated below as Tables 25.1 and 25.2.

Based on the observation from the confusion matrix, it is evident that the Boredom emotion is misclassified as Neutral Emotion and the reason behind this can be dealt with qualitatively as well as quantitatively.

25.5.1   Qualitative Analysis

Qualitatively speaking, neutral emotion means there is no reaction, that is the reaction from the speaker is null or void whereas boredom emotion is one in which the speaker has a lack of interest in doing a particular action and has no feeling of excitement in doing the task which is more or less same as that of neutral emotion.

Table 25.1 Confusion matrix using MFCC and clusters

Emotions	Anger	Sad	Boredom	Neutral
Anger	**64.4**	0	0	35.6
Sad	0	**74.3**	6.9	18.8
Boredom	0	0	**12.9**	87.1
Neutral	0	0	15.8	**84.2**

Table 25.2 Confusion matrix using GTCC and clusters

Emotions	Anger	Sad	Boredom	Neutral
Anger	**91**	0	0	9
Sad	0	**74.3**	6.9	18.8
Boredom	0	0	**9.9**	91.1
Neutral	0	0	5	**95**

25.5.2 *Quantitative Analysis*

Quantitatively speaking, when the same speaker's same utterance is taken in both the emotion, the peak signal to noise ratio (PSNR) computed for each emotion is approximately the same. When the same speaker's same utterance is taken in both the emotion, the misclassification can be dealt with in three ways. They are:

- Feature Vector Analysis
- Histogram Analysis
- Cross-correlation Analysis.

Feature Vector Analysis: When the same speaker's same utterance is taken in both boredom and neutral emotion and feature extraction is carried out for the particular utterance, it is observed that the feature vectors of both the emotions get overlapped with each other. Figure 25.3 indicates the MFCC feature vector corresponding to boredom and neutral emotions. Figure 25.4 indicates the GTCC feature vector corresponding to boredom and neutral emotions.

Histogram Analysis: On extracting the features of a particular utterance from both neutral and boredom emotion, it is observed that the extracted feature vectors have similar sets of vectors for both emotions. Whereas alternatively when a histogram plot is plotted between the feature vectors of anger and boredom emotions there was less resemblance among the feature vectors. This proves that boredom and neutral have alike feature vectors that lead to misclassification. Figure 25.5 indicates the histogram plot between the samples of speech utterances corresponding to boredom and neutral emotions.

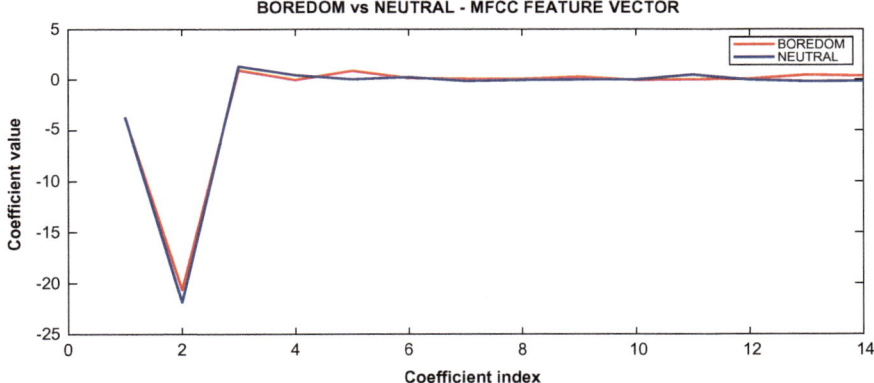

Fig. 25.3 Illustration—MFCC feature variation—boredom and neutral emotions

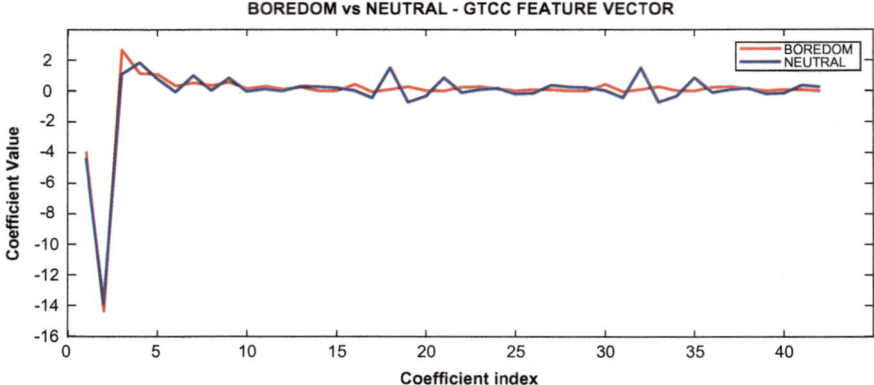

Fig. 25..4 Illustration—GTCC feature variation—boredom and neutral emotions

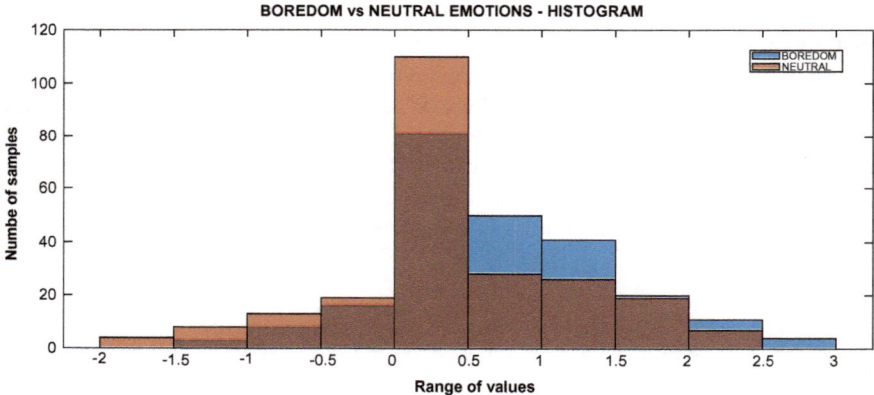

Fig. 25.5 Histogram Plot between speech utterances of boredom and neutral emotions

Cross-Correlation Analysis: When Cross-Correlation is taken between the feature vectors of boredom and neutral emotions, it is observed that the similarity or the relation between both the emotions corresponds to higher cross-correlation value. Alternatively, when cross-correlation is carried out between feature vectors of boredom and anger emotions it results in a lower cross-correlation value (Fig. 25.6).

Hence, these three quantitative analyses give insight into the reasons for misclassification of the two emotions, Neutral and Boredom by extracting features like Mel frequency cepstral coefficients and Gammatone cepstral coefficients thereby serving as the cause for less performance of the overall system.

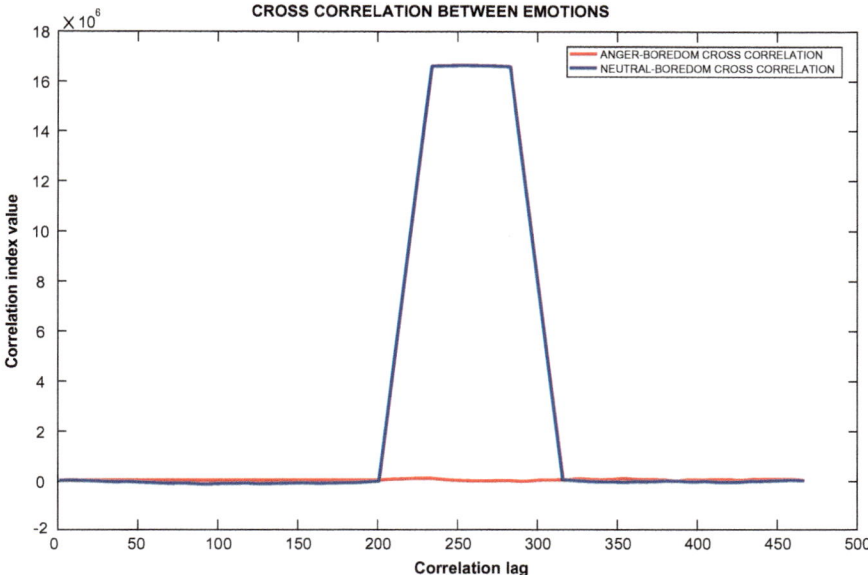

Fig. 25.6 Cross-correlation—(boredom and neutral) and (anger and boredom)

25.6   Conclusions

In this paper, an emotion recognition system is assessed by using MFCC and GFCC features, and VQ based minimum distance classifier is used to classify the emotions from the speech utterances spoken by actors. From the observations, it is evident that the GTCC feature tracks the emotional state precisely as compared to the MFCC feature. Since the GTCC feature captures the movement of the basilar membrane in the cochlea to mimic the human auditory system during the hearing, an emotion is accurately predicted compared to MFCC. The overall accuracy of the system using GFCC is better than MFCC. GTCC provides better overall accuracy of 68% as compared to MFCC whose overall accuracy is 59%. A combination of features can be considered for improving the performance. This emotion recognition system would find applications in the medical field to provide the required treatment to patients with mental disabilities.

Declaration We have taken permission from competent authorities to use the images/data as given in the paper. In case of any dispute in the future, we shall be wholly responsible.

References

1. Revathi, A., Jeyalakshmi, C.: Emotion recognition: different sets of features and models. Int. J. Speech Technol. **22**(3), 473–482 (2019). http://www.scopus.com/inward/record.url?eid=2-s2.0-85049946565&partnerID=MN8TOARS
2. Revathi, A., Jeyalakshmi, C.: Comparative analysis on the use of features and models for validating language identification system. In: Proceedings of the International Conference on Inventive Computing and Informatics, ICICI 2017, pp. 693–698 (2018). http://www.scopus.com/inward/record.url?eid=2-s2.0-85048347705&partnerID=MN8TOARS
3. Dharani, D., Revathi, A.: Singer identification using clustering algorithm. In: International Conference on Communication and Signal Processing, ICCSP 2014—Proceedings, pp. 1927–1931 (2014). http://www.scopus.com/inward/record.url?eid=2-s2.0-84916216617&partnerID=MN8TOARS
4. Revathi, A., Jeyalakshmi, C.: Efficient speech recognition system for hearing impaired children in classical tamil language. Int. J. Biomed. Eng. Technol. **26** (2018). http://www.inderscience.com/offer.php?id=89261
5. Daniel, D., Revathi, A.: Raga identification of carnatic music using iterative clustering approach. In: Proceedings of the International Conference on Computing and Communications Technologies, ICCCT 2015, pp. 19–24 (2015). http://www.scopus.com/inward/record.url?eid=2-s2.0-84961827474&partnerID=MN8TOARS
6. Revathi, A., Sasikaladevi, N., Nagakrishnan, N., Jeyalakshmi, C.: Robust emotion recognition from speech: Gammatone features and models. In: Proceedings of the 2nd International Conference on Communication and Electronics Systems, ICCES 2017 (2018). https://link.springer.com/article/10.1007/s10772-018-9546-1
7. Bombatkar, A., Bhoyar, G., Morjani, K., Gautam, S., Gupta, V.: Emotion recognition using speech processing using K-nearest neighbor algorithm. Int. J. Eng. Res. Appl. (2004). http://ijera.com/special_issue/ICIAC_April_2014/EN/V4/EN-2026871.pdf
8. Revathi, A., Jeyalakshmi, C., Muruganantham, T.: Perceptual features based rapid and robust language identification system for various Indian classical languages. Lect. Notes Comput. Vis. Biomech. (2018). https://link.springer.com/chapter/10.1007%2F978-3-319-71767-8_25
9. Revathi, A., Krishnamurthi, V., Jeyalakshmi, C.: Alphabet model-based short vocabulary speech recognition for the assessment of profoundly deaf and hard of hearing speeches. Int. J. Modell. Identif. Control **23** (2015). http://www.inderscience.com/offer.php?id=69932
10. Rahana, F., Raseena, P.E.: Gammatone cepstral coefficients for speaker identification. Int. J. Adv. Res. Electr. Electron. Instrum. Eng. **2** (2013). https://www.ijareeie.com/upload/2013/dec13-special/63_rahana_fathima.pdf
11. Wang, H., Zhang, C.: The application of gammatone frequency cepstral coefficients for forensic voice comparison under noisy conditions. Austr. J. Foren. Sci. (2019). https://www.tandfonline.com/doi/full/10.1080/00450618.2019.1584830
12. Martinez, J., Perez, H., Escamilla, E., Suzuki, M.M.: Speaker recognition using mel frequency cepstral coefficients (MFCC) and vector quantization (VQ) techniques. In: CONIELECOMP 2012, 22nd International Conference on Electrical Communications and Computers, pp. 248–251, Cholula, Puebla, Mexico (2012)
13. Valero, X., Alias.: Gammatone cepstral coefficients: biologically inspired features for non-speech audio classification. IEEE Trans. Multimed. **14**(6), 1684–1689 (2012). https://ieeexplore.ieee.org/document/6202347
14. Adiga, A., Magimai, M., Seelamantula, C.S.: Gammatone wavelet cepstral coefficients for robust speech recognition. https://core.ac.uk/download/pdf/192527929.pdf

15. Gabrielle, K.L.: Evaluating gammatone frequency cepstral coefficients with neural networks for emotion recognition from speech. Cornell University (2018). https://arxiv.org/abs/1806.09010
16. Ayoub, B., Jamal, K., Arsalane, Z.: Gammatone frequency cepstral coefficients for the speaker identification over VoIP networks. In: 2016 International Conference on Information Technology for Organizations Development (IT4OD), Fez, Morocco, pp. 1–5 (2016). https://ieeexplore.ieee.org/document/7479293

Chapter 26
Healing Blocks: Blockchain Enabled Decentralized Web Application for Securing Medical Records

Shivansh Kumar, Aman Kumar, and Bharti Ruhul Amin

Abstract Cyber-attacks on hospitals in this pandemic are rising day by day and affecting numbers of organizations. Due to the weak IT infrastructure in hospitals nowadays it is very easy for hackers to breach the hospital system. The Interpol Cybercrime Threat division reported a massive surge in ransomware attacks on hospital infrastructure to make the hospitals digitally hostage. Apart from that, there are many cases in which hospitals, medical apps, private medical practitioners fragment the patient's data which is not ideal due to corruption risks. Emerging technologies like blockchain, IPFS are some of the very promising technologies which can bring the paradigm shift in how we store our data. Decentralization is the future and can be a boon for data security in the healthcare industry. In this paper we are proposing a decentralized web application that will use IPFS for decentralized storage for medical data and blockchain will be used to secure the IPFS hashes and for access control. This application can provide the different healthcare organization a SaaS which can not only secure the patient's data but it will also help in proper storage of data as well as management of data.

26.1 Introduction

Cyber attacks on hospitals are affecting different organizations on a vast amount. Providing security to the data of millions of patients is getting harder and harder, Hackers are using the latest technology as well as different types of advanced attacks to breach the data in the medical organizations, and with the pandemic going on

S. Kumar (✉) · A. Kumar · B. R. Amin
DR SPM International Institute of Information, Chattisgarh, India
e-mail: shivansh18100@iiitnr.edu.in

A. Kumar
e-mail: aman18100@iiitnr.edu.in

B. R. Amin
e-mail: ruhul@iiitnr.edu.in

© The Author(s), under exclusive license to Springer Nature Singapore Pte Ltd. 2022   279
V. Bhateja et al. (eds.), *Evolution in Computational Intelligence*,
Smart Innovation, Systems and Technologies 267,
https://doi.org/10.1007/978-981-16-6616-2_26

hospitals are the favorite entity for hackers to attack and harm the victim as well as to create chaos. Most of the hospitals use third-party record management software which is usually not correctly security audited and can be easily breached and if we talk about servers some store the data locally and some use cloud service providers. Most of the time our data is stored in the server that belongs to the hospital or rented by the hospital which is not good. The IT staff of hospitals are not properly trained as well as the software they use is not properly updated. Many hospitals also don't plan proper cybersecurity auditing in their organization to check for any loopholes in the system. So reliable and secure storage of patient data is a very major challenge for healthcare entities. There are many cases of major cyberattacks on healthcare entities, some recent attacks which draw our attention are the attacks on Ambry Genetics, the California based clinic that faced a major email phishing attack which resulted in a data breach of 232,7772 patients, on the other hand, Florida orthopedic institute faced massive ransomware attacks which resulted in loss of data of 640,000 patients. Recently in September 2020, UHS Health services faced a cyber-attack that resulted in the death of 4 patients.

Now If we talk from the perspective of patients they are also facing many challenges, fragmentation of one's medical information across hospitals, private medical practitioners, and medical health apps are very common. Most of the time patients also face many issues while transferring their medical records from one hospital to another. Patients also find difficulties while accessing their vital medical information in case of emergencies and no hospital provides any kind of assistance in the fast track. Data leaks in hospitals, and selling of data to medicine companies are also very common nowadays. Manipulation of data by hospital authorities as well as unauthorized access to private medical data of any patients are very common and a very major issue. One underlying issue that the patients face is the issue while claiming for health insurance and no one talks about it. No hospital stores the data in a proper format so that patients can claim for their insurance easily.

Many researchers around the world are working on the usage of IPFS and Blockchain Technology to take advantage of the traceability and immutability of the Blockchain network as well as the secure distributed mechanism of IPFS to secure healthcare data. IPFS is a key technology which has a very vital role in storing the data in distributed database which results in peer-to-peer storage and hypermedia sharing method. A content-based address is returned after the file is added to the IPFS and the hash is used to retrieve the data. On the other hand,

Blockchains are a distributed ledgers technology which result in peer-to-peer decentralized network which don't require any third party like a bank, etc. The records of transactions are stored in the blocks and they can only be modified by using any consensus mechanism present in the system.

26.2   Technology Used

26.2.1   IPFS

Interplanetary File System (IPFS) is a peer-to-peer hypermedia protocol. This protocol is developed by Juan Benet for decentralized storage system.

A system called DHT (Distributed Hash system) provides a key-value store. The data is stored on the participating nodes and offers excellent performance and scalability. this system is also used in many peer-to-peer systems to maintain the metadata of the system. All the nodes which are participating in the IPFs network have a specified NodeID which is the hash of its public key. The participants nodes use their local storage to store the object data and these objects represent the files stored in the IPFS network. Each node participant has a DHT maintained in their system which is used to find the network address of the other peers in the network and peers which can serve a particular object. DHT helps IPFS to find peers across the network. A system named content addressing is used in IPFS, whenever a file is added in the system the five is converted into chunks of data of the size 256KB. The chunks included some objects and links, which are placed in a Merkle DAG. A single hash which returned from the system is called base CID, it is used to retrieve the file from the IPFS. This hash also makes sure there is no duplication in the network.

26.3   Attacks Prevalent on Current Systems

Hackers and threat actors are using different types of attacks on the current healthcare system. We have discussed some of the attacks further.

26.3.1   Denial of Service (DOS) Attack

In the Cybersecurity domain, a denial of service attack or DOS attack is a type of cyber attack on a machine or network which makes the resources unavailable or unusable temporarily and disrupts the service of the service. DOS attack is typically flooding of malicious requests which overuse the system resources and affect the legitimate user to access the service.

26.3.2   Distributed Denial of Service (DDOS) Attack

Distributed denial of service attack which is known as DDoS attack [11] is a malicious attack on a online service or system to make them unavailable or unreachable to real users by overloading it with requests from multiple malicious sources. It can be done using a group of system called botnet in which an attacker injects malicious code to many systems and order them to send millions of requests to the online services on which he wants to attack. This attack is also widely used by hacktivists for the sake of hacktivism in which numbers of fake requests overload the system and take it down.

26.3.3 Ransomware Attack

A ransomware attack is a very simple yet very effective attack in which an attacker is able to inject a malware which is a malicious software in the victim's system which results in encrypting all his data present in system and other system which are the parts of that network and after that attacker demands ransom to provide the decryption key.

26.3.4 Third-Party Data Breach

Third-party data breach happens when an organization is using a service of any third party for data management is attacked by hackers and the organization is unable to secure the data from their system and their sensitive data gets stolen or leaked through vulnerable software of service providers. This can result in a lot of damage as well as permanent data loss.

26.4 Related Works

There are a variety of researches available that discuss the use of decentralized systems to curb the data theft issues, we have read and analyzed some of the publications, survey papers, and internet articles:

A Telesurgery solution was provided by Rajesh et al. [1], named AaYusH. They have used technologies like 5G-enabled Tactile Internet (TI) for delivering real-time and ultra-responsive surgical services in remote areas without compromising the accuracy and precision. Telesurgery solutions proposed by different researchers have security, privacy, latency, and high cost for storing the data in blockchain networks. So they have presented an approach of solution based on smart contracts (Ethereum) with the power of IPFS technology. The smart contract will play a major role to make sure of the security and privacy issues while the IPFS technology is used for low-cost storage. Using this proposed system gives the advantage that there is no time wasted on paperwork and communication between the surgeon, patients, and caretaker will be reduced and the whole operation becomes more efficient. These smart contracts are self-executable rendering the involved parties cannot oppose the contract as the contract is stored on Ethereum Blockchain Ledger, thus providing a more optimal and efficient form of smart agreements.

The authors in Ref. [2] proposed "Healchain" solution consisting of several layers namely (1) User layer, (2) Blockchain layer, and (3) the Storage layer. In case of the user layer, the vitals information like blood pressure, heart rate, walking distance, etc., are collected by using the wearable devices (a wrist band, smartwatches). After that, it sends using a base station and WiFi access point to some specific servers which are part of a consortium blockchain node. Then in the blockchain layer, miners verify the data integrity that is received from the wifi access points [10] after authenticating the users. The authentication process is utilizing digital signatures signed into healthcare data, digital certificates, public keys of users, etc. Once the healthcare data is authenticated, then it is stored into the IPFS Storage, IPFS works as an off-chain solution for decentralized and secure storage, and the hash that is received after storing the data in IPFS is referenced in the blockchain.

The authors in Ref. [4] proposed a system which includes several layers. These are (1) User layer, (2) Query layer, (3) Data structuring, and (4) Controlling layer and Blockchain network & IPFS-storage layer. The responsibility of the User layer is used by the patient, the physician, the healthcare, and the insurance companies who have authority to store and retrieve data. Then comes the query layer that assigns the address and private key to the users who have successfully registered themselves to the network. The information-query layer is a combination of the database structures accessing, storage, and forwarding or reacting to made system queries. The query layer plays the role of an user interface to all the participants present on the network. Clients can use the query layer to request for the data. The requests are handled by the query system in such a format that is required by the data structuring and other layers to function.

The authors in Ref. [5] proposed "Telemedicine" for the solution of the remote diagnostics and also prescription facility. The advancement in technology is enabling cheaper and good-quality diagnostics and prescriptions in the remote areas where healthcare institutions are absent. In order to provide a better solution, they have developed a blockchain framework that can work efficiently in solving these issues. They are using ethereum smart contracts for developing a transparent and breach-proof framework.

Taras Filatov [6] proposed a system "Dappros", a decentralized encrypted IPFS based patient medical data storage which ensures immunity against server failures. This proposed system stores a database of the patient medical record in a distributed manner across multiple hospitals in private blockchain, the distribution helps the auditors to verify any change in data. This system is not based on any central authority structure so failure of any server does not result in the loss of patient's medical data, as long as there are other nodes present in the network running the IPFS database. As the data is encrypted using multi-signature access lock before storing on the IPFS network, this enables that the patient's data can't be accessed or unlocked unless both the patient's and the healthcare network's private keys are available.

Table 26.1 Literature comparison table

Ref. no.	Proposed solution	Shortcomings
Ref. [6]	The solution is Dappros The medical data is stored between the hospitals present in the blockchain network In a distributed manner using IPFS network protocol for verifying and auditing purposes.Multi-signature access lock is used to enable a secure and permissioned sharing mechanism	No proposal on how the network of hospital is created. No mention of which cryptography system is used. No data is provided regarding data formats
Ref. [7]	The Architecture consists of 3 parts 1st is registration of the entities in Consortium Blockchain network 2nd is the Medical report of patients are uploaded through web app which is verified by the miners 3rd is the verified data is pushed in the IPFS network and received content hash is stored in the blockchain network	The speed of downloading is slow which is not good for emergency. No information provides on how many types of data format this system supports. There is no info regarding we application Security
Ref. [2]	A system is "HealChain" The smart wearable devices are used to collect the medical data Then through verified access points this collected data is uploaded to the Consortium Blockchain network where miners verify the data. After verification the data is pushed in the IPFS network and received content hash is stored in the Consortium Blockchain network	There is no information regarding wifi AP security. registration process in the network is nor discussed in CBN. Increase in number of user can delay the verification process
Ref. [3]	The application generates a RSA key pair, the medical data is encrypted using a random AES key. This AES key is encrypted using the RSA key and is stored in the smart contract whereas the encrypted data is stored in the IPFS network	No information regarding how the network is developed. No retrieval process is proposed if the private key is lost. If a patient got breached then no proper mechanism provided for that too
Ref. [1]	This is a telesurgery solution, which has 3 components 1st component is the surgery robots and different communication mediums like 5G-Tactile Internet 2nd component reduces paperwork with Ethereum Smart Contract and uses IPFS for distributed data storage. 3rd component consists assisting technologies like Augmented Reality- Virtual Reality, 3-D modelling for the surgeons	System is not proposed for private blockchain. smart contract and not to used in the whole infrastructure. Proposed system included many bleeding edge technology which is not feasible in the present scenario
Ref. [4]	The model has 4 layers-user, query, data, and control layers User Layer-Consists of insurance companies, doctors, patients, and healthcare institutions. Query Layer- Generates private keys and addresses for the registered users Data Layer-Operates on the database Control Layer helps patients to control their data	system can be used only on single hospital during the case of emergency no mechanism is provided
Ref. [5]	The model "Telemedicine" enables remote healthcare The IPFS network is used for secure storage and Involvement proof is stored in Ethereum Smart Contracts	no mechanism to share data among different entities. This system can have scalibilty issues as it is only applicable for a single institution

Gururaj et al. [8] proposed a solution which includes technology like IPFS Asymmetric key cryptography, Symmetric key system, and Blockchain technology. In their scheme, it is highlighted that the encrypted file can be decrypted only with the previously generated random key. The key is now encrypted using RSA technique to make it more secure. Now, the encrypted data is sent to the IPFS network and after that the encrypted key is released publicly. The patient can retrieve their data by using their data hash value from the IPFS but to view the data first they need to decrypt the AES key meant for them.

Randhir et al. [9] proposed an architecture in which a consortium blockchain network is developed by registring different healthcare entities and the system provides them a POI which is proof of identity, which is a registration ID. Now, the healthcare entities use a web interface to upload the patients diagnostic reports. After that the local miners present in the network validate the transaction. The validated transaction is then stored in the IPFS database, and the retrieved hash value is stored in the blockchain network. The major advantages of this system are that only the entities which are part of the network can only access the transaction and one can register themselves using the peer registration process. We have also summarized the literature reviews in Table 26.1.

26.5 Proposed System

The architectural design is very important to the overall security of the application. Not only the architecture provides robustness but is also designed to keep all the data access transparent to public where all can see that when their data was requested by some authority and they will have the privilege to share or not. Since the platform is using the IPFS network to store the data there is no issue of data modification as any data on IPFS network is immutable and as it is stored in a distributed manner it is more secure in case of node failure. The major cost of the operation comes when the IPFS hash is stored in the Solidity Smart Contract along with the patient's social security number. The platform is using the social security number of the patients as an unique identifier to map the data to. For hospitals and doctors their registration number is used, not to mention any access should be given after full KYC process. The proposed architecture of our system is presented in Fig. 26.1.

During the sign-up process, all the relevant information and documents are collected, after that e-mail verification and setting up of 2-FA is done. Once all the verification is completed, the user can access their account through the website or the mobile application. For logging in, the user is required to provide their registered e-mail address and the password to their account. Following that, an additional 2-FA authentication is required to fully log in to the user's account.

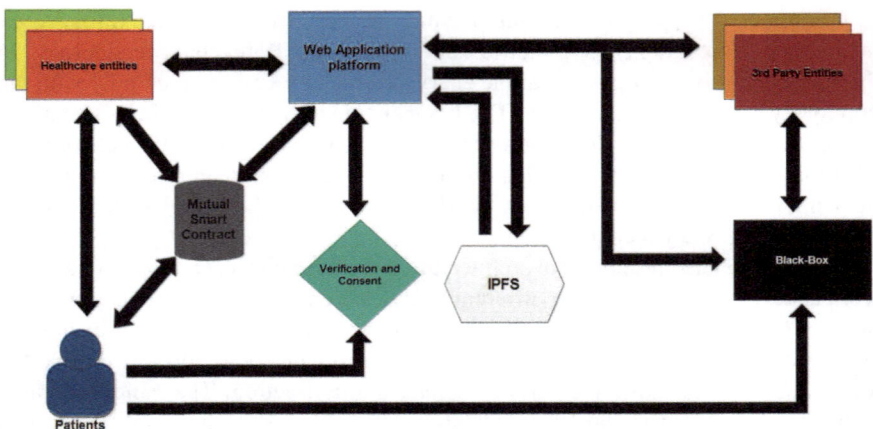

Fig. 26.1 Proposed decentralized web application architecture

First, we will discuss the medical data upload process. If the user wants to add the data they can add it from their dashboard, for doctors and hospitals it is required to first search the patient by using the social security number, and then they can either read the existing data or add new data of the patient. When a request is made to add data, the data is first uploaded to the platform and is encrypted in a predefined manner in which the data is first encrypted using the patient's public key and then the platform's public key is used to encrypt the data, till now the data is encrypted two times. When the data encryption is completed all the files are zipped and then are pushed to the IPFS network which in turn returns the IPFS Hash to the platform servers. Now when the platform has the IPFS hash, it creates a Solidity smart contract. The smart contract contains the social security number of the patient, registration number of the hospital or doctor, and the IPFS hash of the uploaded data, then this information is stored immutably in the smart contract which can be accessed in the future by anyone among whom the smart contract was formed. So the data is stored in the smart contract which is the IPFS hash and that IPFS hash is the address of the patient's data which was encrypted before adding it to the IPFS network.

Now, we must be very clear that the patient's data is stored in the IPFS network in a distributed manner and is immune to mass corruption or deletion as the IPFS network doesn't allow mutability. That is the reason it is also called "the permanent web" and also it doesn't allow duplication either so we are saving the storage cost when someone accidentally pushes the data more than one time.

The patients are only authorized to view their own information, but the doctors and hospitals can search for the other patient's information. To search the other patient's medical records, they have to put the social security number of the patient of interest. The social security number is mapped to the corresponding smart contract in which the IPFS hash is stored. Whenever any entity requests the medical data, the patient to whom this data belongs is informed and they have to approve this transaction with their consent. Now that the patient is given their consent the platform goes to the smart contract in which the information is stored and pays a fee to read the information inside. The IPFS hash is then used to obtain the zipped and encrypted data of the patient, this is moved to container which is referred as a "Black Box" in the architecture. First the data is unzipped and is first decrypted using the platform's private key and then an authorization request is made to the patient who allows the decryption and decrypted the data using their private key. Now the requester of the data is provided the data from the Black Box container.

26.5.1 Algorithms for Proposed System

Algorithm 1: Algorithm for uploading patients data

Result: data upload

//Checking authorization of hospital while data upload//

if *data.sender is not authorized* **then**

  | Return false;

else

  | Data = data.encrypt;

  | Upload document to IPFS network storage return hash;

end

Algorithm 2: Algorithm for mutual smart contract

Result: Mutual smart contract

// Adding necessary information of patients and making a mutual smart contract

Smart-Contract(Platform,Hospital ID,Patient ID, Patient name, Patient info, IPFS-hash, timestamp)

// Adding Smart contract to the blockchain //

Block Id = add contract(Smart-Contract)

Algorithm 3: Algorithm for verifying third-party and hash retrieval

---

Result: Patient data hash

//Third party entity request for patients data//

if *data.query == Patient social security number && Patient Id* **then**

| Return authorization time;

else

| query is wrong or any unusual behaviour Return False;

end

//Verifying the user and taking permission from them to share their data//

Verify.patient = Search(Patient Id, Social security number)

if *Verify.patient == true* **then**

| Ask patient to verify third party and consent for sharing data;;

| **if** *verify.thirdP == true && permission == true* **then**

| | Get the hash from Smart Contract and send it to Black-Box;

| **else**

| | Return False;

| **end**

else

| Return False;

end

Algorithm 4: Algorithm for Black-Box

Result: Patients data

//Decrypting data of the patients and sending it to third party entities//

if *data.hash is present* **then**

| Ask patient for the decryption key;;

| **if** *Patient.key == true* **then**

| | data.decrypt;

| **else**

| | Return False;

| **end**

else

| Return False;

end

26.6 Discussion and Further Research Aspects

This project is using IPFS and Ethereum Smart contracts to protect the patient's data from theft or unauthorized access. All users who are subscribed to the platform's service can access the data at any time without any hassle. IPFS is still in the development phase and can even replace the existing industry-standard- HTTP, some big organizations are already working on migrating to the IPFS ecosystem. This would be a gradual shift, Netflix is working to migrate its database to the IPFS network as it's more fast and robust. This proposed solution is also using the IPFS network for the same reason as to store the data in a distributed manner, also it is a faster and more secure way of storing data. The main cost of operation is the charges to execute the smart contracts in the Ethereum Blockchain. Whenever any new data is added to the patient's account, the IPFS hash would change to some new hash then the platform would make a new smart contract to store the new hash pointing to the newly added data in the IPFS network. This cost of executing the smart contract is not very high but if the data is added more frequently per user account then it can put us in a tough spot. On the other hand, reading the contents in a smart contract is very low as compared to the write. So we can do more read requests which would be the case as before prescribing some medicines or treatment the doctors will review the medical history of the patient. Overall there is some more cost associated with this solution than the currently used systems in place but that is not accounting the cost of employing security staff by the hospitals. If compared to the whole cost of operating in a traditional hospital system this proposed solution is cost-effective and more robust. Needless to say there is a scarcity of security professionals, and periodic system audits can still miss some attack vector. So, instead of hiring new talent and spending money on training them, the organization can opt for our platform where the seasoned security professionals and more secure infrastructure are already deployed in place.

26.7   Concluding Remarks

The main motive of this project is to provide a platform for all the hospitals and their patients where they can access their medical data, which is stored in a distributed manner. To achieve it, this paper proposes a decentralized platform for medical data access and this technology uses the concept of IPFS for storing the data. Further, the same work uses smart contract technology to provide high-level security against data breach. Our future objective on this work is to design a security protocol and its implementation for practical use.

References

1. Gupta, R., Shukla, A. and Tanwar, S.: Aayush: a smart contract-based telesurgery system for healthcare 4.0. In: 2020 IEEE International Conference on Communications Workshops (ICC Workshops), pp. 1–6. IEEE (2020)
2. Ni, W., Huang, X., Zhang, J., Yu, R.: HealChain: a decentralized data management system for mobile healthcare using consortium blockchain. In: 2019 Chinese Control Conference (CCC), pp. 6333–6338. IEEE (2019)
3. Kadam, S., Motwani, D.: Blockchain based E-healthcare record system. In: International Conference on Image Processing and Capsule Networks, pp. 366–380. Springer, Cham (2020)

4. Marangappanavar, R.K., Kiran, M.: Inter-planetary file system enabled blockchain solution for securing healthcare records. In: 2020 third ISEA conference on security and privacy (ISEA-ISAP), pp. 171–178. IEEE

5. Abugabah, A., Nizam, N., Alzubi, A.A.: Decentralized telemedicine framework for a smart healthcare ecosystem. IEEE Access **8**, 166575–166588 (2020)

6. Taras Filatov: https://www.dappros.com/202002/distributed-storage-of-permissioned-access-healthcare-patient-data-using-ipfs-and-blockchain/

7. Kumar, R., Marchang, N., Tripathi, R.: Distributed off-chain storage of patient diagnostic reports in healthcare system using IPFS and blockchain. In: 2020 International Conference on Communication Systems & NETworkS (COMSNETS), pp. 1–5. IEEE (2020)

8. Thimmaiah, C., Disha, S., Nayak, D., Diya, B.B., Gururaj, H.L.: Decentralized electronic medical records. Int. J. Res. Anal. Rev. **6**(1), 199–203 (2019)

9. Kumar, R., Tripathi, R.: Implementation of distributed file storage and access framework using IPFS and blockchain. In: 2019 Fifth International Conference on Image Information Processing (ICIIP), pp. 246–251. IEEE (2019)

10. Kumar, S., Bharti, A.K., Amin, R.: Decentralized secure storage of medical records using Blockchain and IPFS: a comparative analysis with future directions. Secur. Priv. e162 (2021)

11. Kumar, S., Amin, R.: Mitigating distributed denial of service attack: blockchain and software-defined networking based approach, network model with future research challenges. Secur. Priv. e163 (2021)

Chapter 27
Knowledge Graph Based Question-Answering System for Effective Case Law Analysis

Anu Thomas and S. Sangeetha

Abstract Case Law Analysis is a critical step in legal research, as every lawyer has to acquire the skill set to read, understand and apply case laws to augment their arguments and pleadings. We found that the interpretations of case laws provided by existing prominent legal tools are not complete in terms of providing some of the essential details required for effective case law analysis. This paper overcomes the above-mentioned challenge by proposing a judicial knowledge graph based question-answer system. Representing the judicial data in the form of knowledge graph ultimately benefits the legal professionals in analyzing case laws. The proposed approach supports querying the judicial knowledge graph in natural language text and then returns the results directly in natural language. We model the task of answering input queries as a cypher query pattern selection problem. Experimental results are quite satisfying in terms of efficiency and effectiveness in comparison to existing knowledge graph based question answering systems. To the best of the authors' knowledge, this work is the first to offer legal question answering based on knowledge graph.

27.1 Introduction

Knowledge Graphs have been very effective tools to benefit many applications related to Artificial Intelligence, such as websearch, Question & Answer (Q&A) framework, graph analytics, etc. A knowledge graph is a multi-relational graph depicting and storing facts as triples, where entities are represented as nodes and relations as edges. Many enterprises and research groups are now trying to compile their domain information into knowledge graph format, e.g., YAGO [6], DeepDive [11], Knowledge Vault [3], legal knowledge graph [4] and DBpedia [8]. Although these large-scale

A. Thomas (✉) · S. Sangeetha
Text Analytics & NLP Lab, Department of Computer Applications, National Institute of Technology, Tiruchirappalli, Tamil Nadu, India
e-mail: anugraha707@gmail.com

S. Sangeetha
e-mail: sangeetha@nitt.edu

V. Bhateja et al. (eds.), *Evolution in Computational Intelligence*,
Smart Innovation, Systems and Technologies 267,
https://doi.org/10.1007/978-981-16-6616-2_27

knowledge graphs have accumulated a huge amount of reliable world information, many domains are yet to be covered. For instance, the judicial domain. According to [5], the legal research is often performed with the help of a number of information sources namely, primary (case laws or judgments, legislation, and regulations), secondary (law journals), and tertiary sources (Legal citators). While case laws establish the law on a specific topic, it is sometimes difficult to find answers quickly in them. Secondary sources are referred to understand more about the legal principles. Thirdly, the tertiary sources, consisting of systematized sets of law reports, facilitate quicker case search and offer a concise list of the cited cases. However, searching the cases manually through the ever-growing repository can be slow and not always successful.

This paper proposes a question-answer driven approach to extract relevant information from case laws through querying the judicial Knowledge Graph. We model the task of answering input queries as a Cypher Query pattern selection problem. The proposed approach is divided into two sections namely cypher query pattern construction as well as natural language question answering module. For an input question, the proposed methodology constructs the final cypher query to be executed by choosing the matching pattern from the set of cypher query patterns. For creating the cypher query patterns, we used a set of 20 relevant questions that are formulated after consulting with domain experts. Since, domain-specific benchmark question set is not available, we paraphrased these questions and are used as input questions to evaluate the proposed approach. The major contributions of the proposed work are as follows: (1) We propose a novel approach to automatically generate cypher query patterns from a small set of natural language queries rather than relying on large scale of question-answer pairs as training data or by traversing the knowledge graph. (2) A knowledge graph based, Question-Answer driven Judicial Information Extraction system is proposed that can be used for effective Case Law Analysis or as an tertiary source of legal information. Moreover, the proposed work is the first of its kind which focuses on using Knowledge Graphs for better Case Law Analysis.

The rest of the paper is organized as follows. Section 27.2 describes the related works. Section 27.3 explains the proposed approach. In Sect. 27.4, experimental setup and results are discussed and in Sect. 27.5, we conclude the proposed work.

27.2 Related Works

The following section gives an overview of state-of-the-art knowledge graph based question answering systems and legal tools.

27.2.1 Related Works on Knowledge Graphs

Currently, several organizations are trying to assemble the domain knowledge into machine-readable format, such as knowledge graphs. These knowledge graphs vary in a variety of ways, such as domain specificity or methods used to query them. [1] presented a system called 'QUINT' for automatically generating templates that map a question to a triple pattern to query over a knowledge graph. The templates are learned by distant supervision, solely from questions paired with their answer sets as training data. For an input query, the system (i) automatically decomposes the question into its constituent clauses, (ii) computes answers for each constituent using our templates, and (iii) combines these answers to obtain the final answer. [12] proposed a system called 'QATemplate', which used templates to convert unstructured text query to a structured format. They modeled the problem as an uncertain graph join task where given a workload 'D' of SPARQL queries and a set N of natural language questions, the goal is to find some pairs ⟨q, n⟩, for q ∈ D ∧ n ∈ N, where SPARQL query q is the best match for natural language question. [13] proposed 'GAnswer' that generates structured queries by employing the syntactic parser. The proposed approach modeled the RDF Question/Answering to a subgraph matching problem.

27.2.2 Related Works on Legal Tools

There are a variety of legal tools available to search and process court decision files. LexisNexis[1] is a powerful legal research tool with access to court judgments, and other legal resources related to US judiciary system. Casetext[2] is a robust free tool for online legal research, which uses Artificial Intelligence to discover relevant judgments. Fastcase[3] is the legal research service based on American case laws. Fastcase's search software ranks the best cases first and enables the re-sorting of results to help users streamline their legal research. AI-powered research with ROSS Intelligence[4] is like consulting a lawyer. Its aim is to expedite the legal research by providing stronger arguments. ROSS Intelligence allows users to upload a case law and get back a list of cases cited. Westlaw[5] is an online legal research service for lawyers and legal professionals in the United States, the United Kingdom, and Australia, and is a product of Thomson Reuters. Manupatra[6] is a comprehensive database of Indian law. It contains federal Supreme Court decisions, state high court decisions, court rules, decisions of various administrative tribunals. Manupatra is offering a generic legal search based on case laws based on act and section number, or for cases by any combination of judge, appellant/respondent, subject, and case note. OLIS[7] is an initiative to develop an online legal information system for the Indian scenario. OLIS contains case laws of Supreme Court, High Courts, Trial courts, and

---

[1] http://www.lexisnexis.com.

[2] https://casetext.com/.

[3] https://www.fastcase.com/.

[4] https://rossintelligence.com/

[5] https://legal.thomsonreuters.com/en/products/westlaw.

[6] https://www.manupatrafast.com/.

[7] http://www.olisindia.in/.

Tribunals in India. Alike Manupatra, OLIS as well offers a generic search facility on case laws based on case number, subject, section number, or for cases by any combination of judge, appellant, respondent, and attorneys.

From the literature survey, we arrive at the following conclusion:

(1) Although Knowledge Graphs have been used by all types of sectors, the judicial domain remains to be covered.
(2) Interpretations of case laws provided by existing prominent legal tools are not complete in terms of providing some of the essential details required for effective case law analysis. For instance, existing tools search facility permits search by cases and statutes, date, judge, and jurisdiction. But, certain key information from case laws, such as *'cases and rules referred in a judgment', 'final decision in a judgment', 'crimes involved in a case'*, etc., are not covered by current legal tools.

In the light of the above shortcomings, this paper proposes a novel approach to query knowledge graph, which delivers the key information needed for legal decision-making. Judicial Knowledge Graph, that we use in the back end, is constructed and visualized using the Neo4J graph database tool. To store and retrieve data from the graph database, we use cypher query language. So, the proposed approach model the task of answering input natural language queries as a cypher query pattern selection problem. The following section describes the proposed work in detail.

27.3 Proposed Approach

The proposed approach is divided into two phases namely, cypher query pattern generation (offline phase) as well as natural language question answering (online phase). The following subsections describe each of these processes in detail.

27.3.1 Cypher Query Pattern Generation Phase

The proposed algorithm creates Cypher Query patterns corresponding to the natural language queries in the dataset. For any query 'q' in the dataset, algorithm initially formulates the corresponding natural language pattern through identifying its arguments and relation phrases. Arguments represent the named entities (such as names of Person, Organization, etc.), judicial concepts (terms such as petitioner, accused, respondent, complainant, etc.), and judicial facts (e.g., FIR numbers, Section Rules, and Case numbers) appearing in 'q'. For recognizing the judicial concepts such as 'cases', 'rules', petitioner', 'victim', 'accused', etc.) in the input question, proposed methodology utilizes the domain ontology called Judicial Case Ontology (JCO) [2], which depicts a conceptual model for judicial domain. The classes and properties defined in JCO are accessed by importing the ontology in XML format [7]. The

Algorithm 1 Cypher Query Pattern Creation

Input:
NLQ_{List} : List of natural language questions
Output:
CQ_{List} :List of Generic and cypher query patterns

1: Initialize $CQ_{List} \leftarrow \emptyset$
2: **for** each query $q \in NLQ_{List}$ **do**
3:     X $\leftarrow$ Arguments and relation phrases in q
4:     q' $\leftarrow$ Pattern_Creation (q, X)
5:     NLQ $_P \leftarrow \langle q, q' \rangle$
6: **end for**
7: $P_{similar} \leftarrow$ Group_Patterns(NLQ $_P$)
8: **for** each set of similar patterns $q_{similar} \in P_{similar}$ **do**
9:     Pattern$_{Generic} \leftarrow$ Generic_Pattern($q_{similar}$)
10: **end for**
11: **for** each tuple $\langle q_{similar}, P \rangle \in$ Pattern$_{Generic}$ **do**
12:     CQ $\leftarrow$ Cypher_Pattern_Creation (P)
13:     $CQ_{List} \leftarrow (q_{similar}, CQ)$
14: **end for**
15: Return CQ_{List}

---

named entities in 'q' are identified using the hybrid Named Entity Recognition (NER) approach proposed by Thomas et al. [10]. A semi-supervised, knowledge integrated pattern learning approach is followed to identify the judicial facts appearing in 'q'. The relation phrases in 'q' are identified using an Open Information Extraction (OIE) methodology.

Pattern_Creation(q,X): On identifying the arguments 'X', algorithm converts each natural language query 'q' to its corresponding natural language pattern form q' where the arguments and relations are numbered sequentially from left to right.

Group_Patterns(NLQ$_P$): Natural language patterns created in the previous step are grouped on the basis of Question type (such as 'What', 'How many'), and the number of arguments and relations phrases in them. Each group comprises a set of similar natural language queries, $q_{similar}$ and the corresponding natural language patterns, q'.

Generic_Pattern(q$_{similar}$): For each set of similar patterns $q_{similar}$ in $P_{similar}$, algorithm creates the generic natural language pattern by choosing the pattern with the highest number of tokens. Further, the set of similar patterns $q_{similar}$, and the corresponding generic natural language pattern, P are stored in the list of generic patterns, Pattern$_{Generic}$.

Cypher_Pattern_Creation(P): For each set of similar natural language queries $q_{similar}$, the final cypher query pattern, CQ is created from the generic natural language pattern, P by taking account of the question type, argument, and relations list in P. Finally, the set of similar natural language queries $q_{similar}$ and the corresponding cypher query pattern, CQ are stored in the lookup table, CQ_{List} for further use.

27.3.2   Question Answering Phase

Using the cypher query patterns, the question answering phase answers a given input query from the user based on two sub-tasks, namely, Question matching and Final cypher query formulation.

Question Matching: Proposed methodology compares the input query 'q' with the questions in the pattern lookup table to find out a matching question. To find out the most similar question, proposed methodology compares the semantic similarity between the input question and each question in the dataset using an online text similarity as a service (Dandelion API), trained using neural networks on a vast amount of data [9]. The question(s) with similarity score above a particular threshold, 'θ' (here, θ is set as 0.90) is added to the list of semantically similar questions. For each of the semantic similar questions, the corresponding cypher query patterns are also selected for further processing. The set of cypher query patterns and similar questions is passed to the final cypher query formulation phase.

Final Cypher Query Formulation: The proposed system identifies the named entity and judicial fact in the input question 'q' using the methodologies proposed by [10]. Then, it fills up the label slots of the arguments present in the corresponding cypher Query pattern by referring to the *Labels Table*, which contains the labels of entities present in the knowledge graph. For example, in the final cypher query, the terms such as 'Rule', 'Case', represent the labels of entities in the knowledge graph. Also, the labels of named entities (such as Person names) and judicial facts (such as FIR Number, Case Number, Section Rule), are also substituted by referring to the label table. On the other hand, phrases such as 'referred in', 'judged by' depicts the predicates in the knowledge graph. In addition to this, the arguments in the cypher pattern will be substituted with specific entity name or fact identified from query 'q'. At last, the final cypher query is executed on the judicial data stored in Neo4j, and the results are returned to the user.

27.4   Experimental Setup and Results

27.4.1   Dataset Used

The judicial knowledge graph used in this work is made up of judicial facts and relations collected from 1500 e-judgments [8] related to criminal cases. The structured data (such as facts and relations) extracted from these judgments are imported to Neo4j to test the proposed approach. There are 19,398 nodes, 17,435 edges in the graph. Since no standard question sets are available in judicial domain that can be

---

[8] http://hphighcourt.nic.in

Table 27.1 Questions used in cypher query pattern generation

Sl. no	Based on single document	Based on Multiple Documents
1	What are the crimes in Case No. 112/2008?	What are the cases judged by Justice Deepak Sharma?
2	What are the rules referred in Case No. 112/2008?	What are the cases with final decision 'disposed of'?
3	How many witnesses are in Case No. 112/2008?	What are the cases with crime 'murder'?
4	What is the FIR No. of Case No. 102/2009?	What are the cases referring to Section 104 of IPC?
5	Who is the accused in Case No. 102/2009?	What are the cases judged Justice Deepak Sharma with final decision 'disposed of'?
6	Who is the respondent in Case No.102/2001?	What are the rules referred in cases judged by Justice Deepak Mishra?
7	What is the final decision for Case No.102/2001?	What are the cases judged by Justice Deepak Mishra with crime 'murder'?
8	What are the cases cited in Case No. 102/2001?	What are the final decisions of cases judged by Justice Deepak Mishra?
9	Who is the judge in Case No. 102/2009 ?	What are the theft cases with final decision 'disposed of'?
10	Who is the appellant in Case No. 102/2009 ?	What are the cases which cite Case No. 123/2010?

used in the formulation of cypher query patterns, we consulted the legal experts and arrived at set of 20 questions. The questions used for creating cypher query patterns are divided into two categories namely, questions covering single document and multiple documents. Table 27.1 shows the set of questions used in cypher query pattern generation. Later, the proposed approach is tested on questions that are paraphrased from the initial set of 20 questions (Refer Table 27.2).

27.4.2   Results

Effectiveness results. Table 27.3 presents the results on test dataset queries and it is very clear that the proposed approach performs well in answering questions based on single documents and multiple documents. The reason behind the low recall scores of 'GAnswer' and , 'QATemplate' and 'QUINT' is due to the entity linking problem and the failure of the semantic relation extraction. Since GAnswer finds out all feasible matches for the input query, its precision is average. While template generation benefits QATemplate to answer more questions than GAnswer. For 'QUINT', the incorrect results provided by the POS tagger and dependency parser led to framing false queries and this contributed to the low precision score.

Table 27.2 Questions used in evaluating the proposed approach

Sl. no	Based on single document	Based on multiple documents
1	What are the crimes involved in Case No. 102/2009?	For which all cases Mr.Deepak Mishra is the judge?
2	What are the rules cited in Case No. 112/2008?	Which are the cases with final judgment 'disposed of'?
3	How many witnesses are there in Case No. 112/2008?	Which are the murder cases?
4	What is the FIR No. associated with Case No. 102/2009	Which are the cases citing Section 104 of IPC?
5	Case No.102/2001 is defended by whom?	What are the rules mentioned in cases judged by Justice Deepak Mishra?
6	What is the final judgment for Case No.102/2001?	Which all murder cases are judged by Justice Deepak Mishra ?
7	Which are the judgments mentioned in Case No. 102/2001?	What are the final decisions of cases handled by Justice Deepak Mishra?
8	Who has judged Case No. 102/2009 ?	What are the cases with final decision 'disposed of' and related to theft ?
9	Case No. 102/2009 is petitioned by whom ?	Case No. 123/2010 is cited by which all judgments?
10	In Case No.112/2009, who is the accused ?	What are the cases with final decision 'disposed of' and judged by Justice Deepak Sharma?

Table 27.3 Evaluating existing approaches on judicial data

Method	Precision	Recall	FScore
GAnswer	0.90	0.925	0.912
QUINT	0.927	0.931	0.929
QATemplate	0.93	0.928	0.928
Proposed approach	0.98	0.975	0.97

Table 27.4 Response time of each method (in minutes)

Method	Question understanding	Query formulation	Overall time
GAnswer	1.012	2.356	3.368
QUINT	2.961	2.59	5.55
QATemplate	1.326	2.332	4.15
Proposed approach	1.09	1.216	2.306

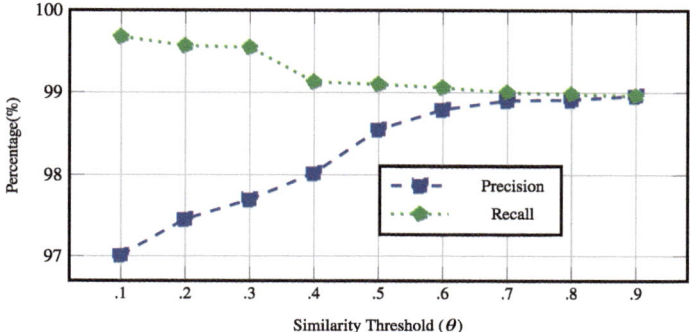

Fig. 27.1 Effect of similarity (θ) on effectiveness

Efficiency Results. We check the response time of each method. Table 27.4 shows that both GAnswer and QATemplate are good at natural language question understanding. For existing systems, the time cost increases for query formulation because of the underlying techniques used (such as graph matching, graph join, question decomposition, etc.). For understanding the input question and to frame the final query, QUINT decomposes the input question into its constituent clauses, followed by computing answers for each constituent using the templates. Finally, it combines these answers to obtain the final answer.

Effect of θ on effectiveness. Mapping an input query to the correct question in the pattern lookup table plays a major role in generating the final Cypher query as well as in the final results. For further analysis, we vary the value of θ from 0.1 to 0.9. Figure 27.1 depicts the results. Here, the precision and recall becomes almost consistent when θ is larger than 0.7. Hence, θ is suggested to be 0.9 (considering the maximum F-Score) in our experiments.

27.5 Conclusion

This paper proposes a knowledge on graph based question-answer driven method to make case law analysis efficient. The proposed system automatically generates cypher query patterns from a small set of natural language questions, based on which an input question from the user is translated into a particular cypher query. The proposed approach excels existing approaches in terms of efficiency and effectiveness, while evaluating on judicial data. In comparison to existing approaches, the proposed approach generates cypher query patterns without relying on a large scale of question-answer pairs or traversing the knowledge graph.

References

1. Abujabal, A., Yahya, M., Riedewald, M., Weikum, G.: Automated template generation for question answering over knowledge graphs. In: Proceedings of the 26th International Conference on World Wide Web, pp. 1191–1200 (2017)
2. Anu, T., Sangeetha, S.: A legal case ontology for extracting domain-specific entity relationships from e-judgments. In: Sixth International Conference on Recent Trends in Information Processing & Computing, pp. 305–312 (2017)
3. Dong, X., Gabrilovich, E., Heitz, G., Horn, W., Lao, N., Murphy, K., Strohmann, T., Sun, S., Zhang, W.: Knowledge vault: a web-scale approach to probabilistic knowledge fusion. In: Proceedings of the 20th ACM SIGKDD International Conference on Knowledge Discovery and Data Mining, pp. 601–610 (2014)
4. Filtz, E.: Building and processing a knowledge-graph for legal data. In: European Semantic Web Conference, pp. 184–194. Springer (2017)
5. Geist, A.: Using citation analysis techniques for computer-assisted legal research in continental jurisdictions. SSRN 1397674 (2009)
6. Hoffart, J., Suchanek, F.M., Berberich, K., Weikum, G.: Yago2: a spatially and temporally enhanced knowledge base from Wikipedia. Artif. Intell. **194**, 28–61 (2013)
7. Lamy, J.B.: Owlready: ontology-oriented programming in python with automatic classification and high level constructs for biomedical ontologies. Artif. Intell. Med. **80**, 11–28 (2017)
8. Lehmann, J., Isele, R., Jakob, M., Jentzsch, A., Kontokostas, D., Mendes, P.N., Hellmann, S., Morsey, M., Van Kleef, P., Auer, S., et al.: Dbpedia-a large-scale, multilingual knowledge base extracted from Wikipedia. Semant. Web **6**(2), 167–195 (2015)
9. Parmesan, S., Scaiella, U., Barbera, M., Tarasova, T.: Dandelion: from raw data to datagems for developers. In: Proceedings of the 2014 International Conference on Developers, vol. 1268, ISWC-DEV'14, p. 1-6. CEUR-WS.org, Aachen, DEU (2014)
10. Thomas, A., Sangeetha, S.: An innovative hybrid approach for extracting named entities from unstructured text data. Comput. Intell. **35**(4), 799–826 (2019)
11. Zhang, C., Ré, C., Cafarella, M., De Sa, C., Ratner, A., Shin, J., Wang, F., Wu, S.: Deepdive: declarative knowledge base construction. Commun. ACM **60**(5), 93–102 (2017)
12. Zheng, W., Zou, L., Lian, X., Yu, J.X., Song, S., Zhao, D.: How to build templates for RDF question/answering: an uncertain graph similarity join approach. In: Proceedings of the 2015 ACM SIGMOD International Conference on Management of Data, pp. 1809–1824 (2015)
13. Zou, L., Huang, R., Wang, H., Yu, J.X., He, W., Zhao, D.: Natural language question answering over RDF: a graph data driven approach. In: Proceedings of the 2014 ACM SIGMOD International Conference on Management of Data, pp. 313–324 (2014)

Chapter 28
SCSF: Supply Chain Sustainability Framework by Bayesian Theory and Markov Model for Risk Analysis

Pratyusa Mukherjee, Sudhansu Shekhar Patra, Swati Samantaray, Lalbihari Barik, and Rabindra Kumar Barik

Abstract Industry 4.0 is a predominantly used terminology to denote the current fourth Industrial Revolution in progress. The use of new technology such as IoT, cloud computing, robotic additive manufacturing and artificial intelligence, and how industry and other working areas are performed facilitates this high-tech metamorphosis. Simultaneously accelerating the changes exponentially, the current technological changeover egresses sustainability as one of the major issues in the domestic and international markets. Sustainability strategies should consider the ambivalence level and the future risks that organizational decisions may impose on the natural and social environments and the obligatory investment costs to make the supply chains more sustainable. The system's performance may be affected by the ripple effect caused by supplier base disruption propagating en route for the supply chain. Although the vulnerability assessments in single-echelon-single-event disruptions are critical for many firms, nevertheless the ripple effect modelling for multi-echelon-correlated events is becoming important. In this paper, two models have been discussed Markov Chain in conjunction with Bayesian Network to quantify the ripple effect.

P. Mukherjee
School of Computer Engineering, KIIT Deemed To Be University, Bhubaneswar, India

S. S. Patra (✉) · R. K. Barik
School of Computer Applications, KIIT Deemed To Be University, Bhubaneswar, India

S. Samantaray
School of Humanities, KIIT Deemed To Be University, Bhubaneswar, India

L. Barik
Department of Information Systems, Faculty of Computing & Information Technology in Rabigh, King Abdulaziz University, Jeddah, Kingdom of Saudi Arabia

© The Author(s), under exclusive license to Springer Nature Singapore Pte Ltd. 2022   301
V. Bhateja et al. (eds.), *Evolution in Computational Intelligence*,
Smart Innovation, Systems and Technologies 267,
https://doi.org/10.1007/978-981-16-6616-2_28

28.1   Introduction

In the contemporary interconnected world, the global supply chain networks (SCNs) play a crucial role in activating international trade and economic growth [1]. According to the protocol, linear supply chains were acquired towards complex systems mainly because of globalization and product specialization, as reflected in Fig. 28.1. Over the past few years, supply chains (SCs) started being progressively disrupted due to the expansion of global sourcing, conglomerated architecture of supply networks, diffusedespousal of demand-driven approach and sloppy production practices. The volatility of these disruptions and the aftermath of their propagation across multiple networks paves the way for a new research domain—the ripple effect in SCs [2, 3]. The ripple effect in SCs emphasizes the structural dynamics and downstream proliferation of the common disruptions and their corresponding consequences. A series of disruptions hamper the demand fulfillment. Detailed analysis of SC disruptions reveals that localization of disruption is practically impossible due to the network interconnections and outsourcing procedures, which drastically impact the performance [4–6].

The rest of the paper is organized as follows. Section 28.2 shows the related work. The Markov chain and Bayesian network for risk assessment in the supply chain are depicted in Sects. 28.3 and 28.4, respectively, and Sect. 28.5 concludes the paper with future research direction.

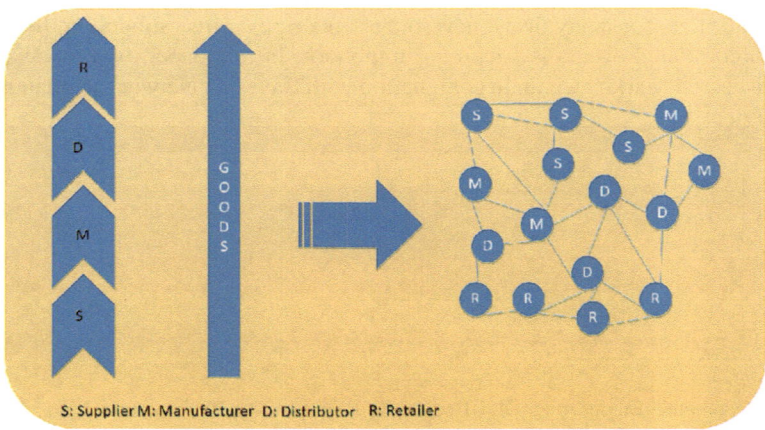

S: Supplier M: Manufacturer D: Distributor R: Retailer

Fig. 28.1   Linear SC to complex SC

28.2 Related Work

In the recent dynamic environment, a disruption or decision was taken by one level of SC systems hierarchy negatively influences the total SC system. The disruption caused by one level spreads quickly to the other levels of the hierarchy. If not appropriately analyzed and not controlled proactively, the total SC system may collapse [7, 8].

There is always a need to measure the effects of disruptions in the SC, but very little work has been done which studies its impact. The articles [9, 10] developed a dynamic model through simulation to analyze disruption's economic effect with disruption lengths as input. In [11], the author has introduced stochastic MILP to find the most suitable flow network to handle the disruption. In [12], the authors developed a Petri net model to study the encroachment of disruption on SC's service level.

Generally, the effect of disruption in SC is not localized, and as it cascades downstream, it degrades the performance of the SC system, and sometimes it halts the system. This causes an increase in delivery time, losing market share and reputation, the profit marginally decreases. In [13], the author proposed a multi-criteria approach to study the disruption in SC. The authors [14] suggested an entropy-based optimization technique to compute the economic loss because of the environmental disruption in SC. In [15], the authors have studied the environmental disruptions and their ripple effects, designed an optimized model, and evaluated the financial losses caused by the disruptions. In [16], the authors have used Bayesian network theory and analyzed and studied the simultaneous disruptions and their effects. It accesses the complex behaviour of risk and its propagation for improved SC.

This paper suggested two methods to study the disruption of SC's downstream using MC and BN.

28.3 Markov Chain for Risk Assessment in SC

Let the process be $Y_0, Y_1, Y_2, \ldots$

- Here the system states are observable and fully autonomous.
- Simplest of all Markov models.
- The state of the MC at instance t is equal to Y_t.

For example, if $Y_t = 5$, then the process is in state 5 during the time t.

The Markov chain's basic property is that Y_{t+1} depends on Y_t, but not upon Y_{t-1}, $Y_{t-2}, Y_{t-3}, \ldots, Y_1, Y_0$.

Mathematically, the Markov property is stated as

$$\pi(Y_{t+1} = j \mid Y_t = i, \ Y_{t-1} = i_1, \ldots \ldots Y_0 = i_0)$$
$$= \pi(Y_{t+1} = j \mid Y_t = i) \tag{28.1}$$

∀ $t = 1,2,3,....$and ∀ states.

Let $\{Y_0, Y_1, Y_2, ...\}$ be a sequence of discrete random variables. Then $\{Y_n, n >= 0\}$ is a MC if it satisfies the Markovian property as given in Eq. (28.1). In a MC, the future depends only on the present and not upon the past.

The matrix describing the Markov Chain is called transition matrix TM. The TM gives the probability of an entity which moves from one state to another within a model. A regular MC is characterized by a primitive TM, π. The primitivity in TM requires $\pi(i, j) < 1$ ∀ states i,

$$\pi = \begin{bmatrix} \pi_{11} & \pi_{12}.....\pi_{1n} \\ \pi_{11} & \pi_{12}.....\pi_{1n} \\ ... & ... & ... \\ \pi_{n1} & \pi_{n2}.....\pi_{nn} \end{bmatrix}.$$

This matrix is said to be TPM if

- It must be a square matrix.
- $\pi_{ij} \geq 0$ for all i and j.
- $\sum \pi_{ij} = 1$ (row-wise).

Entry (i, j) is the conditional probability that the NEXT state $= j$, given that NOW $= i$. In other words, the probability of going FROM state i TO state j

$$\pi_{ij} = \pi\{Y_{n+1} = j | Y_n = i\}. \tag{28.2}$$

Figure 28.2 shows the Markovian chain model, which depicts a disruption for a supplier as the state transition diagram. The behavior can be described using three different states:

1. Fully engage in the operational or functional state without the disruptions (state 0)
2. Fully disrupted state (state 2)

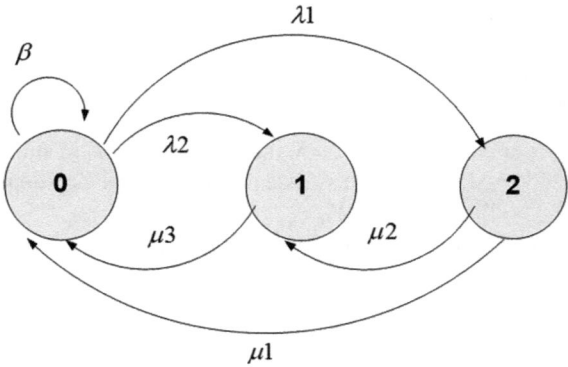

Fig. 28.2 State transition diagram for the supplier capacity disruption

3. The partial-disrupted state (state 1).

As per our assumption, during the partial-disrupted state, the probability of losing the supplier's capacity maybe 0.5.

Let the supplier be working in normal condition, and after the disruption, the supplier reached state 0. Once the supplier is at state 0, the supplier may stay at the same state represented by a self-loop or a 50% lost in its capacity and transferred to state 1, or the supplier lost all its operational capacity and transferred to state 2.

Now let us assume that the supplier is at a partial-disruption state (state 1) at time t. In time $t + 1$, the supplier may restore the capacity and, through backward transition may reach state 0.

Finally, let us take the supplier at a fully disrupted state (state 2) and in time $t + 1$ restore all its capacity and move to state 0, or 50% of the capacity is restored and moved to state 1, and the remaining may be restored in time $t + 2$.

Let π_0, π_1 and π_2 be the probabilities that the supplier at state 0, state 1 and state 2 respectively. The rate of transition is denoted by $\beta, \lambda_1, \lambda_2, \lambda_3, \mu_1, \mu_2, \mu_3$ respectively. The balanced and the normal equations are given by

$$\lambda_1\pi_2 + \lambda_2\pi_1 = (\beta + \mu_1 + \mu_3)\pi_0, \tag{28.3}$$

$$(\lambda_2 + \mu_2)\pi_1 = \mu_3\pi_0, \tag{28.4}$$

$$\sum_{i=0}^{3} \pi_i = 1. \tag{28.5}$$

28.4 Bayesian Network for SC Risk Assessment

28.4.1 Bayesian Network (BN)

BN, also known as belief network, is a directed graphical model called a directed acyclic graph (DAG), a causal graph. A few more graphical models represent certain independence relations or conditional relationships between the different variables in the domain. This helps in making tractable inferences where there are no complete connections or complete dependency among all variables. Let the variables be Y_1, $Y_2, \ldots Y_n$, X is interested in the domain. We also have other variables in that domain, and we are interested in the types of dependencies among them. We can represent the variables by nodes in a graphical model. We use arcs or edges to show the independent relations between the variables. The arcs represent the conditional independence relationships. The edge between nodes should be seen as a cause $\rightarrow$ effect relationship. Among the variables in the domain, some may be causes of the other or effects of other. By the representation of causality, a compact Bayesian network structure is formed. The representation of causality in BN makes the network more efficient.

28.4.2 An Illustration

Here, we will model the supplier disruption by BN. There is a single manufacturer (*M*) who is impacted by the suppliers' *s*1 and *s*2. *M* could face difficulty in its operation if *s*1 and *s*2 failed in supplying the raw materials. Figure 28.3 shows the dependency between manufacturer and supplier by a DAG. The disruption probability of each supplier is 4% means there will get a chance of 4% that each supplier fails to supply the raw materials to the manufacturer. Table 28.1 shows the CPT for the manufacturer disruption. For instance, the manufacturer disruption probability is 99% if the supplier (*s*1) and supplier (*s*2) are disrupted. Similarly, the CPT table shows manufacturer disruption probability is 2% if the supplier (*s*1) and supplier (*s*2) both are functional and so on. Figure 28.4 shows the probability distribution of *s*1, *s*2 and *M*, which is derived using Eq. (28.6).

$$P(M \text{ disruption}) = \sum_{s1,s2} P(M \text{ disruption} | s1, s2)$$

$$= P(M \text{ disruption} | s1 = \text{disruption}, s2 = \text{disruption})$$
$$\times P(s1 = disruption) \times P(s2 = \text{disruption})$$

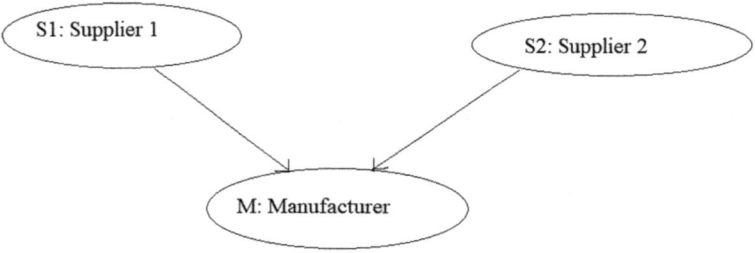

Fig. 28.3 A BN with a manufacturer (*M*) and two suppliers (*s*1 and *s*2)

Table 28.1 CPT for the manufacturer disruption

Supplier 1	Disruption	Functional	Disruption	Functional
Supplier 2	Disruption	Disruption	Functional	Functional
Manufacturer				
Disruption	0.99	0.89	0.86	0.02
Functional	0.01	0.11	0.14	0.98

Fig. 28.4 Probability distribution of $s1$, $s2$ (suppliers) and M (manufacturer)

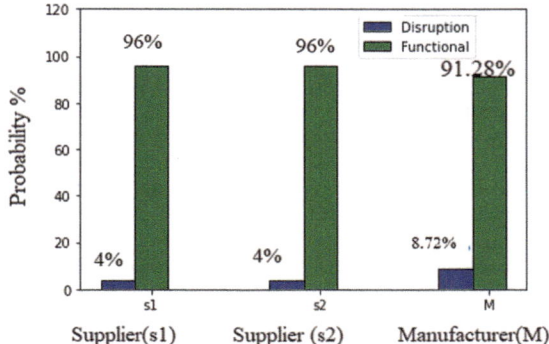

$$+P(M \text{ disruption} \,|s1 = \text{functional}, s2 = \text{disruption})$$
$$\times \ P(s1 = \text{functional}) \times \ P(s2 = \text{disruption})$$
$$+P(M \text{ disruption} \,|s1 = \text{disruption}, s2 = \text{functional})$$
$$\times \ P(s1 = \text{functional}) \times \ P(s2 = \text{disruption})$$
$$+P(M \text{ disruption} \,|s1 = \text{functional}, s2 = \text{functional})$$
$$\times \ P(s1 = \text{functional}) \times \ P(s2 = \text{functional})$$

$$= (0.99)(0.04)(0.04) + (0.89)(0.96)(0.04) + (0.86)(0.04)(0.96)$$
$$+(0.02)(0.96)(0.96)$$
$$= 0.0872 = 8.72\% \tag{28.6}$$

28.5   Conclusion and Future Work

The occurrence of disruption is a probabilistic concept, and hence its impact on the manufacturer followed by the retailers is computed. For any business firm, the disruption analysis in the supply chain illustrates an idea regarding the aftermath of its disruption. This paper first suggested a discrete-time Markov process with three states fully disrupted, semi-disrupted, and functional. Then BN is suggested to handle uncertainty which eases decision-making on CPT efficiently. In the future, the current study can be extended in several ways. The work can be extended for the dynamic measure of supplier disruption propagated from time to time. The second area could consider the hard measures in terms of operational and financial data, and its consequences can be studied, and the risks could be categorized and be prioritized.

References

1. Dolgui, A., Ivanov, D., Rozhkov, M.: Does the ripple effect influence the bullwhip effect? An integrated analysis of structural and operational dynamics in the supply chain. Int. J. Prod. Res. **58**(5), 1285–1301 (2020)
2. Hosseini, S., Ivanov, D.: Bayesian networks for supply chain risk, resilience and ripple effect analysis: a literature review. Expert Syst. Appl. **161**, 113649 (2020)
3. Hosseini, S., Ivanov, D., Dolgui, A.: Review of quantitative methods for supply chain resilience analysis. Transp. Res. Part E: Logist. Transp. Rev. **125**, 285–307 (2019)
4. Hosseini, S., Ivanov, D., Dolgui, A.: Ripple effect modelling of supplier disruption: integrated Markov chain and dynamic Bayesian network approach. Int. J. Prod. Res. **58**(11), 3284–3303 (2020)
5. Hosseini, S., Morshedlou, N., Ivanov, D., Sarder, M.D., Barker, K., Al Khaled, A.: Resilient supplier selection and optimal order allocation under disruption risks. Int. J. Prod. Econ. **213**, 124–137 (2019)
6. Ivanov, D., Dolgui, A.: A digital supply chain twin for managing the disruption risks and resilience in the era of industry 4.0. Prod. Plan. Control 1–14 (2020)
7. Kinra, A., Ivanov, D., Das, A., Dolgui, A.: Ripple effect quantification by supplier risk exposure assessment. Int. J. Prod. Res. **58**(18), 5559–5578 (2020)
8. Kir, H., Erdogan, N.: A knowledge-intensive adaptive business process management framework. Inf. Syst. **95**, 101639 (2021)
9. Scheibe, K.P., Blackhurst, J.: Supply chain disruption propagation: a systemic risk and normal accident theory perspective. Int. J. Prod. Res. **56**(1–2), 43–59 (2018)
10. Shakeri, M., Zarei, A., Azar, A., MalekiMinbashRazgah, M.: Green supply chain risk network management and performance analysis: bayesian belief network modeling. Environ. Energy Econ. Res. **4**(3), 165–183 (2020)
11. Sokolov, B., Ivanov, D., Dolgui, A., Pavlov, A.: Structural quantification of the ripple effect in the supply chain. Int. J. Prod. Res. **54**(1), 152–169 (2016)
12. Tong, Q., Yang, M., Zinetullina, A.: A dynamic Bayesian network-based approach to resilience assessment of engineered systems. J. Loss Preven. Process Ind. **65**, 104152 (2020)
13. Wang, Z., Zeng, S., Guo, J., Che, H.: A Bayesian network for reliability assessment of man-machine phased-mission system considering the phase dependencies of human cognitive error. Reliabil. Eng. Syst. Safety **207**, 107385 (2021)
14. Zhao, K., Kumar, A., Yen, J.: Achieving high robustness in supply distribution networks by rewiring. IEEE Trans. Eng. Manag. **58**(2), 347–362 (2010)
15. Levner, E., Ptuskin, A.: Entropy-based model for the ripple effect: managing environmental risks in supply chains. Int. J. Prod. Res. **56**(7), 2539–2551 (2018)
16. Ojha, R., Ghadge, A., Tiwari, M.K., Bititci, U.S.: Bayesian network modelling for supply chain risk propagation. Int. J. Prod. Res. **56**(17), 5795–5819 (2018)

Chapter 29
QH^2O: Energy Efficient Task Scheduling Using Quasi Reflected Harris Hawks Algorithm in Fog Environment

Lalbihari Barik, Sudhansu Shekhar Patra, Pratyusa Mukherjee, Jnyana Ranjan Mohanty, and Rabindra Kumar Barik

Abstract The volume of the data is increasing exponentially due to the enourmous growth of Industrial Internet of Things (IIoT), digitization with advancement of wireless technologies. Cloud computing was evolved as a technological solution for handling and processing voluminous data. With cloud computing, the next technological advancement known as fog computing is an emerging technology that helps manage a massive volume of IoTs data by delivering their services as well as applications toward nearer to the network edge. There is a rise concern for improving the Quality of Services (QoS) provided to the users all through cloud computing infrastructure. Transferring and communicating all the desired data to the cloud layer and moving back from the cloud layer affect the high latency as well as necessitate high amount of network bandwidth. To consider this situation, we considered the fog computing strategies which are assisting the cloud infrastructure for IIoT. In specific IIoT applications/services, there are sufficient amount of energy needed in the fog computing layer, which is one of the urging area for handling energy minimization scheme by tmany cloud computing service providers. Apart from energy minimization schme, Task scheduling is also crucial factor that adds toward the energy consumptions in fog computing servers. The present research paper proposed an energy-saving task scheduling algorithm based on a meta-heuristic named quasi reflected Harris Hawks optimization technique to achieve energy conservation by attaining the QoS.

L. Barik
Department of Information Systems, Faculty of Computing and Information Technology in Rabigh, King Abdulaziz University, Rabigh, Kingdom of Saudi Arabia

S. S. Patra (✉) · P. Mukherjee · J. R. Mohanty · R. K. Barik
School of Computer Applications, KIIT Deemed to be University, Bhubaneswar, India

© The Author(s), under exclusive license to Springer Nature Singapore Pte Ltd. 2022   309
V. Bhateja et al. (eds.), *Evolution in Computational Intelligence*,
Smart Innovation, Systems and Technologies 267,
https://doi.org/10.1007/978-981-16-6616-2_29

29.1   Introduction

Fog computing is a paradigm which is sandwiched in between the cloud server and IoT devices and renders assistance to appropriate users at the edge networks [1, 2]. The fog servers are geographically scattered and have a resource-constrained association with the cloud servers. It enables them to offer consistent services over a wide area. The fog computing framework enables several service providers to ensure their services at pretty fair rates. Another major merit of the said computing is that the devices are actually located in close proximity to the user rather than the cloud which reduces the latency to great extent [3, 4].The fog computing systems effectively recognize that IoT is enabling computation, control, storage, and networking capabilities. With an enormous rise in the number of data centers, there is a huge increase in energy consumption and CO_2 emission. Task scheduling in fog servers is extremely crucial as an inefficient schedule can cause high energy consumption. This paper proposes the novel Quasi reflected Harris Hawks (QHHO) algorithm [3] that aggregates Harris Hawks algorithm and quasi reflection-based learning system for task consolidation in the fog servers. The drawback of the meta-heuristic HHO algorithm is that it may be stuck at the local optima with iterations. We formulate a QH^2O scheme, a novel Quasi reflected Harris Hawks optimization scheme used to minimize energy consumption.

29.2   Related Work

Over the years, task scheduling is a hot topic among researchers to seamlessly process multiple jobs and design effective task scheduling algorithms. An extensive study suggests that the scheduling algorithms have proportionate impingement on energy consumed as well as resources utilized [1]. It is also highly recommended that proposed techniques conserve energy consumed between the IoT and fog layer. Meta-heuristic algorithms plays an important role in getting the solution for the task scheduling in fog layers [5]. The system's performance is examined on the basis of robust waiting buffer to dynamically scale the VMs and enhance the system's QoS parameters [4, 6]. Suitable algorithms for profit maximization along with conservation of energy and spot allocation can guarantee effective and commendable services in fog assisted cloud environment [4]. Task consolidation to save energy by reducing the unexploited Nano Data Centers in the fog computing ecosystem can gigantically enhance the CPU utilization.

29.3 Task Scheduling in Fog Environment

Fog nodes are available in between the cloud layer and the IoT devices, as in Fig. 29.1.

There is a set of m tasks $\{t_1, t_2, t_3, \ldots t_m\}$. A task ti is assigned a workload w_i. Each w_i can be represented in terms of MI (Million Instructions). Since each wi are different in size, we can say that the tasks are heterogeneous. Each VM_j has a processing speed s_j in terms of MIPS. $F(t_i) = VM_j$, which means a task t_i is assigned to VM_j at an instance. In the ETC matrix, ETC(i,j) shows the execution time of a task i, t_i in VM j VM_j.

$$ETC(i,j) = \frac{w_i}{s_j} \tag{29.1}$$

Four performance measures, makespan, energy consumption, cost, and CO_2 emission are considered.

Makespan is the sum of time taken by the resources in completeing the execution of all tasks.

$$MS = \max ET_j, \ j = 1, 2, \ldots n \ \text{where} ET_j \text{is the total execution time of } VM_j \tag{29.2}$$

Each VM can run either in active or in idle state. The energy consumed by a server is 0.6 times in an idle state. The energy consumption of a VM_j is the energy consumed both in an active and idle state.

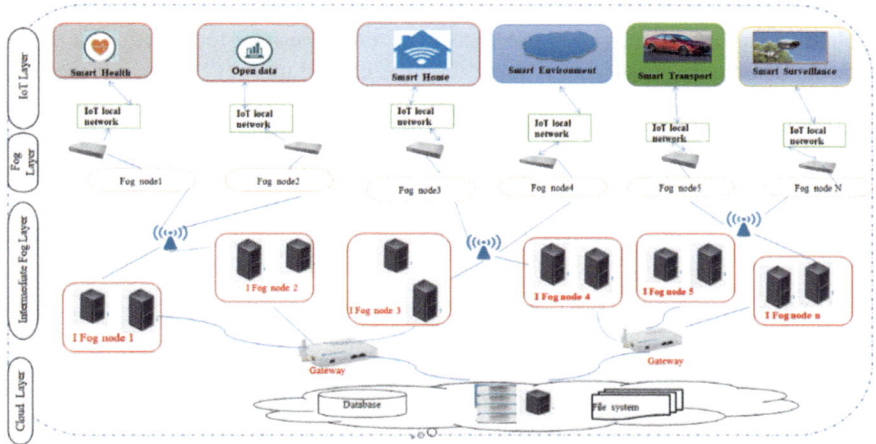

Fig. 29.1 The fog computing architecture [2]

$$\text{Energy} \left(\text{VM}_j \right) = \text{ET}_j * \delta_j + \left(\text{MS} - \text{ET}_j \right) * \gamma_j$$
$$\delta_j = 10^{-8} \text{X} \left(s_j \right)^2 \qquad (29.3)$$
$$\gamma_j = 0.6 * \delta_j$$

Here δj is the energy consumption in terms of joules/Million instructions during the VM's active state.

The sum of the energy consumption in the system is

$$Total\ Energy = \sum_{j=1}^{n} \text{Energy}(\text{VM}_j) \qquad (29.4)$$

Makespan as well as the Energy Consumption in the fog system are the two most important components influencing the other factors such as cost and CO_2 emission; we may use a bi-objective function to evaluate the candidate solution.

$$\text{Fitness} = \tau \text{ X total\_energy} + (1 - \tau) \text{ X MS} \qquad (29.5)$$

Since we are giving more attention toward energy consumption, τ is taken as 0.8.

29.4 The Basic Harris Hawks Optimization (HHO) Algorithm

A nature-inspired HHO is a population-based as well as gradient-free method and is applied to various optimization use cases with the necessary formulation. It works with the behavior of exploration along with the exploitation of the prey. It includes exploration as well as exploitation phases. Harris Hawks carries out several accesses to locate the prey based on the energy shift. The energy of prey minimizes in a noticiable manner. The prey energy is obtained as per the following equation:

$$E = 2E_0(1 - \frac{t}{W}) \qquad (29.6)$$

E and W are the prey's escaping energy and maximum repetitions respectively, and E_0 is the initial energy. E_0 varies between $(-1,1)$ in every iteration. The hawks softly encircle the rabbit and immediately pounced down the rabbit. In every iteration, there will be a decreasing tendency of the value E, which shows that the prey's energy is minimizing in steps in the process. During iteration if the escaping energy $|E|$ *greater than equals* 0.5, the hawks searches the other regions for discovering the location of the prey. The procedure goes to the exploration phase. If $|E|$ *less than* 0.5, the hawk is attacking the prey, and then procedure gets to the exploitation phase. The scenes of all the phases are given in the following subcategories.

29.4.1 Exploration Phase

Harris' hawks represent the population. In each iteration the prey location is the needed candidate solution. At a location randomly, the Harris hawks sits and then awaits for observing the prey with two schemes. Firstly, wait looking on the location of the group fellows with the prey having α less than 0.5, and secondly on randomly selected tall trees with status α greater than equals 0.5.

$$g(t+1) = \begin{cases} g_{randp}(t) - \gamma_1|g_{randp}(t) - 2\gamma_2 g(t)| & \alpha >= 0.5 \\ \\ (g_{rabbit}(t) - g_m(t)) - \gamma_3(LB + \gamma_4(UB - LB)) & \alpha < 0.5 \end{cases} \quad (29.7)$$

Here g(t)= Current location of hawks, g(t+1) = Hawks updated location, $g_{rabbit}(t)$ = prey's current location, $\gamma_{1..4}$ along with α are generated random numbers in range (0,1), $g_m(t)$ is the average location of the hawks calculated using AM.

29.4.2 Exploitation Phase

The prey invariably tries for escaping from the forbiddance position. The hawks have four distinct schemes for furrowing the prey to catch it depending on the prey's escaping behavior. A randomly generated number r is intended to refer the possibility of the prey.

r less than 0.5 prey is escaped with success, and r greater than equals 0.5 prey is escaped without success prior surprise pounce.

$|E|$ greater than equals 0.5 is the condition of soft besiege and, and $|E|$ less than 0.5 is the hard besiege.

29.4.2.1 Soft Besiege

If r greater than equals 0.5 with $|E|$ greater than equals 0.5, the prey have sufficient energy for escaping, therefore softly the hawks encircled the prey, makes the prey fatigued as well as executes surprisal pounce. The modeling is shown as follows:

$$g(t+1) = \Delta g(t) - E|J * g_{rabbit}(t) - g(t)| \quad (29.8)$$

$$\Delta g(t) = g_{rabbit}(t) - g(t) \quad (29.9)$$

Here $\Delta g(t)$ = difference in location in between the prey and hawks in the tth iteration, J equals 2 times $(1-\gamma_5)$ is the prey's jump intensity during the procedure of escape is a random number.

29.4.2.2 Hard Besiege

For r greater than equals 0.5 with $|E|$ less than 0.5, the prey being exhausted with low deflecting energy.

The current positions are updated as

$$g(t + 1) = g_{rabbit}(t) - E|\Delta g(t)| \tag{29.10}$$

29.4.2.3 Soft Besiege with Progressive Rapid Dives

If r less than 0.5 with $|E|$ greater than equals 0.5, the prey's energy is sufficient to escape efficiently. The hawks calculates the further move depending on the following equation:

$$Y = g_{rabbit}(t) - E|Jg_{rabbit}(t) - g(t)| \tag{29.11}$$

Comparing with the previous dive the previous or the current dive is selected by them. If the current one is better compared to previous, the levy flight (LF) is taken for the dive depending on the equation:

$$T = Y + RV \, X \, LF(D) \tag{29.12}$$

Here D is the dimension of the problem, RV is the vector of size 1 X D, LF is the levy flight is computed as

$$LF(x) = 0.01 \, X \, \frac{u \, X \, \sigma}{|v|^{1/\beta}} \tag{29.13}$$

Here u, v are the values in the range (0,1) generated randomly; β equals 1.5 and σ is evaluated as

$$\sigma = \left(\frac{\Gamma(1 + \beta)X \, \sin(\pi\beta/2)}{\Gamma((1 + \beta)/2)X\beta X2^{((\beta-1)/2)}} \right)^{1/\beta} \tag{29.14}$$

$$\text{And finally } g(t + 1) = \begin{cases} Y, & for \ G(g(t)) > G(Y) \\ T, & for \ G(g(t)) \ > G(T) \end{cases} \tag{29.15}$$

In the further iteration, improved position Y or T is selected.

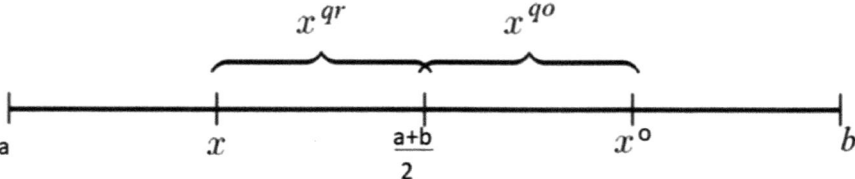

Fig. 29.2 In the domain [a,b] x, xo, xqo, xqr (the pt, opposite pt, quasi-opposite pt and quasi-reflected pt)

29.4.2.4 Hard Besiege Having Progressive Rapid Dives

If r less than 0.5 with $|E|$ less than 0.5, the energy of the prey is not sufficient to escape, and a hard besiege is being constructed.

g(t + 1) is exact equation as (15). Here Y as well as T is derived by:

$$Y = g_{rabbit}(t) - E|Jg_{rabbit}(t) - g_m(t)| \qquad (29.16)$$

and T is computed similar as Eq. (29.12), and $g_m(t)$ is computed as the AM and is the hawks' mean position.

29.4.3 Quasi-reflected Learning

A new learning algorithm, Quasi-Reflection grounded Learning, grounded on opposition-based as well as Quasi Opposition Based; the quasi reflected number x^{qr} is calculated by Eq. (29.17). Figure 29.2 shows all the points x, x^o x^{qo} and x^{qr}.

$$x^{qr} = rand\left(\frac{b+a}{2}, x\right) \qquad (29.17)$$

29.5 Results Analysis

We simulated the proposed scheme using MATLAB R2014a. The dataset is randomly generated. The datacenter is considered with 500 to 1500 tasks having 15 VMs in each case. Figures 29.3 and 29.4 shows the energy consumed Versus the tasks size for the state of art with the proposed algorithm. The graph depicts that the HHMH algorithm outperforms the PCO and TLBO algorithms in terms of energy consumption.

Fig. 29.3 No of tasks versus energy consumption

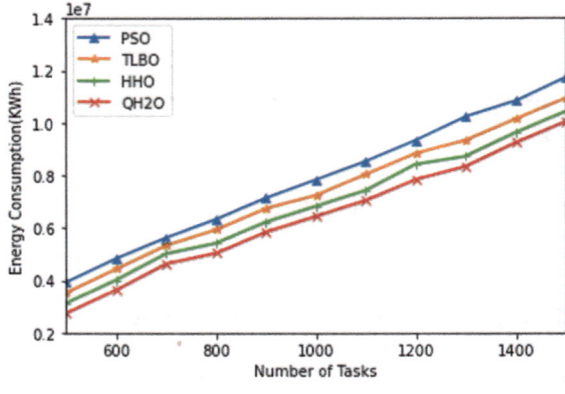

Fig. 29.4 Comparison of various algorithms on the total energy in the average case for 500 tasks and different VMs

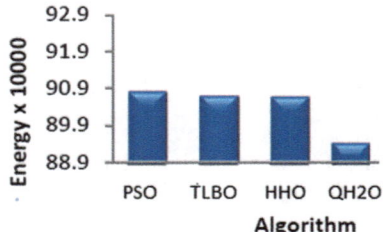

29.6   Conclusion

In this paper, a meta-heuristic algorithm quasi Harris Hawks optimization algorithm is proposed for the task scheduling problem on fog environment for independent tasks. A simulation study has done to compare the performance of the proposed algorithm.

QH^2O with HHO, PSO (Particle Swam optimization) and TLBO (Teaching Learning Based Optimization) techniques. The results of the proposed algorithm outperforms the state of art algorithms in energy-saving and green computing. In future, the proposed algorithm will also test for dependent tasks in a fog environment.

References

1. Al Ahmad, M., Patra, S.S., Barik, R.K.: Energy-Efficient Resource Scheduling in Fog Computing Using SDN Framework. Progress in Computing, Analytics and Networking. Springer, Singapore, pp. 567–578 (2020)
2. Barik, R.K., et al.: FogGIS: Fog Computing for geospatial big data analytics. In: 2016 IEEE Uttar Pradesh Section International Conference on Electrical, Computer and Electronics Engineering (UPCON). IEEE (2016)

3. Goswami, V., Patra, S.S., Mund, G.B.: Performance Analysis of Cloud with Queue-dependent Virtual Machines. In: 2012 1st International Conference on Recent Advances in Information Technology (RAIT), pp. 357–362. IEEE (2012)
4. Rout, S., et al.: Energy Aware Task Consolidation in Fog Computing Environment. Intelligent Data Engineering and Analytics. pp. 195–205. Springer, Singapore (2021)
5. Okay, F.Y., Ozdemir, S.: Routing in fog-enabled IoT platforms: a survey and an SDN-based solution. IEEE Internet J. **5**(6), 4871–4889 (2018)
6. Mukherjee, P., Patra, S.S., Pradhan, C., Barik, R.K.: HHO algorithm for cloud service provider selection. In: 2020 IEEE International Women in Engineering (WIE) Conference on Electrical and Computer Engineering (WIECON-ECE). IEEE, (2020)

Chapter 30
3D CNN Based Emotion Recognition Using Facial Gestures

Kuppa Sai Sri Teja, Thummala Vivekananda Reddy, Mourya Sashank, and A. Revathi

Abstract Emotion is an integral part of everyone's life. With developments in new products in everyday life, it is needed to identify what people think of that in an instant procedure, so better solutions could be provided. For the identification of emotions, it is required to observe the speech and facial gestures of the user. Everywhere in the world is equipped with cameras from laptops to high buildings. With the abundant data of images, it is necessary to build a model that can detect emotion using images or frames without trading for either speed or accuracy. With advancements in computer vision and deep learning, an emotion of the user identity is detected with high accuracy; since computer vision is an automated process, the results are much faster than the regular methods of manual surveys. In this project, the SAVEE database is used, which comprises audio and amp; visual features of seven unique types of emotions; and these emotions are identified by using CNN-based systems exploiting facial gestures of actors. Important features from the faces of the actors in the database are extracted and trained using existing deep learning methods namely 3D convnets. Testing has provided the maximum accuracy of 95.83% for emotion recognition from facial gestures.

30.1 Introduction

Expression in the face is one of the most important ways to express emotional intention. Emotion recognition from computer vision has drawn attention because of its application in human–computer interaction mental health analysis, improving customer service, etc. Especially in the situations of driver fatigue or the intensive emotional state of the person, it is necessary to detect it pre-hand. In the field of enterprises, algorithms are made in a way to determine the review of the product using

K. S. S. Teja · T. V. Reddy · M. Sashank · A. Revathi (✉)
Department of ECE, School of EEE, SASTRA Deemed to be University, Thanjavur, Tamilnadu, India
e-mail: revathi@ece.sastra.edu

© The Author(s), under exclusive license to Springer Nature Singapore Pte Ltd. 2022   319
V. Bhateja et al. (eds.), *Evolution in Computational Intelligence*,
Smart Innovation, Systems and Technologies 267,
https://doi.org/10.1007/978-981-16-6616-2_30

facial analysis and eyeball tracking. With advancements in deep learning and computational power, researchers started experimenting with image data. Facial emotion recognition (FER) systems are classified into two types, static and dynamic FER. We have research works related to detecting facial key points and then identifying the emotion using machine learning techniques, where the results are not that promising. But due to changes with computational technology convolutional neural networks can learn up to some extent to perform, since these can learn from positional invariant data. The real world has a change in emotion during a speech in a rapid way; we can decide just using a single frame to decide it. We have enough SOTA models for static image classification, but in real-time due to problems like occlusions, pose variations, and changes in illumination, these 2D convnets have less ability to perform well. So for a dynamic way of classification, researchers started building 3D convnets and LSTM neural nets to capture temporal details even. But as the model layers increase the memory and computation time increase. We have focused on building a neural network that has minimal layers and without adding computation of finding the key positions in the face to detect the emotion with great accuracy. In this, we have provided a brief overview of advancements in the field of emotion classification using different datasets. For emotion classification, we have various datasets in the formats of speech, static images, and dynamic sequence of frames (videos). Most of the research has progressed in the speech processing domain. Some research works have shown promise by the construction of two separate neural nets consisting of bidirectional LSTM for temporal and CNN for spatial analysis [1, 2]. Datasets and algorithms [3] are analyzed for FER. Emotion recognition from the speech [4] is done using MFCC and wavelet features. Emotion recognition from the speech [5] is performed using SVM. Video-based emotion recognition [6, 7] is done using CNN. Due to high advancements in machine learning algorithms like the random forest, SVM, Adaboost, and Xgboost researchers [8] can achieve high accuracy for the classification of emotion with speech features. Perceptual features and CNN are used [9] for emotion recognition from speech. Local binary patterns are used for facial expression recognition. Many features are extracted from the speech data like ERB-PLPC, MFCC, zero-crossing rate, wavelet features, and spectrograms. Just using just vocal features, even deep neural networks are built and trained for achieving better performance. Competitions such as FER2013 and EmotiQ have provided data with static images. Sequences of images provide a better understanding of emotion rather than a still image. 3D convnets helped to learn both the spatial information and temporal sequence while training. But traditional CNN can observe the spatial features perform well, but when it comes to temporal features LSTM or GRU memory cells are too used. These methods need to extract the facial key points to observe patterns using the LSTM network. This methodology helped to attain a promising result.

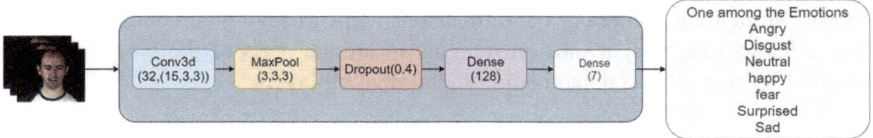

Fig. 30.1 Schematic—3D CNN based emotion recognition using facial gestures *Neural Network Layers*

30.2 Implementation of Emotion Recognition Using Facial Gestures

30.2.1 Dataset and Pre-Processing

The SAVEE database which we used consists of seven unique emotions for classification purposes. It is an Audio-Visual database. We have focused on the visual part alone for this paper. We have split the data into the train (70%), validation (10%), and testing (20%) for each emotion. The average time duration of the video for each emotion is around 3 s. We have used OpenCV for reading frames and resizing. Since we have opted for a 3D convnet as an input for the model, we need a fixed number of frames. With 60fps, in general, the total number of frames we get is around 180(60*3). Due to even computational reasons, we have taken only 96 frames by leaving equidistant frames within frames. This was done using the landscape function in NumPy. Even this, not an ideal way to do it, because we might miss some important frames for the classification process, but in our condition, it did not affect much due to the higher frame rate. Figure 30.1 describes the procedure for the work on 3D CNN based emotion recognition using facial gestures.

30.2.2 Convolution Layers

Convolutional layers are the essential layers that extract the key features in an image. A kernel of matrix 3*3 is convolved with each image in the 3d matrix. When convolved we extract the edges and shapes of a specific emotion. The convolutional operations performed by this layer are matrix multiplications, but we pass them through a nonlinear activation function at the output.

30.2.3 Max Pooling Layer

Max pooling is a robust method in convolutional layers that extract the highest number in a matrix of 2*2 which helps to reduce the dimensions without losing the important features in the data. Reducing the dimensions helps in saving computational costs. This helps to gain insight into movements in unique parts of the face from each emotion. Relu is used as an activation function at all layers.

30.2.4 Dense Layers

The output from the convolutional layers is then flattened to a 1d array and then given to the dense layers. The dense layer is a linear operation that is performed after all the important features that are extracted from the convolutional layers. A series of dense layers of shapes 128, 7 are placed to classify a particular emotion.

30.2.5 Dropout

A dropout of 0.4 is added to avoid the problem of overfitting in the neural network. It makes four out of 10 neurons dead for an epoch and then retains back. Using this way each neuron can learn and produce good results.

30.2.6 Experimentation Settings

We have used SGD optimizer and categorical cross-entropy loss with a learning rate of 1e-4 and have used early stopping. The model is trained on Tesla-K80 GPU, up to around 150 epochs. We have even reduced the learning rate in case validation loss does not decrease up to a patience of 20 epochs. We have done the entire work using python along with the Keras library with TensorFlow as a backend.

Categorical Cross Entropy Loss:

$$Loss = -\sum_{i=1}^{outputsize} y_i . log y_i$$

30.3 Results and Discussions

We have used 3D convnets to capture both spatial and temporal features from the sequence of frames. Inspired by the implementation of this paper we have built the deep learning model. We have tried the implementation but this leads to overfitting

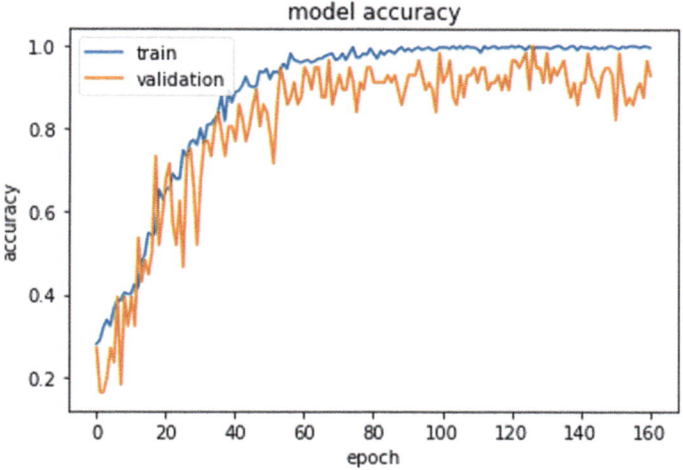

Fig. 30.2 Accuracy versus Epochs—Emotion recognition using facial gestures

and this additional computation for identification of facial points and 2 separate neural networks for processing the data is another problem. We aim to build a faster mechanism to predict without the loss of the Spatio-temporal data. The kernel filter size plays an important role in the high accuracy classification purpose. We have achieved an accuracy of 95.83% on testing data. Table 30.2 tells about the complete classification metrics of all the emotions. In our experiment, the angry emotion is misclassified with disgust and neutral emotion. Emotions like neutral, happy, fear, and sad have the highest accuracy of 100% classification. Figure 30.2 indicates the accuracy versus no. of epochs for the training and validation data to perform emotion recognition from facial gestures.

Figure 30.3 indicates the model loss versus no. of epochs for the proposed 3D CNN-based emotion recognition from facial gestures.

Table 30.1 depicts the comparative analysis between the earlier works and the proposed work.

Table 30.2 indicates the individual performance of the recognition of emotions from facial gestures using CNN.

Table 30.3 indicates the confusion matrix for the proposed 3D CNN-based emotion recognition from facial gestures.

30.4   Conclusions

In this work, a 3D convolutional network is used for recognizing emotions from facial gestures. Video frames of faces considered for training for different emotions are given as input to the CNN and models are created for each emotion. Test video

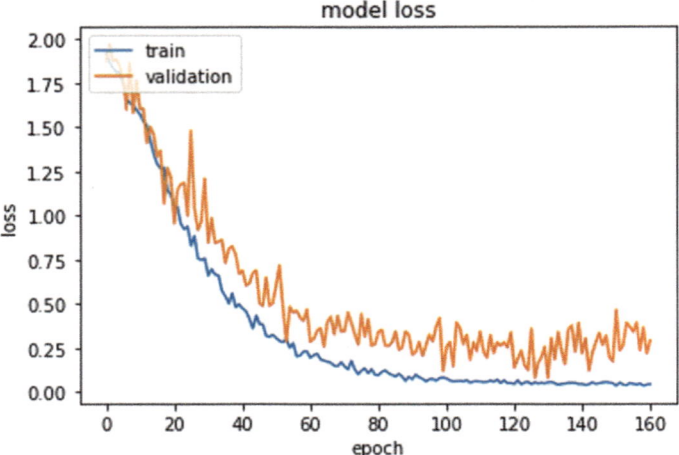

Fig. 30.3 Model loss versus Epochs—Emotion recognition using facial gestures

Table 30.1 Comparative analysis—emotion recognition

Method	Accuracy
Banda and Robinson	95.0
Haq et al.	95.4
CCCNN	93.9
This paper	95.8
Human evaluation	88.0

Table 30.2 Performance assessment of emotion recognition using CNN

Emotion	Precision	Recall	F1-Score
Angry	0.91	0.83	0.87
Disgust	0.92	0.92	0.92
Neutral	0.96	1.00	0.98
Happy	0.92	1.00	0.96
Fear	1.00	1.00	1.00
Surprise	1.00	0.92	0.96
Sad	1.00	1.00	1.00

frames are given to the models, and models are chosen as associated with test images of faces in different emotions. This algorithm has provided 100% accuracy for the emotions neutral, happy, fear, and sadness emotions. However, the performance is relatively low for anger, disgust, and surprise emotions. We have used only facial features for classification. Our future works include adding speech features along with facial features, since in many research works it is proven that vocal has a high

Table 30.3 Confusion matrix—3D CNN based emotion recognition from facial gestures

Emotions	Angry	Disgust	Neutral	Happy	Fear	Surprise	Sad
Angry	84	8	8	0	0	0	0
Disgust	8	92	0	0	0	0	0
Neutral	0	0	100	0	0	0	0
Happy	0	0	0	100	0	0	0
Fear	0	0	0	0	100	0	0
Surprise	0	0	0	8	0	92	0
Sad	0	0	0	0	0	0	100

importance even in classification. We have even fixed a number of frames because of using 3D convnets, and we would expand our research to working by using memory cells like LSTM and GRU layers, since these can use variable lengths of input.

References

1. Zhang, K., Huang, Y., Du, Y., Wang, L.: Facial expression recognition based on deep evolutional spatial-temporal networks IEEE Trans. Image Process **26**(9), 4193–4203 (2017)
2. Jung, H., Lee, S., Yim, J., Park, S., Kim, J.: Joint fine-tuning in deep neural networks for facial expression recognition. In: 2015 IEEE International Conference on Computer Vision (ICCV), pp. 2983–2991. IEEE (2015)
3. Li, S., Deng, W.: Deep facial expression recognition: a survey. In: IEEE Transactions on Affective Computing, (2020) https://doi.org/10.1109/TAFFC.2020.2981446
4. Kishore, K.K., Satish, P.K.: Emotion recognition in speech using MFCC and wavelet features. In: 2013 3rd IEEE International Advance Computing Conference (IACC) Ghaziabad, pp. 842–847 (2013). https://doi.org/10.1109/IAdCC.2013.6514336
5. Sinith, M.S., Aswathi, E., Deepa, T.M., Shameema, C.P., Rajan, S.: Emotion recognition from audio signals using support vector machine. In: 2015 IEEE Recent Advances in Intelligent Computational Systems (RAICS) Trivandrum, pp. 139–144 (2015). ase10.1109/RAICS.2015.7488403
6. Fan, Y., et al.: Video-based emotion recognition using CNN-RNN and C3D hybrid networks. In: Proceedings of the 18th ACM International Conference on Multimodal Interaction (2016)
7. Sharma, S., Shanmugasundaram, K., Ramasamy, S.K.: FAREC—CNN based efficient face recognition technique using Dlib. In: 2016 International Conference on Advanced Communication Control and Computing Technologies (ICACCCT), Ramanathapuram, pp. 192–195 (2016) https://doi.org/10.1109/ICACCCT.2016.7831628.
8. Noroozi, F., Kaminska, D., Sapinski, T., Anbarjafari, G.: Supervised vocal-based emotion recognition using multiclass support vector machine, random forests, and adaboost. J. Audio Eng. Soc. **65**7/8, (2017). https://doi.org/10.17743/jaes.2017.0022
9. Revathi, A., Nagakrishnan, R., Vashista, D.V., Teja, K.S.S., Sasikaladevi, N.: Emotion recognition from speech using perceptual features and convolutional neural networks (2020). https://doi.org/10.1007/978-981-15-3992-3_29

Chapter 31
Breast DCE-MRI Segmentation for Lesion Detection by Multilevel Thresholding Using Arithmetic Optimization Algorithm

Dipak Kumar Patra, Tapas Si, Sukumar Mondal, and Prakash Mukherjee

Abstract It is becoming more common to employ Dynamic Contrast-Enhanced Magnetic Resonance Imaging (DCE-MRI) to diagnose breast disorders. DCE-MRI, which is assisted by Computer-Aided Design (CAD) methods, shows its effectiveness in detecting breast cancer. In image segmentation, multilevel thresholding is an important and easy to implement technique. This paper proposes breast DCE-MRI segmentation by multilevel thresholding using the Arithmetic Optimization Algorithm (AOA). First, the anisotropic diffusion filter is used to denoise MR images, and then, the correction of Intensity Inhomogeneities (IIHs) is performed. The lesions are then retrieved from the segmented images and located in the initial MR images. 50 Sagittal T2-Weighted DCE-MRI images are used to test the suggested approach. Particle Swarm Optimizer (PSO) and Hidden Markov Random Field (HMRF) are compared to the suggested AOA technique. The devised technique achieves a high level of accuracy of 99.80%, sensitivity of 98.06%, and Dice Similarity Coefficient (DSC) of 85.52%. The devised method outperforms the two compared methods.

D. K. Patra (✉)
Research Centre of Natural and Applied Sciences (Department of Computer Science),
Raja Narendralal Khan Women's College (Autonomous), Midnapore 721102, India
e-mail: dpatra11@gmail.com

T. Si
Department of Computer Science and Engineering, Bankura Unnayani Institute of Engineering,
Bankura, India

S. Mondal
Department of Mathematics, Raja Narendralal Khan Women's College (Autonomous),
Midnapore 721102, India

P. Mukherjee
Department of Mathematics, Hijli College, Kharagpur 721306, India

© The Author(s), under exclusive license to Springer Nature Singapore Pte Ltd. 2022   327
V. Bhateja et al. (eds.), *Evolution in Computational Intelligence*,
Smart Innovation, Systems and Technologies 267,
https://doi.org/10.1007/978-981-16-6616-2_31

31.1 Introduction and Background

DCE-MRI is a new method for detecting breast cancer. A dedicated surface coil is used for DCE-MRI. During imaging, the patient is vulnerable, with her breasts lying centrally inside the receiver volumes [1]. Meantime, the scanner captures images before, during, and after the intravenous contrast negotiator is administered. This style of imaging offers far more data than typical imaging techniques (mammography or ultrasound). As a result, DCE-MRI data processing takes longer time and is more difficult. In breast MRI and DCE-MRI, automatic segmentation of the breast area is a major preprocessing phase in the automatic lesion detection [2]. A well-thought-out segmentation strategy could cut down on both the computational burden and the sources of False Positives (FP) [3–5]. The breast, as well as a portion of the patient's chest and a wide area of background, are visible in the DCE-MR images. Due to high-rise noise levels, low contrast, area inhomogeneity, and coil artifacts of the images, segmenting the breast region is difficult. There have been numerous segmentation algorithms created. Shannon thresholding is one of the most common techniques, which is focused on the fact that the sensitivity of the breast area in MRI images is higher than that of other structures. Some researchers suggested a fully automated algorithm for segmenting the breast area based on the anatomical skill of the pectoral muscle properties. It is less susceptible to noise and inhomogeneities in the field than the thresholding procedure [2]. In the current work, AOA [6] is used in Shannon entropy maximization for segmentation of the lesions in breast DCE-MRI. The devised AOA-based method is compared with HMRF [7] and PSO [8] methods. The performance of this technique was evaluated using Accuracy, Specificity, Sensitivity, Geometric mean (G-mean), Precision, False Positive Rate (FPR), F-measure, and DSC. We assessed the overall performance using Multi-Criteria Decision-Making based on the aforementioned criteria. The devised AOA-based method outperforms the compared methods in the experiments.

The remainder of the paper is structured as follows: related work is discussed in Sect. 31.2. Section 31.3 describes the materials and methods. The devised segmentation approach based on the AOA algorithm is presented in this section. The experimental setup is given in Sect. 31.4. Finally, Sect. 31.6 provides a conclusion and recommendations for future work.

31.2 Related Work

Krizhevsky [9] introduced AlexNet with five convolutional and three completely connected layers. The image classification challenges on the ImageNet dataset won by this network. Using a convolution network as a function extractor and a Support Vector Machine (SVM) as a classifier, Arevalo et al. [10] achieved a Receiver Operating Characteristic (ROC) of 86%. Jiao et al. [11] was able to identify tumors as benign and malignant with a precision of 96.7%. They used the Digital Database for Screening Mammography (DDSM) dataset and a Convolutional Neural Network (CNN) as a function extractor and an SVM as a classification system. Rejusha and Kavitha [12] devised a procedure in which they first tried to split the pectoral muscle before attempting to diagnose the lump. In contrast, Jumaat et al. [13] explored the use of active contouring methods for mass segmentation of the breasts in ultrasound images. Lin et al. [14] examined mass segmentation in the form of segmentation and shape scanning. Patra et al. [15] devised for lesion detection of breast DCE-MRI segmentation by thresholding applying student psychological based optimization. The proposed automatic segmentation method has an accuracy of 99.44%, a sensitivity of 96.84%, and a DSC of 93.41%. Si and Mukhopadhyay [16] developed a lesion detection method using Fireworks Algorithm (FWA)-based clustering in breast DCE-MRI. The segmentation approach based on modified hard-clustering with FWA has been developed. The lesions are retrieved from the segmented images. Kar and Si [17] developed for lesion detection using an MVO-based clustering algorithm of breast DCE-MRI segmentation. A modified hard-clustering with MVO is developed for this purpose. Patra et al. [18] developed a Grammatical Fireworks Algorithm (GFWA)-based lesion segmentation method for breast DCE-MRI. The GFWA-based clustering approach is used to segment the preprocessed MR images. From the segmented images, the lesions are retrieved. The proposed approach is being tested with 25 DCE-MRI slices from 5 patients. Ribes et al. [19] devised a statistical approach based on MRF for the automated segmentation of MR images in the breast. Using K-Means clustering and Cuckoo Search Optimization, Arjmand et al. [20] developed breast tumor segmentation. After testing the techniques using the RIDER breast dataset, they found that other techniques outperformed the controlled learning segmentations. Piantadosi et al. [21] suggested U-Net with the Three Time Points (3TP) strategy suitable for improving the performance of lesion segmentation. Si and Mukhopadhyay [22] developed clustering with fireworks algorithm for lesion detection in the segmentation of the breast DCE-MRI images. The use of AOA for breast DCE-MRI segmentation through entropy maximization is devised in this work. The devised method is discussed next.

31.3 Proposed Method

Figure 31.1 depicts the devised AOA-based segmentation method in outline.

31.3.1 Breast MRI Dataset

Total 50 T2-Weighted Sagittal DCE-MRI images of 10 patients are collected from TCGA-BRCA [23, 24]. The size of all MRI slices is 256×256.

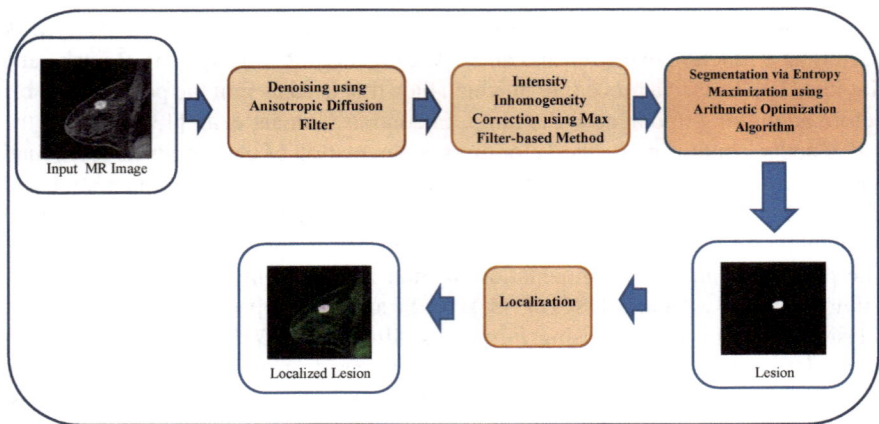

Fig. 31.1 The devised method's outline

31.3.2   Preprocessing

The visual feature of an MRI serves a critical role in appropriately identifying a treatment, which can be harmed by noise present throughout the acquisition procedure. The Anisotropic Diffusion Filter (ADF) [25] is used to denoise the MR images, and the IIHs are rectified using a Max filter-based method [26].

31.3.3   Segmentation

Segmentation is a technique for isolating lesions from the background by splitting the DCE-MRI into nonoverlapping portions. This breast lesions segmentation methodology employs a number of algorithms, including global thresholding, classical methods, and local thresholding based on picture histograms. The work provides pixel-based strategies for area expansion, in which the AOA process, a metaheuristic optimization algorithm, creates seed point and homogeneity parameter thresholds.

Entropy Maximization
Entropy maximization is used to establish segmentation threshold values after the denoised image has been acquired. The goal of Entropy Maximization is to increase the number of homogenous positions between them. To avail the pixel frequency in the image, the image histogram is determined. The entropy value is then calculated using pixel frequencies from histograms. The intended solutions are integer numbers in the [0, 255] span, because the image has gray values between 0 and 255.

Shannon entropy [27] is used in this work. This principle is applied to decide the amount of knowledge showed by any data in a probabilistic manner. Let's say an image has homogeneous $(k + 1)$ regions and k gray threshold measures at $t_1, t_2, t_3, \ldots, t_k$.

$$h(i) = \frac{f_i}{N} \quad i = 0, 1, 2, \ldots, 255$$

where $h(i)$ indicates normalized frequency, f_i indicates the frequency of ith gray level, and The image's gray levels is indicated by N.

The Shannon Entropy Function is defined as follows:

$$H = -\sum_{i=0}^{t_1} P_{1i} \ln(P_{1i}) - \sum_{i=t_1+1}^{t_2} P_{2i} \ln(P_{2i}) - \cdots - \sum_{i=t_k}^{255} P_{ki} \ln(P_{ki}) \qquad (31.1)$$

where,

$$P_{1i} = \frac{h(i)}{\sum_{i=0}^{t_1} h(i)} \qquad \text{for } 0 \leq i \leq t_1,$$

$$P_{2i} = \frac{h(i)}{\sum_{i=t_1+1}^{t_2} h(i)} \qquad \text{for } t_1 + 1 \leq i \leq t_2,$$

$$P_{ki} = \frac{h(i)}{\sum_{i=t_k+1}^{255} h(i)} \qquad \text{for } t_k + 1 \leq i \leq 255.$$

The entropy function H (Eq. (31.1)) is used as an objective function and it is maximized using AOA for searching the optimum threshold regions.

Arithmetic Optimization Algorithm

The following sub-sections of the AOA describe the mathematical exploration and manipulation processes carried out by the (A "+"), Subtraction (S "−"), Division (D "÷"), and Multiplication (M "×"). The pseudocode of AOA is provided in Algorithm 31.1.

The best obtained solution is the best candidate solution in every iteration, and the optimization process of AOA starts with a series of randomly generated candidate solutions (X). Before beginning the work, the AOA should decide on the search process. As a result, the Math Optimizer Accelerated (MOA) is a coefficient manipulated by Eq. 31.2 in the below search steps.

$$a(C_{Iter}) = Min + C_{Iter} \times \left(\frac{Max - Min}{M_{Iter}} \right) \qquad (31.2)$$

where $a(C_{Iter})$ is the function cost at the tth iteration, and C_{Iter} is the current iteration, which is between 1 and (M_{Iter}). The accelerated function's minimum and maximum costs are denoted by Min and Max, respectively.

Algorithm 31.1: AOA

---

1 Initialize the parameters α, μ

2 Start the N solutions' locations haphazardly.

3 **while** $(C_{Iter} < M_{Iter})$ **do**

4 | Find the Fitness Function for the given results

5 | Find the best solution found so far

6 | Renovate a using Eq. 31.2

7 | Renovate p using Eq. 31.4

8 | **for** $(i = 1\ to\ Solutions)$ **do**

9 | | **for** $(j = 1\ to\ Positions)$ **do**

10 | | | Initiate r_1, r_2, and r_3 randomly in [0, 1]

11 | | | **if** $r_1 > a$ **then**

12 | | | | Exploration phase

13 | | | | **if** $r_1 > 0.5$ **then**

14 | | | | | Apply the Division math operator (D " $\div$ ")

15 | | | | | Renovate the ith solutions' locations using the first rule in Eq. 31.3

16 | | | | **else**

17 | | | | | Apply the Multiplication math operator (M " $\times$ ")

18 | | | | | Renovate the ith solutions' locations using the second rule in Eq. 31.3

19 | | | | **end**

20 | | | **else**

21 | | | | Exploitation phase

22 | | | | **if** $r_3 > 0.5$ **then**

23 | | | | | Use the Subtraction math operator (S " $-$ ")

24 | | | | | Renovate the ith solutions' locations using the first rule in Eq. 31.5

25 | | | | **else**

26 | | | | | Apply the Addition math operator (A "$+$ ")

27 | | | | | Renovate the ith solutions' locations using the second rule in Eq. 31.5

28 | | | | **end**

29 | | | **end**

30 | | **end**

31 | **end**

32 | $C_{Iter} = C_{Iter} + 1$

33 **end**

34 Return the best solution (x)

The exploratory conduct of AOA is discussed in this section. Mathematical calculations involving the M or D operators generated high distributed values or decisions, according to the Arithmetic operators. Because of their high dispersion A and S, these operators M and D, unlike others, cannot easily approach the target. A function based on 4 mathematical operations is used to see the effect of the dissimilar operators' distribution values. As a result, after many attempts, the discovery search finds the near-optimal solution. Furthermore, at this stage of optimization, the operators (M and D) are used to assist the other stages of the quest process by improved communication. The AOA exploration operators use two main search strategies (D and M search strategies), which are modeled in Eq. 31.3, to randomly scan several regions in search of a better solution. The a function is used to condition this stage of the quest for the condition $r_1 > a$. The first operator D Eq. 31.3 is conditioned by $r_2 < 0.5$ in this process, and the other operator M is ignored until the current task of this operator is completed. Otherwise, instead of D, the operator M will be assigned to complete the current mission. The factor is given a stochastic scaling coefficient to generate further diversification and explore different areas of the quest space. For the exploration bits, the following position updating equations are devised:

$$x_{i,j}(C_{Iter} + 1) = \begin{cases} x_j^{best} \div (p + \epsilon) \times ((x_j^{max} - x_j^{min}) \times \mu + x_j^{min}) & r_2 < 0.5 \\ x_j^{best} \times p \times ((x_j^{max} - x_j^{min}) \times \mu + x_j^{min}) & otherwise \end{cases}$$

(31.3)

where r_1, r_2 are random number, $x_i(C_{Iter} + 1)$ is the ith solution in the next iteration, $x_{i,j}(C_{Iter})$ is the ith solution in the present iteration, and x_j^{best} is the jth location in the best-acquired solution so far. ϵ is a small integer numeral, and x_j^{max} and x_j^{min} are the upper and lower bound values for the jth location, respectively. μ denotes control parameter for changing the explore process, according to the experiments, and it is set to 0.5. p denotes as a coefficient.

$$p(C_{Iter}) = 1 - \frac{C_{Iter}^{1/\alpha}}{M_{Iter}^{1/\alpha}}$$

(31.4)

where $p(C_{Iter})$ indicates value of function at the jth iteration, C_{Iter} denotes the present iteration, and (M_{Iter}) denotes the highest number of iterations. According to the experiments, α is a sensitive parameter that determines exploitation accuracy across iterations and is put to 5.

The AOA's exploitation strategy is discussed in this section. Mathematical calculations involving S or A yielded high-density performance suggested that the exploitation search mechanism in use was according to the Arithmetic operators. Because of their low dispersion, these operators (A and S) can easily approach the goal. As a consequence, the exploitation scan detects the nearby optimal result after many attempts. The function value for the state of r_1 must be less than the current $a(C_{Iter})$ value (Eq. 31.2) at this point of the search. Using two primary search techniques

(S and A), which are modeled in Eq. 31.5, the AOA exploitation operators (S and A) systematically search the search area across many dense regions to find a better solution.

$$x_{i,j}(C_{Iter}+1) = \begin{cases} x_j^{best} - p \times ((x_j^{\max} - x_j^{\min}) \times \mu + x_j^{\min}) & r_3 < 0.5 \\ x_j^{best} + p \times ((x_j^{\max} - x_j^{\min}) \times \mu + x_j^{\min}) & otherwise \end{cases}$$

$$(31.5)$$

The first operator (S) is conditioned by $r_3 < 0.5$ in this step (Eq. 31.5), an operator A is ignored until the operator completes its current work. Otherwise, the current mission will be accomplished by the second operator A in place of the S. Exploitation operators (A and S) on the other hand, tend to stay out of the local search field as far as possible. This method aids exploration quest strategies in selecting the best solution while reducing the number of candidate solutions. At each iteration, carefully constructed parameters are used to produce a stochastic value, allowing for exploration not only during the first, but also during the last iteration. When local optima stagnation occurs, particularly in the final iterations, this section of the search is especially useful. The search scope's final position, which is within a stochastic range defined by the locations of M, D, A, and S. Other solutions move through the near optimal solution's area stochastically, while M, D, A, and S estimate the location of the nearby optimal result in different ways. The Objective function, i.e., entropy in Eq. (31.1) is used as a fitness function for the individual solution of AOA.

31.4 Experimental Setup

31.4.1 Parameter Settings

The parameters of AOA are set as follows: number of search agents $= 30$, maximum number of iterations $= 100$, $MOP_{Max} = 1$, $MOP_{Min} = 0.2$, $\alpha = 5$, $\mu = 0.499$.

The parameters of PSO are set as follows: swarm size $(N) = 30$, personal cognizance $(C_1) = 1.49618$, social cognizance $(C_2) = 1.49618$, inertia weight $(W) = 0.72984$, maximum number of iterations $= 100$.

The parameters of HMRF are as follows: MAP iterations $= 5$, EM iterations $= 5$.

31.4.2 PC Configuration

1. CPU: Intel® Core™ i3-8130U @ 2.20GHz.
2. RAM: 4.00 GB
3. Operating System: Windows 10 Home Single Language (64-bit)
4. Software: MATLAB 2018a

31.4.3 Performance Measurement

It is a difficult task to design or choose an appropriate effectiveness measure of image segmentation. Information relevant to the task, whether diagnostic or interventional, should be provided by the performance assessment. Accuracy, sensitivity, precision, specificity, FPR, G-mean, DSC, and F-measure are some of the parameters used for the evaluation of this method, and details of these measures can be obtained from [15].

31.5 Results and Discussion

This study presents a breast DCE-MRI lesion segmentation approach. The MR images' noise and IIHs make segmentation more difficult. As a result, an ADF is used to remove noise from MR images, and IIHs are corrected during the preprocessing stage. Ten trials were conducted for a single image because the AOA approach starts with a randomly started population. The quantitative results in terms of the mean value and standard deviation value of performance assessment indicators are evaluated for analysis over 10×50 results. In addition to the existing PSO and HMRF approaches, the new method is assessed and analyzed. The quantitative results of the devised method compared to the other two existing methods are reported in Table 31.1.

The mean classification value of accuracy produced by the suggested technique is higher than that acquired by PSO and HMRF, as shown by the quantitative results in Table 31.1. AOA has an accuracy of 99.80%, which is higher than any other approach. The suggested method's mean sensitivity is 98.06% and its mean specificity is 99.81%, both of which are greater than the PSO and HMRF approach.

Another important element in lesion detection is precision. AOA has a mean precision value of 78.15%, while PSO and HMRF have mean precision values of 68.69%, and 63.09%, respectively.

The devised method has a mean G-mean score of 98.93%, which is higher than PSO, and HMRF. A higher mean of G-mean value indicates better balance classification of lesions and non-lesions.

The devised method's mean F-measure value is 85.52%, which is significantly higher than PSO and HMRF. The F-measure refers to the accuracy with which lesion areas in the breast are classified. The higher this indicator's value, the more likely it is that the AOA classification can accurately identify lesions in breast MRI.

FPR is the proportion of incorrectly marked negative samples to the total number of negative samples. As a result, the specificity adds to it. Changes in data distribution have no effect on FPR. The devised technique has a mean FPR of 0.18%, which is significantly lower than PSO and HMRF.

The devised approach has a mean DSC value of 85.52%, which is greater than the PSO and HMRF. The higher the DSC value, the larger the overlap with the segmen-

Table 31.1 Performance evaluation values (in %) for devised method AOA, existing methods HMRF, and PSO

Performance metrics	AOA	PSO	HMRF
Accuracy	**99.80**	99.06	98.94
	(0.0002)	(0.0004)	(0.0003)
Sensitivity	**98.06**	90.44	90.65
	(0.0848)	(0.1471)	(0.1069)
Specificity	**99.81**	99.35	99.22
	(0.0002)	(0.0965)	(0.0003)
Precision	**78.15**	68.69	63.09
	(0.0351)	(0.1407)	(0.1946)
G-mean	**98.93**	94.40	94.59
	(0.0311)	(0.0711)	(0.0535)
F-measure	**85.52**	72.64.32	67.85
	(0.0422)	(0.0511)	(0.0864)
FPR	**0.18**	0.64	0.78
	(0.0002)	(0.0004)	(0.0003)
DSC	**85.52**	72.64	67.85
	(0.0422)	(0.0511)	(0.0864)

tation lesions' ground truths. The suggested AOA-based segmentation methodology outperforms the PSO and HMRF in the study of the above quantitative results.

It's also important to consider the robustness of lesion detection segmentation approaches. It is calculated over numerous runs in terms of the standard deviation of output variables, with a lower standard deviation indicating higher resilience. When compared to two other approaches, the standard deviation values of AOA in Table 31.1 are minimal, indicating that the suggested segmentation strategy is strong for breast MRI lesion identification.

Statistical Analysis

The performance of the algorithms is evaluated using a nonparametric test. The Wilcoxon Signed-Rank Test [28] is used to compare the suggested AOA-based method with the other algorithms pairwise. Table 31.2 describes the statistical test findings for sensitivity, accuracy, precision, and specificity. Table 31.3 describes F-measure, G-mean, DSC, and FPR, respectively. It has been observed from Table 31.2 that the accuracy of AOA is statistically outperformed PSO and HMRF with a significance level ($\alpha = 0.05$). In Table 31.2, it is reported that the sensitivity and specificity of AOA are statistically better than PSO, and HMRF with a significance level ($\alpha = 0.05$). In Table 31.2, it is reported that the precision of AOA is statistically better than PSO, and HMRF with a significance level ($\alpha = 0.05$). Between AOA and PSO, no statistical difference in precision. Also, from Table 31.3, it is noticed that G-mean, F-measure, FPR, and DSC of AOA are highly statistically significant than HMRF, and PSO with a significance level ($\alpha = 0.05$).

Table 31.2 Statistical test results based on accuracy, sensitivity, specificity, and precision

Sl. no.	Comparison	$p(2 - tailed)$			
		Accuracy	Sensitivity	Specificity	Precision
1	AOA versus PSO	0.000001	0.007682	0.002897	0.229358
2	AOA versus HMRF	0.000000	0.019326	0.000009	0.012926

Table 31.3 Statistical test results based on on G-mean, F-measure, FPR, and DSC

Sl. no.	Comparison	$p(2 - tailed)$			
		G-mean	F-measure	FPR	DSC
1	AOA versus PSO	0.000557	0.000361	0.002897	0.000361
2	AOA versus HMRF	0.000361	0.000231	0.000010	0.000231

Multi-criteria Decision Analysis

The performance is examined in this study utilizing the Technique for Order of Preference by Similarity to Ideal (TOPSIS), a well-known MCDA method [29]. Sensitivity, accuracy, precision, specificity, FPR, G-mean, DSC, and F-measure are some of the criteria used. FPR is in contrast with other criteria since low FPR scores designate that something is better, yet higher values of other categories designate that something is better. The AOA approach is shown to be the most effective. The PSO approach is then followed by the AOA method. It's also worth noting that HMRF is at the bottom of the list.

31.5.1 Visual Results

For the validation of this strategy, a total of 50 images of 10 patients are used. The findings for the 1st image of 1st patients are displayed out of 50 test results due to space constraints. Figure 31.2 for patient-1 shows segmented breast lesions using various approaches. Figures 31.3 for patient-1 show the localized lesion images. In qualitative results (Fig. 31.2) for patient-1, by balancing the ground truth image in Fig. 31.2b and AOA segmented image in Fig. 31.2c for MRI image of patient-1, it can be simply noticed that lesion areas are almost entirely segmented by AOA. When compared to the ground truth image, the suggested AOA-based technique operates well in identifying lesions in the MRI image. It is clear from Fig. 31.2d, for an MRI image of patient-1, that some lesions are not recognized by PSO. When compared to the ground truth image, the PSO technique does not operate well in detecting lesions

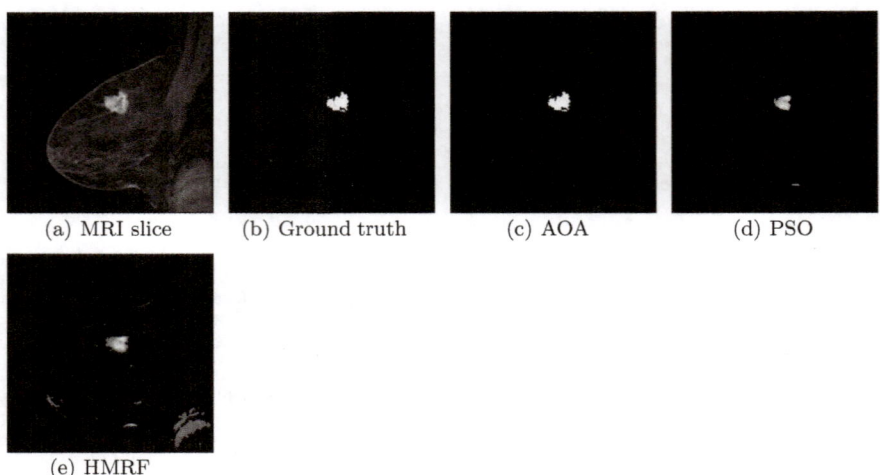

Fig. 31.2 For Patient-1 **a** Original MR image, **b** ground truth, **c** segmented image using AOA, **d** segmented image using PSO, and **e** segmented image using HMRF

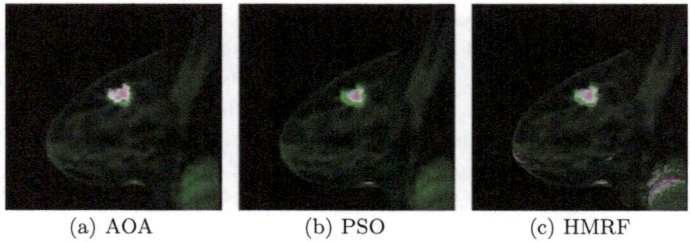

Fig. 31.3 Localized lesions in MR images for patient-1 **a** AOA, **b** PSO, and **c** HMRF

in the image. Figure 31.2e shows how HMRF segments some healthy tissues. When compared to the ground truth image, this approach does not do well in detecting lesions for the images.

The average CPU execution time of AOA is 2.8959 s, according to our results. PSO takes 3.1809 s, and the average execution time for HMRF is 4.1954 s. The devised method takes a lower CPU execution time as compared with all existing compared methods.

According to the above analysis of results, AOA performs better in DCE-MRI segmentation of breast lesions than PSO and HMRF. According to the results of the experiments, AOA is efficient and successful for breast lesion segmentation in DCE-MRI. Because AOA has high searching ability and convergence properties, it is effective at determining appropriate threshold values for segmentation.

31.6 Conclusion with Future Works

The goal of this study is to examine and create a viable method for segmenting breast DCE-MRI to aid radiologists in disease diagnosis and treatment planning. The AOA-based segmentation approach is devised in this work. The experimental findings demonstrate the usefulness of the devised AOA-based scheme, which outperformed the existing PSO and HMRF approaches in terms of sensitivity, accuracy, precision, specificity, FPR, G-mean, DSC, and F-measure. The AOA algorithm will be used to extend entropy maximization to dynamically identify the ideal total of thresholds, which will aid in improved segmentation of breast lesions. In future, an enhanced version of the AOA method could be used to segment breast lesions in MRI scans.

References

1. Turnbull, L.W.: Dynamic contrast-enhanced MRI in the diagnosis and management of breast cancer. Wiley InterScience 28–39 (2008)
2. Giannini, V., Vignati, A., Morra, L., Persano, D., Brizzi, D., Carbonaro, L., Bert, A., Sardanelli, F., Regge, D.: A fully automatic algorithm for segmentation of the breasts in DCE-MR images. In: IEEE Engineering in Medicine and Biology Society Conference Proceedings, New York, pp. 3146–3149 (2010)
3. Vignati, A., Giannini, V., Bert, A., Deluca, M., Morra, L., Persano, D., Martincich, L., Regge, D.: A fully automatic lesion detection method for DCE-MRI fat-suppressed breast images. In: Medical Imaging Computer-Aided Diagnosis, vol. 7260 (2009)
4. Twellmann, T., Lichte, O., Nattkemper, T.W.: An adaptive tissue characterization network for model-free visualization of dynamic contrast-enhanced magnetic resonance image data. IEEE Trans. Med. Imaging $24(10)$, 1256–1266 (2005)
5. Ertas, G., Güçür, H.Ö., Osman, O., Uçan, O.N., Tunaci, M., Dursun, M.: Breast MR Segmentation and lesion detection with cellular neural networks and 3D template matching. Comput. Biol. Med. 116–126 (2008)
6. Abualigah, L., Diabat, A., Mirjalili, S., Elaziz, M.A., Gandomi, A.H.: The arithmetic optimization algorithm. Comput. Methods Appl. Mech. Eng. **376**, 113609 (2021)
7. Chatzis, S.P., Tsechpenakis, G.: The infinite hidden Markov random field model. IEEE Trans. Neural Netw. **21**(6), 1004–1014 (2010)
8. Trelea, I.C.: The particle swarm optimization algorithm: convergence analysis and parameter selection. Inf. Process. Lett. **85**, 317–325 (2002)
9. Krizhevsky, A., Sutskever, I., Hinton, G.E.: Imagenet classification with deep convolutional neural networks. **1**, 1097–1105 (2012)
10. Arevalo, J., Gonzalez, F.A., Ramos-Pollán, R., Oliveira, J.L., Lopez, M.A.G.: Convolutional neural networks for mammography mass lesion classification, pp. 797–800 (2015)
11. Jiao, Z., Gao, X., Wang, Y., Li, J.: A deep feature-based framework for breast masses classification. Neurocomputing **197**, 221–231 (2016)
12. Jumaat, A.K., Zarina, W.E., Rahman, W.A., Ibrahim, A., Mahmud, R.: Segmentation of masses from breast ultrasound images using parametric active contour algorithm. Procedia - Soc. Behav. Sci. **8**, 640–647 (2010)
13. Kavitha, M., Rejusha, M.: Segmentation of pectoral muscle and detection of masses in mammographic images. In: 2015 2nd International Conference on Electronics and Communication Systems (ICECS), pp. 1201–1204 (2015)
14. Liu, X., Xu, X., Liu, J., Tang, J.: Mass classification with level set segmentation and shape analysis for breast cancer diagnosis using mammography. **6839**, 630–637 (2012)

15. Patra, D.K., Si, T., Mondal, S., Mukherjee, P.: Breast DCE-MRI segmentation for lesion detection by multi-level thresholding using student psychological based optimization. Biomed. Signal Process. Control **69**, 102925 (2021)
16. Si, T., Mukhopadhyay, A.: Breast DCE-MRI segmentation for lesion detection using clustering with fireworks algorithm. In: Applications of Artificial Intelligence in Engineering, vol. 1381, pp. 17–35. Springer, Singapore (2021)
17. Kar, B., Si, T.: Breast DCE-MRI Segmentation for Lesion Detection Using Clustering with Multi-verse Optimization Algorithm. In: Advances in Intelligent Systems and Computing, vol. 1381, pp. 265–278. Springer, Singapore (2021)
18. Patra, D.K., Mondal, S., Mukherjee, P.: Grammatical fireworks algorithm method for breast lesion segmentation in DCE-MRI. Int. J. Innov. Technol. Explor. Eng. **10**(7), 170–182 (2021)
19. Ribes, S., laurent, D.D., Decoster, N., Gonneau, E., Risser, L., Feillel, V., Caselles, O.: Automatic segmentation of breast MR images through a Markov random field statistical model. IEEE Trans. Med. Imaging **33**, 1–11 (2014)
20. Arjmand, A., Meshgini, S., Afrouzian, R., Farzamnia, A.: Breast tumor segmentation using K-means clustering and cuckoo search optimization. In: 9th International Conference on Computer and Knowledge Engineering (ICCKE 2019) (2019)
21. Piantadosi, G., Marrone, S., Galli, A., Sansone, M., Sansone, C.: DCE-MRI breast lesions segmentation with a 3TP U-net deep convolutional neural network. In: IEEE 32nd International Symposium on Computer-Based Medical Systems (2019)
22. Si, T., Mukhopadhyay, A.: Breast DCE-MRI segmentation for lesion detection using clustering with fireworks algorithms. In: Gao, X.Z., Kumar, R., Srivastava, S., Soni, B.P. (eds.) Applications of Artificial Intelligence in Engineering. Algorithms for Intelligent Systems. Springer, Singapore (2021). https://doi.org/10.1007/978-981-33-4604-8_2
23. Lingle, W., Erickson, B.J., Zuley, M.L., Jarosz, R., Bonaccio, E., Filippini, J., Gruszauskas, N.: Radiology Data from The Cancer Genome Atlas Breast Invasive Carcinoma Collection [TCGA-BRCA] (2007)
24. Clark, K., Vendt, B., Smith, K., Freymann, J., Kirby, J., Koppel, P., Moore, S., Phillips, S., Maffitt, D., Pringle, M., Tarbox, L., Prior, F.: The cancer imaging archive: maintaining and operating a public information repository. J. Digit. Imaging **26**, 1045–1057 (2013)
25. Bohare, M.D., Cheeran, A.N., Sarode, V.G.: Analysis of breast MRI images using wavelets for detection of cancer. IJCA Spec. Issue Electron. Inf. Commun. Eng. **4**, 1–3 (2011)
26. Balafar, M.A., Ramli, A.R., Mashohor, S.: A new method for MR grayscale inhomogeneity correction. Artif. Intell. Rev. **34**, 195–204 (2010)
27. Shannon, C., Weaver, W.: The Mathematical Theory of Communication. University of Illinois Press, Urbana (1964)
28. Derrac, J., Garcia, S., Molina, D., Herrera, F.: A practical tutorial on the use of nonparametric statistical tests as a methodology for comparing evolutionary and swarm intelligence algorithms. Swarm Evol. Comput. **1**, 3–18 (2001)
29. Brown, S., Tauler, R., Walczak, B.: Comprehensive Chemometrics-Chemical and Biochemical Data Analysis, 2nd edn. Elsevier, Amsterdam (2020)

Chapter 32
Multi Classification of Brain Tumor Detection Using MRI Images: Deep Learning Approach

Rushikesh Bedagkar, Amit D. Joshi, and Suraj T. Sawant

Abstract Brain tumor detection at early stages is very important for successful treatment. The life of the patient can be saved if it is detected in the early stages and also useful for proper and efficient medication. Nowadays, machine learning algorithms are being used for diagnosing purpose in the medical field. Computers always give better result compared to manual diagnostic. Brain tumor detection and segmentation is a very crucial task, as manual processing of medical images leads to the wrong prediction. Magnetic Resonance Imaging scans have proved to be helpful for the diagnosis or segmentation of brain tumors. The image segmentation process is used for the extraction of tumors that are not normal in the brain. With the use of efficient data mining techniques and different classification algorithms, prediction of the disease can be performed at an early stage with better accuracy and effectiveness. In the field of medicine, Machine Learning and Data Mining techniques have proved to be effective and useful for better prediction. Deep learning is one of the subparts of machine learning and recently proved to be of high importance in classification and segmentation-related problems. This work is implemented using a deep learning approach, modeled into Convolutional Neural Network to classify the results into different categories like "Meningioma", "Glioma", "Pituitary" or "TUMOR NOT FOUND".

R. Bedagkar (✉) · A. D. Joshi · S. T. Sawant
College of Engineering Pune, Pune, Maharashtra, India
e-mail: bedagkarra17.comp@coep.ac.in

A. D. Joshi
e-mail: adj.comp@coep.ac.in

S. T. Sawant
e-mail: sts.comp@coep.ac.in

32.1 Introduction

A brain tumor is considered as a growth in brain cells that is abnormal in nature. The tumor may spread into various body organs and affects its normal functionality. Brain tumors can be classified into two categories, viz., primary and secondary. The primary tumor resembles approximately about 70% of all brain tumors and the secondary tumor consists of approximately 30% [1]. The segmentation of brain tumor consists of separation of tumor cells that are not similar in nature compared to the normal brain cells [2]. Gliomas are the brain tumors that first appeared in the glial cells of the brain. Gliomas can be classified into four different grades as per WHO. The different grades of tumor consists of different textures that can be differentiable from each other and visible to normal eyes. Meningioma tumor is formed on the membrane that covers the brain and spinal cord inside the human skull that leads majorly to benign [1].

Data Mining (DM) consists of many techniques, for example, anomaly detection, clustering, regression, and summarization. DM techniques can be used to analyze a large amount of data from different angles. The research using DM techniques helps in building a diagnosis and system of predictions of results that are related to brain tumor [3]. The manual detection of brain tumor is time-consuming, that is why we move toward Computer Aided Diagnosis (CAD) that helps in detecting brain tumor more precisely [4]. Digital image processing is the technique in which an image is processed with help of digital technology. Image processing techniques include anisotropic diffusion. It is a technique of reduction of noise from the image without removing the important parts of the image, hidden markov method, image restoration, linear filtering, neural network, etc.

Image processing is also an important part of brain tumor detection as all the MRI images in the dataset are not equal. Image segmentation is the process of separating images into different parts or segments. The most important image properties are color, texture, and resistance. Image contains many distortions, one of them is noise. The noise is like a bright/dark spot. Increased noise can also significantly distort the quality [5]. With the help of the techniques like image processing, machine learning is used nowadays to process the information further. Clustering and classification are the main concepts used in machine learning [6]. Histogram orientation gradient (HOG) with Extreme Machine Learning (ELM) is used to get a very high rate of recognization called as H-ELM [7]. Graphical Processing Unit (GPU) can be used to enhance the image [8]. The common method that can be used is Convolutional Neural Networks (CNN).

Deep learning is an Artificial Intelligence (AI) function that replicates the human brain as it replicates the functionality of the brain which helps in processing data that can be used in detecting objects, recognizing speech, and making decisions. CNN is one of the classes of deep learning used for analyzing digital images. CNN can be used with the help of different layers such as convolution layers, strides, padding, non linearity, Rectified Linear Unit (ReLU), pooling, and fully connected layer [9].

32.2 Literature Review

Hossam Sultan et al. have used a CNN model with cost function. The proposed model can classify multiple types of brain tumors. The architecture achieved an overall accuracy of 96.13% and 98.7% for the two datasets from Nanfang Hospital and General Hospital, Tianjin Medical University, China, 2005 to 2010 [10] and The Cancer Imaging Archive (TCIA) [1, 11].

Chirodip Choudhury et al. have used images as an array with the help of NumPy and CNN model to predict the outcome. The input image is filtered through various layers such as ReLu function, sigmoid activation function, and hyperbolic tangent function. The achieved accuracy for training is 97.47% in 35 epochs and the training loss is 0.402% [12].

Hemanth et al. have used the CNN model with 3×3 small kernels and filtering algorithm. The proposed model is associated with root mean square error, recall, sensitivity, precision, F-score specificity, Probability of the Misclassification Error (PME), Conditional Random Field (CRF), Support Vector Machine (SVM), Genetic Algorithm (GA). The achieved accuracy is 89% using CRF, 84.5% using SVM, and 91% using CNN [2].

Masoumeh Siar et al. have used CNN and Alexnet architecture. The algorithm uses a duplicate procedure that iteratively attempts to be made so as to obtain points that are in the form of a cluster. The model consists of five convolutional layers and three layers of sub-sampling layers, normalization layers, fully connected layers, and classification layer. The model consists of softmax connected layer classifier, RBF classifier, and the DT classifier. The accuracy with radius-based function (RBF) is 97.34%. The achieved accuracy using CNN and Softmax is 98.67%. The achieved accuracy using CNN and DT is 94.24% [3].

T. Chithambaram et al. have used Artificial Neural Network (ANN) and SVM based model with CAD system for segmentation and classification of Brain tumors on MRI. This method uses segmentation methods, content-based active contour models, static motion field, dynamic motion field, gray level co-occurrence matrix, Rotation invariant circular Gabor features (RICGFs). The achieved accuracy is 94% using ANN and SVM-based model [13].

Raj et al. have used wavelet transformation and multiscale analysis to distribute a given signal using basic functions. The proposed model aims to increase the contrast of MRI images using sub-coding, nonlinear operator, morphological filter, and segmentation. This work has enhanced the cerebral features of MRI [14].

Anuj S. Bhadauria et al. have used a morphological filter that consists of operators such as erosion, dilation, closing and opening for skull stripping that helps in focusing only the skull part. This work has achieved success in retrieving only the skull part of the MRI [15].

32.3 Proposed Solution for Multiple Brain Tumor Detector

MRI images are given as input from the dataset to the model. Each image is then cropped to just focus on the skull part of the MRI image and to get rid of the unwanted parts and then the image is resized to an image size of (224, 244) as all the images in the datasets are not of the same size, then the normalization of image is done to make all the pixels of the images to the same range. Next, the dataset is split into different sets of training, validation, test to check the accuracy and performance of the model. Then the model building starts. Each preprocessed image is taken as an input for the proposed model. Each consecutive step of the model includes a global average pooling 2D, convolutional layer with a defined number of pre-defined sized filter with different number of strides. After this, a dropout layer is added to prevent neural networks and the trained model from overfitting. Consecutively a dense layer has been added that is a deeply connected neural network layer at the end with softmax activation function as shown in Eq. 32.1. Figure 32.1 indicates these different steps included in the model. After all these processes are done to every image in the training and validation dataset, the CNN model starts to build. CNN is flexible and also works well on image data. CNN also uses filters to generate features that are passed on to the next layer. At the end of the completion of model building, the test dataset is provided to check if the model can accurately distinguish between the images containing brain tumor and images without the brain tumor. A brain tumor

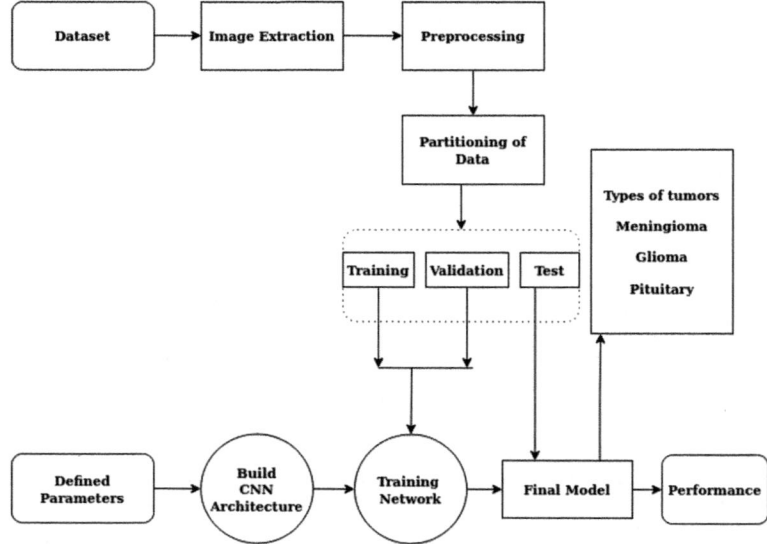

Fig. 32.1 Block diagram of proposed model

classification dataset consists of 3000 MRI images that are used for training and testing.

$$softmax(x)_i = \frac{exp(x_i)}{\sum_j exp(x_j))} \qquad (32.1)$$

32.4 Implementation

When the model starts, it takes MRI images from the provided dataset as input. The image is preprocessed which includes skull cropping to focus only on the skull part of the image. Next step is to resize the image into the same size. The same size of image helps in increasing the computational efficiency of the model. After completion, the data is splitted into parts for training, validation, and testing purposes of the model. After this, the model starts building by adding different layers to the image for the accuracy purpose of all the input images. After the addition of the layers, the CNN model starts to train. With help of the CNN, the model builds a network and creates a prediction model. The model will be used in the prediction of the result when given any MRI image as an input. The results will ultimately be validated with the dataset and then tested with the test dataset created earlier. The model will provide the result as if the tumor is detected (different types of tumor) or not. It will also give the performance report of the model in terms of the accuracy and loss of the model.

32.5 Results and Discussion

Figure 32.2 shows the graph of training accuracy versus validation accuracy. The accuracy is very less for the first few epochs as compared with the increase in epoch size. There is a sudden rise in the accuracy after three epochs and after eight epochs. The accuracy is almost 99.651% at the eighth epoch onwards.

Figure 32.3 shows the graph plot of training loss versus validation loss. It can be observed that the loss is higher at the start. The loss is reduced after four epochs and the line touches almost zero. The loss is very low compared with the earlier one after 12 epochs. The model proposed in the paper is able to detect three types of tumors, viz., Glioma, Meningioma, Pituitary. The model is able to identify the MRI images with no tumor with the accuracy rate of 99.651%.

Figure 32.4 shows the confusion matrix of the proposed model. The Y-axis indicates true labels and X-axis indicates predicted labels. At some places, the model predicts the image with the wrong tumor. Instead of meningioma, the model predicts it as glioma. We can see that rest all the images has been predicted with the accuracy of 99.6%.

It has been observed that the overall accuracy of the proposed model is 99.65% which is better than the existing approach proposed by Hossam Sultan et al. that has an accuracy rate of 98.7% [1].

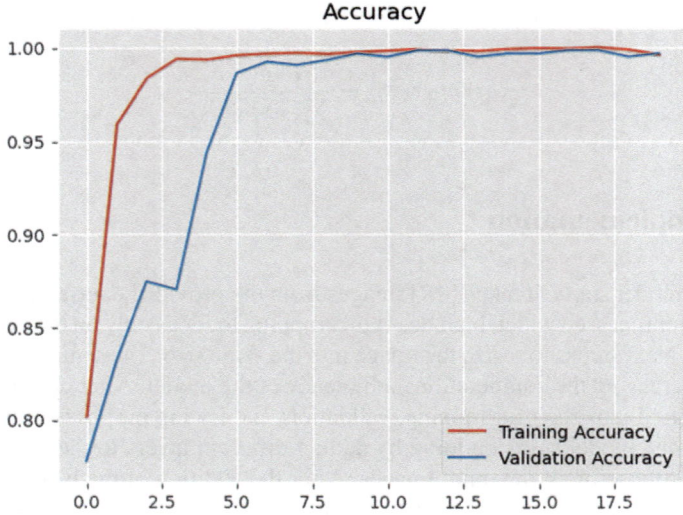

Fig. 32.2 Training accuracy versus validation accuracy

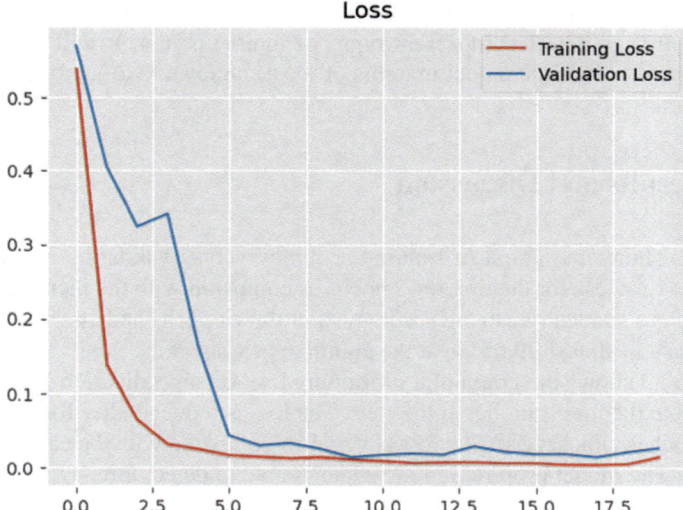

Fig. 32.3 Training loss versus validation loss

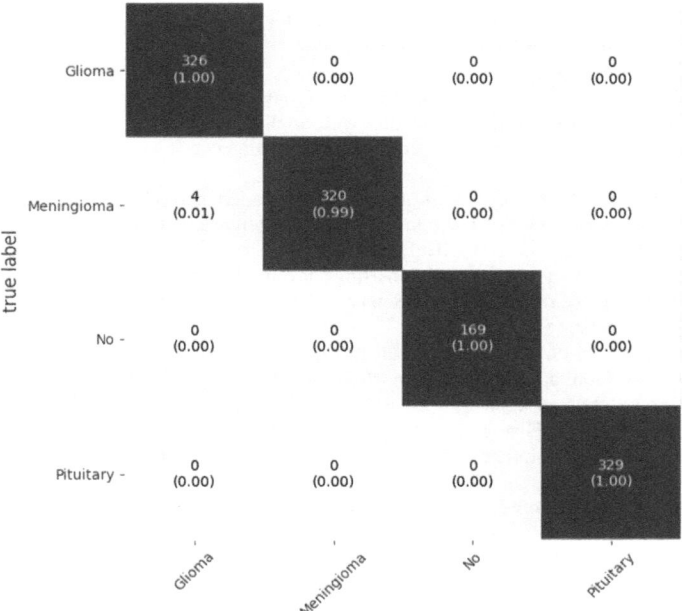

Fig. 32.4 Confusion matrix

32.6   Conclusion

The work proposed a CNN-based model for the detection of three types of tumors, viz., Glioma, Meningioma, and Pituitary using different techniques of image enhancements and various convolutional layers. The model has four distinct layers, the first being an input layer that has an image that is already preprocessed. Then the image goes through a 2D pooling layer, a dropout layer followed by a softmax layer, and a prediction layer that predicts the output. The model that has been proposed in this paper is able to detect the tumors with an accuracy of 99.65% as compared with the existing approach.

References

1. Sultan, H.H., Salem, N.M., Al-Atabany, W.: Multi-classification of brain tumor images using deep neural networks. IEEE Access **7**, 69215–69225 (2019)
2. Hemanth, G., Janardhan, M., Sujihelen, L.: Design and implementing brain tumor detection using machine learning approach. In: 2019 3rd International Conference on Trends in Electronics and Informatics (ICOEI). IEEE (2019)

3. Siar, M., Teshnehlab, M.: Brain tumor detection using deep neural network and machine learn-ing algorithm. In: 2019 9th International Conference on Computer and Knowledge Engineering (ICCKE). IEEE (2019)
4. Gopal, N.N., Karnan, M.: Diagnose brain tumor through MRI using image processing clustering algorithms such as fuzzy C means along with intelligent optimization techniques. In: 2010 IEEE International Conference on Computational Intelligence and Computing Research. IEEE (2010)
5. Goswami, A., Dixit, M.: An analysis of image segmentation methods for brain tumour detection on MRI images. In: 2020 IEEE 9th International Conference on Communication Systems and Network Technologies (CSNT). IEEE (2020)
6. Hossain, T., et al.: Brain tumor detection using convolutional neural network. In: 2019 1st Inter-national Conference on Advances in Science, Engineering and Robotics Technology (ICAS-ERT). IEEE (2019)
7. Phusomsai, W., et al.: Brain tumor cell recognition schemes using image processing with parallel ELM classifications on GPU. In: 2016 13th International Joint Conference on Computer Science and Software Engineering (JCSSE). IEEE (2016)
8. Dalal, N., Triggs, B.: Histograms of oriented gradients for human detection. In: 2005 IEEE Computer Society Conference on Computer Vision and Pattern Recognition (CVPR'05), vol. 1. IEEE (2005)
9. Ker, J., et al.: Deep learning applications in medical image analysis. IEEE Access 6, 9375–9389 (2017)
10. El-Dahshan, E.-S.A., Hosny, T., Salem, A.-B.M.: Hybrid intelligent techniques for MRI brain images classification. Digit. Signal Process. 20(2), 433–441 (2010)
11. Clark, K., et al.: The cancer imaging archive (TCIA): maintaining and operating a public information repository. J. Digit. Imaging 26(6), 1045–1057 (2013)
12. Choudhury, C.L., et al.: Brain tumor detection and classification using convolutional neural network and deep neural network. In: 2020 International Conference on Computer Science, Engineering and Applications (ICCSEA). IEEE (2020). Ker, J., et al.: Deep learning applica-tions in medical image analysis. IEEE Access 6, 9375–9389 (2017)
13. Chithambaram, T., Perumal, K.: Brain tumor segmentation using genetic algorithm and ANN techniques. In: 2017 IEEE International Conference on Power, Control, Signals and Instru-mentation Engineering (ICPCSI). IEEE (2017)
14. Raj, A., Srivastava, A., Bhateja, V.: Computer aided detection of brain tumor in magnetic resonance images. Int. J. Eng. Technol. 3(5), 523 (2011)
15. Bhadauria, A.S., et al.: Skull stripping of brain MRI using mathematical morphology. In: Smart Intelligent Computing and Applications, pp. 775–780. Springer, Singapore (2020)

Chapter 33
Hybrid ANFIS-PSO Model for Monthly Precipitation Forecasting

Subhadipta Chaudhury, Sandeep Samantaray, Abinash Sahoo, Binayini Bhagat, Chinmayee Biswakalyani, and Deba Prakash Satapathy

Abstract Precipitation forecasting is a key constituent of hydrological cycle and is of principal significance in planning and management of water resources, in addition to arrangement of irrigational practices. This study focuses on assessing the potential of hybrid ANFIS-PSO (integrating Adaptive Neuro-Fuzzy Inference System and Particle Swarm Optimization) and simple ANFIS models for precipitation forecasting at Nuapada district of Odisha, India. Evaluations of model performances were studied based on different statistical indices. Results revealed that hybrid ANFIS-PSO model provided superior accurateness compared to standalone ANFIS. In addition, analysis of results revealed that integrating optimisation algorithm with ANFIS can enhance its performance in monthly precipitation forecasting.

33.1 Introduction

Long-term precipitation forecasting is significant for planning and management of water resources due to its extremely precarious situations in various climate conditions. It is also important for application in several hydrological processes, like estimation of water budget and balance of watersheds, water level of rivers, prediction of future streamflows, and computation of long-term inflows for reservoir operation. Changes in precipitation may vary fundamental conditions related to water resources and increases necessity for novel water management strategies and programs specifically in very ambiguous climatic conditions inside semiarid and arid regions [1]. By far, many different models for long-term quantitative forecasting of rainfall have been developed. Among mathematical models, a statistical method, namely time series modelling, is a conventional technique that has been broadly utilised. Even though

S. Chaudhury · B. Bhagat · C. Biswakalyani · D. P. Satapathy
Department of Civil Engineering, CET Bhubaneswar, Bhubaneswar, Odisha, India
e-mail: dpsatapathy@cet.edu.in

S. Samantaray (✉) · A. Sahoo
Department of Civil Engineering, NIT Silchar, Silchar, Assam, India
e-mail: sandeep1139_rs@civil.nits.ac.in

there are many advantages in using statistical modelling, inadequacies of statistical method include usage of nonlinear features of data since statistical models are generally governed by linear correlation of data. It can be conveyed using a correlation coefficient. In recent past, for overcoming inadequacies of statistical approaches, several other models addressing nonlinearity of data were developed, which includes artificial neural networks (ANNs) and fuzzy inference systems (FIS). They are found to be promising techniques which have been broadly employed in many hydrological processes.

Jeong et al. [6] applied ANFIS technique to forecast monthly precipitation in Daejeon weather station, Korea, and compared its results with forecasts provided by Korea Meteorological Administration (KMA). Findings from their study showed a better agreement between both precipitation forecasts. Shamshirband et al. [19] proposed ANFIS and support vector machine (SVM) for estimating precipitation considering data from 29 rain gauge stations located in Serbia. Obtained results demonstrated that enhanced prediction capability and accuracy can be obtained with ANFIS in comparison to SVM technique. Mohammadi et al. [12] applied ANFIS for developing a model to estimate global solar radiation on daily basis. Outcomes highly stimulated usability of ANFIS for estimating solar radiation. Kisi and Sanikhani [8] investigated accuracy of ANN, ANFIS and SVM in long-term prediction of monthly rainfall without the use of climate data in 50 stations throughout Iran. Mohammadi et al. [11] employed ANFIS to identify most important parameters to predict daily dew point temperature (T_{dew}) of two cities located in south and central Iran. They found that ANFIS had higher accurateness in predicting T_{dew} compared to other prediction models. Application of simple ANFIS and its different kinds have been applied in many understanding many hydrological processes [13, 17, 18, 20]. Liu et al. [9] proposed a novel ANFIS-PSO model to forecast weight percentage of oil flocculated asphaltene for broad variety of operation conditions. Their findings revealed that developed model showed good performance in forecasting flocculated weight percentage. Zanganeh [22] employed ANFIS-PSO model for predicting wave parameters at Lake Michigan and assessed the performance of developed model comparing with simple ANFIS and Coastal Engineering Manual. Yaseen et al. [21] proposed hybrid ANFIS-PSO, ANFIS-GA (genetic algorithm), ANFIS-DE (differential evolution) and conventional ANFIS for forecasting rainfall time series. Results specified that for all input combinations all hybrid ANFIS models showed better performance compared to conventional ANFIS. Nguyen et al. [15] applied hybrid ANFIS-PSO model to determine buckling capability of circular steel beam. Obtained results showed that performance of ANFIS-PSO is suitable to determine buckling capability steel beam. From literature it can be observed that ANFIS-PSO model has been applied in many field of engineering studies. Therefore, the objective of this research is to investigate applicability of hybrid ANFIS-PSO model in monthly precipitation forecasting of Nuapada region.

33.2   Study Area

Nuapada district lies in western region of Odisha between 20° 0' N and 21° 5' N latitude and 82° 20' E and 82° 40' E longitude. Its boundaries extend to Mahasamund district of Chhattisgarh in north, south and west direction and in east to Balangir, Kalahandi and Bargarh districts. Geographical area of this district is 3407.5 km^2. The hottest month is May with an average temperature of 35 °C and coldest is January with average temperature of 21 °C. July month is considered the wettest as Nuapada receives the maximum rainfall in this month with average precipitation of 394.7 mm. This district receives an annual precipitation of 1247.0 mm (Fig. 33.1).

33.3   Methodology

33.3.1   Anfis

Jang [4] first introduced a type of ANN based on FIS called the ANFIS, in which fuzzy logic (FL) and ANN balance one another. ANFIS incorporates features of both FL and ANN; hence, it is capable of taking advantage of the two (ANN and FL) in a solitary system [14, 16]). This quality of ANFIS enables it in modelling uncertain situations [3, 5]. Two major categories of FISs are Sugeno and Mamdani systems [7, 10]. In present research, Sugeno FIS was selected since it, (1) effective in computational problems, (2) works well with adaptive, linear and optimisation methods; and (3) suits well in mathematical analysis. Five layers constitute a structure of an adaptive network. Figure 33.2 illustrates the five layers, their nodes, and connection of two inputs into FIS articulated by "x" and "y" and a solitary output "f". As a description about ANFIS configuration, it must be taken into note that two fuzzy 'if–then' rules were used in following Sugeno type FIS:

Rule 1: if $(x = P_1) and y = Q_1$ then $f_1 = m_1x + n_1x + o_1$,

Rule 2: if $(x = P_2) and y = Q_2$ then $f_2 = m_2x + n_2x + o_2$,
    where P and Q respective fuzzy sets, m, n, o resultant model parameters being evaluated in training phase.

33.3.2   Pso

Kennedy and Eberhart [7] introduced PSO algorithm that was modelled after social features of fish and birds. Major benefits of PSO are fewer amounts of computation, instant convergence, and simple procedure, making it appropriate to solve different nonlinear and complex engineering aspects. In brief, PSO deliberates that each particle varies its location in search space regarding preeminent location that

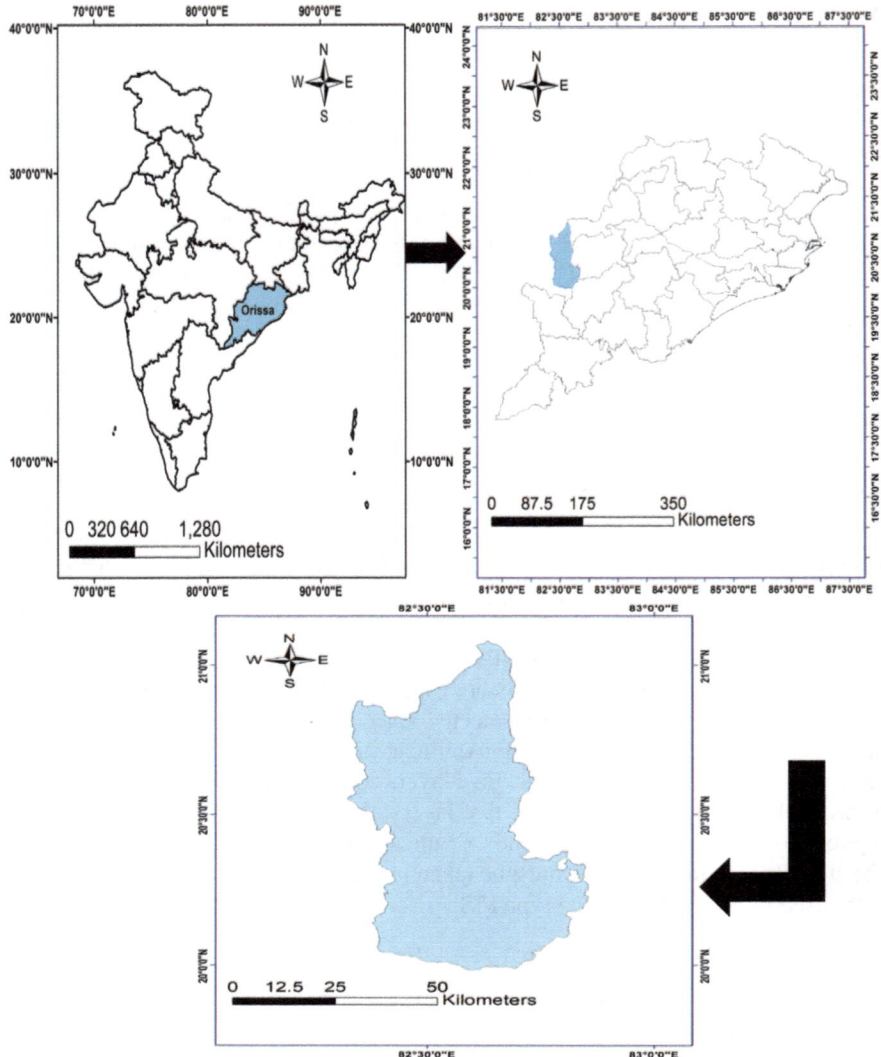

Fig. 33.1 Proposed watershed

it has ever been in and preeminent location nearby its neighbour. PSO works on basis of random probability distribution method which is accompanied by group of population. Primary stage is an arbitrary distribution of particles from search space for joining global optimum system or function of personal best (p_{best}) is preeminent solution that is attained by an element. Value of global best g_{best} is preeminent solution that is attained by swarm having its particle index. Subsequently, during journey of particle over search space having separate time iterations, each particle's

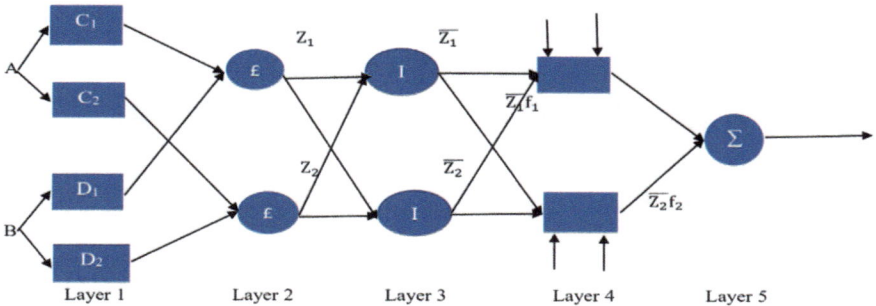

Fig. 33.2 ANFIS model architecture for 2 inputs and 1 output

velocity in following iteration is computed by g_{best} as societal constituent, p_{best}, as perceptive constituent, and its current velocity. Societal and perceptive constituents both arbitrarily supply to location of particle in following iteration [2].

Particle 'p' is expressed as

$$X_{pd}^{iter+1} = X_{pd}^{iter} + v_{pd}^{iter+1}, \tag{33.1}$$

where X_{pi} is value of 'i'th coordinate in D dimension space. Preeminent location of every particle is given by $G = \{g_1, g_2, \ldots, g_D\}$ and best understanding of particle 'p' is expressed by $P_p = \{P_{p1}, P_{p2}, \ldots, P_{pD}\}$. Velocity of change in location i.e. for 'p'particle is given by $V_p = \{V_{p1}, V_{p2}, \ldots, V_{pD}\}$. Based on its velocity, a particle's location is altered. Hence, there is an update in location for every iteration. Particle 'p' changes its location in correspondence to d-coordinate of its velocity. It can be calculated as follows:

$$v_{pd}^{iter+1} = \omega v_{pd}^{iter+1} + C_1 rand(0, 1)\left(P_{pd} - X_{pd}\right) + C_2 rand(0, 1)\left(g_{pd} - X_{pd}\right). \tag{33.2}$$

Particle's new location is attuned by Eq. 33.2. Parameter ω is weight of inertia and is employed for controlling impact of preceding velocities on current velocity. Positive coefficients C_1 and C_2 are signified by learning aspects which help particles for advancing in the direction of more adequate parts of solution space (Fig. 33.3).

33.3.3 Evaluating Standards

Mean monthly precipitation data of 1970–2019 are collected from IMD Bhubaneswar. Based on this study, input and output data (precipitation) were composed from Nuapada station in Odisha and demarcated for machine learning methods. For analysis using ANFIS models, 70% of data (1970–2004) were utilised

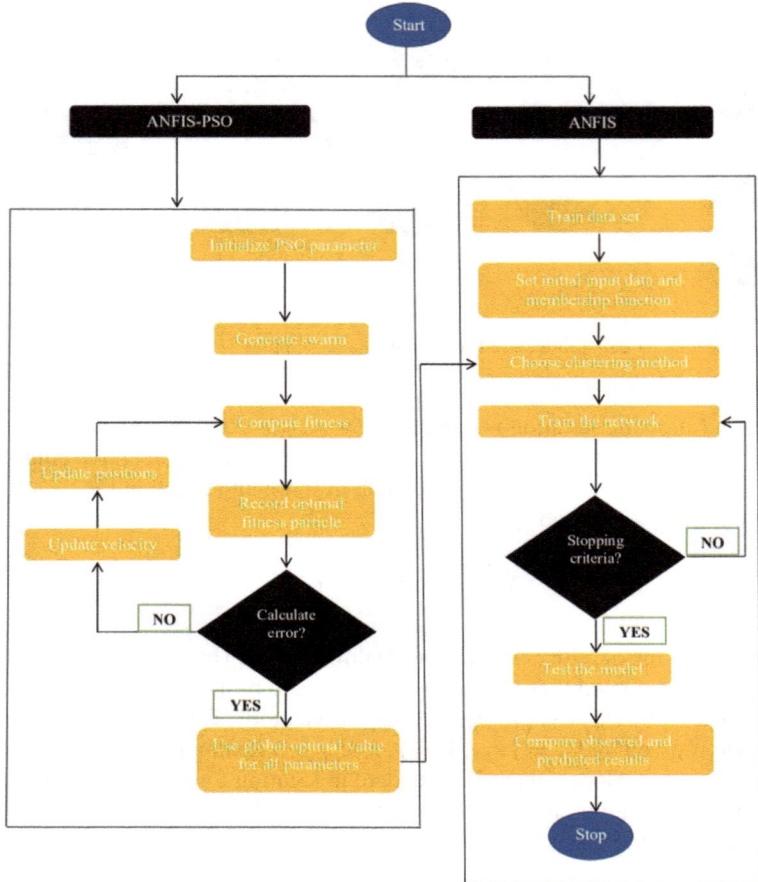

Fig. 33.3 General flowchart of a hybrid ANFIS-PSO model

to train samples and resultant 30% data (2005–2019) served to test the samples. Three constraints Nash–Sutcliffe (E_{NS}), Willmott index (WI) and co-efficient of determination (R^2) are applied to find the best performing model.

$$ENS = 1 - \left[\frac{\sum_{k=1}^{N} \left(x_{obs}^k - x_{comp}^k \right)^2}{\sum_{k=1}^{N} \left(x_{obs}^k - \overline{x_{obs}^k} \right)^2} \right], \tag{33.3}$$

$$R2 = \left(\frac{\sum_{k=1}^{N} (x_{obs}^k - \overline{x_{obs}^k})(x_{comp}^k - \overline{x_{comp}^k})}{\sqrt{\sum_{k=1}^{N} \left(x_{obs}^k - \overline{x_{obs}^k} \right)^2 \sum_{k=1}^{N} \left(x_{comp}^k - \overline{x_{comp}^k} \right)^2}} \right)^2, \tag{33.4}$$

$$\text{WI} = 1 - \left[\frac{\sum_{k=1}^{N} \left(x_{obs}^k - x_{comp}^k \right)^2}{\sum_{k=1}^{N} \left(\left| x_{comp}^k - \overline{x_{comp}^k} \right| + \left| x_{obs}^k - \overline{x_{comp}^k} \right| \right)^2} \right], \qquad (33.5)$$

where x_{comp}^k = Predicted value, x_{obs}^k = Observed value, $\overline{x_{comp}^k}$ = Mean predicted value, and $\overline{x_{obs}^k}$ = Mean observed value.

33.4 Results and Discussion

The potential of ANFIS-PSO and ANFIS models for making sensible forecasts mostly depends on selection of optimal input parameters. Considering suitable aspects controlling the system is thus essential to develop a consistent network. Table 33.1 shows the results of ANFIS and ANFIS-PSO model. The best value of performance of WI, E_{NS}, R^2 are 0.9361, 0.9324, 0.9257 and 0.9302, 0.9287, 0.9219 during training and testing phase for ANFIS technique respectively. Correspondingly for ANFIS-PSO model the value of WI, E_{NS}, R^2 are 0.9945, 0.9903, 0.9894 and 0.9884, 0.9862, 0.9847 for training and testing phase. The result reveals that ANFIS-PSO model shows prominent performance than the ANFIS model.

Sensitivity analysis of various models was performed for gaining assurance from accurateness of certain input data assortment and verification of model. Figure 33.4 depicts the scatter plot of ANFIS and ANFIS-PSO model during testing phase. The time variation of observed versus simulated ANFIS and ANFIS-PSO models are provided in Fig. 33.5. Results illustrate that estimated peak precipitation is

Table 33.1 Performance of model

Technique	Input	E_{NS}	WI	R^2	E_{NS}	WI	R^2
		Training phase			Testing phase		
ANFIS	P_t, P_{t-1}	0.9235	0.9253	0.9138	0.9195	0.9208	0.9093
	P_t, P_{t-1}, P_{t-2}	0.9259	0.9285	0.9152	0.9218	0.9234	0.9125
	P_t, P_{t-1}, P_{t-2}, P_{t-3}	0.9286	0.9317	0.9184	0.9231	0.9257	0.9154
	P_t, P_{t-1}, P_{t-2}, P_{t-3}, P_{t-4}	0.9307	0.9349	0.9215	0.9256	0.9291	0.9187
	P_t, P_{t-1}, P_{t-2}, P_{t-3}, P_{t-4}, P_{t-5}	0.9324	0.9361	0.9257	0.9287	0.9302	0.9219
ANFIS-PSO	P_t, P_{t-1}	0.9828	0.9852	0.9778	0.9794	0.9769	0.9735
	P_t, P_{t-1}, P_{t-2}	0.9846	0.9887	0.9809	0.9813	0.9801	0.9754
	P_t, P_{t-1}, P_{t-2}, P_{t-3}	0.9875	0.9901	0.9837	0.9832	0.9835	0.9786
	P_t, P_{t-1}, P_{t-2}, P_{t-3}, P_{t-4}	0.9882	0.9923	0.9861	0.9857	0.9863	0.9822
	P_t, P_{t-1}, P_{t-2}, P_{t-3}, P_{t-4}, P_{t-5}	0.9903	0.9945	0.9894	0.9862	0.9884	0.9847

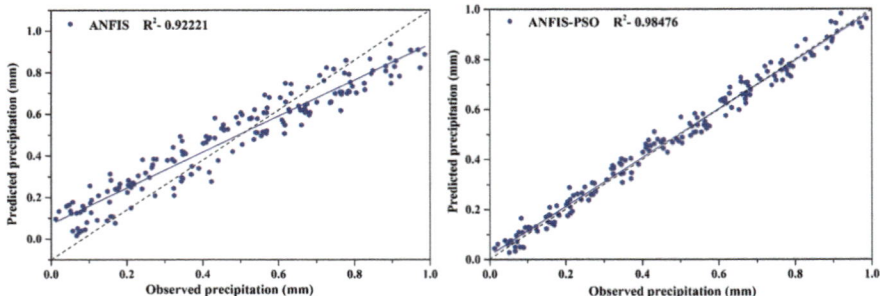

Fig. 33.4 Scatter plot of actual vs predicted precipitation

513.169 mm, 548.022 mm for ANFIS, ANFIS-PSO against actual peak 556.764 mm for the station Nuapada. Comparisons of proposed model in terms of Box plot are presented in Fig. 33.6.

In present study, hybrid ANFIS-PSO and simple ANFIS techniques were employed for estimating precipitation on monthly basis considering precipitation data from selected station as the only input. Fundamentally, precipitation forecasting on basis of month or year provides two benefits. Firstly, there is no reliance to any particular input data like meteorological components, and secondly, any pre-calculation analysis is not necessary. It was investigated that ANFIS-PSO prediction is superior to ANFIS in terms of E_{NS}, WI and R^2 for precipitation forecasting.

33.5 Conclusion

Present study focuses on application of ANFIS-PSO and ANFIS techniques for estimating precipitation in Nuapada district utilising monthly precipitation data from a period of 1970 to 2019. Performances of the applied models were evaluated based on statistical indicators such as E_{NS}, WI and R^2. Based on comparison of results it was found that ANFIS-PSO with $E_{NS} = 0.9903$, WI = 0.9945 and $R^2 = 0.9894$ performed superiorly compared to standalone ANFIS having values of E_{NS}, WI and R^2 as 0.9324, 0.9361, 0.9257 respectively. Robust optimisation algorithm used in this study (PSO) improved results obtained from simple ANFIS model utilised for precipitation forecasting for selected area of study. On a whole, it can be established that PSO algorithm comprised appropriate ability to train and improve performances of ANFIS.

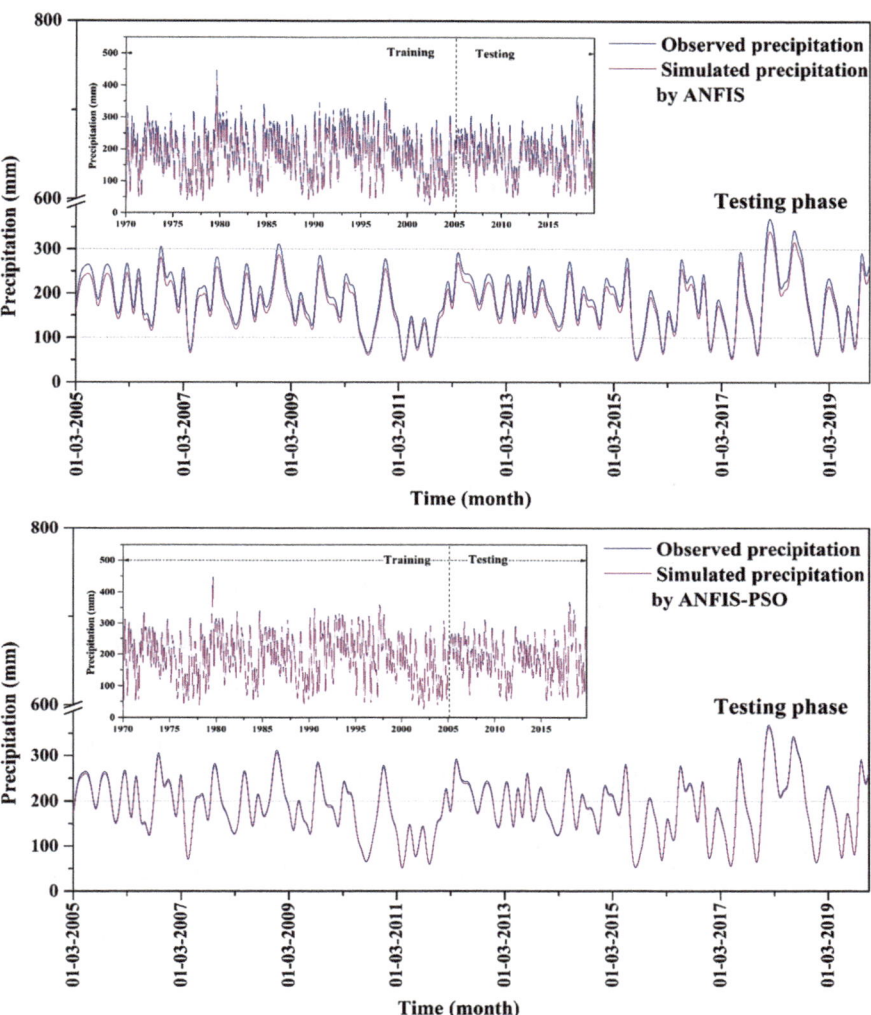

Fig. 33.5 Time series plot of observed and forecasted rainfall

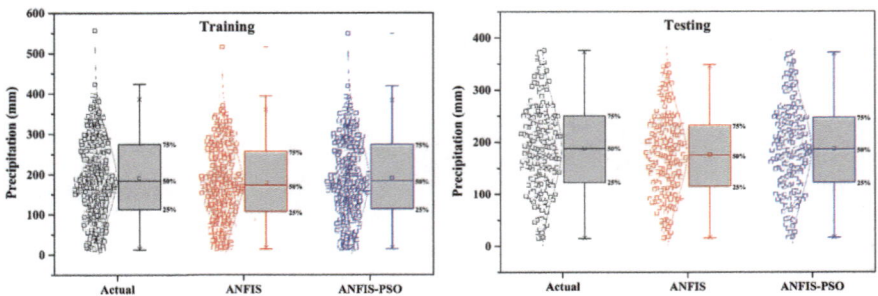

Fig. 33.6 Assessment of Box plot for several models

References

1. Choubin, B., Malekian, A., Samadi, S., Khalighi-Sigaroodi, S., Sajedi-Hosseini, F.: An ensemble forecast of semi-arid rainfall using large-scale climate predictors. Meteorol. Appl. **24**(3), 376–386 (2017)
2. Eberhart, R., Kennedy, J.: A new optimizer using particle swarm theory. In: MHS'95. Proceedings of the Sixth International Symposium on Micro Machine and Human Science (pp. 39–43). IEEE (1995)
3. Jalalkamali, A.: Using of hybrid fuzzy models to predict spatiotemporal groundwater quality parameters. Earth Sci. Inf. **8**(4), 885–894 (2015)
4. Jang, J.S.: ANFIS: adaptive-network-based fuzzy inference system. IEEE Trans. Syst. Man Cybern. **23**(3), 665–685 (1993)
5. Jang, J.S.R., Sun, C.T., Mizutani, E.: Neuro-fuzzy and soft computing-a computational approach to learning and machine intelligence [Book Review]. IEEE Trans. Autom. Control **42**(10), 1482–1484 (1997)
6. Jeong, C., Shin, J.Y., Kim, T., Heo, J.H.: Monthly precipitation forecasting with a neuro-fuzzy model. Water Resour. Manage **26**(15), 4467–4483 (2012)
7. Kennedy, J., Eberhart, R.: Particle swarm optimization. In: Proceedings of the IEEE international conference on neural networks, vol. 4, pp. 1942–1948. (1995)
8. Kisi, O., Sanikhani, H.: Prediction of long-term monthly precipitation using several soft computing methods without climatic data. Int. J. Climatol. **35**(14), 4139–4150 (2015)
9. Liu, Z., Zhang, D., Peng, W.: A Novel ANFIS-PSO Network for forecasting oil flocculated asphaltene weight percentage at wide range of operation conditions. Pet. Sci. Technol. **36**(14), 1044–1050 (2018)
10. Mamdani, E.H., Assilian, S.: An experiment in linguistic synthesis with a fuzzy logic controller. Int. J. Man Mach. Stud. **7**(1), 1–13 (1975)
11. Mohammadi, K., Shamshirband, S., Petković, D., Yee, L., Mansor, Z.: Using ANFIS for selection of more relevant parameters to predict dew point temperature. Appl. Therm. Eng. **96**, 311–319 (2016)
12. Mohammadi, K., Shamshirband, S., Tong, C.W., Alam, K.A., Petković, D.: Potential of adaptive neuro-fuzzy system for prediction of daily global solar radiation by day of the year. Energy Convers. Manage. **93**, 406–413 (2015)
13. Mohanta, N.R., Patel, N., Beck, K., Samantaray, S., Sahoo, A.: Efficiency of river flow prediction in river using Wavelet-CANFIS: a case study. In: Intelligent Data Engineering and Analytics (pp. 435–443). Springer, Singapore (2021a)
14. Mohanta, N.R., Biswal, P., Kumari, S.S., Samantaray, S., Sahoo, A.: Estimation of sediment load using adaptive neuro-fuzzy inference system at indus river basin, India. In: Intelligent Data Engineering and Analytics (pp. 427–434). Springer, Singapore (2021b)

15. Nguyen, Q.H., Ly, H.B., Le, T.T., Nguyen, T.A., Phan, V.H., Tran, V.Q., Pham, B.T.: Parametric investigation of particle swarm optimization to improve the performance of the adaptive neuro-fuzzy inference system in determining the buckling capacity of circular opening steel beams. Materials 13(10), 2210 (2020)
16. Sahoo, A., Samantaray, S., Bankuru, S., Ghose, D.K.: Prediction of flood using adaptive neuro-fuzzy inference systems: a case study. In: Smart Intelligent Computing and Applications (pp. 733–739). Springer, Singapore (2020)
17. Samantaray, S., Sahoo, A., Ghose, D.K.: Infiltration loss affects toward groundwater fluctuation through CANFIS in arid watershed: a case study. In: Smart Intelligent Computing and Applications (pp. 781–789). Springer, Singapore (2020a)
18. Samantaray, S., Sahoo, A., Ghose, D.K.: Prediction of sedimentation in an arid watershed using BPNN and ANFIS. In: ICT Analysis and Applications (pp. 295–302). Springer, Singapore (2020b)
19. Shamshirband, S., Gocić, M., Petković, D., Saboohi, H., Herawan, T., Kiah, M.L.M., Akib, S.: Soft-computing methodologies for precipitation estimation: a case study. IEEE J. Sel. Top. Appl. Earth Observations Remote Sens. 8(3), 1353–1358 (2014)
20. Sridharam, S., Sahoo, A., Samantaray, S., Ghose, D.K.: Assessment of flow discharge in a river basin through CFBPNN, LRNN and CANFIS. In: Communication Software and Networks (pp. 765–773). Springer, Singapore (2021)
21. Yaseen, Z.M., Ebtehaj, I., Kim, S., Sanikhani, H., Asadi, H., Ghareb, M.I., Bonakdari, H., Wan Mohtar, W.H.M., Al-Ansari, N., Shahid, S. Novel hybrid data-intelligence model for forecasting monthly rainfall with uncertainty analysis. Water 11(3), 502 (2019)
22. Zanganeh, M.: Improvement of the ANFIS-based wave predictor models by the Particle Swarm Optimization. J. Ocean Eng. Sci. 5(1), 84–99 (2020)

Chapter 34
Performance of Deconvolution Network and UNET Network for Image Segmentation

Jash Jayesh Kothari, Sai Sandesh Racha, and Joydeep Sengupta

Abstract In this paper, we have discussed the architecture of certain deep learning algorithms namely, Deconvolutional Neural Network and UNET Network. These are compared to the performances of models based on each of them using various mathematical parameters. Briefly, in both of these architectures, the image is passed through several convolution layers, each followed by a rectified linear unit (ReLU) and a max-pooling operation in the contraction path. This enables to capture the feature information. A similar symmetric expanding path helps find spatial information by passing the image through some up-convolution layers. In UNET though, in the expanding path, the spatial information obtained is concatenated with feature information that was obtained from the contraction path. For the CityScapes Dataset, we can see that models based on UNET clearly outperform the prior models based on Deconvolution Network by evaluating and comparing their IOU values. The network was trained on Google Colab's GPU.

34.1 Introduction

Computer vision is a field of science that mainly deals with computers managing to understand high-level information from the input digital images or videos [1, 2]. Its main aim is to replicate the powerful abilities of what a human brain can do. For simplicity, Computer vision can be defined as a complex system that realizes the objects and their characteristics in the image. These characteristics namely shapes, sizes, textures and colors, after being realized by the system, help to study the input image or video in a better way.

J. Jayesh Kothari (✉) · S. S. Racha · J. Sengupta
Visvesvaraya National Institute of Technology (VNIT) Nagpur, Nagpur, India
e-mail: jashkothari35@gmail.com

© The Author(s), under exclusive license to Springer Nature Singapore Pte Ltd. 2022   361
V. Bhateja et al. (eds.), *Evolution in Computational Intelligence*,
Smart Innovation, Systems and Technologies 267,
https://doi.org/10.1007/978-981-16-6616-2_34

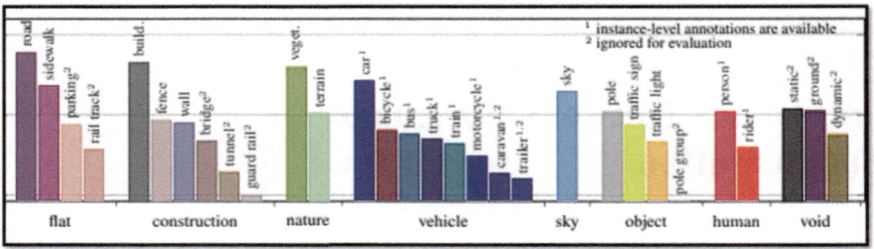

Fig. 34.1 Different colors assigned for different classes [4]

Image classification, object detection and semantic segmentation are the three most important problems that humans have been interested in solving using computer vision [2, 3]. Popularly known as dense prediction, semantic segmentation mainly labels each and every pixel of the input image with the class that it represents. Since the task of prediction of the classes is being done for every pixel in the image, hence it is called dense prediction.

In this study, we are not separating instances of the same class, i.e., when the segmentation map is formed, if the input image consists of two or more objects of the same class then the segmentation map cannot differentiate them as different objects. We are only predicting the class corresponding to each pixel in the image. Even modern semantic segmentation technique [3] is powered by deep learning technology. Its models are useful for a variety of tasks, including autonomous vehicles. Modern cars have to be equipped with these technologies that help sense and understand the environment around them. This will help us integrate self-driving cars into our existing roads. For this, visual understanding of street scenes is a must to help vehicles commute safely on their own.

To address the above problem statement, we use the Cityscapes dataset. It is a dataset that basically comprises a diverse set of street images that have been recorded from 50 different cities [4]. It consists of 5000 high-quality pixel-level annotated images that are further divided into 2975 training image files, 500 validation image files and 1525 test image files. It also has 34 visual classes which are then grouped into 8 major categories:

Flat, Nature, Vehicle, Construction, Object, Human, Sky and Void.

We have only chosen 19 classes for our study as shown in Fig. 34.1. Other classes are too rare and hence they are excluded.

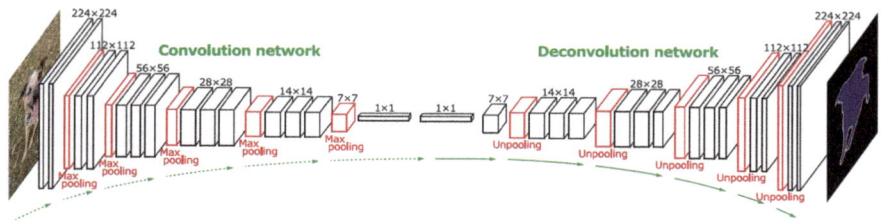

Fig. 34.2 De-CNN architecture [5]

34.2 Training

34.2.1 Deconvolutional Neural Network

A Deconvolutional Neural Network (De-CNN) is a neural network that behaves as an inverse convolution model [5, 6]. They are also called transposed convolutional neural networks or deconvs.

This type of architecture basically makes use of locally connected layers. These layers are convolution layers that are used in the downsampling path and upsampling path. Pooling methods like max-pooling and up-pooling are also used in this type of architecture. We can say that in the deconvolutional neural network, layers are being constructed by running a CNN model in a backward direction. In a simple sense, we can say that we are "reverse engineering" the input parameters of a normal CNN model.

Just like how a CNN model has a downsampling path [7, 8], the deconv network also consists of a down-sampling path that is mainly used to extract and interpret the context of a single-pixel by classifying it among various classes. We can say that this path helps to capture the "What of the pixel". Adding to that, the De-CNN model has an upsampling path that consists of transposed convolutional layers [5, 9]. This helps for the predictions to have a one-to-one correspondence with the input image in the spatial domain. This layer mainly deals with localization, that is, it helps to capture the "Where of the pixel". This means that this layer helps to recover the spatial information around the pixel. The above architecture can be represented as shown in Fig. 34.2.

A transposed convolutional model can be easily made by using back-propagation and reverse filtering. Padding and Striding can also be used to build a model with greater efficiency.

Fig. 34.3 UNET
architecture

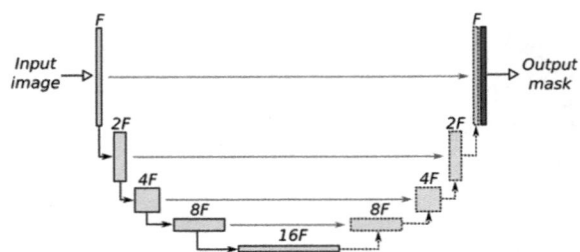

34.2.2 UNET

UNet architecture is a Convolutional Neural Network that was developed by Olaf
Ronneberger et al. for Biomedical Image Segmentation [10]. This network consists
of two paths, namely a contraction path and an expansion path, which gives it a
U-shaped architecture, hence named at UNet [11].

Contraction Path: This path is also called the encoder path. The main function
of the contraction path is to capture the context of the pixel in the image just similar
to that in the De-CNN model. The contraction path is a convolutional network where
the image is passed through a repeated number of convolutions layers, each followed
by a rectified linear unit (ReLU) and then this is passed through a max-pooling
operation. The number of feature channels is doubled in every step. The contraction
path helps to increase the feature information and decrease the spatial information
of the image.

Expansion Path: This path is also called the decoder path. The main function of
the expansion path is to combine the spatial information to the feature information
obtained in the downsampling path through several up-convolutions layers thereby
decreasing the number of channels by 2. Concatenations with the corresponding high-
resolution features from the contracting path are done in the UNET architecture that
make it different from the De-CNN.

The above architecture can be represented as shown in Fig. 34.3.

34.3 Results

The models discussed above were tested with the test dataset and the results are
shown in Figs. 34.4, 34.5 and 34.6.

Fig. 34.4 Original image [4]

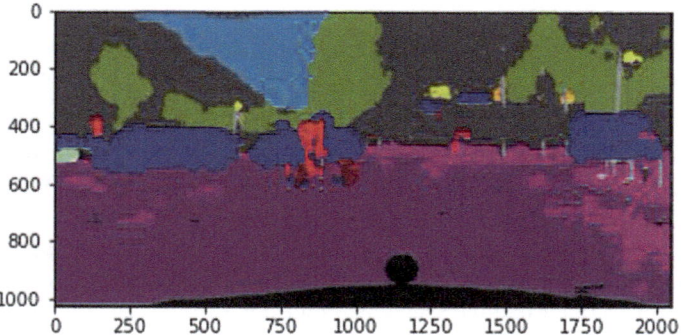

Fig. 34.5 Segmented image using De-CNN

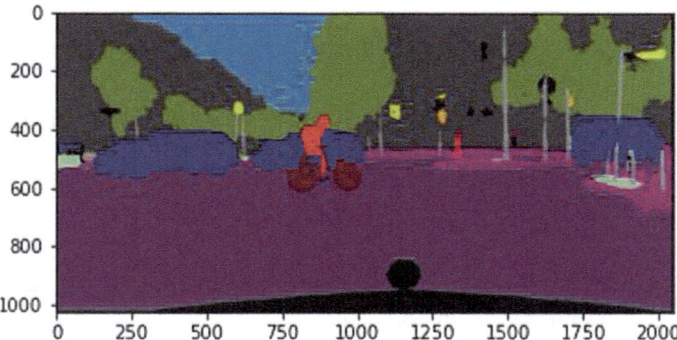

Fig. 34.6 Segmented image using UNET

Epoch No	Train Loss	Validation Loss	Avg Pixel Accuracy	Avg IOU
1	3.0164629	0.02685575	39.13788795	0.227693
5	2.9772706	0.02091291	62.39209175	0.380966
7	1.8818187	0.008726799	73.6093998	0.668281
8	1.6304272	0.011434129	90.71044922	0.575111
13	1.5037672	0.009094001	86.10086441	0.651523
14	1.3394839	0.007486659	79.88996506	0.671265
15	0.7272205	0.010424027	82.02514648	0.603628
18	0.564506	0.007647223	89.54043388	0.694425
22	0.4825182	0.008224421	92.55332947	0.68868
30	0.4053306	0.008936689	69.01197433	0.700909
32	0.3689286	0.009191109	81.83197975	0.634313

Fig. 34.7 Tabulated data for De-CNN

34.3.1 *Mathematical Computations*

The above results cannot be differentiated visually so we have used Average Pixel accuracy values and IOU (intersection over union) metric to different the segmented output image from both the models and comment on them.

Pixel accuracy is defined as the ratio of the number of correctly predicted pixels by our model to the total number of pixels in the image. But high pixel accuracy does not always confer that the segmented image has been predicted correctly. This problem is called a class imbalance. It basically occurs when few classes dominate the entire image while the other classes only constitute a small portion of the image. So, to evaluate our model more accurately, we have used the IOU metric along with pixel accuracy values. This helps us to have a better understanding of the segmented images.

Intersection over Union (IOU) [12] is defined as the overlap between the area of the predicted image and the ground truth image divided by the union of the area of the predicted image and the area of the ground truth image.

$$IOU = \frac{Area\,of\,Intersection}{Area\,of\,Union}$$

IoU metric helps us to understand how much similar a predicted box is to the box in the ground truth image. By seeing the formula, we can say that the prediction result is poor for a low value of IoU and if the IOU value is close to 1, it means the prediction result is accurate enough. The ideal IOU value is 1.

These values were formulated along with the training losses and validation losses for both the models till 32 epochs owing to the in-availability of a GPU. The training and validation losses were evaluated to keep a check on the model that it does not overfit at any time. These results have been presented separately in Figs. 34.7 and 34.8.

Epoch No	Train Loss	Validation Loss	Avg Pixel Accuracy	Avg IOU
1	2.0782367	0.021851443	35.55088043	0.402744
2	1.1474725	0.011131913	80.31687737	0.588733
3	0.7915547	0.013057315	61.91029549	0.55593
8	0.5169051	0.009195452	57.04298019	0.636423
9	0.4932101	0.006973442	86.74879074	0.730583
15	0.3848666	0.005853947	89.19682503	0.796484
19	0.360206	0.005925882	81.32209778	0.781359
25	0.4535975	0.00647835	93.31684113	0.767415
26	0.2964574	0.004315377	85.61353683	0.838641
31	0.2560886	0.006266952	91.69726372	0.772369
32	0.2497416	0.004180761	82.75270462	0.834771

Fig. 34.8 Tabulated data for UNET

Here we can see that for De-CNN architecture after 32 epochs, the average pixel accuracy found was 81.83% and the average IOU value is found to be 0.634. Further, we have tabulated similar data for the UNET architecture.

Here we can see that for UNET architecture after 32 epochs, the average pixel accuracy found was 82.75% and the average IOU value is found to be 0.834.

34.4   Conclusion

As mentioned earlier, only pixel accuracy is not enough to decide the efficiency of a model. Thus, the IOU values of two different model architectures are compared as shown in the tables above. The ideal value of IOU being 1, the model with its IOU value close to 1 is considered to have predicted the test data in a better way.

– The expansion path and the contraction path in De-CNN are mutually independent of each other. This causes a loss of information in the expansion path.
– As it can be seen in the results above, for the Cityscapes Dataset, the final IOU value after 32 epochs for UNET Model is 0.834 and for De-CNN architecture is 0.634.
– The average IOU value and the average pixel accuracy found were lesser than the values of the UNET model showing the lesser efficiency of the model.
– Hence it can be concluded mathematically that the UNET architecture predicts the dataset more efficiently. This is due to the concatenations of the spatial information with the feature information from the contraction path in the upsampling path.
– It can also be seen from the above tables that both the training losses as well as the validation losses decrease with every epoch and both the losses 8 are lesser for the model with UNET architecture than the model with De-CNN architecture for the same number of epochs trained.

Hence, owing to our study above, we have concluded that UNET Architecture is better than De-CNN for the Cityscapes Dataset by verifying it using the two most important metrics that are average pixel accuracy metric and the average IOU Metric.

References

1. Wiley, V., Lucas, T.: Computer vision and image processing: a paper review. Int. J. Artif. Intell. Res. **2**, 29–36 (2018)
2. Barik, D., Mondal, M.: Object identification for computer vision using image segmentation. In: 2010 2nd International Conference on Education Technology and Computer, pp. V2-170–V2-172 (2010). https://doi.org/10.1109/ICETC.2010.5529412
3. Wang, L., Chen, X., Hu, L., Li, H.: Overview of image semantic segmentation technology. In: 2020 IEEE 9th Joint International Information Technology and Artificial Intelligence Conference (ITAIC), pp. 19–26 (2020)
4. Cordts, M., Omran, M., Ramos, S., Rehfeld, T., Enzweiler, M., Benenson, R., Franke, U., Roth, S., Schiele, B.: The Cityscapes dataset for semantic urban scene understanding. In: IEEE Conference on Computer Vision and Pattern Recognition (CVPR) (2016)
5. Noh, H., Hong, S., Han, B.: Learning deconvolution network for semantic segmentation. In: 2015 IEEE International Conference on Computer Vision, pp. 1520-1528 (2015)
6. Zeiler, M.D., Krishnan, D., Taylor, G.W., Fergus, R.: Deconvolutional networks. In: IEEE Computer Society Conference on Computer Vision and Pattern Recognition, pp. 2528–2535 (2010). https://doi.org/10.1109/CVPR.2010.5539957
7. Albawi, S., Mohammed, T.A., Al-Zawi, S.: Understanding of a convolutional neural network. In: International Conference on Engineering and Technology (ICET), pp. 1–6 (2017). https://doi.org/10.1109/ICEngTechnol.2017.8308186
8. Chauhan, R., Ghanshala, K.K., Joshi, R.C.: Convolutional neural network (CNN) for image detection and recognition. In: 1st International Conference on Secure Cyber Computing and Communication (ICSCCC), pp. 278–282 (2018). https://doi.org/10.1109/ICSCCC.2018.8703316
9. "Review: DeconvNet — Unpooling Layer (Semantic Segmentation)", Towardsdatascience by Sik-Ho Tsang. Referred from https://towardsdatascience.com/review-deconvnet-unpooling-layer-semantic-segmentation-55cf8a6e380e. Accessed 8 Oct 2018
10. Ronneberger, O., Fischer, P., Brox, T.: U-net: convolutional networks for biomedical image segmentation. In: Medical Image Computing and Computer-Assisted Intervention – MICCAI, vol. 9351. Springer (2015)
11. "UNet — Line by Line Explanation", Towardsdatascience by Jeremy Zhang. Referred from https://towardsdatascience.com/unet-line-by-line-explanation-9b191c76baf5. Accessed 18 Oct 2019
12. Rezatofighi, H., Tsoi, N., Gwak, J., Sadeghian, A., Reid, I., Savarese, S.: Generalized intersection over union: a metric and a loss for bounding box regression. In: IEEE/CVF Conference on Computer Vision and Pattern Recognition (CVPR), pp. 658–666 (2019). https://doi.org/10.1109/CVPR.2019.00075

Chapter 35
On Correlations Between Feedback on Project Proposals for External Financier Support and Final Marking

Sylvia Encheva

Abstract High education organizations world while are strongly encouraged to obtain substantial amounts of external support for their research activities. Employees invest an enormous amount of time and effort in proposal writing while granting entities use large amounts of resources, both human and financial, to administrate, evaluate, and finally select applications that are to be granted. One of the many open questions in this complicated process is to which degree feedback from evaluation committees assists researchers in their further work on improving proposals that have not received support and in writing new ones. How the application of modern technologies can assist in the search for good answers to this question is the main focus of this work.

35.1 Introduction

Employees at higher education organizations are expected to deliver excellent teaching, publish high-quality research results, disseminate their findings in a way understandable to the general public, and obtain substantial amounts of external funding from both national and international organizations for their research and development activities. Granting organizations use a lot of resources to administer and evaluate such applications, universities increase considerably the number of employees at their research departments to facilitate the development of grant applications and assist in project management, and last but not least come researchers doing their utmost best to prepare competitive proposals and handle granted applications on usually very tight budgets and short deadlines. A relatively little discussed but very important question is how can modern technologies be applied to assist and enhance the work of granting institutions, researchers, and administrators at writing applications and working on projects that have received financial support.

S. Encheva (✉)
Western Norway University of Applied Sciences, Bergen, Norway
e-mail: sbe@hvl.no

© The Author(s), under exclusive license to Springer Nature Singapore Pte Ltd. 2022   369
V. Bhateja et al. (eds.), *Evolution in Computational Intelligence*,
Smart Innovation, Systems and Technologies 267,
https://doi.org/10.1007/978-981-16-6616-2_35

Do granting organizations have a systematic approach to control and raise the quality of provided feedback to applicants? Applicants are provided with clear responses on whether their proposals are granted or not, along with overall marks, and marks for each predefined criterion. The latter ones however are far too often followed by a short text with very little practical meaning if at all. While submissions of formal complaints are allowed, requests for more detailed feedbacks are not granted. Shortcomings of conventional project evaluation practice are discussed in [6].

There is an obvious need for extracting knowledge from previous experiences to assist the hard work of all parties involved in the process of both developing applications for external funding and project management.

35.2   Text Mining Tools

Text mining techniques, [1] can be used to assist applicants in their efforts to write better proposals.

Bag of words is a popular technique to visualize the occurrence of words within a document, [10]. Sentiment analysis is a special case of text mining used to analyze attitudes, evaluations, and opinions, and depict the sentiments behind it, [11].

A silhouette plot illustrates the level of similarity between an instance and the cluster it belongs to in comparison to other clusters, [2]. Numerically the level of similarity is presented by a number in the interval from -1 to $+1$ where $+1$ indicates complete similarity.

In data science, nomograms provide a visual representation of outcomes obtained by applying various classifiers. For further details on the subject see [3, 4]. Dynamic nomograms are proposed as translational tools which can accommodate models of increased complexity in [5]. Nomograms in Orange, [8] enable Naive Bayes and Logistic Regression classifiers.

Among the numerous machine learning open source tools it is worth looking at Orange, [8] and Weka, [9]. All figures in this work are drawn using Orange [8].

35.3   Evaluation Criteria and Final Outcomes

The process of obtaining external funding for research projects is complicated and requires problem-solving skills at many levels. Problems can be due to poorly designed goals, lack of competence in management, low-level domain knowledge among team members, insufficient administrative support, etc. While some researchers attend seminars and workshops where they exchange experience from project management, others prefer personal contacts in order to get advises on how to handle current challenges. Employment of modern technologies to assist scientists in the development of project proposals is however far too little considered by the research community.

Applicants need to know how to interpret the provided referee comments in order to improve the quality of their future applications and granting organizations should know how to advise potential evaluators in order to assure clarity of feedback to grant seekers.

Note that in general, granting organizations employ different scales for giving marks to submitted proposals.

In this study, we will use a scale from 1 to 7 where 7 is the highest and 1 is the lowest. Such a scale is used by RCN, [7]. Nearly all proposals that are given an overall mark 7 are granted, a substantial number with given mark 6 receive support as well, in some areas those that are given mark 5 or less are not considered to be given financial support, in other areas the same is valid for applications that are given mark 4 or less. Very seldom a proposal with a mark 3 is returned to the project team with a request for improvement and a resubmitted version is evaluated where the outcome is either granted or rejected. Marks 1 and 2 are hardly known to be given in practice.

Proposals are assessed on three main criteria: excellence, impact, and implementation.

From this point on we use data about research proposals presented in Table 35.1, which is a toy example meant to illustrate tendencies from real world data.

Table 35.1 Applications, criteria, overall marks, and final outcomes

Proposals (P)	Excellence	Impact	Implementation	Overall mark	Final outcome
P 1	6	5	6	6	Granted
P 2	5	5	4	5	Rejected
P 3	5	6	7	6	Rejected
P 4	3	6	5	4	Rejected
P 5	5	4	4	4	Rejected
P 6	5	5	5	5	Rejected
P 7	5	5	5	5	Granted
P 8	4	5	4	4	Rejected
P 9	6	6	6	6	Granted
P 10	5	4	5	5	Granted
P 11	4	4	5	5	Rejected
P 12	6	5	3	4	Rejected
P 13	5	6	4	5	Rejected
P 14	5	6	6	6	Granted
P 15	7	4	4	5	Rejected
P 16	6	5	5	5	Rejected
P 17	6	6	4	6	Rejected

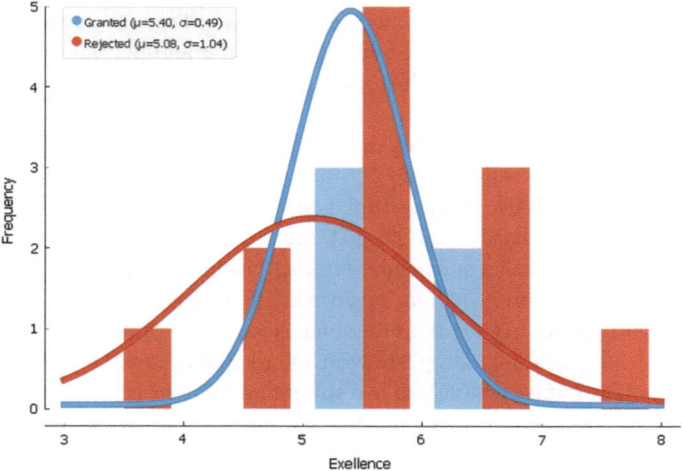

Fig. 35.1 Normal distributions of marks for excellence

Normal distributions of marks for the two criteria excellence and implementation are shown in Figs. 35.1 and 35.2, where blue and red colors indicate granted and rejected proposals respectively. This color code is applied for drawing corresponding curves and bar diagrams. Horizontal axes represent criteria and vertical axes represent frequencies.

In both cases, a number of granted and rejected proposals are given similar marks on the two criteria but there is no clear indication why some of them have received support while the others have not. Applicants with rejected proposals are sometimes advised to consider the degree to which their suggestions fit the corresponding call. Sentiment analysis on proposal evaluations can be applied to discover hidden issues that need improvements. Fine-grained sentiment, aspect-based, and intent analysis are recommended for discovering somewhat hidden hints.

Nomograms in Figs. 35.3 and 35.4 indicate which feature has the highest importance for the final decision. Nomograms in Orange are interactive and provide support for prediction of class probabilities.

Figure 35.3 thus shows that about one third of the proposals listed in Table 35.1 are granted where the marks for criteria excellence, impact, and implementation are either 5 or 6. This indicates that granted proposals are well thought of, all criteria are seriously considered and properly addressed.

Figure 35.4 reveals that some of the proposals in Table 35.1 are rejected even though their marks for criterion excellence can be between 5 and 7. This is in contrast to the popular belief among applicants that excellence is the most important criterion in decision-making for granting a research proposal.

The silhouette plot in Fig. 35.5 clearly shows that projects P1, P9, P14, and P7 are granted according to given overall marks while project P10 is granted for a reason not entirely based on the given overall mark. The case with rejected proposals points

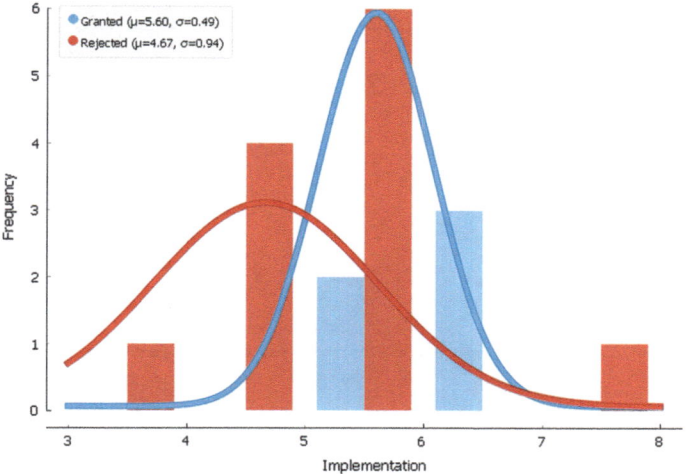

Fig. 35.2 Normal distributions of marks for implementation

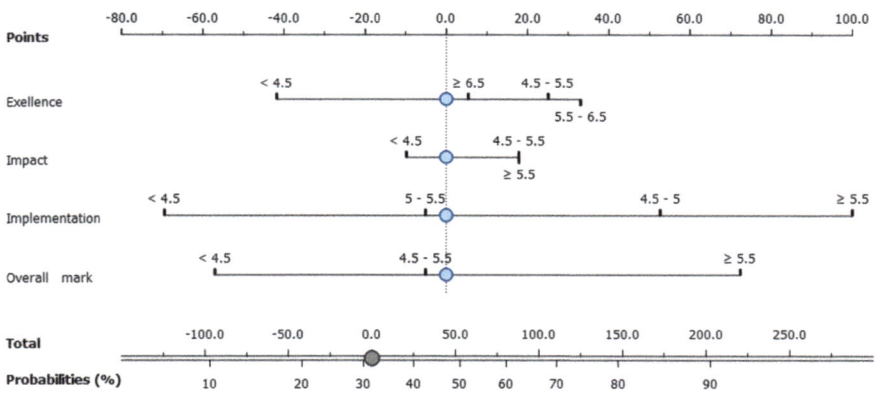

Fig. 35.3 Nomogram for granted proposals

out that five of them, i.e. P8, P12, P5, P4, and P13 are rejected based on given overall marks and seven of them P11, P15, P17, P2, P16, P6, and P3 are not granted for reasons unrelated to given overall marks.

35.3.1 Granting Organizations

Sentiment analysis on granted and rejected proposals can help granting organizations to find out to which degree their comments are helpful to those who work on developing proposals. This way they can obtain higher quality feedback to applicants

S. Encheva

374

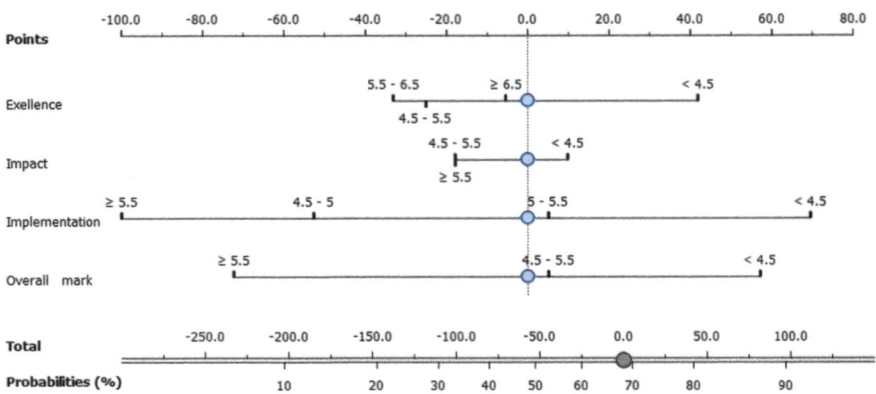

Fig. 35.4 Nomogram for rejected proposals

Fig. 35.5 Silhouette plot

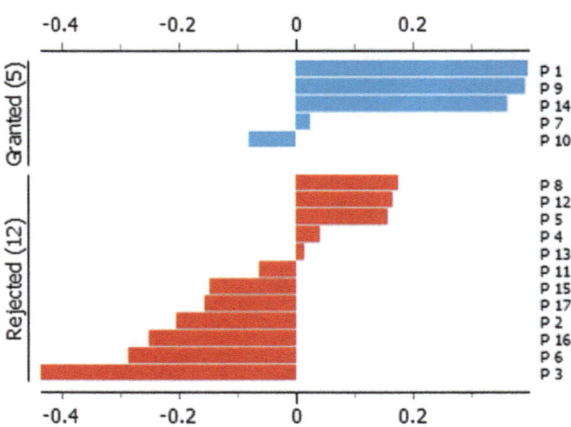

and help particularly good ideas be supported in future applications. Considering the amount of time and efforts researchers spent on developing applications for external support it is really necessary for granting institutions to do their best to provide meaningful and helpful comments.

Granting organizations can apply text mining techniques to extract knowledge about the quality of the work of their evaluation committees. This can be done by f. ex. comparing feedback a single committee is providing to all proposals assigned to that committee. It is also beneficial to compare the text following each evaluation criterion and the text following the overall mark. In a similar fashion one can study the work performance among committees.

Another interesting aspect is to find out what is the correlation between the committee's evaluations and final outcomes, i.e. how many and which of the highest ranked by a committee proposals are actually granted. Does it make sense to keep a committee providing evaluations that are not followed by the corresponding granting organization? If this is to be taken for further considerations then where is the threshold for acceptable differences between a committee's ranking and granting organizations' final decisions?

Bag of words is very useful for providing information about words used in evaluation reports. Suppose one proposal is given a mark 6 for the criterion excellence and another one is given 5 for the same criterion and in both cases, the actual feedback is 'very well formulated scientific contribution'. In such cases, the respective committees ought to be contacted with a request to provide meaningful feedback strongly related to each proposal.

Evaluation committees are usually employed for a period of several years. If their work is found to be unsatisfactory based on previously described comparisons, granting organizations can take proper actions to improve the committees' deliverables.

35.3.2 Discussions

Granting organizations can send questionnaires to applicants inquiring about the clarity and usability of provided feedback. Text mining techniques like sentiment analysis and hash distances may reveal useful information in terms of inconsistencies, vague remarks, etc.

Consider projects given the same overall mark where some of them are granted while others are rejected. How is the final outcome related to different applying organizations? The ratio between granted and rejected proposals at national level is usually quite clear but is there any correspondence between this ratio and the ratios for different organizations, are there any clusters of organizations?

Project coordinators use a lot of time and effort to deliver intermediate reports. While such activities may help granting institutions to have better control over how resources are used by single projects, there is a need to extract knowledge based on a large population of data gathered from a large number of projects granted to different research organizations and thus been able to draw conclusions about general tendencies, positive outcomes, and eventual needs for improvements.

In future work, we will investigate whether overall proposals' marks are related to timely deliveries and final project outcomes applying supervised learning.

35.4 Conclusion

In this paper, we propose a way to extract knowledge from previous applications in order to improve new applications and increase the number of high-quality results obtained from new projects. This way researchers will be encouraged to work on improving rejected applications and granting institutions will be supported in electing evaluators not only based on scientific merits but also on abilities to provide meaningful and useful feedback to applicants.

References

1. Bengfort, B., Bilbro, R., Ojeda, T.: Applied Text Analysis with Python: Enabling Language-Aware Data Products with Machine Learning. O'Reilly Media, Newton (2018)
2. Ciaburro, C.: MATLAB for Machine Learning. Packt Publishing, Birmingham (2017)
3. Evesham, H.A.: The History and Development of Nomography. Docent Press, Boston (2010)
4. Gronwall, T.H., Doerfler, R., Gluchoff, A., Guthery, S.: Calculating Curves: The Mathematics, History, and Aesthetic Appeal of T. H. Gronwall's Nomographic Work. Docent Press, Boston (2012)
5. Jalali, A., Alvarez-Iglesias, A., Roshan, D., Newell, J.: Visualising statistical models using dynamic nomograms. PLoS One **14**(11) (2019)
6. Haass, O., Guzman, G.: Understanding project evaluation a review and reconceptualization. Int. J. Manag. Proj. Bus. **13**(3), 573–599 (2019)
7. https://www.forskningsradet.no/
8. https://orange.biolab.si/
9. https://www.cs.waikato.ac.nz/ml/weka/
10. Sarkar, D.: Text Analytics with Python: A Practitioner's Guide to Natural Language Processing. Apress, Berkley (2019)
11. Zizka, J., Darena, F., Svoboda, A.: Text Mining with Machine Learning Principles and Techniques. CRC Press, Boca Raton (2020)

Chapter 36
Evaluation of Older Adults Fitness Level and Accomplishments with Fuzzy Similarity Relations and Data Analytics Techniques

Sylvia Encheva

Abstract Increasing amount of aging population worldwide is a matter of concern to politicians, healthcare providers, economists, and researchers among many others. A number of studies emphasize the importance of physical fitness for keeping older adults healthy and strong in their retirement life. Among the many open questions here are those related to how to evaluate their physical fitness at the beginning of a training program, recommend appropriate set of exercise, and provide a reasonable estimation of what an individual can expect to accomplish in a relatively short period of time, like for example in terms of months. Such estimations may contribute to building stronger motivation for more physical activities among older adults and contribute to a better quality of life.

36.1 Introduction

Medical practitioners strongly recommend an improvement of physical fitness to most older adults. Advises on types, amount, and intensity of exercises however are usually obtained from fitness instructors and to some extent from physiotherapists. On the one hand, such recommendations are more often than not based on relatively small-scale case studies and general knowledge. On the other hand, it is a common knowledge that many people, although for different reasons, prefer to minimize the number of times they take medical tests. There is a need for a systematic approach both tailored to individual needs and at the same time not requiring medical examinations.

The intention of this study is to detect cases where people's physical fitness is lower compared to what is common for their age. This can be accomplished by considering their daily activities and personal opinions. It is of utmost importance to take into account how they feel and draw conclusions afterwards.

S. Encheva (✉)
Western Norway University of Applied Sciences, Bergen, Norway
e-mail: sbe@hvl.no

We employ similarity degrees [4] among data sets calculated by fuzzy functions for finding correlations between biological age and physical fitness. For the purpose of predictive analytics we use the open source machine learning and data visualization tool Orange, [3]. Decision trees and hierarchy clusterings are further on employed for compact graphical representation of complex data and detecting otherwise difficult to notice dependencies.

Once presented with such comparisons, older adults can be more motivated to increase the amount of activities by considering their current achievements and what other people their age are capable of doing.

36.2 Related Theories

Data analytics analyzes data and extracts information from it. For a guide to the principles and methods of data analysis that does not require knowledge of statistics or programming see [6]. Data analytics and its applications are well presented in [9, 11] where special attention is paid to four types of data analytics: descriptive, diagnostic, predictive, and prescriptive analytics.

Outliers in data mining are commonly understood to be data points placed far from the known clusters. While they may appear due to incorrect measurements or an error in classification they may still indicate significant events that have to be considered separately.

It is a well known fact that real life data comes with missing values. While most available data analytics tools provide ways to fill in missing values a special care is needed in order to select a method providing the best outcomes, [12].

According to fuzzy set theory an object may partially belong to a set with a given degree. A fuzzy set in X is an $X \to [0, 1]$ mapping, while a fuzzy relation in X is a fuzzy set in $X \times X$.

Fuzzy similarity relations are studied in [4, 5].

A weak fuzzy similarity relation is a mapping, $S : U \times U \to [0, 1]$, such that for $x, y, z \in U$,

(a) $S(x, x) = 1$, reflexivity
(b) if $S(x, y) > 0$ then $S(y, x) > 0$, conditional symmetry
(c) if $S(y, x) \geq S(x, y) > 0$, and $S(z, y) \geq S(y, z) > 0$ then $S(z, x) > S(x, z)$
      conditional transitivity, [5].

For importance of physical activities for a healthier life see [2]. The impact of mobility limitations on health outcomes among older adults is presented in [8].

36.3 Physical Fitness

For those individuals, who have not been involved in sport professionally or at least being accustomed to regular sport activities, it might be quite a challenge to begin with and maintain health-related fitness activities in their retirement age. More often than not, people expect a quick fix and drop out of otherwise good fitness programs because they do not notice positive changes in a short period of time and they do not receive information on what is actually possible to accomplish in their particular cases. One way to motivate them is to be able to predict what to expect in terms of achievements based on current fitness level, types, amount, and intensity of physical activities.

In order to illustrate the approach we take a small sample of simulated data. Using a large amount of data is a part of another project that we are planning to work on in the near future.

Let's first have a look at what is the meaning of the terms used in this study.

- Preliminary self-assessment—ability to get up from the floor without using hands as support, number of stands up from a chair in 30 s without support, etc.
- Types of exercises—walking, swimming, pilates, etc., [10]. Due to space limitations we use only one column for that in Table 36.1. A working model will contain a significantly large amount of related attributes.
- Intensity—high, moderate, low.
- Frequency—irregular, once a week, twice a week, three times a week, …, daily, see [7].

Results refer to self-evaluation—'insignificant' improvement, 'positive' change is noticed, 'real' gain is observed.

Table 36.1 is constructed based on simulated data.

Notations in Table 36.1 are as follows. 'G' stands for 'Group' while 'I' denotes a single individual. Groups are formed based on outcomes of applying clustering techniques [12] for extracting information from given datasets. The number of groups will significantly increase once a large amount of real data is inserted. For older adults a year or two extra have much more effect on physical performance compared to younger adults. The same goes for the rest of the attributes.

Empty cells in Table 36.1 indicate missing information.

We use a fuzzy function to make comparisons based on individual data.

A fuzzy information Table 36.2 is constructed based on data from Table 36.1 where numerical and text values are converted to numbers between 0 and 1. These numbers indicate the extent to which a group satisfies listed conditions. The degree of similarity between two objects can be calculated by a conditional probability relation on fuzzy sets, as in [4].

Let $x = \{\chi_1^x/u_1, \ldots, \chi_n^x/u_n\}$ and $y = \{\chi_1^y/u_1, \ldots, \chi_n^y/u_n\}$ are two fuzzy sets over $U = \{u_1, \ldots, u_n\}$, $s_j : D_j \times D_j \to [0, 1]$, such that for $x, y \in D_j$,

Table 36.1 Data set

Group	Gender	Age	Preliminary self assessment	Exercise impact	Intensity	Weekly based frequency	Results in a month	Results in three months
G1	Female	60–65	Low	High	Moderate	Irregular	Positive change	Positive change
G2	Female	66–70	High	None	Low	Once	Positive change	Insignificant change
G3	Female	71–75	Middle	Low	High	Twice	Insignificant change	Real gain
G4	Male	60–65	Middle		High	Three	Positive change	Real gain
G5	Male	66–70	Low	Low	Low		Insignificant change	Insignificant change
G6	Male	71–75	High	High	Moderate	Daily	Positive change	Real gain
G7	Male	76–80	Middle	Low	Low	Twice	Positive change	Real gain
G8	Female	76–80	Low	None	Low	Once	Insignificant change	Positive change
I	Female	60–65	Middle	Low	Low	Irregular	Insignificant change	Insignificant change

Table 36.2 Fuzzy information table extracted from Table 36.1

Group	Gender	Age	Preliminary self assessment	Exercise impact	Intensity	Weekly based frequency	Results in a month	Results in three months
G1	Female	60–65	0.3	1	0.6	0.1	0.6	0.6
G2	Female	66–70	1	0.2	0.3	0.3	0.6	0.1
G3	Female	71–75	0.6	0.7	1	0.4	0.1	1
G4	Male	60–65	0.6		1	0.5	0.6	1
G5	Male	66–70	0.3	0.7	0.3		0.1	0.1
G6	Male	71–75	1	1	0.6	1	0.6	1
G7	Male	76–80	0.6	0.7	0.3	0.4	0.6	1
G8	Female	76–80	0.3	0.2	0.3	0.3	0.1	0.6
I	female	60–65	0.6	0.7	0.3	0.1	0.1	0.1

$$s_j(x|y) = \frac{\sum_{i=1}^{n} \min\{\chi_i^x, \chi_i^y\}}{\sum_{i=1}^{n} \chi_i^y} \tag{36.1}$$

where $s_j(x|y)$ means level similarity of x given y, [4].

The last row in Table 36.1 contains data provided by a woman of age between 60 and 65. Calculating the similarity level between her data and the ones in Group 1 and Group 2 we obtain $s_1(G1|Ex) = 0.61$ and $s_2(G2|Ex) = 0.73$ respectively. Her results are closer to ones provided by an older group of people than her self. This indicates a need for a change in her physical activities. The level of similarity may also be used to provide an indication as to what the physical condition of a person may be later on in life depending on current level of physical activities. Knowing how much current efforts spent on physical activities could contribute to a better physical fitness in a decade or two, may also contribute to a stronger motivation for increasing the amount of current physical activities.

Predictive analytics can be used to identify possible outcomes for longer periods of time for individuals in general. Prescriptive analytics can afterwards be employed to facilitate choices and recommendations of new activities supporting healthier living. Here it might be worthwhile to have a closer look at outliers and see whether they are results of some kind of errors or they indicate cases that have to be considered separately.

A decision tree shown in Fig. 36.1 is built based on data from Table 36.1, using Orange, [3]. It illustrates groups' distribution according to listed features. In an attempt to place the individual denoted as 'I' in Table 36.1 in one of the leaves of the tree in Fig. 36.1 we can see that again the fitness level of that individual, does not correspond to the fitness level of persons of the same age and gender in a way that her results are closer to those that are older than her. Decision trees tend to become big when visualising large data sets. Pythagoras trees can be used instead since they present a more compact picture of depicted data sets. Pythagoras trees are fractals that can be used to depict binary hierarchies, [1].

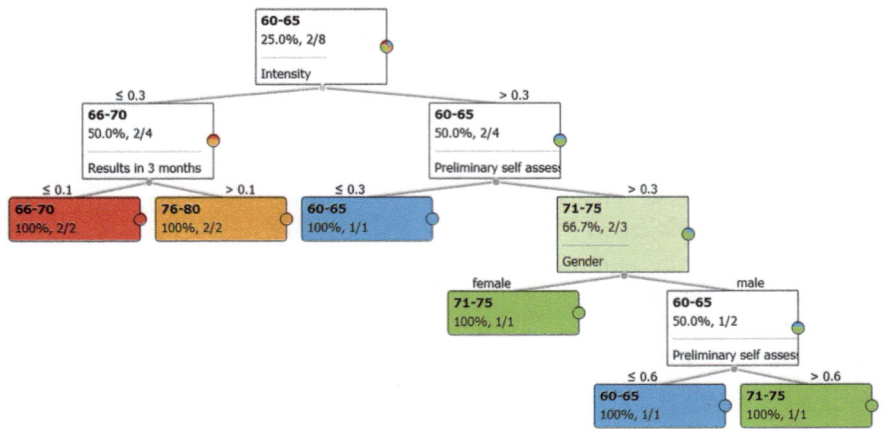

Fig. 36.1 A decision tree

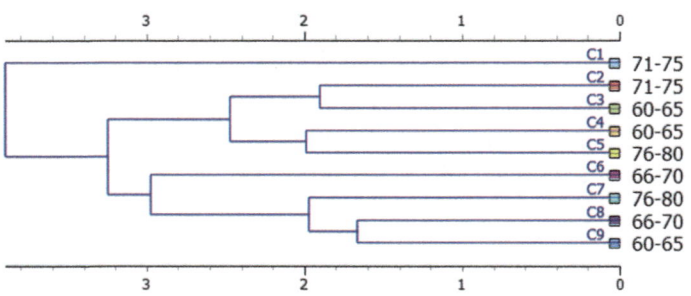

Fig. 36.2 Hierarchy clustering

Another way to deduce to which group the individual, denoted as 'I' in the last row of Table 36.1, is close to, is to apply hierarchy clustering. By looking at Fig. 36.2 it is easy to see that our conclusion is conformed. The hierarchy clustering also indicates that there are different age groups that have similar level of physical fitness, i.e. the younger ones are under performing.

While the similarity function allows finer tuning of the decision process, methods from predictive analytics are much more suitable for working with large data sets. A combination of the two might be a sensible approach, i.e. apply predictive analytics first to divide the data in smaller sets and then use the similarity function for detailed comparisons.

Considering the importance of motivating older adults to increase their physical activities, it is worth taking a closer look at to which groups that have experienced 'insignificant' improvement, 'positive' change, or 'real' gain. This can be used to advise individuals not only on types and amount of exercises but also on where to look for inspiration. Figure 36.3 f. ex. implies that the person 'I' we described in the last row of Table 36.1 should look for inspiration at those that are older than her self.

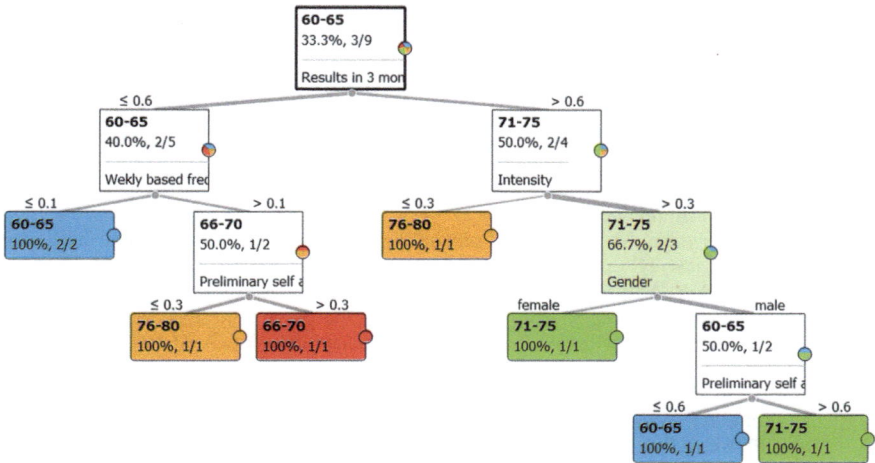

Fig. 36.3 Results in 3 months

Remark In a following study we plan to introduce expected results from physical activities after six months and after a year of training. This can give encouragement for continuing spending time and efforts for doing more physical exercises.

The person 'I' we have considered can also appear to be an outlier. In order to draw conclusions about similar occurrences it will be necessary to work with much larger data sets.

Information about weight in terms of body mass index (BMI) will be included as well. Older adults have difficulties losing weight and they definitely need extra motivation to work for it. Recommendations and advises on what to do in order to maintain optimal weight come from medical doctors, fitness gurus, and media, to name a few. What effect losing weight has on physical fitness however is relatively little emphasised. Therefore it is worth involving BMI as a part of a future study.

36.4 Conclusion

Both short and long time evaluations can be used to motivate people to increase the amount of their physical activities and thus have a healthier aging.

Additional research is needed in order to adjust the proposed approach to individual needs and to make it easier accessible to anyone who would be interested in using it. Another important question we intend to consider in future research is whether it is possible to predict the effect of exercise in a long period of time in terms of years.

References

1. Beck, F., Burch, M., Munz, T., Di Silvestro, L., Weiskopf, D.: Generalized Pythagoras trees for visualizing hierarchies. In: Proceedings of the 5th International Conference on Information Visualization Theory and Applications - Volume 1: IVAPP (VISIGRAPP 2014). ISBN 978-989-758-005-5 (2014), pp. 17–28
2. https://www.who.int/dietphysicalactivity/factsheet-adults/en/
3. https://orangedatamining.com/
4. Intan, R., Mukaidono, M.: Conditional probability relations in fuzzy relational database. In: Proceedings of RSCTC'00. Lecture Notes in Computer Science, pp.251–260 (2001)
5. Intan, R., Yao, Y.Y., Mukaidono, M.: Generalization of rough sets using weak fuzzy similarity relations. Rough Set Theory and Granular Computing, pp. 37–46. Springer, Berlin (2003)
6. Moreira, J., Carvalho, A., Horvath, T.: A General Introduction to Data Analytics. Wiley, Hoboken (2018)
7. Musich, S., Wang, S.S., Hawkins, K., Greame, C.: Frequency and health benefits of physical activity for older adults. Popul. Health Manag. **20**(3), 199–207 (2017)
8. Musich, S., Wang, S.S., Hawkins, K., Greame, C.: The impact of mobility limitations on health outcomes among older adults. Geriatr. Nurs. **39**(2), 162–169 (2018)
9. Provost, F., Fawset, T.: Data Science for Business: What You Need to Know About Data Mining and Data-Analytic Thinking, pp. 44–48. O'Reilly Media, Sebastopol (2013)
10. Roberts, C.E., Phillips, L.H., Cooper, C.L., Gray, S., Allan, J.L.: Effect of different types of physical activity on activities of daily living in older adults: systematic review and meta-analysis. J. Aging Phys. Act. **25**(4), 653–670 (2017)
11. Siegel, E.: Predictive Analytics: The Power to Predict Who Will Click, Buy, Lie, or Die, pp. 17–37. Wiley, New Jersy (2016)
12. Witten, I., Frank, E., Hall, M., Pal, C.: Data Mining: Practical Machine Learning Tools and Techniques, pp. 17–37. Morgan Kaufmann, Cambridge (2016)

Chapter 37
Multimodal Biometrics Recognition Using Soft Computing

U. Ramani, M. Thilagaraj, and V. Mueeswaran

Abstract The acknowledgment of human from the Gait, Footprint, and Voice was useful in security purposes. The ID of the people by the framework needs some extraordinary instruments and will be helpful in PC vision measure. Perceiving the people utilizing the component esteems extricated from Footprint and Gait is more solid since the impressions of similar people have the same examples. The voice of the people was likewise more solid in the ID of the individual people. This technique is tried by different preparing tests with the yield shows that the right characterization rate is high for the mix of Gait and impression contrasted with ID utilizing walk and impression alone. The new system produces high accuracy when compared to the previous system algorithms which indicates that the misclassifications are cheap to a better extent.

37.1 Introduction

The Biometric scheme is especially a sample-recognition scheme that perceives an individual dependent on a feature vector got from a physiological characteristic that the person possesses. Based on the application situation, a biometric scheme ordinarily works in one of two modes: check and distinguishing proof.

Multimodal biometric scheme utilizes one of the important physiological or individual person characteristics for enrollment, confirmation, and recognition process as shown in Fig. 37.1. It is also used in many security applications such as in the country borders, person identification, and network security and so on. They are

U. Ramani
K.Ramakrishnan College of Engineering, Trichy, India

M. Thilagaraj (✉)
Karpagam College of Engineering, Coimbatore, India

V. Mueeswaran
Kalasalingam Academy of Research and Education, Krishnankoil, India
e-mail: v.munneswaran@klu.ac.in

© The Author(s), under exclusive license to Springer Nature Singapore Pte Ltd. 2022   385
V. Bhateja et al. (eds.), *Evolution in Computational Intelligence*,
Smart Innovation, Systems and Technologies 267,
https://doi.org/10.1007/978-981-16-6616-2_37

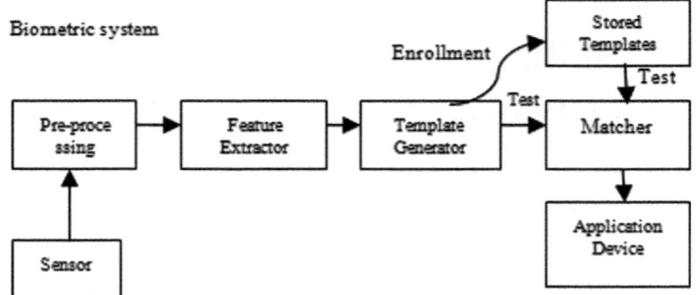

Fig. 37.1 Biometric system block diagram

used for reducing false non-match and false match rates. The main advantage of multimodal biometrics is if any failures occur in any one of the biometrics it will be overcome by the other biometrics, which will be used for deciding the accuracy of the scheme.

Here this thesis we consider the problems of identifying the person by means of Gait, Footprint, and voice. The gait is taken in the form of silhouettes [1, 2] footprint as images and then these images are converted into Gait Energy Image (GEI). The voice inputs are taken in the form of signals.

The Voice, Gait, and the Footprint datasets were gathered and they were prepared by extricating the PCA and LDA highlights from the pictures and statistical features from voice and labeling the footprint and gait [3]. Step portrayals are one of the unequivocal elements in the presentation of stride acknowledgment. GEI is a useful spatio-temporal step portrayal that has a very superior perceptive power with also has strength adjacent to segmental mistakes. Compared with the binary silhouette series Gender Equity Index is a smaller amount perceptive to noise and stores the storage space and calculation time [4]. Footprint identification is a promising biometric method to recognize persons from footprint images and forms. The footprint also has high discriminate power for the identification of humans. These are analyzed by ANFIS classifier for better authentication and identification process.

37.2 Procedure

The following project consists of four processes they are as follows:

(A) Preprocessing (B) Feature Extraction (C) Recognition (D) Performance Measures.

37.2.1 Preprocessing

The fingerprint and Gait images were first re-sized to a particular size. The blare in the images reduces the excellence of the images. In order to progress the excellence of the images we usually occupy some filtering techniques. The center channel considers each pixel in the image along these lines and sees its near to neighbors to pick whether it is illustrative of its ecological components. The center is controlled by first orchestrating all the pixel regards from the enveloping neighborhood into numerical solicitation and subsequently replacing the pixel being considered with the middle pixel regard. The training diagram is shown in Fig. 37.2.

The proposed strain is a nonlinear digital filtering method and it is worn to remove the unwanted noise. This unwanted noise decrease is a classic preprocessing step to get better the outcome of shortly handing out and will preserve edges while edge detection step. The major concept of the proposed filter is to sprint throughout the image entrance by entrance, replacing every entrance with the proposed filter of adjacent entries. The model of adjacent is called the "window". It is a more powerful technique than the customary direct separating, on the grounds that it protects the sharp edges. Middle channel is a spatial separating activity. It is also same as that of Gaussian filter in case of smothering technique. Since proposed method is very generally worn in digital image processing technique.

Fig. 37.2 Training diagram

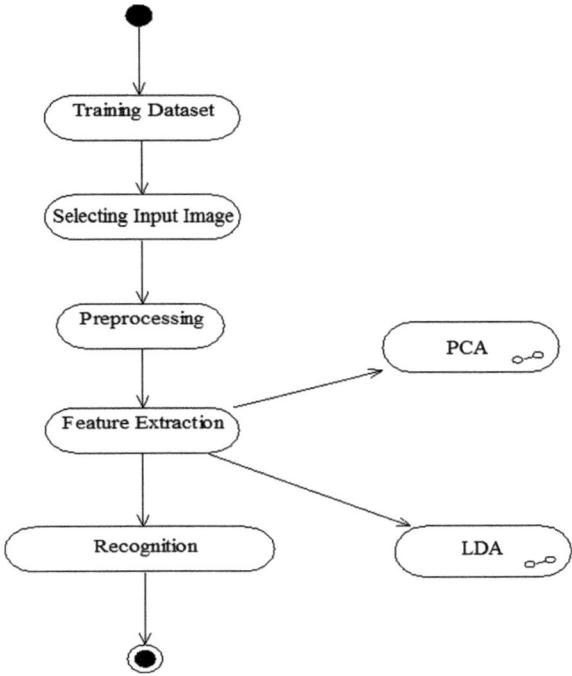

37.2.2 Feature Extraction

For inputs Gait and Footprint the features were extracted by LDA and PCA and for Voice the statistical features like Mean, Standard Deviation, and Entropy were extracted. LDA (Linear Discriminant Analysis) works when the measurements made on independent variables for each observation are continuous quantities. PCA (Principle Component Analysis) [5] which is also referred by eigenspace projection uses eigenvalues and eigenvectors for dimensionality reduction of the images.

LDA and PCA features are used to reduce the similarities in the combination of biometrics as shown in Fig. 37.3.

LDA and PCA features are used to reduce the similarities in the combination of biometrics. Statistical features are calculated by the following formulas.

Mean:

$$\overline{X} = \frac{\Sigma X}{n} \tag{37.1}$$

Standard Deviation:

$$s = \sqrt{\frac{\Sigma(x - \overline{x})^2}{n - 1}} \tag{37.2}$$

Entropy:

$$H = -\sum p(x) \log p(x) \tag{37.3}$$

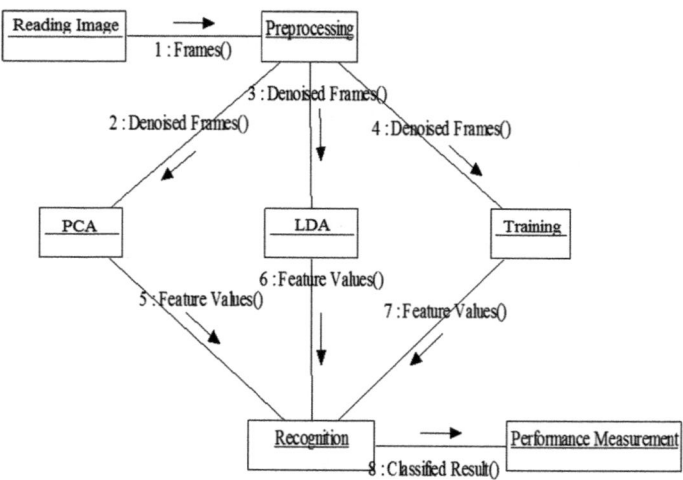

Fig. 37.3 Collaboration diagram

37.2.3   General Architecture of ANFIS

ANFIS consists of 5 layers and 27 and also some rules, couples of inputs and outputs and the architecture is shown in Fig. 37.4.

The problem-solving methodology is shown below:

The inspiration of ANFIS model as shown in Fig. 37.5 is to prepare the FIS utilizing neural organization design. We begin with an underlying FIS to prepare [6, 7].

As expressed above, we start by producing an underlying FIS, which might be utilized for preparing the framework. In the event that the underlying FIS isn't produced, we may utilize a haphazardly per input [8, 9]. The logical operation of ANFIS is shown in Fig. 37.6.

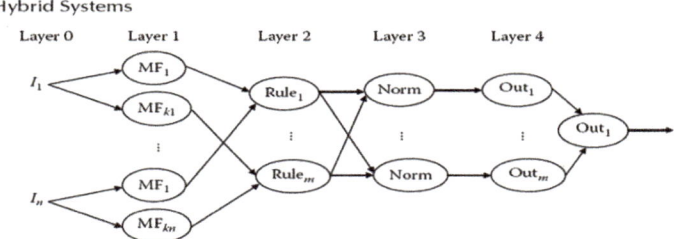

Fig. 37.4   Architecture of ANFIS classifier

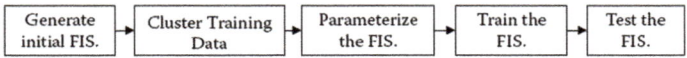

Fig. 37.5   ANFIS Problem-solving method

Fig. 37.6 Logical operation of ANFIS

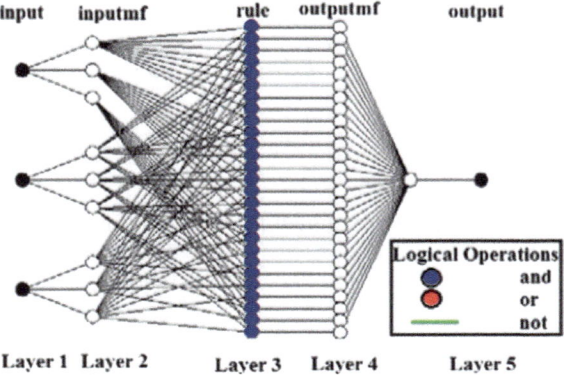

The layer 1, all the nodes are adaptive. The yields of this layer are the fuzzy membership of the data sources, which are given by the following condition:

$$O_i^1 = M_i(x_i)$$

where $i = 1, 2, 3,..., P$; xi indicates the ith contribution of ANFIS and is the yield of hub I. In the subsequent layer, the hubs are fixed hubs. They are named with M, demonstrating that they proceed as a basic multiplier. Here, to figure the yield of the layer, AND (min) activity is utilized:

$$O_t^1 = M_i(x_i) AND M_j(x_i)$$

In the third layer, each hub is a fixed hub, too. Examination between terminating strength of the principles and the amount of all terminating strength is done in this layer. The yields of this layer can be determined as: $O_i^3 = \frac{O_i^2}{\sum_i O_i^2}$.

In the fourth layer, each hub is a versatile hub. The yield of every hub in this layer is essentially the result of the standardized terminating strength and a first request polynomial (for a first request Sugeno model). In this way, the yields of fourth layer are given by $O_i^4 = O_i^3 \sum_{i=1}^{p} P_j x_j + c_j$.

The parameters P1, P2, ...,Pn and c1, c2, ..., cn are consequent parameters set.

In the fifth layer, there is just one single fixed hub that figures the general yield as the summation of every approaching sign. Consequently, the yields of layer 4 are accumulated: $O_i^5 = \sum_i O_i^4$.

In the learning algorith of the ANFIS, the task of the learning count for this designing is to tune all the modifiable limits, specifically and to make the ANFIS yield arrange the readiness data. Exactly when the explanation limits ai, bi, and ci of the cooperation limits are fixed, the yield of the ANFIS model can be formed as

$$f = \frac{w_1}{w_1 + w_2} f_1 + \frac{w_2}{w_1 + w_2} f_2 \tag{37.4}$$

$$f = \overline{w_1}(p_1 + q_1 y + r_1) + \overline{w_2}(p_2 x + q_2 y + r_2) \tag{37.5}$$

$$f = \overline{w}_1(p_1 + q_1 y + r_1) + \overline{w_2}(p_2 x + q_2 y + r_2) \tag{37.6}$$

which is a linear combination of the modifiable consequent parameters p1, q1, r1, p2, q2, and r2.

ANFIS advantages are as follows:

- Refining fluffy in the event that rules to portray the conduct of an intricate framework.
- Not needing earlier human ability that is frequently required in fluffy frameworks, and it may not generally be accessible.
- Presenting more prominent decision of enrollment capacity to utilize.
- Bringing exceptionally quick union time.

37.2.4 Measuring Performance

The introduction of the framework is exact by computing the Correct Classification Rate. The Correct Classification Rate of the classifier addresses to which broaden the classifier arranges the pictures dependent on the given name. (i.e.,) the quantity of names precisely returned by the classifier. The exhibition for acknowledgment with Gait Energy Image alone, Footprint alone and consolidating both Footprint and Gait Energy Image were estimated for various number of preparing tests. True positive = correctly identified:

False positive = incorrectly identified.
True negative = correctly rejected;
False negative = incorrectly rejected

$$\text{CCR} = \frac{(TP + TN)}{(FP + TN) + (TP + FN)} \qquad (37.7)$$

Using the over formula the presentation of the procedure is calculated. In a classification task, the precision for a class is the amount of certifiable positives (for instance, the amount of things viably named as having a spot with the positive class) divided by irrefutably the quantity of segments named as having a spot with the positive class (for instance, the measure of veritable positives and counterfeit positives, which are things mistakenly named as having a spot with the class). Audit in this setting is portrayed as the amount of certifiable positives confined by the total number of segments that truly have a spot with the positive class. For characterization assignments, the terms genuine positives, genuine negatives, bogus positives, and bogus negatives (see additionally Type I and type II blunders) look at the aftereffects of the classifier under test with confided in outer decisions. The terms positive and negative allude to the classifier's forecast (now and again known as the assumption), and the terms valid and bogus allude to whether that expectation compares to the outer judgment. The sequence diagram is shown in Fig. 37.7.

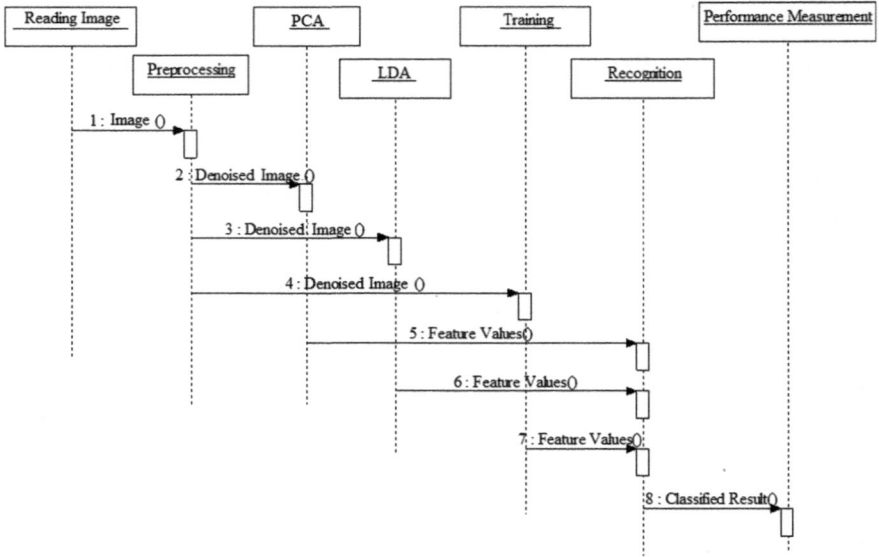

Fig. 37.7 Sequence Diagram

37.3 Conclusion

The new technique is straightforward and compelling. The proposed framework
perceives the people from the walk energy picture and impression dependent on the
highlights extricated utilizing PCA and LDA. The separated highlights depend on
the PCA and LDA with the goal that the component extraction measure is more
compelling. The separated enormous number of the LDA Eigenvectors and qualities
were proficiently improved by the dimensionality decrease of the Principal Component Analysis. Henceforth the proposed strategy is liberated from complexities and
it is all around advanced. The acknowledgment of the individual is finished utilizing
the distance measurements determined utilizing Euclidean distance. The proposed
structure gives accuracy which is higher than the current estimations which perceive
that the misclassifications are diminished to a more imperative growth. The right
characterization rate is high for the mix of the impression and step successions.
This shows that the detriments in steps were standardized with the assistance of the
impression. The proposed framework is certifiably not a totally computerized framework to upgrade the framework can be mechanized by dodging managed classifiers.
The directed classifiers need the preparation of the framework and manual marks to
group the classifier. The framework can upgrade by including some extra element
extraction calculations, for example, surface and article surface-based highlights and
some other measurable highlights which can portray more data's about the items. The
solo methodologies may incorporate some grouping strategies followed by improvement procedures that can make the interaction liberated from preparing. Some other

additional biometrics such as iris and fingerprint can be employed which makes the process more reliable and authenticated.

References

1. Wang, L., Tan, T., Ning, H., Hu, W.: Silhouette analysis-based gait recognition for human identification. IEEE Trans. Pattern Anal. Mach. Intell. **25**(12), 1505–1518 (2003)
2. Nixon, M.S., Carter, J.N.: Automatic recognition by gait. Proc. IEEE **94**(11), 2013–2024 (2006)
3. Gavrila, D.: The visual analysis of human movement: a survey. Comput. Vis. Image Underst. **73**(1), 82–89 (1999)
4. Jung, J., Bien, Z., Sato, T.: Person recognition method using sequential walking footprints via overlapped foot shape and center-of-pressure trajectory. IEICE Trans. Fundam Electron Commun Comput Sci **87**(6), 1393–1400 (2004)
5. Belhumeur, P.N., Hespanha, J.P., Kriegman, D.J.: Eigenfaces vs. Fisherfaces: recognition using class specific linear projection. IEEE Trans. Pattern Anal. Mach. Intell. **19**(7), 711–720 (1997)
6. Wang, Y., Tan, T., Jain, A.K.: Combining face and iris biometrics for identity verification. In: Proceeding of the Audio- and Video-Based Biometric Person Authentication, Guildford, UK, pp. 805–813 (2003)
7. Ross, A., Jain, A.K.: Multimodal biometrics: an overview. In: 2004 12th European Signal Processing Conference (pp. 1221–1224). IEEE (2004)
8. Sanjekar, P.S., Patil, J.B.: An overview of multimodal biometrics. Signal Image Proc. **4**(1), 57 (2013)
9. Kumar, G.H., Imran, M.: Research avenues in multimodal biometrics. IJCA Special Issue on "Recent Trends in Image Processing and Pattern Recognition", RTIPPR, 3, 1–8 (2010)

Chapter 38
Feature Based Transfer Learning for Kinship Verification

Rupali Kute, Alwin Anuse, and Pranav Bhat

Abstract Kinship verification based on facial images is a recent and challenging problem in computer vision and machine learning. It has many real-world potential applications including tracing of missing children cases, social media image analysis, and image annotation. The existing methods do not focus on probability distribution differences between kinship facial images; that is why the performance of these methods is poor. The deep learning methods required a large amount of data. Practically, parents and their children share common features. To take into account this transfer learning based framework is proposed. The extracted hand-crafted features are given to a transfer learning framework that transfers the knowledge acquired from facial images of parents to children and vice versa. We evaluated the proposed methods on standard kinship datasets and experimental results showed that the proposed method outperforms in terms of accuracy and computational efficiency as compared to state of art methods.

38.1 Introduction

The main aim of kinship verification is to determine whether there is a kin relation between the pair of persons under consideration. It is an upcoming problem in the field of computer vision and has applications to trace missing children and social media image analysis. Compared to traditional face recognition methods, kinship verification is a challenging task because of large age and gender differences among father-daughter, father-son, mother-daughter and mother-son face images.

R. Kute (✉) · A. Anuse
Vishwanath Karad MIT–World Peace University, Pune, Maharashtra, India
e-mail: rupali.kute@mitwpu.edu.in

A. Anuse
e-mail: alwin.anuse@mitwpu.edu.in

P. Bhat
Maharashtra Institute of Technology, Savitribai Phule Pune University, Pune, Maharashtra, India

© The Author(s), under exclusive license to Springer Nature Singapore Pte Ltd. 2022   395
V. Bhateja et al. (eds.), *Evolution in Computational Intelligence*,
Smart Innovation, Systems and Technologies 267,
https://doi.org/10.1007/978-981-16-6616-2_38

Traditional kinship verification methods are feature-based methods. Histogram of gradient (HOG) [6], local Binary patterns (LBP) [7], local descriptor [9], and gabor gradient [8] are main approaches which focus on feature-based methods. Some kinship verification methods are based on subspace learning [3, 4, 10, 11] and metric learning [1, 2]. Researchers have focused on extracting the features using handcrafted features-based methods HOG and LBP. These features are then combined using transfer learning.

This research proposes to use handcrafted features extracted from the photos of children and their respective parents along with a transfer learning framework which uses Linear Discriminant Analysis (LDA) to leverage the power of transfer learning. The children and their respective parents share some common features. Using Bregman-divergence based regularization it is possible to bring these features in the common subspace. The pair of children and respective parents are given the same label. Their probability distribution is used to bring them together in the same space. While bringing them together LDA is used to discriminate between different pairs of children and respective parents. This will help in increasing the accuracy required for the task of Kinship verification. Using LDA this research also shows how accuracy depends on the number of dimensions selected.

38.2 Proposed System

Figure 38.1 shows the overview of the proposed methodology of this research. Children and Parents are assigned the same label. The features are extracted using handcrafted features HOG and LBP methods. These features are given to LDA and then combined using the Bregman-divergence based transfer learning method.

The KinFaceW and UBKinFace dataset both have a set of photos of corresponding children and parents. Features extracted from these pairs are used for transfer learning along with the LDA model for the task of Kinship verification. The transfer learning framework is tested in combination with the HOG and LBP feature descriptors and corresponding results are obtained.

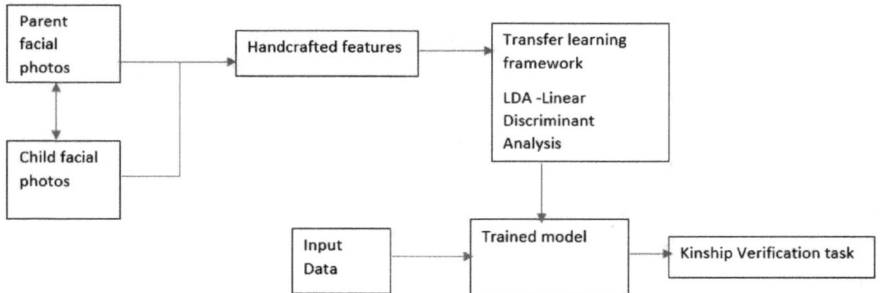

Fig. 38.1 Proposed methodology of kinship verification

Linear discriminant analysis (LDA), normal discriminant analysis (NDA), or discriminant function analysis is a generalization of Fisher's linear discriminant, a method used in statistics and other fields, to find a linear combination of features that characterizes or separates two or more classes of objects or events [6]. Fisher discriminant does not make some of the assumptions of the generally referred LDA such as equal class covariances and normally distributed classes. LDA with transfer learning is mentioned as TrFLDA.

The FLDA subspace algorithm is represented using following mathematical steps: The 'within-class scatter' matrix S_w,

$$S_W = \sum_{i=1}^{C} \sum_{j=1}^{n} (X_j - m_i)(X_j - m_i)^T. \tag{38.1}$$

The 'between-class scatter' matrix S_B,

$$S_B = \sum_{i=1}^{C} n_i (m_i - m)(m_i - m)^T. \tag{38.2}$$

Here, C is represented as the class number, n_i is the total number of samples of class i, m_i is the mean of class i, and m is the overall mean.

The objective function $F(W)$ is given in Eq. (38.3),

$$F(W) = tr^{-1}(W^T S_B W) tr(W^T S_W W). \tag{38.3}$$

38.3 Experimentation and Results Discussion

There are many popular datasets currently available for Kinship verification tasks like KinFaceW [1, 2] and UB KinFace [3, 4, 5]. Most of the research works either propose feature learning or metric learning for the task of Kinship verification on KinFaceW-I and II and UBKinFace datasets. This research uses the Local Binary Pattern (LPB) and Histogram of Oriented Gradients (HOG) feature descriptors which are used by all the baseline models and combines it with the transfer learning model.

The verification accuracy results obtained for the KinFaceW-I, KinFaceW-II, and UB KinFace dataset along with the respective CPU time required for each cases and dataset are as given in the below tables.

Tables 38.1, 38.2, and 38.3 show the verification accuracy (%) of TrFLDA on different subsets of the KinFaceW-I dataset, KinFaceW-II dataset, and the UB KinFace dataset respectively. It depicts that HOG + TrFLDA method performs better on all the different children and respective parent pair as compared to LBP + TrFLDA method.

Table 38.1 Verification accuracy (%) of TrFLDA on different subsets of the KinFaceW-I dataset

Method	F-D	D-F	F-S	S-F	M-D	D-M	M-S	S-M	Mean
LBP + TrFLDA	96.2	97.01	89.74	91.02	98.42	96.85	100	99.13	96.04
HOG + TrFLDA	99.25	95.52	99.35	96.79	99.21	98.42	100	100	98.56

Table 38.2 Verification accuracy (%) of TrFLDA on different subsets of the KinFaceW-II dataset

Method	F-D	D-F	F-S	S-F	M-D	D-M	M-S	S-M	Mean
LBP + TrFLDA	78.4	81.6	69.6	66.8	88.4	86.4	78	84.8	79.25
HOG + TrFLDA	92.8	92.4	94	93.2	95.6	93.6	94.8	91.6	93.5

Table 38.3 Verification accuracy (%) of TrFLDA on different subsets of the UBKinFace dataset

Method	Children-Old parent	Old Parent-Children	Children-Young Parent	Young Parent-Children	Mean
LBP + TrFLDA	74.5	77.5	88.5	85.5	81.5
HOG + TrFLDA	87	87	89	88	87.75

Tables 38.4, 38.5, and 38.6 show the CPU time required for different subsets of the KinFaceW-I dataset KinFaceW-II dataset, and the UB KinFace dataset respectively. CPU time required for HOG + TrFLDA method is less as compared to LBP + TrFLDA method.

Figure 38.2 gives a visual representation of the relation between the results in terms of accuracy and the dimension selected for the UBKinFace dataset. It shows that as number of dimensions increases accuracy decreases.

Table 38.4 CPU time required in seconds for the KinFaceW-I dataset

Method	F-D	D-F	F-S	S-F	M-D	D-M	M-S	S-M	Mean
LBP + TrFLDA	1.10	1.62	1.1	1.71	1.57	1.1	1.31	1.019	1.3161
HOG + TrFLDA	0.93	0.95	1.13	1.01	0.951	0.904	0.826	0.764	0.936

Table 38.5 CPU time required in seconds for the KinFaceW-II dataset

Method	F-D	D-F	F-S	S-F	M-D	D-M	M-S	S-M	Mean
LBP + TrFLDA	2.41	2.29	3.18	1.76	1.95	1.76	1.74	1.59	2.085
HOG + TrFLDA	1.46	1.51	1.54	1.41	1.48	1.48	1.29	1.4	1.4462

Table 38.6 CPU time required in seconds for the UBKinFace dataset

Method	Children	Young Parent	Old Parent	Mean
LBP + TrFLDA	14.16	16.84	25.39	14.0975
HOG + TrFLDA	2.69	2.7	3.18	2.1425

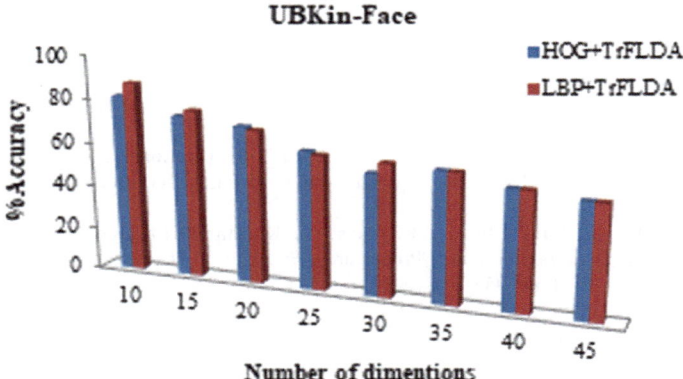

Fig. 38.2 Accuracy versus number of dimensions for UBKin-Face

38.4   Conclusion

The approaches used in the previous research works propose either based on descriptor level or transfer learning level. This research used handcrafted features and transfer learning approach by making use of the availability of data and extracted knowledge of the facial photos of both the parent and child for kinship verification of each other. As seen in Tables 38.1, 38.2, and 38.3, on testing on KinFaceW-I, KinFaceW-II, and UBKinFace datasets it is seen that the transfer learning approach implemented in this research gives better results than the previous proposed state of the art models. It is seen from Tables 38.1, 38.2, and 38.3 that the transfer learning framework gives better results along with HOG feature descriptors for all the three datasets. Also, the CPU time shown in Tables 38.4, 38.5, and 38.6 shows that the transfer learning framework along with HOG descriptors takes much less time as compared with LBP. Figure 38.2 shows that accuracy increases as the number of dimensions selected decreases and the highest accuracy is recorded for 10 dimensions.

References

1. Lu, J., Tan, Y.-P., Shang, Y.: Neighborhood repulsed metric learning for kinship verification. IEEE Trans. Pattern Anal. Mach. Intell. (PAMI) **36**(2), 331–345 (2014)
2. Lu, J., Hu, J., Shang, Y., Tan, Y.P., Wang, G.: Neighborhood repulsed metric learning for kinship verification. IEEE Conf. Comput. Vision Pattern Recogn. (CVPR) 2594–2601 (2012)
3. Ming, J.L., Fu, Y.: Understanding Kin relationships in a photo. IEEE Trans. Multimedia (T-MM) **14**(4), 1046–1056 (2012)
4. Ming, J.L., Fu, Y.: Kinship verification through transfer learning. Int. Joint Conf. Artif. Intell. (IJCAI) 2539–2544 (2011)
5. Ming, J.L., Fu, Y.: Genealogical face recognition based on UB KinFace database. IEEE CVPR Workshop (BIOM) (2011)
6. Fang, R., Tang, K.D., Snavely, N., Chen, T.: Towards computational models of kinship verification. In: 2010 17th IEEE International Conference on Image Processing (ICIP), IEEE. 1577–1580 (2010)
7. Zhou, X., Hu, J., Lu, J., Shang, Y., Guan, Y.: Kinship verification from facial images under uncontrolled conditions. In: Proceedings of the 19th ACM International Conference on Multimedia, ACM. pp. 953–956 (2011)
8. Zhou, X., Lu, J., Hu, J., Shang, Y.: Gabor-based gradient orientation pyramid for kinship verification under uncontrolled environments, in: Proceedings of the 20th ACM International Conference on Multimedia, ACM. pp. 725–728 (2012)
9. Qin, X., Liu, D., Gui, B., Wang, D.: Neighborhood min distance descriptor for kinship verification. Appl. Soft Comput. **95**, 106569 (2020)
10. Dornaika, F., Arganda-Carreras, I., Serradilla, O.: Transfer learning and feature fusion for kinship verification. Neural Comput Appl **32**, 7139–7151 (2020)
11. Yan, H., Song, C.: Multi-scale deep relational reasoning for facial kinship verification. Pattern Recogn. **110**, 1–11 (2021)

Chapter 39
Unsupervised Document Binarization Via Disentangled Representation

K. H. Salman and Chakravarthy Bhagvati

Abstract Binarization of document is considered the first key step in many document processing tasks. In this paper, we try to reformulate the problem as an image-to-image translation. Most of the existing learning methods for document binarization make use of supervised approach, but obtaining ground truth for binarized documents is difficult. Here we developed an unsupervised adversarial training procedure for binarization. We use disentangling of style and content from a binarized document and transfer the binarized style to the input document. Our results indicate that this approach works on par with many other results published in literature.

39.1 Introduction

Document image binarization is considered as the first step for most document image processing tasks like document page segmentation, OCR, etc. Being the first step of the whole document image processing pipeline, the quality of binarization is very important for higher-level document process and analysis tasks. A significant amount of semantic information will be available by separating text from the background. A binary image is produced by quantization of the image gray values to binary values, usually 0 and 1 [1].

Although binarization seems an easy task and many algorithms have been proposed, in a real-world environment, it can become challenging because of multiple factors like document degradation, lighting condition, ink stains, warping effect, non-uniform variation of intensity during scanning, etc. Besides, the complexity of image contents and the varying scales of characters make it extremely challenging in some cases to reliably distinguish the foregrounds of a document image from its backgrounds and noises [2].

Classical binarization algorithms relied on global and or local thresholds based on image statistics [1, 3]. Although better binarization is possible with local thresh-

K. H. Salman (✉) · C. Bhagvati
University of Hyderabad, Hyderabad, India
e-mail: chakravarthybhagvati@uohyd.ac.in

© The Author(s), under exclusive license to Springer Nature Singapore Pte Ltd. 2022   401
V. Bhateja et al. (eds.), *Evolution in Computational Intelligence*,
Smart Innovation, Systems and Technologies 267,
https://doi.org/10.1007/978-981-16-6616-2_39

olding, the presence of more empirical parameters increases the complexity of the method. Tuning these parameters was done heuristically. Over time several learning algorithms were developed in order to alleviate the issue of tuning the parameters. Most of the classical methods were also focused on the specific problem, like bleed-through correction, degradations on historical documents, etc.

Convolutional Neural Networks (CNN) have dramatically improved state of the art in many tasks such as image, video and speech processing [4]. It has also been applied for document binarization [5, 6]. Deep learning methods using CNN give far better results than the previous methods. But almost all the state-of-the-art methods require ground truth for the binarized data to train.

In this paper, we are proposing an unsupervised technique for binarization using disentangled representation. Here we frame the problem as an image-to-image translation problem. We used convolutional auto encoders to encode content and style of both original documents and binarized documents into different latent spaces. For generating the required binarized document, we use a generator that takes content and style as input. Our method is completely unsupervised and does not require any kind of paired image samples.

39.2 Proposed Method

In this paper, we propose a document binarization technique by formulating it as an image-to-image translation problem. Image-to-image translation aims to learn the mapping between different visual domains without paired training data. Here being one domain as normal document image represented as X and the other as binarized image represented as $\mathcal{Y}$. Our method aims to disentangle style and content from the input image and learn the mapping of content to a domain-invariant latent space and style to a domain-specific latent space. We used content encoders $\{E_X^c, E_y^c\}$ to encode the content to a domain-invariant shared latent space C and style encoders $\{E_X^s, E_y^s\}$ to encode style to a specific different latent space S_X, S_y. In order to learn a domain-invariant mapping of content, we used weight sharing of last layer of encoders E_X^c and E_y^c with the first layer of generators G_X and G_y, respectively similar to [7, 8], and a content discriminator D^c which tries to distinguish the content encoding between both domains. For effective style representation we made use of style-specific encoder introduced in [9] and domain discriminators $\{D_X, D_y\}$. These discriminators try to discriminate between real image and generated image by the corresponding generator. The generators $\{G_X, G_y\}$ generate images conditioned on both content and style vector (see Fig. 39.1). We used Least Squared GAN loss instead of conventional GAN loss for both content and domain discriminators as it does better density estimation and generates better quality images than vanilla GAN networks [10]. In order to avoid the need for an unpaired binarized example while at the test time, the style vector is regularized and made similar to normal distribution using KL-divergence at training period. So while testing, the generator can generate the required stylized image by sampling the style from a normal distribution.

Fig. 39.1 The encoder
$E_X = (E_X^c, E_X^s)$ and
$E_Y = (E_Y^c, E_Y^s)$ take
images from their
corresponding domains,
disentangle content and style
into separate latent spaces (C
and S_X, S_Y, respectively)

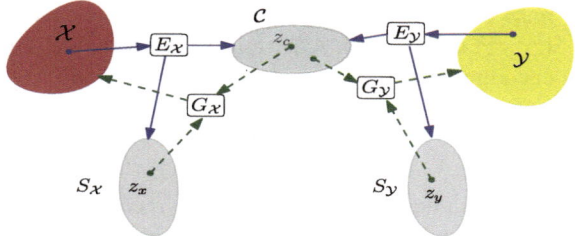

39.2.1 Disentangling Content and Style

For an image $x \in X$ and $y \in Y$, the content encoder (E_X^c, E_Y^c) maps the input image to shared content space C and style encoder (E_X^s, E_Y^s) maps to respective style space S_X, S_Y. The disentanglement of style and content has been achieved using the technique mentioned in [8]. Since both domains contain the document image information, it makes sense to map their contents into one shared content space. In addition, since both have two distinct style information, we will map the style information to two different latent spaces.

$$\{z_x^s, z_x^c\} = \{E_X^s(x), E_X^c(x)\} \qquad z_X^s \in S_X, z_X^c \in C \tag{39.1}$$

$$\{z_y^s, z_y^c\} = \{E_Y^s(y), E_Y^c(y)\} \qquad z_Y^s \in S_Y, z_Y^c \in C \tag{39.2}$$

We achieve this by weight sharing and content discriminator. As mentioned in Sect. 39.2, we share the last layer of both content encoders E_X^c and E_Y^c and the first layer of generator G_X and G_Y. On the other hand, only sharing the high-level features does not fully guarantee the shared encoding. Hence a content discriminator D^c is used. D^c which forces the encoders to learn to encode, whose domain membership that which is not recognized by D^c. We used Least Squared GAN loss for this.

$$L_{adv}^{content} = \tfrac{1}{2}\mathbb{E}_x[D^c(E_X^c(x) - 1)^2] + \tfrac{1}{2}\mathbb{E}_x[(D^c(E_X^c(x))^2] + \tag{39.3}$$

$$\tfrac{1}{2}\mathbb{E}_y[D^c(E_Y^c(y) - 1)^2] + \tfrac{1}{2}\mathbb{E}_y[(D^c(E_Y^c(y))^2] \tag{39.4}$$

39.2.2 Cyclic-Consistency

Unlike the conventional cyclic-consistency loss used in [11], which learns a one-to-one mapping between two domains $X \to Y \to X$, here we leverage disentangled content and style space representation for cyclic reconstruction. The cycle consists of two passes.

– **pass 1:** Given a non-corresponding pair of images x and y, we encode them into $\{z_x^c, z_x^s\}$ and $\{z_y^c, z_y^s\}$. After this, we perform the translation by swapping the style encoding to generate u, v. such that $u \in X$ and $v \in Y$, i.e.,

$$u = G_X(z_y^c, z_x^s) \qquad v = G_Y(z_x^c, z_y^s) \tag{39.5}$$

– **pass 2:** In the second pass we try to reconstruct the input image. We achieve this by encoding the fake images u, v, generated in previous pass, i.e.,

$$\{z_u^c, z_u^s\} = \{E_X^c(u), E_X^s(u)\} \tag{39.6}$$
$$\{z_v^c, z_v^s\} = \{E_Y^c(v), E_Y^s(v)\} \tag{39.7}$$

After encoding u, v into $\{z_u^c, z_u^s\}$ and $\{z_v^c, z_v^s\}$, respectively, a second translation is performed as follows.

$$\hat{x} = G_X(z_v^c, z_u^s) \qquad \hat{y} = G_Y(z_u^c, z_v^s) \tag{39.8}$$

After the second translation, the generated images should be as close as the input images x and y. To enforce this constraint, we used L_1 loss as shown in Fig. 39.2

$$L_1^{cc} = \mathbb{E}_{x,y}[\|\hat{x} - x\|_1 + \|\hat{y} - y\|_1] \tag{39.9}$$

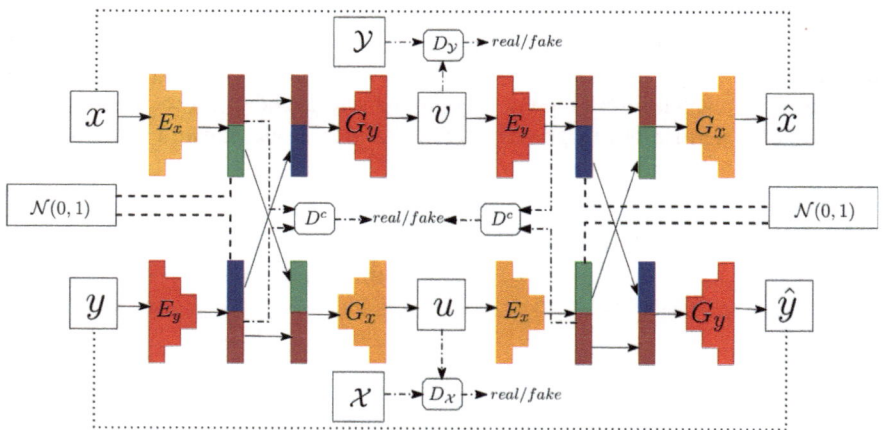

Fig. 39.2 Our two-pass image translation model: forward translation comprising an encoding and decoding where the style vector is swapped between two domains; second translation, the images again encoded and decoded with swapped style vector to reproduce the original input image

39.2.3 Other Loss Functions

Apart from the loss function mentioned above, we have used other loss terms also to facilitate the network training (Fig. 39.3).

Adversarial Loss:
In order to discriminate between real and generated images, we used a domain adversarial loss L_{adv}^{domain} where D_X and D_Y are multiscale discriminators. While G_X and G_Y attempt to generate images that look realistic, D_X and D_Y attempt to discriminate them. Here also we resorted to Least Squared GAN loss for both generator and discriminator.

Reconstruction loss:
We also enforced a reconstruction loss (L_1^{rec}) at stage 1 itself. With the encoded style and content, z_x^s, z_x^c and z_y^s, z_y^c, the decoders G_X and G_Y should be able to reconstruct the original input x and y. i.e., $x' = G_X(z_x^c, z_x^s)$ and $y' = G_Y(z_y^c, z_y^s)$.

KL-Loss:
At test time, in order to avoid the need of an unpaired binarized example to extract style from, we encourage the style attribute representation to be as close to a prior Gaussian distribution. It has also shown in [12] that KL-divergence can be used to leverage content-style disentanglement. Hence we apply

$$L_{KL} = \mathbb{E}[D_{KL}(z_a \parallel N(0, 1))] \qquad (39.10)$$

$$\text{where } D_{KL}(p \parallel q) = -\int p(z) \, log \frac{p(z)}{q(z)} \, dz \qquad (39.11)$$

Style code regression loss:
In order for latent style vector that is sampled from prior Gaussian distribution to learn the style mapping, we additionally used an L_1 latent regression loss for style latent encoding which is similar to [13]. We draw a latent vector z from $N(0, 1)$ as style representation and try to reconstruct it as following $\hat{z} = E_X^s(G_X(z_x^c, z))$ and $\hat{z} = E_Y^s(G_Y(z_y^c, z))$

Fig. 39.3 Legend

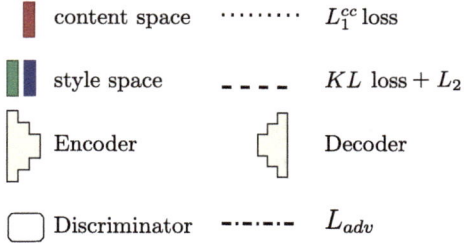

content space	L_1^{cc} loss
style space	KL loss + L_2
Encoder	Decoder
Discriminator	L_{adv}

So the total loss function will be

$$\min_{G,E^s,E^c} \max_{D,D^c} \quad \lambda_{adv}^{content} L_{adv}^{content} + \lambda_1^{cc} L_1^{cc} + \lambda_{adv}^{domain} L_{adv}^{domain} \qquad (39.12)$$

$$+\lambda_1^{rec} L_1^{rec} + \lambda_{KL} L_{KL} + \lambda_{latent} L_1^{latent} \qquad (39.13)$$

39.3 Experiment

39.3.1 Model Details

Encoders:
It consists of a content encoder which has four stridden convolution layers to down-sample the input and four residual blocks to further process the features learned. All convolution layers are followed by an Instance normalization layer. Whereas for the style encoder, it has five stridden convolutional layers with relu activation, followed by a global average pooling and a fully connected layer. Since Instance normalization removes the original features mean and variance that represents the important style information [14], it is not used for encoding style features.

Decoders:
In [15], Kazemi et al.; developed a general framework for decoders that can be used in style transfer. The decoder takes two inputs, the content code and style code. The decoder first processes the content features via four upsampling layers and resnet block with adaptive instance normalization block. Inspired by recent works that use affine transformation parameters in the normalization layer for representing style [14], we used Adaptive instance normalization (AdaIN) in residual blocks whose parameters are dynamically generated by multilayer perceptron (MLP) from style code.

Discriminators:
We used Least squared objective since it generates better quality images and is better at density estimation. We also used multiscale discriminators for fine grained output.

We implemented the model in PyTorch. We created a 256×256 image patched with 25% overlap from the training set as input to the models. The style vector size $z^s \in \mathbb{R}^{10}$. For training we used Adam optimizer with a learning rate of 0.0001 and batch size of 8. We set the hyper-parameters values as the following $\lambda_{adv}^{content} = 1, \lambda_1^{cc} = 10, \lambda_{adv}^{domain} = 1, \lambda_1^{rec} = 10, \lambda_{latent} = 10, \lambda_{KL} = 0.01$. We evaluated our model both qualitatively and quantitatively. At test time we used randomly sampled vector from style space to binarize the given input. See Fig. 39.4 for testing mode.

Fig. 39.4 Testing with style being sampled from random normal distribution

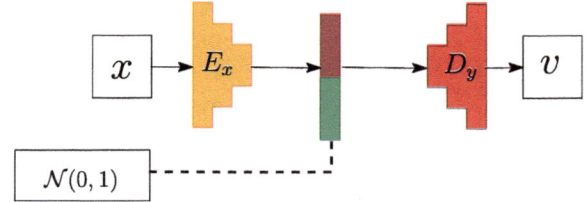

39.3.1.1 Dataset:

We used document datasets which are available in public such as DIBCO 2009, DIBCO 2011, H-DIBCO 2010, H-DIBCO 2012, H-DIBCO 2014 for training and evaluation and we used DIBCO 2013 for tests. We compare our results with the following baseline models.

Pix2pix:
Introduced by Isola et al. [16] is a popular image-to-image translation model. It trained using adversarial loss and L_2 loss in a paired setting.

CycleGAN:
CycleGAN is an unpaired image-to-image translation model [17]. To the best of our knowledge, CycleGAN is the first model to make use of cyclic-consistent loss function for image translation.

39.4 Results

We compare our results with several state-of-the-art classical and learning-based methods. Here we can see our results are performing better when compared to other *unsupervised* technique like CycleGAN. We show empirically in Table 39.1 that our method works better than conventional unsupervised training mechanism in all the metrics (Fig. 39.5).

39.5 Conclusion and Future Work

It has been shown that disentangling content and style representation and using the learned latent space can be successfully adopted for document binarization and is performing well than canonical unsupervised neural network-based style transfer. It is worth investigating that for what number of training samples are required for mapping the concept (content and style) to latent spaces? Since different documents

Table 39.1 Please write your table caption here

Method	PSNR	DRD	F-measure	F_{ps}[a]-measure
Otsu	16.6	11.0	83.9	86.5
Niblack	13.6	24.9	72.8	72.2
Sauvola	16.9	7.6	85.0	89.8
Gatos	17.1	9.5	83.4	87.0
BERN	10.1	62.2	52.6	52.8
Su	19.6	4.2	87.7	88.3
DSN	21.4	1.8	94.4	96.0
Pix2pix	20.8	2.7	94.8	97.0
CycleGAN	12.5	17.6	66.8	70.1
Ours	15.1	10.1	72.8	76.7

[a] F_{ps}: pseudo f-measure

Fig. 39.5 Results from test set. Qualitative comparison against test image, CycleGAN, pix2pix, our model and ground truth, respectively

have different deformities, earlier specific binarization techniques had to be tailored. It will also be interesting to investigate that disentangling content and style will work across different deformities of documents.

References

1. Otsu, N.: A threshold selection method from gray-level histograms. IEEE Trans. Syst. Man Cybern. **9**(1), 62–66 (1979)
2. Meng, G., Yuan, K., Wu, Y., Xiang, S., Pan, C.: Deep networks for degraded document image binarization through pyramid reconstruction. In: 2017 14th IAPR International Conference on Document Analysis and Recognition (ICDAR), vol. 1, pp. 727–732. IEEE (2017)
3. Sauvola, J., Pietikäinen, M.: Adaptive document image binarization. Pattern Recognit. **33**(2), 225–236 (2000)
4. LeCun, Y., Bengio, Y., Hinton, G.: Deep learning. Nature **521**(7553), 436 (2015)
5. Vo, Q.N., Kim, S.H., Yang, H.J., Lee, G.: Binarization of degraded document images based on hierarchical deep supervised network. Pattern Recognit. **74**, 568–586 (2018)
6. He, S., Schomaker, L.: Deepotsu: document enhancement and binarization using iterative deep learning. Pattern Recognit. **91**, 379–390 (2019)
7. Huang, X., Liu, M.Y., Belongie, S., Kautz, J.: Multimodal unsupervised image-to-image translation. In: Proceedings of the European Conference on Computer Vision (ECCV), pp. 172–189 (2018)
8. Lee, H.Y., Tseng, H.Y., Huang, J.B., Singh, M.K., Yang, M.H.: Diverse image-to-image translation via disentangled representations. In: European Conference on Computer Vision (2018)
9. Li, X., Chen, L., Wang, L., Wu, P., Tong, W.: Scgan: disentangled representation learning by adding similarity constraint on generative adversarial nets. IEEE Access **7**, 147928–147938 (2019)
10. Mao, X., Li, Q., Xie, H., Lau, R.Y.K., Wang, Z., Smolley, S.P.: Least squares generative adversarial networks. In: 2017 IEEE International Conference on Computer Vision (ICCV), pp. 2813–2821 (2017)
11. Zhu, J.Y., Park, T., Isola, P., Efros, A.A.: Unpaired image-to-image translation using cycle-consistent adversarial networks. In: 2017 IEEE International Conference on Computer Vision (ICCV) (2017)
12. Bao, J., Chen, D., Wen, F., Li, H., Hua, G.: Towards open-set identity preserving face synthesis. In: Proceedings of the IEEE Conference on Computer Vision and Pattern Recognition, pp. 6713–6722 (2018)
13. Zhu, J.Y., Zhang, R., Pathak, D., Darrell, T., Efros, A.A., Wang, O., Shechtman, E.: Toward multimodal image-to-image translation (2017)
14. Huang, X., Belongie, S.: Arbitrary style transfer in real-time with adaptive instance normalization. In: Proceedings of the IEEE International Conference on Computer Vision, pp. 1501–1510 (2017)
15. Kazemi, H., Iranmanesh, S.M., Nasrabadi, N.: Style and content disentanglement in generative adversarial networks. In: 2019 IEEE Winter Conference on Applications of Computer Vision (WACV), pp. 848–856. IEEE (2019)
16. Isola, P., Zhu, J.Y., Zhou, T., Efros, A.A.: Image-to-image translation with conditional adversarial networks. CVPR (2017)
17. Zhu, J.Y., Park, T., Isola, P., Efros, A.A.: Unpaired image-to-image translation using cycle-consistent adversarial networks. In: Proceedings of the IEEE International Conference on Computer Vision, pp. 2223–2232 (2017)

Chapter 40
Detection of Online Hate in Social Media Platforms for Twitter Data: A Prefatory Step

D. Venkata Swetha Ramana⬤ and **T. Hanumantha Reddy**⬤

Abstract Online social media (OSM) plays a major role in today's life by providing people with a platform to share individual contents such as photos, ideas, opinions, and videos. However, at the same time, there are many negative carry-overs on social media users, which lead to uncontrolled and offensive communication called online hate. It has become a major concern for all online social media (OSM) users as it creates nothing but a negative influence on the users. Hence, developing a better generalized online hate classifier and network analyzer across domains is the challenge. This article highlights a novel 8-stage methodology that is proposed to develop an efficient generalized online hate classifier and network analyzer both which can help in overcoming the challenges. As a preliminary step toward 8-stage process proposed, the dataset related to women is collected from Twitter and analyzed to find out the frequent words used, number of retweets increased are negative tweets.

40.1 Introduction

OSM is defined to be a communication platform for people to share individual contents such as photos, videos, opinions, and career interests. Some are interested in businesses, crisis communication, disaster management, journalism, political communication, etc. Examples of some social media sites are Facebook, Twitter, LinkedIn, WhatsApp, etc. Nowadays, social media is very popular as it is efficient, low cost, and omnipresent. As the usage of social media is increasing day by day, the accumulation of data volume is also rising exponentially, which is known as social media big data. The OSM analytic field aims to examine social media data by integrating, expanding, and modifying methods. It allows us to gain deep insights

D. V. S. Ramana (✉) · T. H. Reddy
Rao Bahadur Y Mahabaleswarappa Engineering College, Ballari, Karnataka, India
e-mail: swetharamana@rymec.in

T. H. Reddy
e-mail: thrbly@rymec.in

© The Author(s), under exclusive license to Springer Nature Singapore Pte Ltd. 2022   411
V. Bhateja et al. (eds.), *Evolution in Computational Intelligence*,
Smart Innovation, Systems and Technologies 267,
https://doi.org/10.1007/978-981-16-6616-2_40

into mass communication and derive useful information, such as trends, influential actors, and communication behaviors. [1].

The key features of OSM data are the 4 "V"s: Volume, Velocity, Variety, and Veracity, which can be better handled by machine learning. As machine learning algorithms are used to identify patterns in data and classify them, they perfectly suit unstructured data because social media posts do not follow any particular rules.

OSM data also carries some negative influence. A large number of social media users often involve in uncontrolled and offensive communication, which is called online hate. This kind of behavior of social media users is a threat and even detected as a drawback in multiple OSM platforms. Therefore, analysis of social media is important to detect this online hate behavior easily. Manually online hate can be flagged, but only 12% of adults have reported others for online harassment. To overcome manual technique an automated system which uses machine learning models to automatically detect online hate called as online hate classifiers have been gaining popularity. But these online hate classifiers are concentrating on focus on single platform but not multiple platforms.

In this article, gaps identified in the present literature are addressed and an efficient and generalized online hate classification methodology consisting of 8 stages is proposed. As a preliminary step for the proposed methodology, Twitter data on women is collected and an algorithm is implemented to analyze those data, preprocess them by removing stop words and special characters, and identify negativity in tweets and plot wordcloud.

40.2   Literature Survey

In Ref. [2], three dimensions of hate content to be considered by the researchers are proposed—"What the hate aims at?" "Who takes the hate?," and "How the hate is communicated?" The authors have identified the limitations of the existing work. According to them, there are five linguistic difficulties associated with classification errors in previous work:

1.   Humorous or sarcastic abusive content is usually observed as a cause of classification error.
2.   In social media, spelling variations are ubiquitous and cannot be detected.
3.   Polysemy is a challenge.
4.   Existing research have focused only on short posts.
5.   The syntax, grammar, and lexicons of language are unpredictable as they mutate.

In Ref. [3], comments were labeled as a toxic, severe toxic, obscene, threat, and insult using Kaggle's toxic comment dataset and trained the deep learning model to identity hate. They applied deep learning techniques such as ANN, CNN, LSTM, Glove and CNN, Glove and LSTM, GLOVE, and LSTM and CNN to analyze better models in the classification of comments. In Ref. [4], in order to detect hate automatically in online communities, the authors proposed a fusion method by combining content and graph-based features to improve the results. But computational time and small datasets are found to be the limitations of this work.

In Ref. [5], the authors introduce KNLPEDNN (Killer Natural Language Processing Optimization Ensemble Deep Learning approach), a method that is a fusion of natural language processing and machine learning for the purpose of detecting and classifying hate speech on OSM platforms. Tweets gathered from the storm front and crowd flower dataset are processed using the NLP approach. NLP tokenization process, NLP feature extraction process, and ensemble deep learning classification processes are used.

In Ref. [6], the authors manually labeled 5413 hateful expressions collected from videos posted on YouTube and Facebook and applied machine learning models to detect and classify hate. They found the best performing model is linear SVM. They also observed four challenges associated with it—interpretation problem, linguistic variety, the danger of over-moderation, and limits of automation.

In Ref. [7], the authors used Logistic Regression, Naïve Bayes, Support-Vector Machines, XGBoost, and Neural Network to identify online hate. They identified XGBoost as the best classifier and BERT as the powerful representation of online hate.

The authors in Ref. [8] mentioned that outright offensive terms are very few, so a complete storyline needs to be analyzed. In this context, they proposed a learning approach with many layers to analyze the storyline. In Ref. [9], the data is collected from Kaggle and preprocessed to remove punctuation marks and duplicate texts. A bird's eye view on the detection of hate speech is given in Ref. [10]. Different approaches to discover hate speech, preprocessing techniques, analysis techniques, and evaluation metrics are mentioned.

In Ref. [11], cyberhate is measured by collecting data at adolescent schools by carrying out a questionnaire survey. And frequency rates, correlations, and descriptive statistics were used to analyze cyberhate victimization. Data of 19,000 comments are collected from YouTube which is labeled as bullying and nonbullying utilizing machine learning. A model is created through preprocessing feature extraction and classification. Three algorithms SVM, NB, and CNN are used for the classification and are evaluated using precision, recall, and f-score. SVM and NB had an average accuracy, and also the authors in Ref. [12] concluded that categorizing in one topic will increase the results of the classifier.

A topic-oriented model is used in Ref. [13] to observe densely interacted communities using real datasets. English and Dutch data accumulated by GNU Wget software and annotated by a scheme of two levels in Ref. [14]. In the first level the participant role of authors is considered and in the second level fine-grained cyberbullying and victims are detected by binary classification implementing kernel SVM. Moreover, the authors recommended few future directions such as identification of cyberbullying through irony, deep representation learning, and detection of fine-grained cyberbullying. Annotated datasets are classified to establish lexical baselines. The original dataset is added with emotional information using NLP techniques for machine learning classification in Ref. [15]. Challenges in hate detection are also mentioned such as the definition of hateful content is changed regularly and new expressions are updated frequently.

40.3   Problem Statement

From the above-mentioned literature survey, the following gaps are identified:

- Many researchers have developed models that are nongeneralizable across domains; hence, a generalized hate classifier is needed to develop.
- Analyze the network which is influenced or influences such online hate.
- Spelling variations are very common, especially in social media. (e.g., oh to ohhh, goodnight to gn, ok to k, etc.). These spelling variations increase the number of errors as many vocabulary terms are created.

  We can address the above-mentioned gaps identified by the following objectives:

- To collect datasets from various platforms of social media, the hate split percentage of which is higher than non–hate split percentage and preprocess the data collected from social media.
- To develop a better model of online hate classifier using deep learning techniques which automatically detect and classify hateful comments.
- Developing a social network analyzer to know the influences.

40.4   Methodology

A methodology consisting of eight stages, as shown in Fig. 40.1, is proposed to achieve the above-mentioned objectives: social media data collection, data preprocessing, feature extraction, word embedding, classification, social network analyzer, hate lexicon analyzer, and hate target identification.

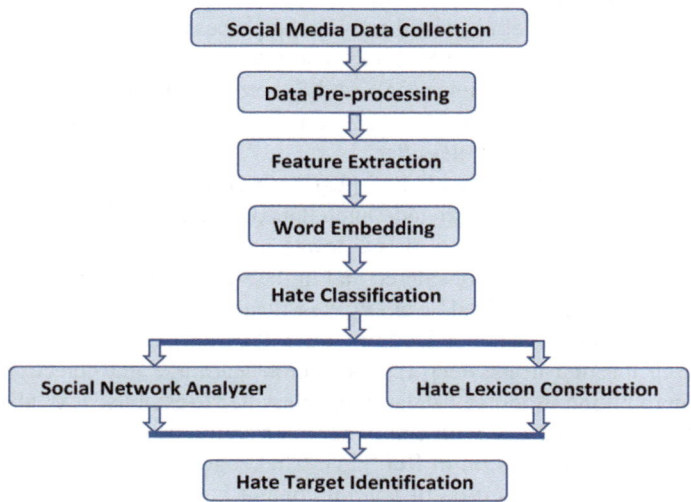

Fig. 40.1 Proposed methodology to discover online hate

Social Media Data Collection: Data is collected from various sources such as Facebook, Twitter, Reddit, YouTube, Wikipedia, Instagram, after analyzing hate split percentage.

Data Preprocessing: Data is preprocessed to remove URLs, numbers, hashtags, punctuations, date, time, email, emoticons, and elongated words to avoid the problems of codification.

Feature Extraction: In this stage, features are extracted based on the ranking of input features showing their prominence in the classification process. Linguistic features of posts and comments were studied well to form a feature space for later use in classification.

Word Embedding: A sentence can be denoted as w_1, w_2, …, w_n, where w_1 is a real-valued vector. This stage maps each tweet, represented as a sequence of integer indexes, to an n-dimension vector space using a pretrained model to detect syntactic and semantic word relations.

Hate Classification: Features attained from the extraction stage are stored to train and classify online social media data as hate speech and nonhate speech using deep learning techniques. Then the proposed method will be evaluated using evaluation metrics such as recall, precision, accuracy, and F1 score.

Social Network Analyzer: The most common hate speech with frequencies and places that are appeared is mentioned in a directed graph with nodes representing users and arc showing the speech between them.

Hate Lexicon Construction: A full list of hate words is obtained by applying top K similar word-finding techniques on online hate detected. This list can be referred to find the most dominant words in the corpus of posts to generate a lexicon based on words that are in the annotated dataset, in the place of a predefined dictionary. To train a classifier, the frequencies of words present in the text, manually annotated as hate speech, are used as features.

Hate Target Identification: In this stage, the list of hate words will be nominated as the initial lexicon and then enlarged by the most similar co-occurring word vectors. Then, the target can be identified by counting the tallied words in the lexicon.

40.5   Implementation

As the first step of the proposed method to discover online hate, we tried to implement the following algorithm using python. This algorithm is used to analyze the data collected from Twitter, to learn how to preprocess the data by removing stopwords and special characters, and to plot histograms and wordcloud.

Algorithm

Begin

   Step 1: Collect datasets from Twitter.

   Step 2: Import all the necessary libraries in python like numpy, pandas, nltk, matplotlib, etc.

   Step 3: Read the csv file resultwomen.csv.

   Step 4: Preprocess the tweet data on women.

     Step 4.1: Access only the tweet text from the csv file.

     Step 4.2: Drop duplicates from the tweet text.

     Step 4.3: Clean the text by removing stopwords using the remove_stopwords() function.

     Step 4.4: Convert all the tweet text to lower case.

     Step 4.5: Tokenize the words using word_tokenize() function.

     Step 4.6: Remove few repeated meaningless words using regular expressions.

   Step 5: Print the total number of words, which was around 530.

   Step 6: Plot histograms for friends, followers, and retweets.

   Step 7: Join the tokenized words using the join() function.

   Step 8: Identify percentage of positive, negative and neutral tweets.

   Step 9: Plot piechart for the above data.

   Step 10: Identify and display top 10 sentences which frequently occur.

   Step 11: Plot wordcloud figure using WordCloud.generate() function.

End

40.6 Experimental Results

Result 1: Dataset is collected from Twitter using Twitter developer accounts Twitter API and using python tweepy function. Datasets comprise total 4000 tweets in which 2000 tweets on keyword "Woman" and another 2000 tweets on keyword "Female" as on May 28, 2021. The data is analyzed by plotting histograms on retweets, as shown in Fig. 40.2.

In Fig. 40.2, the histogram presents that out of 2000 tweets, 1480 had 0 to 100 retweets, 192 had 101 to 500 retweets, 116 had 501 to 1000 retweets, 128 had 1001 to 1500, 13 had 1501 to 2000 retweets, and only 6 had 2001 to 2500 retweets.

By analyzing Fig. 40.3, wordcloud on the data collected from Twitter on the topic women, we can see that "woman" is the most occurred word and, similarly, other negative words such as kill, rape, stop, and drunk were also more frequently occurred words.

Result 2: Sentiment analysis was done on the data collected to analyze percentage of positive, negative, and neutral tweets. As a result it was found that maximum tweets with a percentage of 43.7 were neutral and 23.3% of tweets were negative tweets.

Result 3: Out of 2000 tweets, 212 tweets had retweets more than 1000. This data is analyzed to find maximum of 39.6 tweets are negative. This implies that more retweets happen for negative tweets and such data need to be analyzed to get more insights (Figs. 40.4 and 40.5).

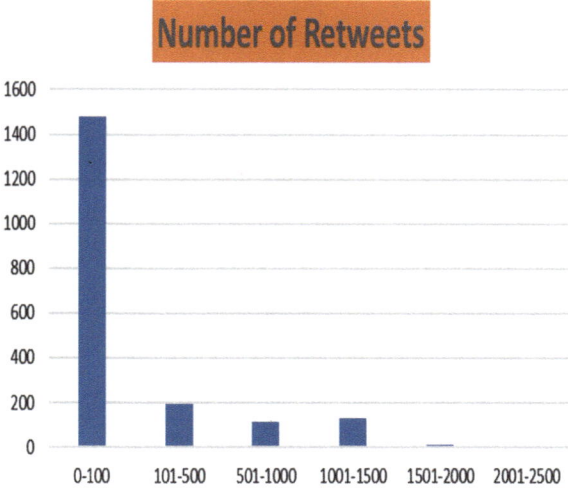

Fig. 40.2 Number of Retweets

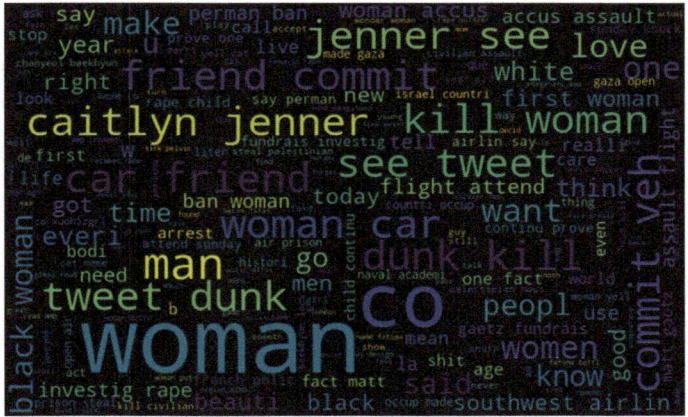

Fig. 40.3 Wordcloud of the data collected

40.7 Conclusion

In the online social world of today, it is important to detect and classify online hate as early as possible to stop its spreading. Developing a generalized online hate classifier and social network analyzer is a major challenge. To overcome this challenge, an 8-step process which is more generalized is proposed. In the first step, a dataset from twitter consisting of 4000 tweets was collected and analyzed. Data visualization techniques were implemented to show histograms and scatter plots to find out the

Fig. 40.4 Pie chart for
sentiment analysis of tweets

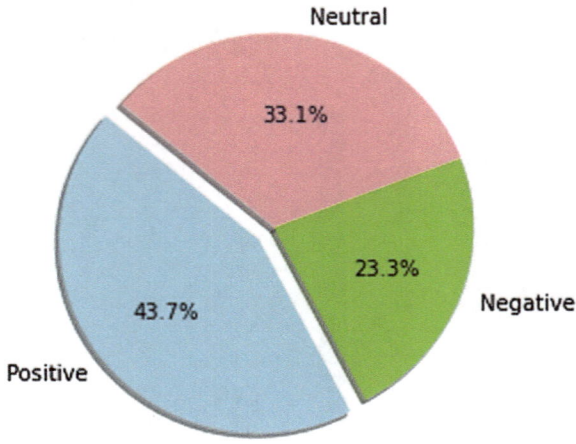

Fig. 40.5 Pie chart for
sentiment analysis on
maximum retweets

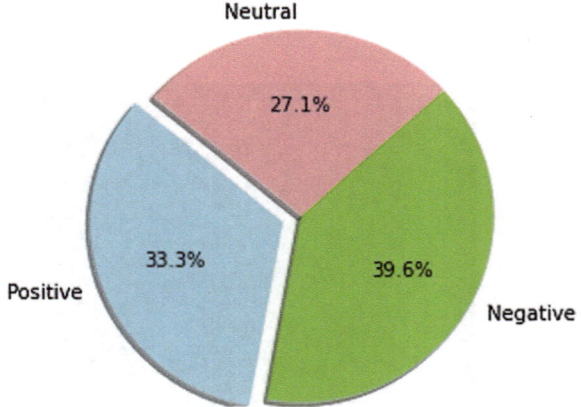

range of retweets with different bin values, a wordcloud to display frequent words used in the tweets, and pie charts for sentiment analysis with positive, negative and neutral percentages in 2000 tweets. The results show that there is a need to analyze the tweets which have more than 1000 retweets to get more insights on negative.

References

1. Zeng, D., Chen, H., Lusch, R., Li, S.: Social media analytics and intelligence. IEEE Intell. Syst. **25**(6), 13–16 (2010)
2. Vidgen, B., Harris, A., Nguyen, D., Tromble, R., Hale, S., Margetts, H.: Challenges and frontiers in abusive content detection. In: Proceedings of the Third Workshop on Abusive Language Online (2019). https://doi.org/10.18653/v1/W19-3509

3. Anand, M., Eswari, R.: Classification of abusive comments in social media using deep learning. In: 2019 3rd International Conference on Computing Methodologies and Communication (ICCMC), Erode, India, pp. 974–977 (2019)
4. Cécillon, N., Labatut, V., Dufour, R., Linarès, G.: Abusive language detection in online conversations by combining content- and graph-based features. Front. Big Data **2**, 8 (2019). https://doi.org/10.3389/fdata.2019.00008
5. Al-Makhadmeh, Z., Tolba, A.: Automatic hate speech detection using killer natural language processing optimizing ensemble deep learning approach. Computing **102**, 501–522 (2020). https://doi.org/10.1007/s00607-019-00745-0
6. Salminen, J., Almerekhi, H., Milenković, M., Jung, S., An, J., Kwak, H., Jansen, B.J.: Anatomy of online hate: developing a taxonomy and machine learning models for identifying and classifying hate in online news media. In: Proceedings of the International AAAI Conference on Web and Social Media (ICWSM 2018), San Francisco, CA, USA (2018)
7. Salminen, J., Hopf, M., Chowdhury, S.A. et al.: Developing an online hate classifier for multiple social media platforms. Hum. Cent. Comput. Inf. Sci. **10**, 1 (2020). https://doi.org/10.1186/s13673-019-0205-6
8. Englmeier K., Mothe J.: Application-oriented approach for detecting cyberaggression in social media. In: Ahram, T. (ed) Advances in Artificial Intelligence, Software and Systems Engineering. AHFE 2020. Advances in Intelligent Systems and Computing, vol. 1213. Springer, Cham (2021). https://doi.org/10.1007/978-3-030-51328-3_19
9. Shibly F.H.A., Sharma U., Naleer H.M.M.: Classifying and measuring hate speech in twitter using topic classifier of sentiment analysis. In: Gupta D., Khanna A., Bhattacharyya S., Hassanien A.E., Anand S., Jaiswal A. (eds.) International Conference on Innovative Computing and Communications. Advances in Intelligent Systems and Computing, vol. 1165. Springer, Singapore (2021). https://doi.org/10.1007/978-981-15-5113-0_54
10. Adesara, A., Tanna, P., Joshi, H.: Hate speech detection: a bird's-eye view. In: Kotecha K., Piuri V., Shah H., Patel R. (eds.) Data Science and Intelligent Applications. Lecture Notes on Data Engineering and Communications Technologies, vol. 52. Springer, Singapore (2021). https://doi.org/10.1007/978-981-15-4474-3_26
11. Wachs, S., Costello, M., Wright, M.F., Flora, K., Daskalou, V., Maziridou, E., Kwon, Y., Na, E.-Y., Sittichai, R., Biswal, R., Singh, R., Almendros, C., Gámez-Guadix, M., Görzig, A., Hong, J.S.: "DNT LET 'EM H8 U!": Applying the routine activity framework to understand cyberhate victimization among adolescents across eight countries, Computers & Education, vol. 160. 104026. ISSN **0360–1315**, (2021). https://doi.org/10.1016/j.compedu.2020.104026
12. Trana, R.E., Gomez, C.E., Adler, R.F.: Fighting cyberbullying: an analysis of algorithms used to detect harassing text found on YouTube. In: International Conference on Applied Human Factors and Ergonomics (pp. 9–15). Springer, Cham (2020)
13. Das, S., Anwar, M.M.: Discovering topic oriented highly interactive online community. Front. Big Data **2**, 10 (2019)
14. Van Hee, C., Jacobs, G., Emmery, C., Desmet, B., Lefever, E., Verhoeven, B., Hoste, V.: Automatic detection of cyberbullying in social media text. PloS one **13**(10), e0203794 (2018)
15. Martins, R., Gomes, M., Almeida, J. J., Novais, P., Henriques, P.: Hate speech classification in social media using emotional analysis. In: 2018 7th Brazilian Conference on Intelligent Systems (BRACIS) (pp. 61–66). IEEE (2018)

Chapter 41
Exponential Similarity Measure for Spherical Fuzzy Sets and Its Application in Pattern Recognition

D. Ajay and **P. Pon Hidaya David**

Abstract The aim of this paper is to introduce an alternative Pattern Recognizing technique using similarity measure in the spherical fuzzy set environment. In many applications of pattern recognition, the structural data is intrinsically ambiguous. It is essential to boost the descriptive power of pattern recognition in such applications, and the fuzzy technique is often used. So, a fuzzy logic (FL)-based similarity measure (SM) called Exponential similarity measure for Spherical fuzzy sets (SFSs) is applied in the field of Pattern Recognizing. In this paper similarity measures known as Exponential Similarity Measure (ESM), Weighted Exponential Similarity Measure (W-ESM), Weighted Average Exponential Similarity Measure (WA-ESM), Weighted Generalized Average Exponential Similarity Measure (WGA-ESM) are developed. Spherical fuzzy set is extended from fuzzy sets to cope with uncertain situations more accurately. Further, a problem on Intrusion Detection and Risk evaluation is chosen as an application of pattern recognition and is solved using similarity measures proposed in this paper.

41.1 Introduction

Recognition and classification of pattern problems in the real world are fraught with fuzziness that connects to a number of information about human cognition. A pattern is an object's physical or abstract structure. It is distinguished from others by a range of characteristics known as features, which together form a pattern. "A search for structure in data" is known as Pattern recognition (PR) [1].

For instance, a right cylindrical pattern can be recognized by following features (i) curved surface area with height. (ii) Base is of circle shaped flat bottom. (iii) Both bases with same diameter. (iv) radius makes 90° at the center of base. With these features one can recognize that it is a right cylindrical Pattern. In general, the pattern recognition process can be represented as a series of steps, including (i) data collecting, (ii) selecting features, and (iii) categorization procedure [2].

D. Ajay (✉) · P. Pon Hidaya David
Sacred Heart College (Autonomous), Tirupattur 635 601, India

© The Author(s), under exclusive license to Springer Nature Singapore Pte Ltd. 2022   421
V. Bhateja et al. (eds.), *Evolution in Computational Intelligence*,
Smart Innovation, Systems and Technologies 267,
https://doi.org/10.1007/978-981-16-6616-2_41

In accordance with the above steps, it is evident that there exist several sources of ambiguity. Couple of them are of evident interest (i) ambiguity concerned with a character of recognition space (ii) ambiguity resulting from a process of labeling. In Classical Pattern recognition the recognizing patterns are mostly Boolean which interprets whether the pattern belongs to a class or not which sometimes overlaps the legitimate Patterns resulting in miscalculation. Therefore, uncertainty theory can be implemented in overcoming those errors. A number of uncertainty theories have been put forth to deal with uncertainty and ambiguity.

Theory of probability, fuzzy set theory [3], rough sets [4], and so on are widely in usage to overcome the uncertainty and ambiguity. Fuzzy set theory has proven to be efficient over time. Fuzzy set theory [3] was introduced by Zadeh to deal with uncertainties but falsity values are not discussed. IFS [5] contains grades for truth and falsehood, whose sum should not exceed 1. And if the sum of grades for truth and falsehood is greater than 1 there fails IFS. Yager introduced Pythagorean fuzzy sets [6] which is a generalization of IFSs. The limitation of PyFS is that the sum of squares of grades for truth and falsehood should not exceed 1.

Later, to address abstinence along with truth and falsity grades Cuong and Kreinovich introduced picture fuzzy sets (PFS). The constraint for PFS is that the sum of grades should not exceed 1. The spherical fuzzy sets (SFS) [7] were established by Gundogdu et al. as a generalizing tool to deal with uncertain information effectively.

Keeping the positive impacts of the SFS in mind, SFSs-based similarity measures such as Exponential Similarity Measure (ESM), Weighted Exponential Similarity Measure (W-ESM), Weighted Average Exponential Similarity Measure (WA-ESM), and Weighted Generalized Average Exponential Similarity Measure (WGA-ESM) are introduced and some basic properties of similarity measures are discussed. Further, its applications in pattern recognition are studied in detail and numerical illustrations are presented. Further, a problem on Intrusion Detection and Risk evaluation is chosen as an application of pattern recognition and solved using similarity measures proposed in this paper.

41.2 Preliminaries

Definition 41.1 Pattern Recognition is defined as a search for structure in data. Thus recognition of pattern requires determining the nearest match of an unknown structure with a set of known structure of data. Pattern recognition is applied in Face ID, Intrusion detection system, Disease Categorizing, etc.

Definition 41.2 ([3]) A fuzzy set $\tilde{A}$ in Υ is characterized by a membership value $\mu_{\tilde{A}}$ which takes the value in the interval $[0, 1]$, i.e.,

$$\mu_{\tilde{A}} : \Upsilon \longrightarrow [0, 1] \tag{41.1}$$

The value $\mu_{\tilde{A}}$ at $\wp \in \Upsilon$, denote by $\mu_{\tilde{A}}(\wp)$, represents the grade of membership of $\wp$ in $\tilde{A}$ and is a point of $[0, 1]$. Then $\tilde{A}$ is given by

$$\tilde{A} = \sum_i \mu_{\tilde{A}}(\wp_i)/\wp_i, \ \wp_i \in \Upsilon. \tag{41.2}$$

Definition 41.3 ([7]) A Spherical fuzzy set $\mathbb{S}_{F(\tilde{A})}$ is of the form,

$$\mathbb{S}_{F(\tilde{A})} = \left\{ \langle \wp, \tau_{\tilde{A}}(\wp), \sigma_{\tilde{A}}(\wp), \xi_{\tilde{A}}(\wp) \rangle /\wp \in \Upsilon \right\} \tag{41.3}$$

provided that, $0 \leq \tau_{\tilde{A}}^2(\wp) + \sigma_{\tilde{A}}^2(\wp) + \xi_{\tilde{A}}^2(\wp) \leq 1$. $\mathcal{E}A(\wp) = \left(1 - \left(\tau_{\tilde{A}}^2(\wp) + \sigma_{\tilde{A}}^2(\wp) + \xi_{\tilde{A}}^2(\wp)\right)\right)^{1/2}$ is the degree of negation of $\wp \in X$ in $\mathbb{S}_{F(\tilde{A})}$.

For any two SFSs, $\mathbb{S}_{F(\tilde{A})} = \left\{ \tau_{\tilde{A}}(\wp), \sigma_{\tilde{A}}(\wp), \xi_{\tilde{A}}(\wp) \right\}$, $\mathbb{S}_{F(\tilde{B})} = \left\{ \tau_{\tilde{B}}(\wp), \sigma_{\tilde{B}}(\wp), \xi_{\tilde{B}}(\wp) \right\}$ their operations are defined as,

1. $\mathbb{S}_{F(\tilde{A})} \subseteq \mathbb{S}_{F(\tilde{B})}$ iff $\tau_{\tilde{A}}(\wp) \leq \tau_{\tilde{B}}(\wp), \sigma_{\tilde{A}}(\wp) \leq \sigma_{\tilde{B}}(\wp), \xi_{\tilde{A}}(\wp) \geq \xi_{\tilde{B}}(\wp), \forall \wp \in X$
2. $\mathbb{S}_{F(\tilde{A})} \cup \mathbb{S}_{F(\tilde{B})} = \left\{ \wp, max\left(\tau_{\tilde{A}}(\wp), \tau_{\tilde{B}}(\wp)\right), max\left(\sigma_{\tilde{A}}(\wp), \sigma_{\tilde{B}}(\wp)\right), min\left(\xi_{\tilde{A}}(\wp), \xi_{\tilde{B}}(\wp)\right) \right\}$
3. $\mathbb{S}_{F(\tilde{A})} \cap \mathbb{S}_{F(\tilde{B})} = \left\{ \wp, min\left(\tau_{\tilde{A}}(\wp), \tau_{\tilde{B}}(\wp)\right), min\left(\sigma_{\tilde{A}}(\wp), \sigma_{\tilde{B}}(\wp)\right), max\left(\xi_{\tilde{A}}(\wp), \xi_{\tilde{B}}(\wp)\right) \right\}$
4. $\mathbb{S}_{F(\tilde{A})}^c = \left\{ \langle \wp, \xi_{\tilde{A}}(\wp), \sigma_{\tilde{A}}(\wp), \tau_{\tilde{A}}(\wp) \rangle /\wp \in \Upsilon \right\}$

Definition 41.4 ([8]) For two spherical fuzzy sets $\mathbb{S}_{F(\tilde{U})} = \left(\tau_{\tilde{U}}(\wp), \sigma_{\tilde{U}}(\wp), \xi_{\tilde{U}}(\wp)\right)$ and $\mathbb{S}_{F(\tilde{v})} = \left(\tau_{\tilde{V}}(\wp), \sigma_{\tilde{V}}(\wp), \xi_{\tilde{V}}(\wp)\right)$ the Cosine Similarity Measure(CSM) is given by,

$$CSM_{SFS}(\mathbb{S}_{F(\tilde{U})}, \mathbb{S}_{F(\tilde{v})}) = \frac{1/t \sum_{i=1}^{t} \left(\tau_{\tilde{U}}^2(\wp_i) \cdot \tau_{\tilde{V}}^2(\wp_i) + \sigma_{\tilde{U}}^2(\wp_i) \cdot \sigma_{\tilde{V}}^2(\wp_i) + \xi_{\tilde{U}}^2(\wp_i) \cdot \xi_{\tilde{V}}^2(\wp_i)\right)}{\sqrt{\left(\tau_{\tilde{U}}^2(\wp_i)\right)^2 + \left(\sigma_{\tilde{U}}^2(\wp_i)\right)^2 + \left(\xi_{\tilde{U}}^2(\wp_i)\right)^2} \sqrt{\left(\tau_{\tilde{V}}^2(\wp_i)\right)^2 + \left(\sigma_{\tilde{V}}^2(\wp_i)\right)^2 + \left(\xi_{\tilde{V}}^2(\wp_i)\right)^2}} \tag{41.4}$$

which holds the following conditions,

1. $0 \leq CSM_{SFS}(\mathbb{S}_{F(\tilde{v})}, \mathbb{S}_{F(\tilde{v})}) \leq 1$
2. $CSM_{SFS}(\mathbb{S}_{F(\tilde{v})}, \mathbb{S}_{F(\tilde{v})}) = CSM_{SFS}(\mathbb{S}_{F(\tilde{v})}, \mathbb{S}_{F(\tilde{v})})$
3. $CSM_{SFS}(\mathbb{S}_{F(\tilde{v})}, \mathbb{S}_{F(\tilde{v})}) = 1$ if $\mathbb{S}_{F(\tilde{v})} = \mathbb{S}_{F(\tilde{v})}$ and $i = 1, 2, \ldots t$
4. If $\mathbb{S}_{F(\tilde{v})} \subseteq \mathbb{S}_{F(\tilde{v})} \subseteq \mathbb{S}_{F(\tilde{w})}$ then $CSM_{SFS}(\mathbb{S}_{F(\tilde{v})}, \mathbb{S}_{F(\tilde{w})}) \leq CSM_{SFS}(\mathbb{S}_{F(\tilde{v})}, \mathbb{S}_{F(\tilde{v})})$ and $CSM_{SFS}(\mathbb{S}_{F(\tilde{v})}, \mathbb{S}_{F(\tilde{w})}) \leq CSM_{SFS}(\mathbb{S}_{F(\tilde{v})}, \mathbb{S}_{F(\tilde{w})})$

Definition 41.5 ([8]) For two spherical fuzzy sets $\mathbb{S}_{F(\tilde{U})} = \left(\tau_{\tilde{U}}(\wp), \sigma_{\tilde{U}}(\wp), \xi_{\tilde{U}}(\wp)\right)$ and $\mathbb{S}_{F(\tilde{v})} = \left(\tau_{\tilde{V}}(\wp), \sigma_{\tilde{V}}(\wp), \xi_{\tilde{V}}(\wp)\right)$ the Weighted Cosine Similarity Measure (W-CSM) is given by,

$$CSM_{SFS}(\mathbb{S}_{F(\tilde{U})}, \mathbb{S}_{F(\tilde{v})}) = \frac{1/t \sum_{i=1}^{t} \Psi_i \left(\tau_{\tilde{U}}^2(\wp_i) \cdot \tau_{\tilde{V}}^2(\wp_i) + \sigma_{\tilde{U}}^2(\wp_i) \cdot \sigma_{\tilde{V}}^2(\wp_i) + \xi_{\tilde{U}}^2(\wp_i) \cdot \xi_{\tilde{V}}^2(\wp_i)\right)}{\sqrt{\left(\tau_{\tilde{U}}^2(\wp_i)\right)^2 + \left(\sigma_{\tilde{U}}^2(\wp_i)\right)^2 + \left(\xi_{\tilde{U}}^2(\wp_i)\right)^2} \cdot \sqrt{\left(\tau_{\tilde{V}}^2(\wp_i)\right)^2 + \left(\sigma_{\tilde{V}}^2(\wp_i)\right)^2 + \left(\xi_{\tilde{V}}^2(\wp_i)\right)^2}}$$
(41.5)

where $\Psi = (\Psi_1, \Psi_2, \Psi_3, \ldots, \Psi_t)^T$ represents a weight vector such that $\Psi_i \in [0, 1]$ and $\sum_{i=1}^{t} \Psi_i = 1$ which holds the following conditions,

1. $0 \le WCSM_{SFS}(\mathbb{S}_{F(\tilde{U})}, \mathbb{S}_{F(\tilde{v})}) \le 1$
2. $WCSM_{SFS}(\mathbb{S}_{F(\tilde{U})}, \mathbb{S}_{F(\tilde{v})}) = WCSM_{SFS}(\mathbb{S}_{F(\tilde{v})}, \mathbb{S}_{F(\tilde{U})})$
3. $WCSM_{SFS}(\mathbb{S}_{F(\tilde{U})}, \mathbb{S}_{F(\tilde{v})}) = 1$ if $\mathbb{S}_{F(\tilde{U})} = \mathbb{S}_{F(\tilde{v})}$ and $i = 1, 2, \ldots t$
4. If $\mathbb{S}_{F(\tilde{U})} \subseteq \mathbb{S}_{F(\tilde{v})} \subseteq \mathbb{S}_{F(\tilde{w})}$ then $WCSM_{SFS}(\mathbb{S}_{F(\tilde{U})}, \mathbb{S}_{F(\tilde{w})}) \le WCSM_{SFS}(\mathbb{S}_{F(\tilde{U})}, \mathbb{S}_{F(\tilde{v})})$ and $WCSM_{SFS}(\mathbb{S}_{F(\tilde{U})}, \mathbb{S}_{F(\tilde{w})}) \le WCSM_{SFS}(\mathbb{S}_{F(\tilde{v})}, \mathbb{S}_{F(\tilde{w})})$

41.3 Exponential Similarity Measure Based on Spherical Fuzzy Sets

In this section we establish SFSs-based ESM and W-ESM.

Definition 41.6 For two Spherical Fuzzy Sets $\mathbb{S}_{F(\tilde{U})} = \{\langle \wp_i, \tau_{\tilde{U}}(\wp_i), \sigma_{\tilde{U}}(\wp_i), \xi_{\tilde{U}}(\wp_i)\rangle / \wp_i \in \Upsilon\}$ $\mathbb{S}_{F(\tilde{v})} = \{\langle \wp_i, \tau_{\tilde{V}}(\wp_i), \sigma_{\tilde{V}}(\wp_i), \xi_{\tilde{V}}(\wp_i)\rangle / \wp_i \in \Upsilon\}$ the exponential functions are defined as

$$E_{SFS}^{\tau}(\mathbb{S}_{F(\tilde{U})}, \mathbb{S}_{F(\tilde{v})}) = \exp^{-\left|\tau_{\tilde{U}}(\wp_i)^2 - \tau_{\tilde{V}}(\wp_i)^2\right|}$$

$$E_{SFS}^{\sigma}(\mathbb{S}_{F(\tilde{U})}, \mathbb{S}_{F(\tilde{v})}) = \exp^{-\left|\sigma_{\tilde{U}}(\wp_i)^2 - \sigma_{\tilde{V}}(\wp_i)^2\right|}$$

$$E_{SFS}^{\xi}(\mathbb{S}_{F(\tilde{U})}, \mathbb{S}_{F(\tilde{v})}) = \exp^{-\left|\xi_{\tilde{U}}(\wp_i)^2 - \xi_{\tilde{V}}(\wp_i)^2\right|}$$

Theorem 41.1 If $E_{SFS}^{\tau}(\mathbb{S}_{F(\tilde{U})}, \mathbb{S}_{F(\tilde{v})})$, $E_{SFS}^{\sigma}(\mathbb{S}_{F(\tilde{U})}, \mathbb{S}_{F(\tilde{v})})$, $E_{SFS}^{\xi}(\mathbb{S}_{F(\tilde{U})}, \mathbb{S}_{F(\tilde{v})})$ are exponential functions for SFSs $\mathbb{S}_{F(\tilde{U})}$ and $\mathbb{S}_{F(\tilde{v})}$ then the following conditions holds:

1. $0 \le E_{SFS}^{\tau}(\mathbb{S}_{F(\tilde{U})}, \mathbb{S}_{F(\tilde{v})}), E_{SFS}^{\sigma}(\mathbb{S}_{F(\tilde{U})}, \mathbb{S}_{F(\tilde{v})}), E_{SFS}^{\xi}(\mathbb{S}_{F(\tilde{U})}, \mathbb{S}_{F(\tilde{v})}) \le 1$
2. $E_{SFS}^{\tau}(\mathbb{S}_{F(\tilde{U})}, \mathbb{S}_{F(\tilde{v})}) = E_{SFS}^{\tau}(\mathbb{S}_{F(\tilde{v})}, \mathbb{S}_{F(\tilde{U})})$, $E_{SFS}^{\sigma}(\mathbb{S}_{F(\tilde{U})}, \mathbb{S}_{F(\tilde{v})}) = E_{SFS}^{\sigma}(\mathbb{S}_{F(\tilde{v})}, \mathbb{S}_{F(\tilde{U})})$, $E_{SFS}^{\xi}(\mathbb{S}_{F(\tilde{U})}, \mathbb{S}_{F(\tilde{v})}) = E_{SFS}^{\xi}(\mathbb{S}_{F(\tilde{v})}, \mathbb{S}_{F(\tilde{U})})$
3. $E_{SFS}^{\tau}(\mathbb{S}_{F(\tilde{U})}, \mathbb{S}_{F(\tilde{v})}) = E_{SFS}^{\sigma}(\mathbb{S}_{F(\tilde{U})}, \mathbb{S}_{F(\tilde{v})}) = E_{SFS}^{\xi}(\mathbb{S}_{F(\tilde{U})}, \mathbb{S}_{F(\tilde{v})})$ iff $\mathbb{S}_{F(\tilde{U})} = \mathbb{S}_{F(\tilde{v})}$
4. If $\mathbb{S}_{F(\tilde{U})} \subseteq \mathbb{S}_{F(\tilde{v})} \subseteq \mathbb{S}_{F(\tilde{w})}$ then

$$E_{SFS}^{\tau}(\mathbb{S}_{F(\tilde{U})}, \mathbb{S}_{F(\tilde{w})}) \le min\left\{E_{SFS}^{\tau}(\mathbb{S}_{F(\tilde{U})}, \mathbb{S}_{F(\tilde{v})}), E_{SFS}^{\tau}(\mathbb{S}_{F(\tilde{v})}, \mathbb{S}_{F(\tilde{w})})\right\},$$

$$E_{SFS}^{\sigma}(\mathbb{S}_{F(\tilde{U})}, \mathbb{S}_{F(\tilde{w})}) \le min\left\{E_{SFS}^{\sigma}(\mathbb{S}_{F(\tilde{U})}, \mathbb{S}_{F(\tilde{v})}), E_{SFS}^{\sigma}(\mathbb{S}_{F(\tilde{v})}, \mathbb{S}_{F(\tilde{w})})\right\},$$

$$E_{SFS}^{\xi}(\mathbb{S}_{F(\tilde{U})}, \mathbb{S}_{F(\tilde{w})}) \le min\left\{E_{SFS}^{\xi}(\mathbb{S}_{F(\tilde{U})}, \mathbb{S}_{F(\tilde{v})}), E_{SFS}^{\xi}(\mathbb{S}_{F(\tilde{v})}, \mathbb{S}_{F(\tilde{w})})\right\}.$$

Proof 1, 2, 3 are obvious.

4. $\mathbb{S}_{F(\tilde{U})} \subseteq \mathbb{S}_{F(\tilde{v})} \subseteq \mathbb{S}_{F(\tilde{W})}$ then we have

$0 \le \tau_{\tilde{U}}(\wp_i) \le \tau_{\tilde{V}}(\wp_i) \le \tau_{\tilde{W}}(\wp_i) \le 1, 0 \le \sigma_{\tilde{U}}(\wp_i) \le \sigma_{\tilde{V}}(\wp_i) \le \sigma_{\tilde{W}}(\wp_i) \le 1$
and $1 \ge \xi_{\tilde{U}}(\wp_i) \ge \xi_{\tilde{V}}(\wp_i) \ge \xi_{\tilde{W}}(\wp_i) \ge 0$

$0 \le \tau_{\tilde{U}}(\wp_i)^2 \le \tau_{\tilde{V}}(\wp_i)^2 \le \tau_{\tilde{W}}(\wp_i)^2 \le 1, 0 \le \sigma_{\tilde{U}}(\wp_i)^2 \le \sigma_{\tilde{V}}(\wp_i)^2 \le \sigma_{\tilde{W}}(\wp_i)^2 \le 1$
and $1 \ge \xi_{\tilde{U}}(\wp_i)^2 \ge \xi_{\tilde{V}}(\wp_i)^2 \ge \xi_{\tilde{W}}(\wp_i)^2 \ge 0$

Hence,

$$- \left| \tau_{\tilde{U}}(\wp_i)^2 - \tau_{\tilde{W}}(\wp_i)^2 \right| \le min \left\{ - \left| \tau_{\tilde{U}}(\wp_i)^2 - \tau_{\tilde{V}}(\wp_i)^2 \right|, - \left| \tau_{\tilde{V}}(\wp_i)^2 - \tau_{\tilde{W}}(\wp_i)^2 \right| \right\}$$

$$- \left| \sigma_{\tilde{U}}(\wp_i)^2 - \sigma_{\tilde{W}}(\wp_i)^2 \right| \le min \left\{ - \left| \sigma_{\tilde{U}}(\wp_i)^2 - \sigma_{\tilde{V}}(\wp_i)^2 \right|, - \left| \sigma_{\tilde{V}}(\wp_i)^2 - \sigma_{\tilde{W}}(\wp_i)^2 \right| \right\}$$

$$- \left| \xi_{\tilde{U}}(\wp_i)^2 - \xi_{\tilde{W}}(\wp_i)^2 \right| \le min \left\{ - \left| \xi_{\tilde{U}}(\wp_i)^2 - \xi_{\tilde{V}}(\wp_i)^2 \right|, - \left| \xi_{\tilde{V}}(\wp_i)^2 - \xi_{\tilde{W}}(\wp_i)^2 \right| \right\}$$

It implies that $E^\tau_{SFS}(\mathbb{S}_{F(\tilde{U})}, \mathbb{S}_{F(\tilde{W})}) \le min \left\{ E^\tau_{SFS}(\mathbb{S}_{F(\tilde{U})}, \mathbb{S}_{F(\tilde{v})}), E^\tau_{SFS}(\mathbb{S}_{F(\tilde{v})}, \mathbb{S}_{F(\tilde{W})}) \right\}$
Similarly, $E^\sigma_{SFS}(\mathbb{S}_{F(\tilde{U})}, \mathbb{S}_{F(\tilde{W})}) \le min \left\{ E^\sigma_{SFS}(\mathbb{S}_{F(\tilde{U})}, \mathbb{S}_{F(\tilde{v})}), E^\sigma_{SFS}(\mathbb{S}_{F(\tilde{v})}, \mathbb{S}_{F(\tilde{W})}) \right\}$

$$E^\xi_{SFS}(\mathbb{S}_{F(\tilde{U})}, \mathbb{S}_{F(\tilde{W})}) \le min \left\{ E^\xi_{SFS}(\mathbb{S}_{F(\tilde{U})}, \mathbb{S}_{F(\tilde{v})}), E^\xi_{SFS}(\mathbb{S}_{F(\tilde{v})}, \mathbb{S}_{F(\tilde{W})}) \right\}$$

Thus the proof of 4.

Now we define the Weighted Similarity measures using Exponential Similarity Values.

Definition 41.7 Let $\mathbb{S}_{F(\tilde{v})}, \mathbb{S}_{F(\tilde{v})}$ be two SFS defined over Υ and $\Psi_i > 0$ be the weight element of Υ which satisfies the condition $\sum_{i=1}^{t} \Psi_i = 1$. Then a weighted similarity measure between $\mathbb{S}_{F(\tilde{v})}, \mathbb{S}_{F(\tilde{v})}$ is defined as,

$$E^1_{SFS}(\mathbb{S}_{F(\tilde{v})}, \mathbb{S}_{F(\tilde{v})}) = \sum_{i=1}^{t} \Psi_i \cdot E^\tau_{SFS}(\mathbb{S}_{F(\tilde{v})}, \mathbb{S}_{F(\tilde{v})}) \cdot E^\sigma_{SFS}(\mathbb{S}_{F(\tilde{v})}, \mathbb{S}_{F(\tilde{v})}) \cdot E^\xi_{SFS}(\mathbb{S}_{F(\tilde{v})}, \mathbb{S}_{F(\tilde{v})})$$

(41.6)

Theorem 41.2 *The measure given in (41.7) is a valid Exponential Similarity Measure for Spherical Fuzzy sets.*

Proof The theorem's proof is identical to that of Theorem 41.1.

Definition 41.8 Let $\mathbb{S}_{F(\tilde{v})}, \mathbb{S}_{F(\tilde{v})}$ be two SFS defined over Υ and $\Psi_i > 0$ is the weight element of Υ which satisfies the condition $\sum_{i=1}^{t} \Psi_i = 1$. Then the weighted average Similarity measure of the function $E^\tau_{SFS}(\mathbb{S}_{F(\tilde{v})}, \mathbb{S}_{F(\tilde{v})}), E^\sigma_{SFS}(\mathbb{S}_{F(\tilde{v})}, \mathbb{S}_{F(\tilde{v})}), E^\xi_{SFS}(\mathbb{S}_{F(\tilde{v})}, \mathbb{S}_{F(\tilde{v})})$ is defined as,

$$E^2_{SFS}(\mathbb{S}_{F(\tilde{v})}, \mathbb{S}_{F(\tilde{v})}) = \sum_{i=1}^{t} \Psi_i \left(\frac{E^\tau_{SFS}(\mathbb{S}_{F(\tilde{v})}\mathbb{S}_{F(\tilde{v})}) + E^\sigma_{SFS}(\mathbb{S}_{F(\tilde{v})}, \mathbb{S}_{F(\tilde{v})}) + E^\xi_{SFS}(\mathbb{S}_{F(\tilde{v})}, \mathbb{S}_{F(\tilde{v})})}{3} \right)$$

(41.7)

Theorem 41.3 *The measure given in (41.8) is valid Exponential Similarity Measure for Spherical Fuzzy sets.*

Proof The theorem's proof is identical to that of Theorem 41.1.

Definition 41.9 Let $\mathbb{S}_{F(\tilde{U})}, \mathbb{S}_{F(\tilde{v})}$ be two SFS defined over Υ and $\Psi_i > 0$ is the weight element of Υ which satisfies the condition $\sum_{i=1}^{t} \Psi_i = 1$. Then the generalized weighted average Similarity measure of the function $E_{SFS}^{\tau}(\mathbb{S}_{F(\tilde{U})}, \mathbb{S}_{F(\tilde{v})})$, $E_{SFS}^{\sigma}(\mathbb{S}_{F(\tilde{U})}, \mathbb{S}_{F(\tilde{v})})$, $E_{SFS}^{\xi}(\mathbb{S}_{F(\tilde{U})}, \mathbb{S}_{F(\tilde{v})})$ is defined as,

$$E_{SFS}^{2}(\mathbb{S}_{F(\tilde{U})}, \mathbb{S}_{F(\tilde{v})}) = \sum_{i=1}^{t} \Psi_i \left(\frac{\sqrt[g]{E_{SFS}^{\tau}(\mathbb{S}_{F(\tilde{U})}, \mathbb{S}_{F(\tilde{v})})^g + E_{SFS}^{\sigma}(\mathbb{S}_{F(\tilde{U})}, \mathbb{S}_{F(\tilde{v})})^g + E_{SFS}^{\xi}(\mathbb{S}_{F(\tilde{U})}, \mathbb{S}_{F(\tilde{v})})^g}}{3} \right).$$

(41.8)

$\forall g = \{1, 2, 3 \ldots\} \in N^*$

Theorem 41.4 *The measure given in (41.9) is valid Exponential Similarity Measure for Spherical Fuzzy sets.*

Proof The theorem's proof is identical to that of Theorem 41.1.

41.4 Application of Spherical Fuzzy Set-Based Exponential Similarity Measure in Pattern Recognition

This section will discuss the advantages of the proposed SMs in terms of solving pattern recognition problem. Pattern recognition has been implemented in various fields. In this section we choose Intrusion Detection and Risk evaluation and apply SM's proposed in this paper to recognize the pattern.

41.4.1 Intrusion Detection and Risk Evaluation

Data confidentiality, data and communications integrity, and assurance against denial-of-service are all services that a protected computer or network system must provide. These services are sometimes exploited by intruders [9]. To prevent this from occurring many intrusion detection systems are available. Pattern recognition is being widely used in Intrusion detection system. Here we use SFS-based ESM to detect the high risk intruder and rank according to their risk behavior using the proposed SMs.

Let us assume an intrusion scenario for smart phone application permissions.

Let $\mathcal{P}_1 \rightarrow$ Permission for system audio, $\mathcal{P}_2 \rightarrow$ Permission for system camera, $\mathcal{P}_3 \rightarrow$ Permission for system media, $\mathcal{P}_4 \rightarrow$ Permission for system Location, $\mathcal{P}_5 \rightarrow$ permission for messages and contacts are recognized patterns of permissions.

Consider the known permission patterns $\mathcal{P}_1$, $\mathcal{P}_2$, $\mathcal{P}_3$, $\mathcal{P}_4$, $\mathcal{P}_5$ whose characteristics are represented in terms of SFSs over the universe of discourse $\Upsilon = \{\wp_1, \wp_2, \wp_3, \wp_4\}$ where $\wp_1 \rightarrow$ Mobile bot detection, $\wp_2 \rightarrow$ Encryption of known store of user data, $\wp_3 \rightarrow$ Shielding the app code-base, and $\wp_4 \rightarrow$ Software security.

$\mathcal{P}_1 = \{\langle \wp_1, 0.2, 0.2, 0.3\rangle, \langle \wp_2, 0.3, 0.3, 0.4\rangle, \langle \wp_3, 0.4, 0.4, 0.5\rangle, \langle \wp_4, 0.5, 0.5, 0.6\rangle\}$
$\mathcal{P}_2 = \{\langle \wp_1, 0.3, 0.3, 0.4\rangle, \langle \wp_2, 0.4, 0.4, 0.5\rangle, \langle \wp_3, 0.5, 0.5, 0.6\rangle, \langle \wp_4, 0.6, 0.6, 0.7\rangle\}$
$\mathcal{P}_3 = \{\langle \wp_1, 0.2, 0.3, 0.4\rangle, \langle \wp_2, 0.1, 0.2, 0.3\rangle, \langle \wp_3, 0.3, 0.4, 0.5\rangle, \langle \wp_4, 0.2, 0.1, 0.1\rangle\}$
$\mathcal{P}_4 = \{\langle \wp_1, 0.1, 0.3, 0.5\rangle, \langle \wp_2, 0.2, 0.4, 0.6\rangle, \langle \wp_3, 0.7, 0.2, 0.1\rangle, \langle \wp_4, 0.1, 0.3, 0.4\rangle\}$
$\mathcal{P}_5 = \{\langle \wp_1, 0.2, 0.3, 0.8\rangle, \langle \wp_2, 0.3, 0.5, 0.7\rangle, \langle \wp_3, 0.1, 0.5, 0.4\rangle, \langle \wp_4, 0.2, 0.4, 0.1\rangle\}$

Let $\mathcal{U}_1$, $\mathcal{U}_2$, $\mathcal{U}_3$ be three unknown intruders that attack $\mathcal{P}_1$, $\mathcal{P}_2$, $\mathcal{P}_3$, $\mathcal{P}_4$, $\mathcal{P}_5$. We detect the high risk intruder and rank according to their risk behavior using the proposed SMs.

$\mathcal{U}_1 = \{\langle \wp_1, 0.1, 0.1, 0.1\rangle, \langle \wp_2, 0.3, 0.2, 0.2\rangle, \langle \wp_3, 0.2, 0.3, 0.4\rangle, \langle \wp_4, 0.3, 0.2, 0.4\rangle\}$
$\mathcal{U}_2 = \{\langle \wp_1, 0.7, 0.5, 0.2\rangle, \langle \wp_2, 0.3, 0.1, 0.7\rangle, \langle \wp_3, 0.2, 0.6, 0.1\rangle, \langle \wp_4, 0.2, 0.5, 0.1\rangle\}$
$\mathcal{U}_3 = \{\langle \wp_1, 0.2, 0.3, 0.5\rangle, \langle \wp_2, 0.3, 0.4, 0.6\rangle, \langle \wp_3, 0.6, 0.2, 0.1\rangle, \langle \wp_4, 0.1, 0.2, 0.4\rangle\}$

Let the weight vector be $\Psi = \{0.2, 0.3, 0.4, 0.1\}$

CASE 1: In Table 41.1 we find the values of $E^1_{SFS}(\mathcal{P}_i, \mathcal{U}_1)$, $E^2_{SFS}(\mathcal{P}_i, \mathcal{U}_1)$, $E^3_{SFS}(\mathcal{P}_i, \mathcal{U}_1)$. Further, we rank them according to their risk behavior.
CASE 2: In Table 41.2 we find the values of $E^1_{SFS}(\mathcal{P}_i, \mathcal{U}_2)$, $E^2_{SFS}(\mathcal{P}_i, \mathcal{U}_2)$, $E^3_{SFS}(\mathcal{P}_i, \mathcal{U}_2)$. Further, we rank them according to their risk behavior.
CASE 3: In Table 41.3 we find the values of $E^1_{SFS}(\mathcal{P}_i, \mathcal{U}_3)$, $E^2_{SFS}(\mathcal{P}_i, \mathcal{U}_3)$, $E^3_{SFS}(\mathcal{P}_i, \mathcal{U}_3)$. Further, we rank them according to their risk behavior.

Table 41.1 Similarity measures of $\mathcal{P}_i$ *with* $\mathcal{U}_1$

	$(\mathcal{P}_1, \mathcal{U}_1)$	$(\mathcal{P}_2, \mathcal{U}_1)$	$(\mathcal{P}_3, \mathcal{U}_1)$	$(\mathcal{P}_4, \mathcal{U}_1)$	$(\mathcal{P}_5, \mathcal{U}_1)$	Risk ranking
E^1_{SFS}	0.785837	0.582409	0.634626	0.768145	0.653877	$\mathcal{P}_2 > \mathcal{P}_3 > \mathcal{P}_5 > \mathcal{P}_4 > \mathcal{P}_1$
E^2_{SFS}	0.921897	0.847929	0.862468	0.8595	0.872872	$\mathcal{P}_2 > \mathcal{P}_4 > \mathcal{P}_3 > \mathcal{P}_5 > \mathcal{P}_1$
E^3_{SFS}	0.532539	0.489917	0.501449	0.500105	0.509514	$\mathcal{P}_2 > \mathcal{P}_4 > \mathcal{P}_3 > \mathcal{P}_5 > \mathcal{P}_1$

Table 41.2 Similarity measures of P_i *with* $\mathcal{U}_2$

	$(\mathcal{P}_1, \mathcal{U}_2)$	$(\mathcal{P}_2, \mathcal{U}_2)$	$(\mathcal{P}_3, \mathcal{U}_2)$	$(\mathcal{P}_4, \mathcal{U}_2)$	$(\mathcal{P}_5, \mathcal{U}_2)$	Risk ranking
E^1_{SFS}	0.354545	0.535644	0.568441	0.557542	0.583028	$\mathcal{P}_1 > \mathcal{P}_2 > \mathcal{P}_4 > \mathcal{P}_3 > \mathcal{P}_5$
E^2_{SFS}	0.840313	0.814772	0.832063	0.825718	0.840313	$\mathcal{P}_2 > \mathcal{P}_4 > \mathcal{P}_3 > \mathcal{P}_1 > \mathcal{P}_5$
E^3_{SFS}	0.488466	0.472787	0.484969	0.481205	0.488466	$\mathcal{P}_2 > \mathcal{P}_4 > \mathcal{P}_3 > \mathcal{P}_1 > \mathcal{P}_5$

Table 41.3 Similarity measures of $\mathcal{P}_i$ *with* $\mathcal{U}_3$

	$(\mathcal{P}_1, \mathcal{U}_3)$	$(\mathcal{P}_2, \mathcal{U}_3)$	$(\mathcal{P}_3, \mathcal{U}_3)$	$(\mathcal{P}_4, \mathcal{U}_3)$	$(\mathcal{P}_5, \mathcal{U}_3)$	Risk ranking
E^1_{SFS}	0.443335	0.665924	0.914516	0.925819	0.646907	$\mathcal{P}_1 > \mathcal{P}_5 > \mathcal{P}_2 > \mathcal{P}_3 > \mathcal{P}_4$
E^2_{SFS}	0.874839	0.866973	0.971505	0.975273	0.865073	$\mathcal{P}_5 > \mathcal{P}_2 > \mathcal{P}_1 > \mathcal{P}_3 > \mathcal{P}_4$
E^3_{SFS}	0.506099	0.501673	0.561456	0.563545	0.501868	$\mathcal{P}_5 > \mathcal{P}_2 > \mathcal{P}_1 > \mathcal{P}_3 > \mathcal{P}_4$

From the above tables we have detected and prioritized the intrusion risk of unknown intruders for the smart phone App permissions. Hence, the system will now be able to identify the high risk intruder and act accordingly.

41.5 Conclusion

In this paper some new similarity measures in Spherical Fuzzy set domain such as ESM, W-ESM, WA-ESM, and WGA-ESM are introduced and some of their properties are discussed. A Pattern Recognition problem in intrusion detection and risk evaluation for smart phone App permissions has been solved in three cases. In future the proposed similarity measures can be extended in different fuzzy domain and also for other applications of pattern recognition.

References

1. Bezdek, J.C.: Fuzzy mathematics in pattern classification. Ph.D. Thesis, Applied Mathematics Center, Cornell University, Ithaca (1973)
2. Fuzzy pattern recognition. Computational Intelligence. Springer, Berlin (2005). https://doi.org/10.1007/3-540-27335-2_5
3. Zadeh, L.: Fuzzy sets. Inf. Control **8**, 338–353 (1965)
4. Pawlak, Z.: Rough set theory and its applications to data analysis. Cybern. Syst. **29**, 661–688 (1998)
5. Atanassov, K.: Intuitionistic fuzzy sets. Fuzzy Sets Syst. **20**, 87–96 (1983)
6. Yager, R.: Pythagorean fuzzy subsets. In: 2013 Joint IFSA World Congress and NAFIPS Annual Meeting (IFSA/NAFIPS), pp. 57–61 (2013)
7. Mahmood, T., Ullah, K., Khan, Q., Jan, N.: An approach toward decision-making and medical diagnosis problems using the concept of spherical fuzzy sets. Neural Comput. Appl. **31**(11), 7041–7053 (2019)
8. Mahmood, T., Ilyas, M., Ali, Z., Gumaei, A.: Spherical fuzzy sets-based cosine similarity and information measures for pattern recognition and medical diagnosis. IEEE Access **9**, 25835–25842 (2021)
9. National Computer Security Center.: Department of Defense, Trusted Computer System Evaluation Criteria, DOD 5200.28-STD. Dec. 1985 (Orange Book)

Chapter 42
Machine Learning Techniques to Analyze Pandemic-Induced Economic Outliers

Anindita Desarkar and Ajanta Das

Abstract Outliers, or outlying observations, are values in data, which appear unusual. It is quite essential to analyze various unexpected events or anomalies in economic domain like sudden crash of stock market, mismatch between country's per capita incomes and overall development, abrupt change in unemployment rate and steep falling of bank interest to find the insights for the benefit of humankind. These situations can arise due to several reasons, out of which pandemic is a major one. The present COVID-19 pandemic also disrupted the global economy largely as various countries faced various types of difficulties. This motivates the present researchers to identify a few such difficult areas in economic domain, arises due to the pandemic situation and identify the countries, which are affected most under each bucket. Two well-known machine-learning techniques DBSCAN (density based clustering approach) and Z-score (statistical technique) are utilized in this analysis. The results can be used as suggestive measures to the administrative bodies, which show the effectiveness of the study.

42.1 Introduction

An outlier is a data point, which significantly varies from other observations present in a set. One burning example of economic outlier is the economic shock, which the global community faced in the context of the present COVID-19 scenario. Several studies were performed to understand the pandemic effect on global economy, out of which two have been utilized in this research. The first one yielded a dataset, which reveals the pandemic's social and economic impact on 170 individual countries of the world [1], renamed as Dataset 1 in the present work. The second involves a dataset, which divulges the present unemployment status prevailing in 40 countries across

A. Desarkar (✉)
Department of Computer Science and Engineering, Jadavpur University, Kolkata, India

A. Das
Amity Institute of Information Technology, Amity University, Kolkata, India

© The Author(s), under exclusive license to Springer Nature Singapore Pte Ltd. 2022
V. Bhateja et al. (eds.), *Evolution in Computational Intelligence*,
Smart Innovation, Systems and Technologies 267,
https://doi.org/10.1007/978-981-16-6616-2_42

the globe [2] and renamed as Dataset 2 here. Dataset 1 contains various aspects of the pandemic effect such as the daily frequency of infection and death, stringency index, human development index (HDI), and gross domestic product (GDP) per capita whereas Dataset 2 consists mainly country-wise unemployment related data.

The general *motivation* behind the current research work is to judge whether the factors used to measure country's development are truly reflecting the same. Two factors are considered here: GDP per capita and HDI. GDP refers the total value of goods produced and services provided in a country during one year whereas GDP per capita is a measure of a country's economic output, which accounts for its number of people. HDI is a summary measure of average achievement in key dimensions of human development: a long and healthy life, being knowledgeable and have a decent standard of living. HDI is developed to emphasize that people and their capabilities should be the ultimate criteria for measuring the development of a country, not economic growth alone. So the present research is motivated to analyze whether HDI and GDP per capita are strongly correlated, any deviations exists and countries with strong HDI are really able to manage the pandemic situation more efficiently compared to other countries.

The overall *objective* is to explore the economic correlations between entities and discover the outliers by identifying the anomalous countries from the two source data sets. Two outlier detection techniques are applied: DBSCAN (a well-known density-based clustering algorithm) and Z-score (a statistical measure of deviation from the standard).

Clustering algorithm, an unsupervised type, is a good fit here as no training dataset is available. On the other hand, Z-score—a well-known statistical technique chosen as it is already utilized before for many similar areas across different domains.

It is felt that pointing out the discrepancies may raise a global alert to review and ponder on feasible solutions by the respective outlier countries' government. The results may also have far-reaching effects on the general policy making Supremos' decisions regarding how such a crisis can be handled in future. Another objective is to analyze the performance of India in this global scenario to understand, to what extend it has handled the situation.

Section 42.2 highlights existing research works in related areas. The outline of the proposed framework and brief overview of outlier detection techniques are presented in Sect. 42.3. Section 42.4 discusses the source data acquisition details. It is followed by Sect. 42.5, which lists down the experimental results and inferences drawn accordingly. Conclusion is depicted in Sect. 42.6.

42.2   Related Work

A few recent studies are listed here, which show how the current pandemic has disrupted economy in all countries to some extent. Beyer et al. showed in their work that daily electricity consumption and monthly nighttime light intensity could be a good indicator of the economic activities in India [8]. In another study by Mele et al., it was evident that a relationship exists between pollution emissions, economic growth, and COVID-19 deaths in India [6]. Unidirectional causality between economic growth and pollution was found after performing Stationarity and Toda-Yamamoto causality tests on the dataset between 1980 and 2018. The underlying hypothesis is that a predetermined pollution concentration, due to economic growth, could foster COVID-19 by making the respiratory system more susceptible to infection. A detailed study was conducted by Chetty et al. on the economic impacts of COVID-19 pandemic which includes of building a public database that tracks economic activity at a granular level in real time using anonymized data from private companies [5]. Polyakova et al. had discussed why initial economic damage from the COVID-19 pandemic in the United States is more intense across ages and geographies than initial mortality impacts [3]. Stojkoski et al. explored various socio-economic determinants of the COVID-19 pandemic [4]. Total 31 determinants were selected which describes a diverse ensemble of social and economic factors, including healthcare infrastructure, societal characteristics, economic performance, demographic structure, etc. Bonaccorsi et al. performed an in-depth analysis on real time Italian data to analyze how economy is affected through lockdown strategies [7].

42.3 Proposed Outlier Detection Techniques

42.3.1 Brief Overview of Outlier Detection Techniques

Several approaches are available for detecting the outliers in different domains. Out of them, two are utilized in this research, which are briefly described in the following.

(a) Outlier Detection with Clustering—DBSCAN: It is a well-known clustering algorithm that is commonly used in data mining and machine learning domain and uses two parameters [10, 11]. Here, one big advantage is automatic detection of cluster numbers based on the given input data and parameter. Also, outlier detection can be done easily as points which do not reside within any cluster, labelled as outliers.

eps (ε): A distance measure that will be used to locate the points about any point. eps value is considered as **0.8** in the current research work.

minPts: The minimum number of points (a threshold) clustered together for a region to be considered dense. The value is taken as **5** here.

(b) Outlier Detection with Statistical Technique—Z-Score: Z-score is probably the simplest algorithm that can rapidly screen candidates for further analysis to determine whether anomalies are present [12]. It is a statistical measure, which represents dispersion of data for a given observation is from the mean. The formula for calculating a Z-score: $z = (x - \mu)/\sigma$, where x is the raw score, μ is the population mean, and σ is the population standard deviation. Standard deviation function is used to identify outliers and compared with the mean. If the value is over a certain number of standard deviations away from the mean, it can be classified as an outlier. This certain number is called the threshold, and often the default number is 3.

42.3.2 Identifying Attributes for Finding Correlations

In the experimental results, correlation is analyzed among the four pair of columns along with outlier detection. First two cases are included to understand whether HDI is a true indicator of country's development and any defaulter exists. In the present scenario, various stringency measures like lockdown are applied in almost everywhere to reduce the infection rate. So the third case will show the efficacy of the same. Additionally, the correlation between GDP per capita and HDI is also analyzed, as it is one of our objectives.

- Case 1: Total_Death_Percentage (Y-axis) and HDI (X-axis)
- Case 2: Unemployment Value (Y-axis) and HDI (X-axis)
- Case 3: Total_Infection_Percentage (Y-axis) and Stringency_index (X-axis)
- Case 4: GDP per Capita (Y-axis) and HDI (X-axis)

42.4 Source Data Acquisition/Experimental Set-Up

42.4.1 Source File Description

Two real datasets, Dataset 1[1] and Dataset 2 [2], are utilized in the present research work and attributes according to the source are mentioned below.

Metadata for Dataset 1 (*Impact of Covid-19 Pandemic on the Global Economy* [1]):

- **iso_code**: Code of the country (As example, For Afghanistan – AFG)
- **Location**: Country name (As example, Afghanistan)
- **Date**: Date with day, month and year
- **total_cases**: Total number of COVID 19 active cases on that date
- **total_deaths**: Total number of COVID 19 Deaths reported on that date
- **stringency_index** (SI): These indicators examine containment policies such as school and workplace closings, public events, public transport, stay-at-home policies. A higher index score indicates a higher level of stringency.
- **Population**: Population of the specific country
- **gdp_per_capita**: Already mentioned in the "Introduction" section
- **human_development_index**: Already mentioned in the "Introduction" section

Metadata for Dataset 2: (*OECD Unemployment rate (2020)* [2]):

- **LOCATION**: Country Name, same as iso_code column of Dataset 1
- **TIME**: Specific time in terms of month and year
- **Value**: Refers value of Unemployment (Unempt)—This indicator is measured in numbers of unemployed people as a percentage of the labor force and it is seasonally adjusted. The labor force is defined as the total number of unemployed people plus those in employment.
- **INDICATOR, SUBJECT, MEASURE, FREQUENCY, Flag Codes**—Not utilized in present research

42.4.2 Data Pre-Processing

- Truncating the unutilized columns from Dataset 2
- Each dataset is aggregated by their date or time column so that one record becomes available for each country in both the datasets
- Merging the two datasets based on country name (Dataset1.iso_code = Dataset2. LOCATION)
- Null values are identified in the merged dataset and replaced by zero in case of numerical columns.
- Two columns are derived: Total_Death_Percentage (Tot_Dth_Pct) and Total_Infection_Percentage (Tot_Infct_Pct). The percentage is calculated based on the total population of the respective country.

Implementation Details: The complete application is developed in Python 3 accessed through Google Colab [13].

42.5 Experimental Results and Inferences

Following COVID-imprints 1–4 describe four scenarios where correlation is investigated along with the outliers through two proposed techniques. First scatter plot is built based on the Z-score values of the identified columns whereas second one represents the actual values. The "X" sign indicates the outliers in each figure.

COVID-imprints 1: **Correlation between Total_Death_Percentage and HDI.**

Following Figs. 42.1 and 42.2 present the Z-score and DBSCAN results respectively.

Inference: The two countries Belgium and Peru are true outliers because their death rate is very high compared to other countries with the same HDI. However, the position of India (shown as pentagon in Fig. 42.1) is not bad with respect to other countries.

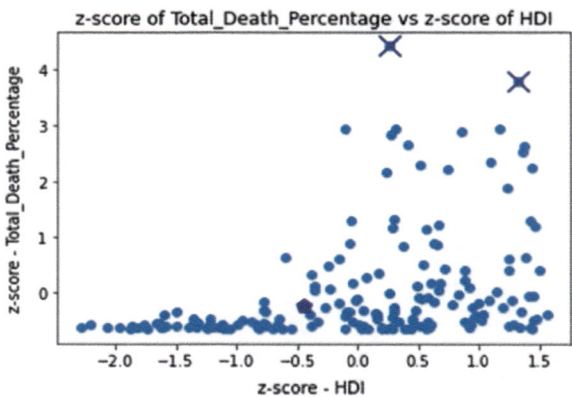

Fig. 42.1 Z-Score—Tot_Dth_Pct & HDI

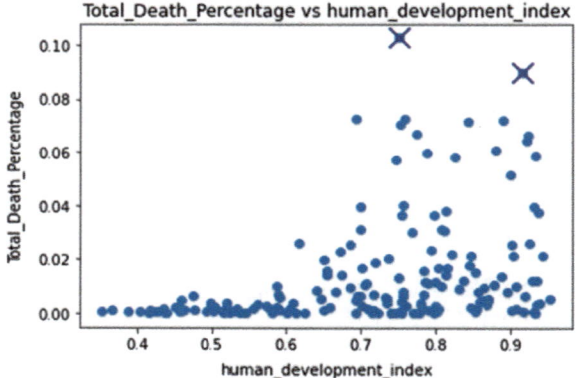

Fig. 42.2 DBSCAN—Tot_Dth_Pct & HDI

COVID-imprints 2: Correlation between Unemployment Value and HDI.

Following Figs. 42.3 and 42.4 present the Z-score and DBSCAN results respectively. However, no outlier is identified through Z-score technique here.

Inference: The first three countries Colombia, Mexico, and Turkey are not true outliers as their HDI is not very high. So, it is obvious that, unemployment will be an issue in these cases. However, the other two, Spain and Greece (placed in the right side of the graph) can be considered as outliers as their unemployment rate is very high compared to other countries with same HDI.

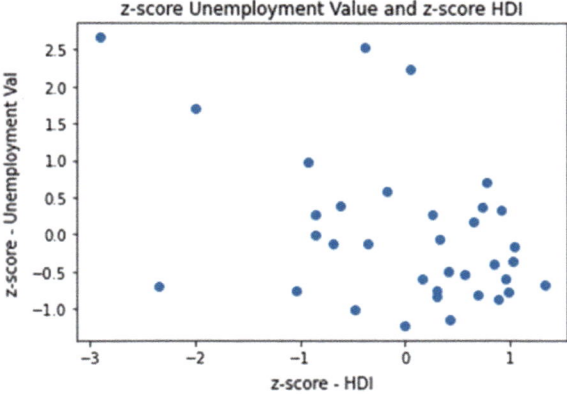

Fig. 42.3 Z-Score—Unempt & HDI

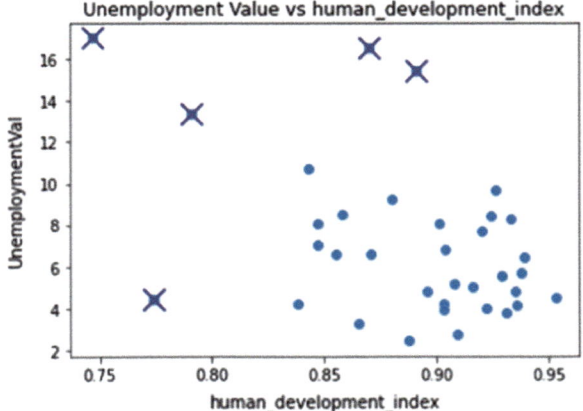

Fig. 42.4 DBSCAN—Unempt & HDI

COVID-imprints 3: Correlation between Total_Infection_Percentage and Stringency_index.

The objective of implementing stringency related measures is to reduce the Total_Infection_Percentage. Hence, the relationship between these two is analyzed here and following Figs. 42.5 and 42.6 present the Z-score and DBSCAN results respectively.

Inference: The three countries Bahrain, Israel, and Qatar can be treated as outliers because their total infection percentage is very high compared to other countries with the same stringency index. The other two countries are Armenia and Montenegro, which are coming through DBSCAN technique only, and can be ignored as stringency index is zero in these cases. So little bit higher infection percentage is obvious. The

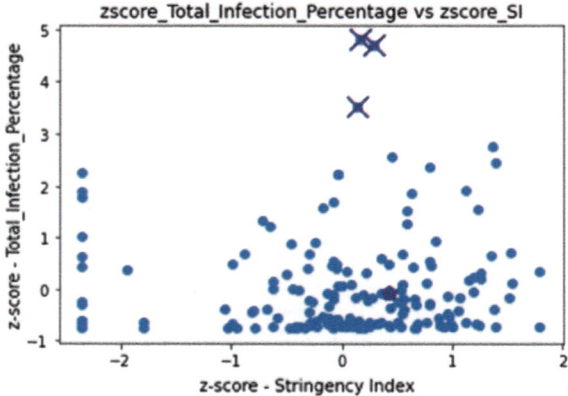

Fig. 42.5 Z-Score—Tot_Infct_Pct & SI

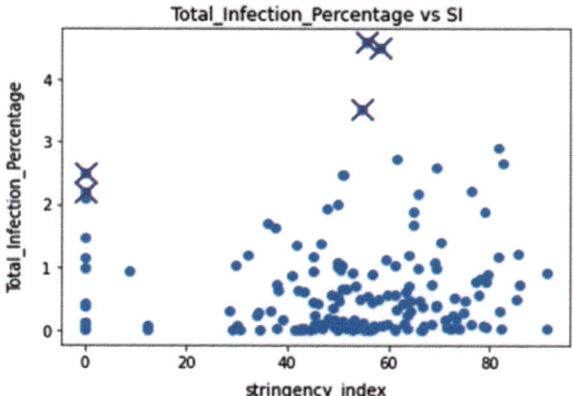

Fig. 42.6 DBSCAN—Tot_Infct_Pct & SI

position of India is highlighted in Fig. 42.5 as pentagon, which shows the infection rate is quite controllable.

COVID-imprints 4: Finding Relationship between GDP per Capita and HDI.

HDI measures economic conditions using GDP per capita. So, there should be a strong relationship between GDP per Capita and HDI. The same is evaluated here by finding correlation between the above two variables. Hence, Figs. 42.7 and 42.8 present the Z-score and DBSCAN results respectively. The position of India is highlighted as pentagon sign in below Fig. 42.7.

Inference: It is evident from the above figures that a strong correlation exists between the two features GDP per capita and HDI. In this scenario, the three countries Luxembourg, Qatar, and Singapore are identified as outliers through both the techniques,

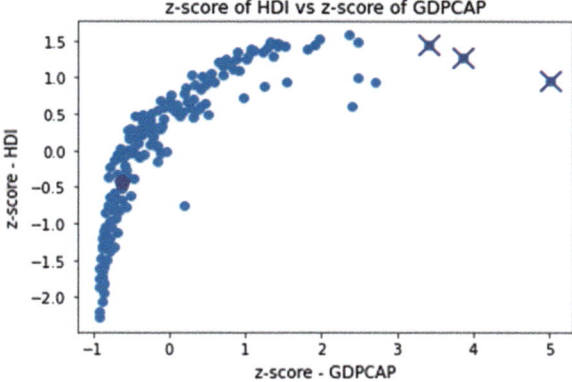

Fig. 42.7 Z-Score—GDP & HDI

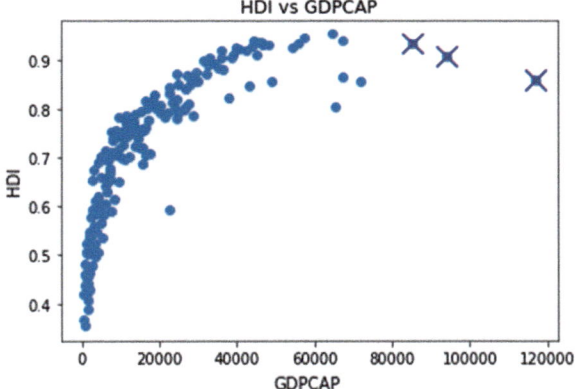

Fig. 42.8 DBSCAN—GDP & HDI

which are true anomalies. Because they are attaining the same HDI like a few other countries where the GDP per capita is much lesser than these three countries.

Summary of Observation/Knowledge Discovery

Table 42.1 presents the outlier countries, which are appearing in the above four cases.

Inference

- Total 14 countries are present which are anomalous in at least any of the four cases.
- One country Qatar is appearing in two cases because of its excessive infection percentage and GDP-HDI mismatch issues.
- COVID-imprints 1 and COVID-imprints 4—DBSCAN and Z-score results are completely matching.

Table 42.1 Summary of outlier countries for all COVID-imprints 1, 2, 3, 4

Outlier country	Tot_Dth_Pct & HDI	Unempt & HDI	Tot_Infct_Pct & SI	HDI & GDP
Luxembourg	–	–	–	DBSCAN & Z-score
Qatar	–	–	DBSCAN & Z-score	DBSCAN & Z-score
Singapore	–	–	–	DBSCAN & Z-score
Belgium	DBSCAN & Z-score	–	–	–
Peru	DBSCAN & Z-score	–	–	–
Colombia	–	DBSCAN	–	–
Spain	–	DBSCAN	–	–
Greece	–	DBSCAN	–	–
Mexico	–	DBSCAN	–	–
Turkey	–	DBSCAN	–	–
Armenia	–	–	DBSCAN	–
Bahrain	–	–	DBSCAN & Z-score	–
Israel	–	–	DBSCAN & Z-score	–
Montenegro	–	–	DBSCAN	–

- COVID-imprints 2—Five outliers are identified through DBSCAN only.
- COVID-imprints 3—Three are common anomalous countries but DBSCAN identifies extra two countries.

42.6 Conclusion

Total four scenarios are analyzed using two machine learning techniques, DBSCAN, and Z-score, in the current research where first two find the relationship between two COVID related parameters like Tot_Dth_Pct and Unempt vs HDI respectively. First scenario reveals that two outlier countries Belgium and Peru have higher death rate compared to other countries having same HDI, which indicates they require special attention in this context. Five countries (mentioned in Table 42.1) are listed through DBSCAN technique in the second scenario, which identifies that unemployment is a big concern in these places in the current COVID situation. The next case shows how various stringency measures helped the world to fight against the infection rate. Three countries, Qatar, Bahrain, and Israel, are identified as anomalies by both the techniques as they face a high infection rate in spite of moderate stringency measures.

In the last scenario, the correlation between HDI and GDP is analyzed to understand whether any loophole exists because ideally positive correlation should exist here. However, three countries (Luxembourg, Qatar, and Singapore) are identified by both the techniques, which reveals the gap present in these countries from economic perspective.

The position of India is analyzed in three scenarios out of the four mentioned ones as unemployment related data is not received for it. Fortunately, it is not listed as outlier in none of the cases and the position indicates that it has handled the pandemic situation.

References

1. Vitenu-Sackey, P.A.: The impact of Covid-19 pandemic on the global economy: emphasis on poverty alleviation and economic growth. Mendeley Data **V1**,(2020). https://doi.org/10.17632/b2wvnbnpj9.1
2. OECD (2020) Unemployment rate (indicator). https://doi.org/10.1787/52570002-en. Accessed 07 Dec 2020
3. Polyakova, M., Kocks, G., Udalova, V., Finkelstein, A.: Initial economic damage from the COVID-19 pandemic in the United States is more widespread across ages and geographies than initial mortality impacts. Proc. Natl. Acad. Sci. **117**(45), 27934–27939 (2020)
4. Stojkoski, V., Utkovski, Z., Jolakoski, P., Tevdovski, D., & Kocarev, L.: The socio-economic determinants of the coronavirus disease (COVID-19) pandemic (2020). arXiv:2004.07947
5. Chetty, R., Friedman, J., Hendren, N., Stepner, M.: The economic impacts of COVID-19: Evidence from a new public database built from private sector data. Opportunity Insights (2020)
6. Mele, M., Magazzino, C.: Pollution, economic growth, and COVID-19 deaths in India: a machine learning evidence. Environ. Sci. Pollut. Res. 1–9 (2020)
7. Bonaccorsi, G., Pierri, F., Cinelli, M., Porcelli, F., Galeazzi, A., Flori, A., Pammolli, F.: Evidence of economic segregation from mobility lockdown during covid-19 epidemic. (2020) Available at SSRN 3573609
8. Beyer, R. C., Franco-Bedoya, S., Galdo, V.: Examining the economic impact of COVID-19 in India through daily electricity consumption and nighttime light intensity. World Dev. 105287 (2020)
9. Albulescu, C.T.: COVID-19 and the United States financial markets' volatility. Finance Res. Lett. **38**, 101699 (2021)
10. Behera, S., Rani, R.: Comparative analysis of density based outlier detection techniques on breast cancer data using Hadoop and MapReduce. In: International Conference on Inventive Computation Technologies (ICICT), (vol. 2, pp. 1–4). IEEE (2016)
11. Berkhin, P.: 'A survey of clustering data mining techniques', Grouping multidimensional data, pp. 25–71. Springer, Berlin, Heidelberg (2006)
12. Kannan, K.S., Manoj, K., Arumugam, S.: Labeling methods for identifying outliers. Int. J. Stat. Syst. **10**(2), 231–238 (2015)
13. Accessed Google Colab from https://colab.research.google.com/notebooks/intro.ipynb

Chapter 43
MABAC Method for Assessment of Cyber Security Technologies Under Fermatean Fuzzy Sets

J. Aldring and **D. Ajay**

Abstract Cyber security is a basic need for every digital device such as computers, mobiles, networks, servers and electronic items. It protects digital devices from various malicious cyber-attacks or threats created over the Internet. In the digitized world, cyber security threats have been an emerging problem. In the current scenario the growth of digital devices is very high, and the risks of cyber security technologies (CST) are serious issues. These types of threats should be considered when choosing the best among cyber security technologies. For this purpose, we approach mathematically to determine the safest CST. In this paper, first we study CST and their risk factors. Then, Fermatean fuzzy sets (FFSs) are used to develop Fermatean fuzzy multi attribute border approximation area comparison (FF-MABAC) method. This method handles more uncertain decision-making problems by computing distance between each alternative and bordered approximation area (BAA). At the end of the paper, an application has been given to CST and their computing approaches are demonstrated numerically. Finally, a sensitivity analysis of results is conducted and direction for future research is provided.

43.1 Introduction

We are living in a digital world that is networked together which acknowledges that our private data are more sensitive than ever before. Sensitive information, whether personal information, intellectual property, financial data or other forms of data for which unauthorized access or disclosure could have significant effects, may form part of that data. Cyber-attack is now a global issue and has given several warnings of hacks and certain security threats could even threaten the global economy. Cyber security provides the protection for Internet-connected systems or internet of things, which include software, hardware and data from cyber threats. The purpose of information security is significant and it is assured by cyber security technologies (CST).

J. Aldring · D. Ajay (✉)
Sacred Heart College (Autonomous), 635601 Tirupattur District, Tamilnadu, India

© The Author(s), under exclusive license to Springer Nature Singapore Pte Ltd. 2022   441
V. Bhateja et al. (eds.), *Evolution in Computational Intelligence*,
Smart Innovation, Systems and Technologies 267,
https://doi.org/10.1007/978-981-16-6616-2_43

Nowadays many cyber-attacks or threats have taken place on CST and it is becoming a crucial problem [1]. To take an action on cyber-attacks a global assessment of CST is needed. In this context, our research made an effort to conduct a study on CST and their risk factors in order to rank and select the most secure technology. For this purpose a fuzzy multi criteria decision making (MCDM) method has been developed using Fermatean fuzzy sets and it is utilized to rank the CST. It is evident from the literature that many distinguished MCDM models are available with fuzzy sets [2] and its extensions intuitionistic fuzzy sets (IFSs) [3], Pythagorean fuzzy sets (PFSs) [4] and so on. But in some ideal situations a particular fuzzy set cannot handle the uncertain information properly so the fuzzy sets moved to the next level. Fermatean fuzzy sets (FFSs) [5] are another extension of fuzzy sets which replaced IFSs and PFSs. The developed FF-MABAC model predicts the alternatives more effectively with the help of decision makers or experts information in the field of CST.

The rest of the paper has been organized as follows: Sect. 43.2, briefly discusses the basic fuzzy notion which is used in this paper. Section 43.3 explains the proposed methodology based on FFSs. Section 43.4 shows the risk analysis of CST using the proposed FF-MABAC method. The sensitivity analysis is presented in Sect. 43.5. Finally, Sect. 43.6 represents the conclusion of the study and includes the scope for future research.

43.2 Preliminaries

Definition 1 [5] Let $\mathbb{R}$ be a universe of discourse. A FFS $\mathbb{F}$ in $\mathbb{R}$ is defined by the following form

$$\mathbb{F} = \{\langle r, \xi_{\mathbb{F}}(r), \psi_{\mathbb{F}}(r)\rangle : r \in \mathbb{R}\}$$

where $\xi_{\mathbb{F}}(r) : \mathbb{R} \to [0, 1]$ and $\psi_{\mathbb{F}}(r) : \mathbb{R} \to [0, 1]$, with the condition that $0 \leq (\xi_{\mathbb{F}}(r))^3 + (\psi_{\mathbb{F}}(r))^3 \leq 1, \forall r \in \mathbb{R}$ The fuzzy numbers $\xi_{\mathbb{F}}(r)$ and $\psi_{\mathbb{F}}(r)$ denote the degree of membership and the degree of nonmembership of the element r in the set $\mathbb{F}$, respectively. For any FFSs $\mathbb{F}$ and $r \in \mathbb{R}$, the degree of indeterminacy of r to $\mathbb{F}$ is defined as $\pi_{\mathbb{F}}(r) = \sqrt[3]{1 - (\xi_{\mathbb{F}}(r))^3 - (\psi_{\mathbb{F}}(r))^3}$. In the interest of simplicity, a Fermatean fuzzy set is denoted as $\mathbb{F} = \langle \xi_{\mathbb{F}}, \psi_{\mathbb{F}}\rangle$.

Definition 2 [5] Let $\mathbb{F}_v = \{\langle r, \xi_{\mathbb{F}_v}(r), \psi_{\mathbb{F}_v}(r)\rangle : r \in \mathbb{R}\}$ be a Fermatean fuzzy set in $\mathbb{R}$. The Fermatean fuzzy score function is given by the following equation:

$$S\left(\mathbb{F}_v\right) = \frac{1}{2}\left(1 + \xi_{\mathbb{F}_v}^3 - \psi_{\mathbb{F}_v}^3\right), \, S\left(\mathbb{F}_v\right) \in [0, 1] \tag{43.1}$$

43.2.1 Aggregations Operators of FFSs

Garg et al. [6] developed Yager's weighted aggregation operators under FF environment and the operators are given as follows:

Definition 3 Let $\mathbb{F}_v = \left\{ \langle r, \xi_{\mathbb{F}_v}(r), \psi_{\mathbb{F}_v}(r) \rangle : r \in \mathbb{R} \right\}$ $(v = 1, 2, 3, \ldots \Bbbk)$ be a group of FFSs. Then, the Fermatean fuzzy yager weighted arithmetic operator $(FFYWA_\alpha)$ is a function $\mathbb{Q}^\Bbbk \rightarrow \mathbb{Q}$ such that

$$FFYWA_\alpha \left(\mathbb{F}_1, \mathbb{F}_2, \mathbb{F}_3, \ldots, \mathbb{F}_\Bbbk \right) = \bigoplus_{v=1}^{\Bbbk} (\alpha_v \mathbb{F}_v)$$

$$= \left\langle \sqrt[3]{\min \left(1, \left(\sum_{v=1}^{\Bbbk} \left(\alpha_v \xi_{\mathbb{F}_v}^{3\eta} \right) \right)^{\frac{1}{\eta}} \right)}, \sqrt[3]{1 - \min \left(1, \left(\sum_{v=1}^{\Bbbk} \left(\alpha_v \left(1 - \psi_{\mathbb{F}_v}^3 \right)^\eta \right) \right)^{\frac{1}{\eta}} \right)} \right\rangle$$

$$(43.2)$$

where the weight vector (WV) of $\mathbb{F}_v$ is $\alpha = (\alpha_1, \alpha_2, \alpha_3, \ldots, \alpha_\Bbbk)^T$ with $\alpha_v > 0$ and $\sum_{v=1}^{\Bbbk} \alpha_v = 1$.

The $FFYWG_\alpha$ (Fermatean fuzzy yager weighted geometric) operator is given as follows;

$$FFYWG_\alpha \left(\mathbb{F}_1, \mathbb{F}_2, \mathbb{F}_3, \ldots, \mathbb{F}_\Bbbk \right) = \bigotimes_{v=1}^{\Bbbk} (\mathbb{F}_v)^{\alpha_v}$$

$$= \left\langle \sqrt[3]{1 - \min \left(1, \left(\sum_{v=1}^{\Bbbk} \left(\alpha_v \left(1 - \xi_{\mathbb{F}_v}^3 \right)^\eta \right) \right)^{\frac{1}{\eta}} \right)}, \sqrt[3]{\min \left(1, \left(\sum_{v=1}^{\Bbbk} \left(\alpha_v \psi_{\mathbb{F}_v}^{3\eta} \right) \right)^{\frac{1}{\eta}} \right)} \right\rangle$$

43.2.2 Distance Measures of Fermatean Fuzzy Sets

Some distance measures between two Fermatean fuzzy numbers and the formulae are discussed below.

Definition 4 Let $\mathbb{F}_v = \left\{ \langle r, \xi_{\mathbb{F}_v}(r), \psi_{\mathbb{F}_v}(r) \rangle : r \in \mathbb{R} \right\}$ $(v = 1, 2, 3, \ldots \Bbbk)$ be a group of FFSs. If the weight of different elements r_j is $\alpha_j = (\alpha_1, \alpha_2, \alpha_3, \ldots, \alpha_\Bbbk)^T$, with the condition that $0 \leq \alpha_j \leq 1$ and $\sum_{j=1}^{\Bbbk} \alpha_j = 1$, then the Fermatean fuzzy weighted generalized distance measures between any two Fermatean fuzzy sets or numbers can be defined as follows:

$$\Delta_{\alpha_G} (\mathbb{F}_1, \mathbb{F}_2) = \left[\sum_{j=1}^{k} \frac{\alpha_j}{2} \left(\left| \xi_{\mathbb{F}_1}^3 (r_j) - \xi_{\mathbb{F}_2}^3 (r_j) \right|^\lambda + \left| \psi_{\mathbb{F}_1}^3 (r_j) - \psi_{\mathbb{F}_2}^3 (r_j) \right|^\lambda + \right. \right.$$

$$\left. \left. \left| \pi_{\mathbb{F}_1}^3 (r_j) - \pi_{\mathbb{F}_2}^3 (r_j) \right|^\lambda \right) \right]^{\frac{1}{\lambda}}$$

(43.3)

where $\lambda > 0$.

Remark 1 If $\lambda = 1$, then the generalized weighted Fermatean fuzzy distance measure is reduced to Fermatean fuzzy hamming distance measure Δ_{α_H}. If $\lambda = 2$, then the $\Delta_{\alpha_G} (\mathbb{F}_1, \mathbb{F}_2)$ is reduced to the Fermatean fuzzy weighted Euclidean distance measure Δ_{α_E}.

Also, the generalized weighted Fermatean fuzzy distance measure $\Delta_{\alpha_G} (\mathbb{F}_1, \mathbb{F}_2)$ satisfies the following properties:

1. $\Delta_{\alpha_G} (\mathbb{F}_1, \mathbb{F}_2) \geq 0$;
2. $\Delta_{\alpha_G} (\mathbb{F}_1, \mathbb{F}_2) = \Delta_{\alpha_G} (\mathbb{F}_2, \mathbb{F}_1)$;
3. $\Delta_{\alpha_G} (\mathbb{F}_1, \mathbb{F}_2) = 0 \Longleftrightarrow \mathbb{F}_1 = \mathbb{F}_2$.

43.3   The MABAC Model with Fermatean Fuzzy Numbers

Fuzzy have been extended to Fermatean fuzzy sets and they are very effective to solve many uncertain problems through scientific ways. In this paper, the conventional MABAC method [7] is built based on the Fermatean fuzzy sets. In the process of decision-making, consider m number of alternatives $(Ł_1, Ł_2, Ł_3, \dots, Ł_m)$ and n number of attributes $(\beta_1, \beta_2, \beta_3, \dots, \beta_n)$ with the weight vectors of criteria $\alpha_j (j = 1, 2, 3, \dots, n)$. E_κ indicates that κ number of experts are utilized to evaluate the alternative over criteria and their weight vectors are denoted as $\varpi_\kappa (\kappa = 1, 2, 3, \dots n)$. Then, the Fermatean fuzzy (FF)-MABAC method can be described as follows.

Step 1 Frame the Fermatean fuzzy decision matrix $DM = \left[e_{ij}^\kappa \right]_{m \times n} = (\xi_\mathbb{F}^\kappa, \psi_\mathbb{F}^\kappa)$ $\langle i = 1, 2, 3, \dots, m \rangle$; $\langle j = 1, 2, 3, \dots, n \rangle$ as shown below.

$$DM = \left[e_{ij}^\kappa \right]_{m \times n} = \begin{array}{c} \\ Ł_1 \\ Ł_2 \\ \vdots \\ Ł_m \end{array} \begin{array}{cccc} \beta_1 & \beta_2 & \dots & \beta_n \end{array} \left(\begin{array}{cccc} e_{11}^\kappa & e_{12}^\kappa & \dots & e_{1n}^\kappa \\ e_{21}^\kappa & e_{22}^\kappa & \dots & e_{2n}^\kappa \\ \vdots & \vdots & \ddots & \vdots \\ e_{m1}^\kappa & e_{m2}^\kappa & \dots & e_{mn}^\kappa \end{array} \right)$$

where $e_{ij} = \left(\xi_{\mathbb{F}_{ij}}, \psi_{\mathbb{F}_{ij}}\right)$ ($\forall i = 1, 2, 3, \ldots, m; \, j = 1, 2, 3, \ldots, n$) denotes the Fermatean fuzzy values of alternative $\mathbb{L}_i$ on $\mathbb{B}_j$ given by experts E_κ.

Step 2 Using Fermatean fuzzy aggregation operators, we can fuse overall $\left[e_{ij}^\kappa\right]_{m \times n} = \left(\xi_{\mathbb{F}}^\kappa, \psi_{\mathbb{F}}^\kappa\right)$ to $\left[e_{ij}\right]_{m \times n} = \left(\xi_{\mathbb{F}}, \psi_{\mathbb{F}}\right)$, then the fused Fermatean fuzzy decision matrix is shown as follows;

$$
\left[e_{ij}\right]_{m \times n} = \begin{array}{c} \\ \mathbb{L}_1 \\ \mathbb{L}_2 \\ \vdots \\ \mathbb{L}_m \end{array}
\begin{pmatrix}
\overset{\mathbb{B}_1}{\left(\xi_{\mathbb{F}_{11}}, \psi_{\mathbb{F}_{11}}\right)} & \overset{\mathbb{B}_2}{\left(\xi_{\mathbb{F}_{12}}, \psi_{\mathbb{F}_{12}}\right)} & \overset{\cdots}{\cdots} & \overset{\mathbb{B}_n}{\left(\xi_{\mathbb{F}_{1n}}, \psi_{\mathbb{F}_{1n}}\right)} \\
\left(\xi_{\mathbb{F}_{21}}, \psi_{\mathbb{F}_{21}}\right) & \left(\xi_{\mathbb{F}_{22}}, \psi_{\mathbb{F}_{22}}\right) & \cdots & \left(\xi_{\mathbb{F}_{2n}}, \psi_{\mathbb{F}_{2n}}\right) \\
\vdots & \vdots & \ddots & \vdots \\
\left(\xi_{\mathbb{F}_{m1}}, \psi_{\mathbb{F}_{m1}}\right) & \left(\xi_{\mathbb{F}_{m2}}, \psi_{\mathbb{F}_{m2}}\right) & \cdots & \left(\xi_{\mathbb{F}_{mn}}, \psi_{\mathbb{F}_{mn}}\right)
\end{pmatrix}
$$

Step 3 Normalize the Fermatean fuzzy decision matrix $DM = \left[e_{ij}\right]_{m \times n}$ based on the type of each attribute $\mathbb{B}_j$ using the following conditions;

$$
\text{For benefit attributes:} N_{ij} = \left[e_{ij}\right]_{m \times n} = \left(\xi_{\mathbb{F}_{ij}}, \psi_{\mathbb{F}_{ij}}\right)_{m \times n}
$$

$$
\text{For cost attributes:} N_{ij} = \left[e_{ij}\right]_{m \times n}^c = \left(\psi_{\mathbb{F}_{ij}}, \xi_{\mathbb{F}_{ij}}\right)_{m \times n}
$$

Step 4 Based on the normalized Fermatean fuzzy decision matrix $N_{ij} = \left[e_{ij}\right]_{m \times n}$ and attribute's weighting vector $\alpha_j (j = 1, 2, 3, \ldots, n)$, the weighted normalized Fermatean fuzzy decision matrix $WN_{ij} = \left(\xi'_{\mathbb{F}_{ij}}, \psi'_{\mathbb{F}_{ij}}\right)_{m \times n}$ can be determined by

$$
WN_{ij} = \alpha_j \otimes N_{ij} = \left(\sqrt[3]{1 - \left(1 - \xi_{\mathbb{F}_{ij}}^3\right)^{\alpha_j}}, \psi_{\mathbb{F}_{ij}}^{\alpha_j}\right) \tag{43.4}
$$

Step 5 Compute BAA (Border Approximation Area) values and form the BAA matrix $B = \left[\beta_j\right]_{1 \times n}$ by the following Eq. 43.5

$$
\beta_j = \left(\prod_{i=1}^m WN_{ij}\right)^{\frac{1}{m}} = \left\{\left(\prod_{i=1}^m \xi'_{\mathbb{F}_{ij}}\right)^{\frac{1}{m}}, \sqrt[3]{1 - \prod_{i=1}^m \left(1 - \xi_{\mathbb{F}_{ij}}^3\right)^{\frac{1}{m}}}\right\} \tag{43.5}
$$

Step 6 Calculate the values of $\Delta_{ij} = \left[\Delta_{ij}\right]_{m \times n}$ between alternatives and BAA matrix (β_j) using the following equation;

$$
\Delta_{ij} = \begin{cases} \Delta\left(WN_{ij}, \beta_j\right), & \text{If } WN_{ij} > \beta_j \\ 0, & \text{If } WN_{ij} = \beta_j \\ -\Delta\left(WN_{ij}, \beta_j\right), & \text{If } WN_{ij} < \beta_j \end{cases} \tag{43.6}
$$

where $\Delta\left(WN_{ij}, \beta_j\right)$ represents the distance from WN_{ij} to β_j which can be calculated by Eq. 43.3. According to the values of Δ_{ij}, we define the following:

1. If $\Delta_{ij} > 0$, then the chosen alternative lies in $\Delta^+(UAA)$ (Upper Approximation Area) ;
2. If $\Delta_{ij} = 0$, then the chosen alternative lies in $\Delta(BAA)$ (Border Approximation Area) ;
3. If $\Delta_{ij} < 0$, then the chosen alternative lies in $\Delta^-(LAA)$ (Lower Approximation Area).

Consequently, the greatest and worst alternatives are in the area of $\Delta^+(UAA)$ and $\Delta^-(LAA)$, respectively.

Step 7 Find the values of $\eth_i$ using the following equation;

$$\eth_i = \sum_{j=1}^{n} \Delta_{ij} \tag{43.7}$$

43.4 Risk Assessment of Cyber Security Technologies

Cyber security vulnerabilities have been identified as a rising issue for networks and computers in recent years. In this scenario, it is very important for cyber security technology to implement risk analysis. The following are the privacy issues involved in cyber security technologies in the literature [8]: β_1—Data exposure, β_2—Level of identification, β_3—Data sensitivity, β_4—Level of user control. Also, as provided in the research [9], we have taken the weights of criteria, respectively, as follows 0.253, 0.330, 0.276 and 0.140. From this perspective, we carried out a risk-based prioritization analysis for cyber security technologies and a list of CST are chosen as alternatives as determined by Gartner [10]: $(Ł_1)$—Cloud Workload Protection Platforms, $(Ł_2)$—Remote Browser, $(Ł_3)$—Deception, $(Ł_4)$—Network Traffic Analysis, $(Ł_5)$—Managed Detection and Response, $(Ł_6)$—Micro segmentation, $(Ł_7)$—Software-Defined Perimeters, $(Ł_8)$—Cloud Access Security Brokers, $(Ł_9)$—OSS Security Scanning and Software Composition Analysis for DevSecOps, $(Ł_{10})$—Container Security. After the criterion-alternative determination, the proposed FF-MABAC theoretical approach is carried out as follows:

Step 1 The preference relations or decision matrices are collected from the experts $(E_1 \& E_2)$ and their weight vectors are 0.6 and 0.4, respectively, according to their levels of experience, who provided the assessment values of alternatives over criteria based on Fermatean fuzzy numbers as shown in Table 43.1.

Table 43.1 Decision matrix by the experts E_1 and E_2

E_1	β_1	β_2	β_3	β_4	E_2	β_1	β_2	β_3	β_4
Ł$_1$	⟨0.8, 0.4⟩	⟨0.9, 0.3⟩	⟨0.9, 0.2⟩	⟨0.9, 0.2⟩		⟨0.7, 0.3⟩	⟨0.9, 0.1⟩	⟨0.7, 0.5⟩	⟨0.8, 0.1⟩
Ł$_2$	⟨0.8, 0.5⟩	⟨0.7, 0.5⟩	⟨0.8, 0.5⟩	⟨0.9, 0.3⟩		⟨0.7, 0.3⟩	⟨0.8, 0.6⟩	⟨0.9, 0.5⟩	⟨0.9, 0.4⟩
Ł$_3$	⟨0.6, 0.3⟩	⟨0.6, 0.2⟩	⟨0.7, 0.5⟩	⟨0.7, 0.5⟩		⟨0.7, 0.4⟩	⟨0.8, 0.5⟩	⟨0.7, 0.4⟩	⟨0.8, 0.3⟩
Ł$_4$	⟨0.7, 0.6⟩	⟨0.9, 0.2⟩	⟨0.6, 0.5⟩	⟨0.8, 0.2⟩		⟨0.8, 0.3⟩	⟨0.8, 0.3⟩	⟨0.7, 0.6⟩	⟨0.7, 0.1⟩
Ł$_5$	⟨0.8, 0.4⟩	⟨0.8, 0.3⟩	⟨0.8, 0.4⟩	⟨0.7, 0.2⟩		⟨0.7, 0.3⟩	⟨0.7, 0.2⟩	⟨0.7, 0.5⟩	⟨0.8, 0.5⟩
Ł$_6$	⟨0.8, 0.6⟩	⟨0.9, 0.4⟩	⟨0.7, 0.5⟩	⟨0.8, 0.2⟩		⟨0.8, 0.5⟩	⟨0.9, 0.2⟩	⟨0.6, 0.6⟩	⟨0.7, 0.2⟩
Ł$_7$	⟨0.9, 0.1⟩	⟨0.9, 0.3⟩	⟨0.9, 0.5⟩	⟨0.9, 0.5⟩		⟨0.9, 0.3⟩	⟨0.8, 0.5⟩	⟨0.8, 0.3⟩	⟨0.9, 0.6⟩
Ł$_8$	⟨0.9, 0.3⟩	⟨0.8, 0.5⟩	⟨0.8, 0.4⟩	⟨0.9, 0.1⟩		⟨0.9, 0.4⟩	⟨0.7, 0.6⟩	⟨0.7, 0.6⟩	⟨0.9, 0.2⟩
Ł$_9$	⟨0.9, 0.5⟩	⟨0.7, 0.5⟩	⟨0.7, 0.3⟩	⟨0.7, 0.3⟩		⟨0.9, 0.3⟩	⟨0.8, 0.6⟩	⟨0.6, 0.8⟩	⟨0.8, 0.5⟩
Ł$_{10}$	⟨0.9, 0.4⟩	⟨0.9, 0.4⟩	⟨0.8, 0.6⟩	⟨0.8, 0.4⟩		⟨0.9, 0.6⟩	⟨0.8, 0.2⟩	⟨0.9, 0.5⟩	⟨0.7, 0.3⟩

Table 43.2 The fused Fermatean fuzzy decision matrix

$DM = [e_{ij}]$	β_1	β_2	β_3	β_4
Ł$_1$	⟨0.7675, 0.3660⟩	⟨0.9000, 0.2547⟩	⟨0.8457, 0.3758⟩	⟨0.8667, 0.1732⟩
Ł$_2$	⟨0.7675, 0.4389⟩	⟨0.7489, 0.5431⟩	⟨0.8472, 0.5000⟩	⟨0.9000, 0.3466⟩
⋮	⋮	⋮	⋮	⋮
Ł$_{10}$	⟨0.9000, 0.4955⟩	⟨0.8667, 0.3454⟩	⟨0.8472, 0.5629⟩	⟨0.7675, 0.3660⟩

Step 2 Using aggregation operators Eq. (43.2), we can fuse overall $\left[e_{ij}^{\kappa}\right]_{m \times n} = \left(\xi_{\mathbb{F}}^{\kappa}, \psi_{\mathbb{F}}^{\kappa}\right)$ to $\left[e_{ij}\right]_{m \times n} = (\xi_{\mathbb{F}}, \psi_{\mathbb{F}})$, then the fused Fermatean fuzzy decision matrix is shown in Table 43.2 when the value of $\eta = 2$.

Step 3 The provided attributes β_j; $(j = 1, 2, 3, \ldots, n)$ are of the same kind and hence the normalized Fermatean fuzzy decision matrix $DM = N_{ij} = \left[e_{ij}\right]_{m \times n}$ is same as mentioned in Table 43.2.

Step 4 The normalized Fermatean fuzzy decision matrix $N_{ij} = \left[e_{ij}\right]_{m \times n}$ and attribute's weighting vector $\alpha_j (j = 1, 2, 3, 4) = (0.253, 0.330, 0.276, 0.140)^T$ are used to compute the Fermatean fuzzy weighted normalized decision matrices $WN_{ij} = \left(\xi'_{\mathbb{F}_{ij}}, \psi'_{\mathbb{F}_{ij}}\right)_{m \times n}$ by Eq. 43.4 as follows;

$$WN_{11} = \alpha_1 \otimes N_{11} = \left(\sqrt[3]{1 - \left(1 - 0.7675^3\right)^{0.253}}, 0.3660^{0.253}\right) = \langle 0.5207, 0.775 \rangle$$

Similarly, we can derive the values of WN_{ij} as shown in Table 43.3

Step 5 Eq. 43.5 can be used to calculate BAA values and the BAA matrix $B = \left[\beta_j\right]_{1 \times n}$ as follows;

Table 43.3 The weighted normalized Fermatean fuzzy decision matrix

WN_{ij}	β_1	β_2	β_3	β_4
Ł₁	⟨0.5207, 0.7755⟩	⟨0.7048, 0.6368⟩	⟨0.6092, 0.7633⟩	⟨0.5156, 0.7823⟩
Ł₂	⟨0.5207, 0.8119⟩	⟨0.5473, 0.8175⟩	⟨0.6107, 0.8259⟩	⟨0.5507, 0.8621⟩
⋮	⋮	⋮	⋮	⋮
Ł₁₀	⟨0.6552, 0.8372⟩	⟨0.6646, 0.7041⟩	⟨0.6107, 0.8533⟩	⟨0.4323, 0.8687⟩

Table 43.4 The Fermatean fuzzy hamming distance between $\Delta_{\alpha_{Hij}}$ and β_j

$\Delta_{\alpha_{Hij}}$	β_1	β_2	β_3	β_4
Ł₁	0.0178	0.0472	0.0343	0.0198
Ł₂	−0.0073	−0.0481	0.0163	0.0089
⋮	⋮	⋮	⋮	⋮
Ł₁₀	0.0356	0.0199	0.0305	−0.0058

$$\beta_1 = \left\{ \left(\prod_{i=1}^{10} \xi'_{\mathbb{F}_{i1}} \right)^{\frac{1}{10}}, \sqrt[3]{1 - \prod_{i=1}^{10} \left(1 - \xi^3_{\mathbb{F}_{i1}}\right)^{\frac{1}{10}}} \right\} = \langle 0.5609, 0.8009 \rangle$$

Similarly, $\beta_2 = \langle 0.6107, 0.7378 \rangle$, $\beta_3 = \langle 0.5308, 0.8245 \rangle$ and $\beta_4 = \langle 0.4685, 0.8525 \rangle$

Step 6 Calculate the values of $\Delta_{ij} = \left[\Delta_{ij} \right]_{m \times n}$ between alternatives and the values of BAA using Eq. 43.6; For example, the Fermatean fuzzy hamming distance measure $\Delta_{\alpha_{H11}}$ is shown below

$$\Delta_{\alpha_{H11}} (WN_{11}, \beta_1) = \left[\frac{0.253}{2} \left(\left|0.5207^3 - 0.5609^3\right|^\lambda + \left|0.775^3 - 0.8009^3\right|^\lambda + \left|0.8152^3 - 0.8432^3\right|^\lambda \right) \right]^{\frac{1}{\lambda}}$$

where $\lambda = 1$, then $\Delta_{\alpha_{H11}} (WN_{11}, \beta_1) = 0.0178$. Similarly, all other values calculated and are shown in Table 43.4.

Step 7 The values of $\eth_i$ are obtained using Eq. 43.7

$$\eth_1 = 0.0178 + 0.0472 + 0.0343 + 0.0198 = 0.1190.$$

Similarly, $\eth_2 = -0.0302$, $\eth_3 = -0.0940$, $\eth_4 = 0.0267$, $\eth_5 = 0.0750$, $\eth_6 = 0.0053$, $\eth_7 = 0.0633$, $\eth_8 = -0.0003$, $\eth_9 = -0.0556$ and $\eth_{10} = 0.0802$. Obviously, the best and the worst alternatives lie in the area of $\Delta^+(UAA) = 0.1190$ and $\Delta^-(LAA) = -0.0940$, respectively.

43.5   Sensitive Analysis

The performance of the proposed FF-MABAC model is checked with $FFYWG_\alpha$ aggregation operator and we have calculated the final results as follows: $\eth_1 = 0.1150$, $\eth_2 = -0.0294$, $\eth_3 = -0.0560$, $\eth_4 = 0.0292$, $\eth_5 = 0.0813$, $\eth_6 = 0.0178$, $\eth_7 = 0.0633$, $\eth_8 = 0.0044$, $\eth_9 = -0.0728$ and $\eth_{10} = 0.0758$. The rank of alternatives by Fermatean fuzzy aggregation operators are listed in Table 43.5. We can see from Fig. 43.1 and Table 43.5 that the rank of the most alternatives are same and slight variation in results exists due to the preference of experts' values. Here the best alternative is $Ł_1$ and it remains the same when using $FFYWA_\alpha$ and $FFYWG_\alpha$ operators. Also, this sensitive analysis implies that the proposed FF-MABAC model is authentic and validates the results obtained.

Table 43.5 The rank of the alternatives

Methods	The ranking order
FF-MABAC—$FFYWA_\alpha$	$Ł_1 > Ł_{10} > Ł_5 > Ł_7 > Ł_4 > Ł_6 > Ł_8 > Ł_2 > Ł_9 > Ł_3$
FF-MABAC—$FFYWG_\alpha$	$Ł_1 > Ł_5 > Ł_{10} > Ł_7 > Ł_4 > Ł_6 > Ł_8 > Ł_2 > Ł_3 > Ł_9$

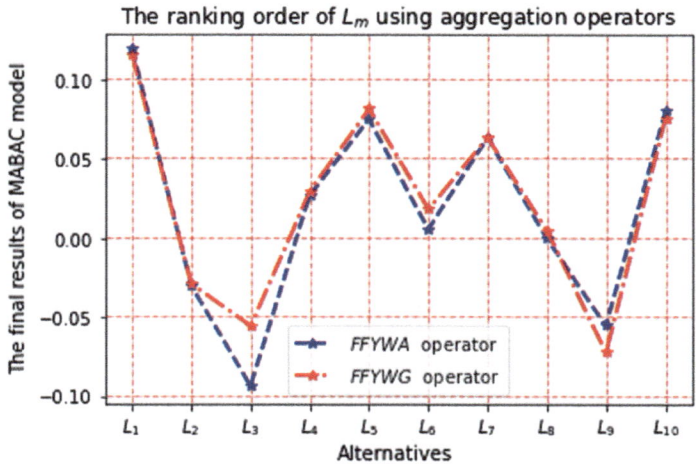

Fig. 43.1 Comparison results of FF-MABAC model using aggregation operators

43.6   Conclusion

In this article, we have developed an MCDM model called FF-MABAC and it is used in the process of selecting the most secure and best cyber security technologies. In order to give best cyber security, our proposed FF-MABAC model attempts to fulfill the need in the fields of information security. Also the scientific computing of the model is explained with all its supplements. The real-time application of FF-MABAC in the risk assessment of CST is demonstrated numerically. From the sensitive analysis, the proposed model is strongly validated and the results obtained are authenticated. In future, unsupervised techniques could be used to extract the expert values from the fields of respective domains and further it can be utilized to find new decision-making model with the help of fuzzy sets.

References

1. Alali, M., Almogren, A., Hassan, M.M., Rassan, I.A.L., Bhuiyan, M.Z.A.: Improving risk assessment model of cyber security using fuzzy logic inference system. Comput. Secur. **74**, 323–339 (2018)
2. Zadeh, L.A.: Fuzzy sets. Inf. Control **8**, 338–353 (1965)
3. Atanassov, K.T.: Intuitionistic fuzzy sets. Fuzzy Sets Syst. **20**(1), 8796 (1986)
4. Yager, R.R.: Pythagorean fuzzy subsets. Joint IFSA World Congress and NAFIPS Annual Meeting, Edmonton, Canada (2013)
5. Senapati, T., Yager, R.R.: Fermatean fuzzy sets. J Ambient Intell Human Comput **11**, 663–674 (2020). https://doi.org/10.1007/s12652-019-01377-0
6. Garg, H., Shahzadi, G., Akram, M.: Decision-making analysis based on fermatean fuzzy yager aggregation operators with application in COVID-19 testing facility. Mathematical Problems in Engineering (2020)
7. Wang, J., Wei, G., Wei, C., Wei, Y.: MABAC method for multiple attribute group decision making under q-rung orthopair fuzzy environment. Defence Technology 16(1), (2020). https://doi.org/10.1016/j.dt.2019.06.019
8. Toch, E., Bettini, C., Shmueli, E., Radaelli, L., Lanzi, A., Riboni, D., Lepri, B.: The privacy implications of cyber security systems: a technological survey. ACM Comput. Surv. **51**(2), 1–27 (2018)
9. Erdogan, M., Karasan, A., Kaya, I., Budak, A., Çolak, M.: A Fuzzy Based MCDM Methodology for Risk Evaluation of Cyber Security Technologies. In International Conference on Intelligent and Fuzzy Systems (pp. 1042-1049). Springer, Cham (2019, July). https://doi.org/10.1007/978-3-030-23756-1-123
10. Gartner Identifies the Top Technologies for Security in 2017. https://www.gartner.com/en/newsroom/press-releases/2017-06-14-gartner-identifies-the-top-technologies-for-security-in

Chapter 44
Skin Cancer Image Classification Using Deep Neural Network Models

Mayank Upadhyay, Jyoti Rawat, and Srabanti Maji

Abstract Skin cancer is among the most prominent types of cancer, which if not detected in the initial stages can metastasize and become fatal. As the number of Melanoma cases surges, the interest in using CAD (computer-aided diagnosis) for skin cancer prognosis is also rising. In this work, the proposed methodology is applied to HAM 10,000 dataset containing skin lesion images. This paper aims to implement neural network architectures on pre-processed segmented lesions for labeling the lesion image as benign or malignant. The images from the dataset are first pre-processed to remove the noise present, and then segmentation is performed to separate skin lesion from the background. Textural features are extracted from the segmented lesion, which are later used by perceptron and multilayer perceptron (MLP) for classification. Pre-processed images are classified using transfer learning models (ResNet-50, Inception-V3, MobileNet, Inception-ResNet-V2, and DenseNet201), where DenseNet201 has given the best performance and achieved an accuracy of 93.24% and AUC of 0.932.

44.1 Introduction

The skin lesion is described as an abnormal growth of cell on the skin surface. It can be benign which rarely metastasize or malignant which usually metastasize, i.e., it can spread to other organs [1]. There are different classes of skin lesion, some of them are benign keratosis (BKL), dermatofibroma (DF), nevus (NV), and melanoma of which BKL, NV, and DF are benign, i.e., non-cancerous and melanoma is malignant, i.e., cancerous. Melanoma is considered the most precarious type as it can easily metastasize. The main role of melanocytes cell is to create melanin which gives skin its dark color and protects it from UV rays, and irregular growth of melanocytes cell causes melanoma [2]. In the US alone around 207,396 cases of melanoma are estimated with 7,180 deaths in 2021 [3]. Most of the melanoma cases are initially

M. Upadhyay · J. Rawat · S. Maji (✉)
School of Computing, DIT University, Dehradun, India
e-mail: srabanti.maji@dituniversity.edu.in

detected by non-specialist or through self-diagnosis making it vital to raise attention toward skin cancer [4]. There are several methods for diagnosing skin lesion but differentiating benign lesion from malignant ones remains a challenging task as the true negative rate of identifying malignant lesion by an expert dermatologist is around only 59% [5].

In recent years deep-learning techniques like convolution neural network (CNN) have been extensively used in image classification problems due to the advancement in computer hardware, and they have performed exceptionally well in visual tasks like pattern recognition and object classification. CNN is the most conventional and common deep-learning multilayer neural network architecture [6]. Various CNN models have more accurately classified skin lesion images than trained professionals. Techniques like transfer learning where the framework is pre-trained on a large dataset have further improved the classification performance of CNN models.

DenseNet201 is a transfer learning CNN model that is prepared on millions of images belonging to the ImageNet database. The model extends 201 layers, which is capable of labeling objects into 1000 categories. The model takes images of size 244×244.

44.2 Literature Survey

Classification of skin lesion as benign or malignant can be achieved by labeling the data based on extracted features from the skin lesion. One of the challenges in skin cancer prognosis is the presence of noise in the image. In [7], for hair detection median filter and bottom hat filtering were applied on grayscale image and region filling for removal of detected hair. For image segmentation, Otsu's thresholding followed by Geodesic Active Contours (GAC) was applied. In the classification step, SVM using features extracted by HOG and GLCM gave the best result with 97.8% accuracy.

The benefit of CNNs is that they can extract features automatically but the disadvantage of CNNs is the requirement of a large data sample. Many of the papers implemented architectures that were pre-trained on a large dataset (transfer learning models), Sae-Lim et al., together with using modified MobileNet architecture which is a pre-trained model also applied data augmentation and data-up sampling in pre-processing to deal with problems of imbalanced data and model overfitting. MobileNet provides a quid pro quo between less computational time and some accuracy lost. They proved that their model classifies skin cancer better than the original MobileNet model and also showed that data pre-processing increases the accuracy of the model [8].

A computer-aided detection system can be used for the prognosis of skin cancer. Aminur Rab Ratul et al. showed that compared to traditional CNNs, dilated convolution provides better accuracy with the same computational complexity. Dataset used was HAM10000 which is imbalanced, so they used the stratified method for splitting the dataset. For classification implemented Inception-V3, MobileNet, VGG16, and

VGG19 models with dilated convolution, out of which Inception-V3 outperformed all the other three models in all criteria [9].

44.3 Proposed Methodology

This portion explains the procedure involved in the classification of lesion images as benign or malignant, also shown in Fig. 44.1.

44.3.1 Data Acquisition

The dataset used is HAM 10,000 (Human Against Machine), which can be accessed online [10] and was gathered by Philipp et al. in a duration of 20 years from MedUni Vienna and C Rosendahl's skin cancer practice. More than half of the data was validated by the means of histopathology, and the remaining by either confocal, consensus, or 3 follow-ups in a duration of 1.5 years [2].

The dataset contains metadata file and skin lesion images. The metadata file had records regarding each image data with attributes like age, sex, location of the lesion, type of lesion (dx), and method used to identify the type of lesion (dx_type). The image dataset had 10,015 skin lesion images with 600 × 450 dimensions in JPG format, comprised of seven classes defined in the introduction section. Images of nevus and melanoma classes were taken from the dataset.

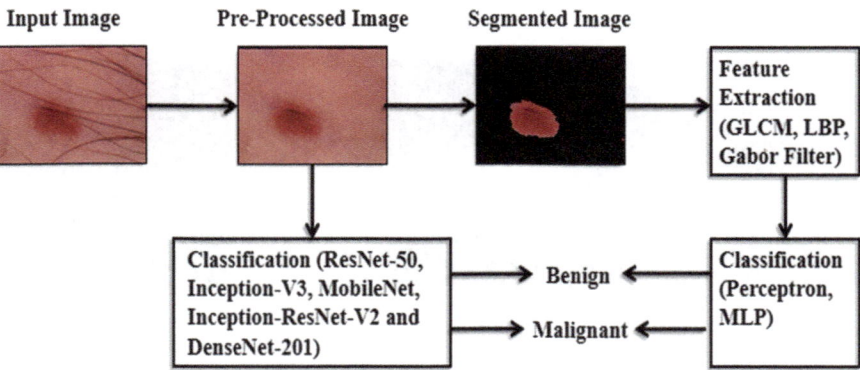

Fig. 44.1 The procedure involved in classifying the skin lesion image dataset

44.3.2 Pre-processing

The first step in skin lesion image classification is pre-processing, which is done to remove unwanted information from the image and highlight the skin lesion [11, 12]. In this step artefacts like hair, air bubble, blood vessel, and skin line are removed as their presence can result in reduced classification accuracy [13]. The captured image might not be properly illuminated so image enhancement techniques are implemented to improve the contrast between the lesion border and surrounding. The acquired images are resized into 244 × 244 dimension and, a blackhat operation is applied on the monochrome of the image to define hair contour, then binary thresholding was applied to intensify the obtained hair contour, after which the value of each pixel in contour is replaced by the normalized weighted sum of the neighborhood pixels [14].

44.3.3 Segmentation

The second phase in skin lesion image classification is segmentation, where the pre-processed image is taken and the region of interest (ROI) is extracted [15, 16]. For segmentation of the skin lesion, applied Otsu's thresholding on the pre-processed image to obtain the binarized image. This binarized image is applied with the morphological closing operation for smoothening the boundary. To define the border of the lesion used GAC (Geodesic Active Contours), which measures the highest variation in the overall skin tumor. Figure 44.2 shows the result after segmentation steps are applied to pre-processed image.

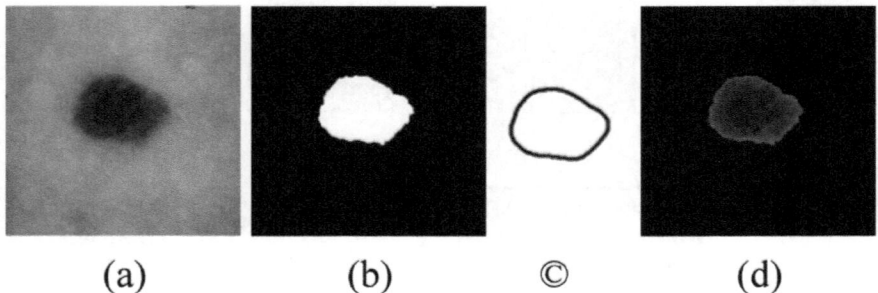

(a)   (b)   ©   (d)

Fig. 44.2 Extracted ROI from a lesion image **a** Pre-processed image **b** Segmented lesion area **c** GAC **d** Final output

44.3.4 Feature Extraction

The third step in skin lesion image classification is feature extraction. The extracted features from the segmented lesion are textural features. The feature extraction techniques employed were: GLCM, LBP, and Gabor Filter and in total 9 features were extracted from each image.

44.3.4.1 GLCM (Gray Level Co-occurrence Matrix)

GLCM is used to obtain textural features, which are variation in pixel intensity at the local level. GLCM measures the occurrence of a grey level pixel having intensity value k with another pixel of intensity value l. Features obtained using GLCM are correlation, energy, homogeneity, dissimilarity, and contrast [17, 18].

44.3.4.2 LBP (Local Binary Patterns)

LBP is a structural textural feature extraction technique predominantly used on the greyscale image. It works by thresholding the pixel intensity with its neighbor's intensity resulting in an eight-bit (let's assume 3×3 window) binary code, called LBPs. The histogram is calculated and normalized over this cell. The LBP features extracted were energy and entropy.

44.3.4.3 Gabor Filter

Gabor filter is beneficial in several domains of biometric image processing like fingerprint identification, texture classification, and edge detection. They are used for textural feature extraction. The features extracted were energy and entropy.

44.3.5 Classification

The final step in skin lesion image classification is this phase; the information obtained from the previous steps is utilized for skin lesion prognosis.

44.3.5.1 Perceptron Model

A perceptron is the basic unit of an artificial neural network. The model used is a 10-class classification with one-hot encoding, i.e., total of 10 perceptrons are present at the output layer and at a time only one neuron will have a value closer to one and the rest of the neurons will have values closer to zero with the final layer being SoftMax which is used for binary classification, with the learning rate value as 1. From each image 9 features were extracted and stored into a pickle file which was later fed to the perceptron model for classification.

44.3.5.2 Multilayer Perceptron Model

MLP is a combination of single-layer perceptrons. The model used has 10 input nodes and one hidden layer where activation function RELU is used and SoftMax for binary classification in the output layer. For any given value v the activation function RELU provides output as $MAX(0, v)$, thus resolving the problem like vanishing gradient.

44.3.5.3 Transfer Learning

The models of transfer learning are employed for the skin lesion image classification. The transfer learning models are pre-trained on a large dataset, thus acquiring knowledge in the process and so the learning process does not start from scratch when working on a new classification problem. Transfer learning is particularly significant in the domain of medical image processing due to the lack of image data available that is required to train the deep-learning architectures. The transfer learning models used were—ResNet-50, Inception-V3, MobileNet, Inception-ResNet-V2, and DenseNet201 for the present work.

The pre-processed images were utilized, and the top layer of the transfer learning models were replaced with average pooling and finally SoftMax to classify an image as either begin or malignant. The learning rate was set to 0.0001 and Adam optimizer was used with the dropout rate set to 60%.

44.3.6 Experimental Results

The presented classification approach is applied to images from HAM 10,000 dataset. From the dataset, 1200 images of melanoma class and 1200 images of nevus class were taken, out of which 222 images (equal nevus and melanoma class) were used for testing the model.

As explained in the proposed methodology section the images were first pre-processed to remove noise and segmentation was applied to obtain ROI, then texture features were extracted using GLCM, LBP, and Gabor Filter which were stored into a pickle file. Several neural network models were applied to this dataset like Perceptron, MLP, Inception-V2, Inception-ResNet-V2, ResNet-50, MobileNet, and DenseNet201. To improve performance, the dropout rate was set to 60% and batch normalization was applied. For loss function, binary cross entropy was employed. In the training phase, hyperparameters like the optimizer, batch size, learning rate, and the number of epochs were changed to achieve better accuracy. Evaluation metrics like accuracy, precision, recall, F1-score, and ROC (Receiver Operating Characteristic) curve are utilized to measure the efficiency of the classification model.

The classification accuracy of 93.24% and AUC of 0.932 is attained from DenseNet201 in contrast to other models. The batch size is defined as no. of images taken in one iteration and training the model with different batch size provides different classification accuracy see Table 44.1. Learning rate is another hyperparameter that needs to be adjusted as it decides how fast global minima is reached. Table 44.2 provides a comparison between different learning rates. Optimizers are used to update weight in order to reduce loss, moreover different optimizer converge differently. Table 44.3 shows the accuracy achieved by the classification model using different optimizers. Also, perceptron achieved 85.806% accuracy for 50 epoch and MLP achieved 73.02% accuracy.

To examine the learning and performance of the DenseNet201 model, Fig. 44.3 (a) accuracy on the training and validation set are plotted on the number of epoch vs accuracy graph. The model displays an excellent learning rate as the training and validation accuracy increases with the number of epochs (b) training and validation loss are plotted showing them downward sloping to point of stability. The reduced

Table 44.1 Comparison of accuracy with different batch size

Batch Size	Accuracy
8	0.88
16	**0.93**

Table 44.2 Comparison of accuracy with different learning rate

Learning rate	Accuracy
0.00001	0.88
0.0001	**0.93**

Table 44.3 Comparison of accuracy with different optimizers

Optimizers	Accuracy
SGD (Stochastic gradient descent)	0.81
Adam (Adaptive Moment Estimation)	**0.93**

Table 44.4 Comparison of classification models

Model	Precision	Recall	F1-score	Accuracy	AUC
Inception-V3	0.93	0.79	0.85	0.86	0.865
Inception- ResNet- V2	0.93	0.87	0.90	0.91	0.905
ResNet-50	0.92	0.86	0.89	0.89	0.892
MobileNet	0.89	0.91	0.90	0.90	0.896
DenseNet201	**0.93**	**0.92**	**0.93**	**0.93**	**0.932**

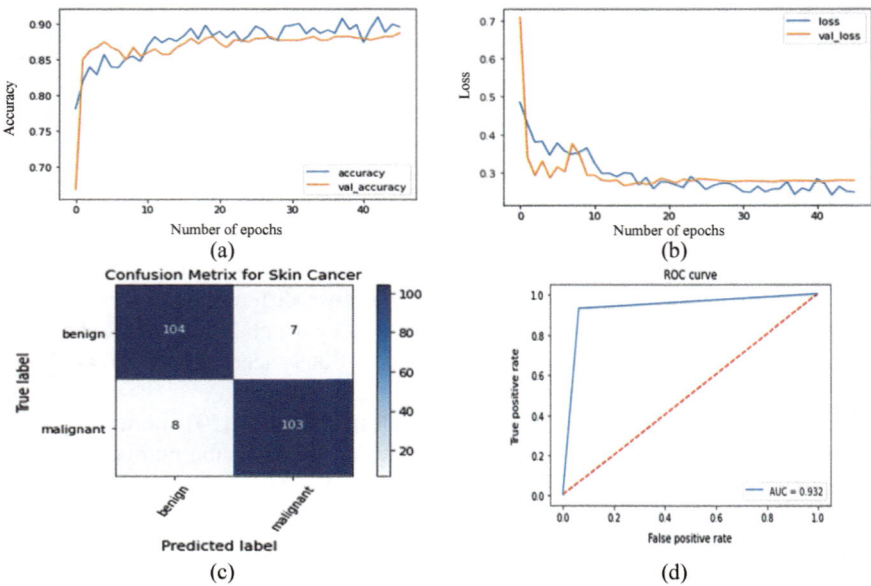

Fig. 44.3 DenseNet201 performance on skin lesion image classification **a** test accuracy and validation accuracy **b** test loss and validation loss **c** confusion matrix **d** ROC curve

gap between training and validation loss shows that the model can classify new images with high accuracy.

AUROC (Area Under ROC curve) curve is a plot between True Positive Rate (TPR) and False Positive Rate (FPR) used to measure the performance of a binary classifier. In the ROC curve threshold value between 0 and 1 is taken to classify output as benign or malignant. ROC curve in particular is advantageous in the field of medical image processing as it presents a graphical tradeoff between TPR and FRP. The more area under the curve (AUC), the better the model.

44.4 Conclusion

In this work neural network and transfer learning models were employed to classify the skin lesion images as either benign or malignant. The proposed method was first to eliminate noise present in the dataset and obtain the skin lesion, from the segmented image textural feature were extracted using GLCM, LBP, and Gabor. The extracted features were used by Perceptron and MLP. The pre-processed images were classified as benign or malignant using transfer learning architectures. DenseNet201 gave better performance than any other model with an overall accuracy of 93%. The weighted average of precision, recall, and f1-score were 93%, 93%, and 93%, respectively. Future research includes utilizing all the images in the dataset and applying data augmentation techniques to deal with the imbalance in the dataset for developing a multiclass skin lesion classifier.

References

1. Cancer. World Health Organization. https://www.who.int/health-topics/cancer#tab=tab_1. Accessed July 06, 2020
2. Tschandl, P., Rosendahl, C., Kittler, H.: Data descriptor: The HAM10000 dataset, a large collection of multi-source dermatoscopic images of common pigmented skin lesions. Sci. Data, vol. 5 (2018). https://doi.org/10.1038/sdata.2018.161
3. "Cancer Facts and Figures. American Cancer Society, 2021. https://www.cancer.org/content/dam/cancer-org/research/cancer-facts-and-statistics/annual-cancer-facts-and-figures/2021/cancer-facts-and-figures-2021.pdf. Accessed Feb. 23, 2021
4. Titus, L.J., et al.: Recent skin self-examination and doctor visits in relation to melanoma risk and tumour depth. Br. J. Dermatol. **168**(3), 571–576 (2013). https://doi.org/10.1111/bjd.12003
5. Bhattacharya, A., Young, A., Wong, A., Stalling, S., Wei, M., Hadley, D.: Precision Diagnosis of Melanoma And Other Skin Lesions From Digital Images.," AMIA Jt. Summits Transl. Sci. proceedings. AMIA Jt. Summits Transl. Sci. vol. 2017, pp. 220–22 (2017). Accessed: Feb. 13, 2021. http://www.ncbi.nlm.nih.gov/pubmed/28815132.
6. Xin, M., Wang, Y.: Research on image classification model based on deep convolution neural network. Eurasip J. Image Video Process. **2019**(1), 1–11 (2019). https://doi.org/10.1186/s13640-019-0417-8
7. V. Ma and M. V. Karki, "Skin Cancer Detection using Machine Learning Techniques," Jul. 2020. https://doi.org/10.1109/CONECCT50063.2020.9198489.
8. W. Sae-Lim, W. Wettayaprasit, and P. Aiyarak, "Convolutional Neural Networks Using MobileNet for Skin Lesion Classification," in JCSSE 2019 - 16th International Joint Conference on Computer Science and Software Engineering: Knowledge Evolution Towards Singularity of Man-Machine Intelligence, Jul. 2019, pp. 242–247. https://doi.org/10.1109/JCSSE.2019.8864155.
9. Ratul, M.A.R., Mozaffari, M.H., Lee, W.S., Parimbelli, E (2019) Skin lesions classification using deep learning based on dilated convolution. *bioRxiv*. bioRxiv, p. 860700. https://doi.org/10.1101/860700.
10. "The HAM10000 dataset, a large collection of multi-source dermatoscopic images of common pigmented skin lesions - ViDIR Dataverse. https://doi.org/10.7910/DVN/DBW86T. Aaccessed Feb. 14, 2021.
11. Rawat, J., Singh, A., Bhadauria, H.S., Virmani, J., Devgun, J.S.: Classification of acute lymphoblastic leukaemia using hybrid hierarchical classifiers. Multimed. Tools Appl. **76**(18), 19057–19085 (Sep. 2017). https://doi.org/10.1007/s11042-017-4478-3
12. Rajinikanth, V., Satapathy, S.C., Dey, N., Fernandes, S.L., Manic, K.S.: Skin melanoma assessment using Kapur's entropy and level set—A study with bat algorithm. Smart Innovation, Systems and Technologies **104**, 193–202 (2019). https://doi.org/10.1007/978-981-13-1921-1_19
13. Ruela, M., Barata, C., Marques, J.S., Rozeira, J.: A system for the detection of melanomas in dermoscopy images using shape and symmetry features. Comput. Methods Biomech. Biomed. Eng. Imaging Vis. **5**(2), 127–137 (Mar. 2017). https://doi.org/10.1080/21681163.2015.1029080
14. Telea, A.: An Image Inpainting Technique Based on the Fast Marching Method. J. Graph. Tools **9**(1), 23–34 (Jan. 2004). https://doi.org/10.1080/10867651.2004.10487596
15. Rawat, J., Virmani, J., Singh, A., Bhadauria, H.S., Kumar, I., Devgan, J.S.: FAB classification of acute leukemia using an ensemble of neural networks. Evol. Intell. (2020). https://doi.org/10.1007/s12065-020-00491-9

16. Rawat, J., Singh, A., Bhadauria, H.S., Virmani, J., Devgun, J.S.: Computer assisted classification framework for prediction of acute lymphoblastic and acute myeloblastic leukemia. Biocybern. Biomed. Eng. vol. 37, no. 4, pp. 637–654 (2017). https://doi.org/10.1016/j.bbe.2017.07.003.
17. Haralick, R.M., Shanmugam, K., Dinstein, I.H.: Textural Features for Image Classification. IEEE Trans. Syst. Man Cybern. SMC-3(6), 610–621 (1973). https://doi.org/10.1109/TSMC.1973.4309314.
18. Rawat, J., Singh, A., Bhadauria, H.S., Virmani, J., Devgun, J.S.: Leukocyte Classification using Adaptive Neuro-Fuzzy Inference System in Microscopic Blood Images. Arab. J. Sci. Eng. 43(12), 7041–7058 (2018). https://doi.org/10.1007/s13369-017-2959-3

Chapter 45
Application of Deep Learning in Detection of Covid-19 Face Mask

Anuja Jana Naik and M. T. Gopalakrishna

Abstract With the increase in the number of Covid-19 cases throughout the globe wearing face masks has proved to be effective in the prevention of the virus. In this work, we have originated a method that can detect if people are violating the rule of wearing a mask outdoors using a two-stage deep learning system. The first stage of the system detects different faces present in the input image using YOLO (You Only Look Once) model trained for the face detection and returns face ROIs. In the second stage extracted face ROI is passed through face mask detector model trained using MobileNetV2 which in turn classifies it as Mask or No mask. The dataset used for training the mask detector model is Real-World Masked Face Dataset (RMFD) and for Face Detection model is the WIDER dataset. The proposed method gives 98% accuracy for mask detection. The promising results derived from the proposed model demonstrate that the deployment of the model can be done in real-time systems.

45.1 Introduction

Video 2019-nCoV is identified as a new coronavirus spreading like a wildfire across the world. It is caused by severe acute respiratory syndrome coronavirus 2 (SARS-CoV-2). The natural home of virus origin is suspected to be a specific breed of the bat which might have transfused to humans via some intermediate host. Every single day there is an addition to the number of cases detected throughout the world of novel coronavirus which actually originated from the Chinese city of Wuhan in the month of December. Research in the "The Lancet" predicts the rate of transmission to be 2 to 3 persons per positive case [1]. As per Worldometer in India, the total count of positive cases has surpassed 3.33 lakhs making its place among the top 5 nations in the world for the highest cases. As of now, there is no drug approved by

A. J. Naik (✉)
Department of Computer Science and Engineering, S.J.B.Institute of Technology, Bengaluru, India

M. T. Gopalakrishna
Visvesavaraya Technological University, Belgavi, Karnataka, India

© The Author(s), under exclusive license to Springer Nature Singapore Pte Ltd. 2022   461
V. Bhateja et al. (eds.), *Evolution in Computational Intelligence*,
Smart Innovation, Systems and Technologies 267,
https://doi.org/10.1007/978-981-16-6616-2_45

the Food and Drug Administration (FDA) for Covid-19. However, a lot of studies and investigations are being carried out for Covid-19 treatment around the world.

As a preventive measure, a nationwide lockdown was enforced by the Indian government and is currently in its fifth phase with partial Unlocks. Many of the states with lockdown still on have trajectories dropping which indicates preventive measures like social distancing can help fight the pandemic. With the ease in the lockdown, people should be even more cautious and continue to maintain physical distancing, wear masks, and avoid going outdoors until unavoidable.

As per the World Health Organization (WHO) guidelines [2] the most effective practices to avoid the spread of Covid-19 disease are:

- In public places especially in the areas having active coronavirus cases avoid direct unprotected contact with anyone by maintaining social distancing.
- Always cover mouth and nose with a mask while outdoors to avoid any droplets or virus entering your body through nose or mouth.
- Wash hands frequently using sanitizer.
- Incase of health worker wearing shields to protect eyes is a must.

As per the research published by [3] more than social distancing or imposing lockdowns, wearing masks is important to prevent the spread of the virus. Infection trends shifted dramatically when mask-wearing rules were implemented as shown in Fig. 45.1. R0 is a reproduction number which is an epidemiological measure of spread. It basically determines the rate of growth, with a super-linear effect in simple terms the number of cases increased from one infected case. Initially, R0 is 2.4 which is gradually changing as indicated by different colors.

In this paper, the problem of monitoring if people are wearing face masks when outdoors is monitored. The Covid-19 Face Mask detector system uses a two-stage Deep Learning model to check if the person is wearing a mask. A Face Mask detector

Fig. 45.1 Ramification of wearing a mask in public under the full range of mask adherence and efficacy scenarios [3]

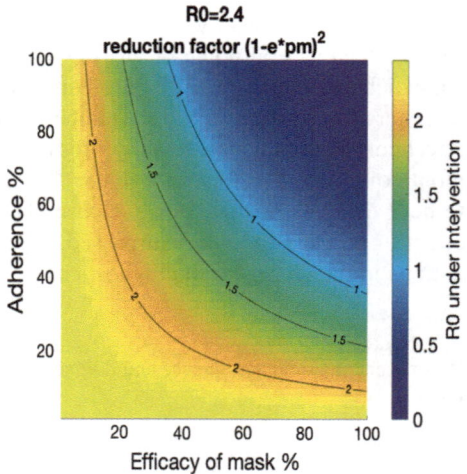

system has direct applicability in the surveillance domain. It can be installed at the entry point of the airports, railway stations, or bus stations to detect if commuters or staff are without masks. It can be also deployed in offices, schools, or health care firms to monitor the staff breaching the guidelines especially in the quarantine wards if patients are wearing masks or not. An appropriate alert can be sent to the concerned authority and required action can be taken.

The paper is organized as follows. The existing work related to facts about wearing mask and face detection algorithms are discussed in Sect. 45.2. The description about the proposed method is given in Sect. 45.3. Experimental results are summarized in Sect. 45.4 with outputs in Sect. 45.5 and conclusion in Sect. 45.6.

45.2 Related Work

45.2.1 Covid-19

On March 11, 2020, a novel health crisis emerged due to the outbreak of the severe acute respiratory syndrome coronavirus 2 (SARS-CoV-2) or novel coronavirus (2019-nCoV). The disease is circulated by inhalation of droplets present in the air which are spread by coughing or sneezing of an infected person. The viral infection moves from the nasal passage and throat to the lungs [4]. As per recent literature from various researchers on wearing masks in public, it is proposed that wearing a face mask can lead to a reduction of transmission of viral infections. Community transmissions can be further reduced if infected people who are Symptomatic as well as Asymptomatic wear masks. However, the effect might be reduced if not used correctly or can even lead to an increase in transmissions as people tend to touch their faces multiple times. A mask that is worn for a long time can become moist and touching such a contaminated mask can help spread the virus.

Due to the shortage of surgical masks people are encouraged to make homemade masks but there is no certainty whether these homemade masks can protect humans from the virus. A plethora of reviews have been undertaken on whether wearing face masks are beneficial or not. In a study by [5] efficacy of three types of masks in blocking avian influenza virus (AIV) in aerosols and the efficacy of instant hand wiping in removing AIV from hands is evaluated. AIV is used as a substitute for SARS-Cov-2 since both are enveloped and pleomorphic spherical viruses with a diameter of around 80–20 nm. Results by [6] suggested face masks should be used along with other non-pharmaceutical practices such as following social distancing. Another review by [7] assessed the effect of mask usage on a healthy population of the community.

Most researchers concluded their studies by proposing that wearing a mask is important until an effective vaccine against coronavirus is marketed. The study also indicated homemade masks can be used on a temporary basis to reduce severe and fatal cases of the virus overburdening health care workers.

45.2.2 Face Detection Methods

The efficiency of the Face Mask Detection system depends on how accurately it can detect faces in the image. As per the current scenario, deep learning methods have superseded the traditional methods [9–11]. Deep Learning methods are able to achieve high accuracy due to their ability to learn features from training data containing the real-world variations. Deep learning methods are inspired by state-of-the-art networks achieving good accuracy in ImageNet Large Scale Visual Recognition Challenge (ILSVRC). There are two kinds of networks explored for object detection based on deep learning: region-based and regression-based. Region with Convolutional Neural network (R-CNN), Fast R-CNN [12], Faster R-CNN [13] are region-based methods. Regression-based methods are Single Shot Multi Box Detector (SSD) [14] and You only look once (YOLO) [15].Region-based methods do not consider entire image rather consider only those regions with high probabilities for locating the object. YOLO model considers entire image to locate the object and to calculate their class probabilities. Regression-based methods are considered to be faster without affecting accuracy of the system. In [16] faster version of YOLO named YOLO V3 is proposed. In the proposed work, YOLO V3 is adopted for performing face detection in images.

45.3 Proposed Method

Covid-19 Face Mask Detector system consists of two stages: Face Detection and Mask Detection. The proposed network accepts a raw image as input and outputs whether each detected person is wearing a mask or not as shown in Fig. 45.2.

45.3.1 Face Detection

The face Detection model is implemented using the YOLO V3 (You Only Look Once) model. The input image is directly passed to the trained YOLO model. The

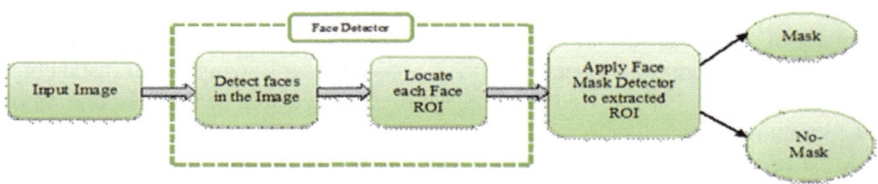

Fig. 45.2 The proposed architecture for Face Mask detection system comprising of two stages Face Detector and Mask Detector

SxS → SxSxB → **Face Detection**
Detection Frame   NMS

Fig. 45.3 Face detection framework using YOLO model depicting detection of face in an image

YOLO model partitions each image into a grid of $S1 \times S1$ ($S1 = 7$). If the center of the human face ground truth is in a grid, then that particular grid detects the Human Face. Each grid predicts five values with respect to each bounding box $B1$. These values are $X1$, $Y1$, $W1$, $H1$, and $C1$ where $X1$, $Y1$ is the center of $B1$ and $W1$, $H1$ are the width and height, respectively, and $C1$ indicates the confidence of $B1$. If a grid indicates Human Face then it also predicts its conditional probability. YOLO model uses non-maximum suppression (NMS) to finalize the bounding box in case of multiple same face detections. Originally YOLO model has been trained for object detection and is able to recognize 80 different classes of objects. YOLO model is a single neural network that considers the entire image at a time and is able to predict confidence of the bounding box and classify the desired object in one go. In real-time detection, the YOLO model outperforms other existing models such as Faster R-CNN [13] since it considers detection as a regression problem. The YOLO Face detection model is shown in Fig. 45.3.

45.3.2 Mask Detection

For Face Mask Detection phase transfer learning using imagenet weights on a lightweight architecture MobileNet V2 [17] is performed on RMFRD [18] dataset. The input given to this model is Face ROI extracted from the input image in the first phase. Based on the detected faces model classifies each Human face as Mask or NoMask.

The system also displays the total number of faces detected and the number of faces non-adhering wearing masks. MobileNet V2 is the enrichment of MobilenetV1 [19]. The model is computationally efficient and fast due to the use of depthwise separable convolutions thus reducing the total computations involved. MobileNetV2 uses depthwise separable convolutions with a different structure than that of MobileNetV1. In MobileNetV2 there are three convolution layers: an expansion layer in which the number of channels is expanded by 1×1 convolutional layer, in the second layer named as depthwise convolution layer input is filtered and in the third layer called projection layer these refined values are combined by pointwise convolution (1×1) to form new features. These layers together perform normal CNN but at a much faster speed without compromising with accuracy.

A total of 17 building blocks are present in the model. Gradients flow through the network with the help of residual connections. These layers are followed by batch normalization and the ReLU6 activation function except for the projection layer. Lastly, there is a global average pooling layer succeeded by a fully connected layer and a softmax. MobileNetV2 performs 300 million MACs which are the multiply-accumulate operations for an RGB image of 224×224 and have 20% fewer parameters compared to its predecessor. The model is more appropriate for applications with limited memory and computation power.

45.4 Experimental Results

This section discusses the results of our proposed method. Experimental environment: the processor used is Intel Core i7-8550, the CPU is 1.80GHZ*8, 8.00 GB RAM memory. The operating system is Windows 10, 64-bit.

45.4.1 Dataset

In this section, we provide a quick overview of the datasets used in our work, RMFRD [18] and WIDER Face Dataset [20].

- Face Detection: For the face detection model dataset used for training and testing model is the WIDER FACE dataset. It is a face detection benchmark dataset comprising of 32,203 images with variability in scale, occlusion with 393,703 faces labeled. There are a total of 61 event classes out of which random split of 40:10:50 is done for training, validation, and testing of images.
- Face mask Detection: For the face mask detection model only dataset available is Real-world Masked Face Recognition Dataset (RMFRD). The dataset comprises of 525 different subject images out of which 5,000 are while wearing masks, and 90,000 without masks. Sample images are shown in Fig. 45.4.

45.4.2 Performance Metrics

For the experiments, our model is trained with a learning rate of 0.0005 as an initial learning rate, batch size of 20 for CPU training with Adam as an optimizer, and a Dropout of 0.5. After trying freezing of different layers we learned that freezing first

No-
Mask

Mask

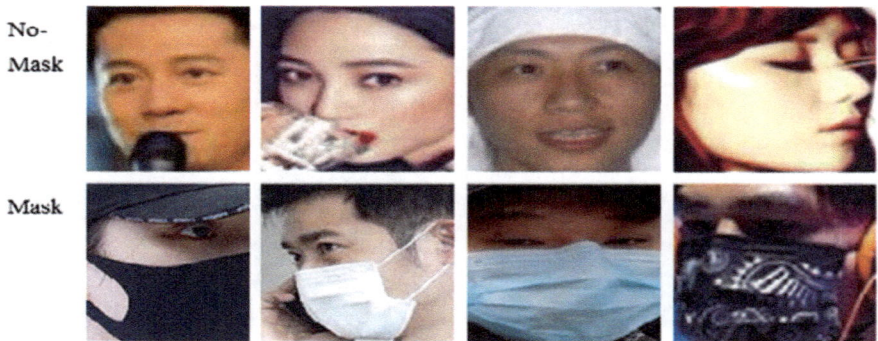

Fig. 45.4 RMFRD dataset: Sample images depicting with Mask and without Mask images

eighty layers led to comparable learning of features from the dataset. After training the face mask detection model on the RMFRD dataset, an accuracy of 98% was achieved due to exclusive set of images fed as input to network as shown in Fig. 45.5. The model can be generalized easily as witnessed from the results. Figure 45.6 presents the confusion matrix derived after testing the model on RMFRD dataset. The precision, recall, and f1-score per class are given in Table 45.1 and key performance measures of the classifier in Table 45.2. Different metrics used to measure the performance of our model are listed below:

- Accuracy: Accuracy in a classification problem is the number of correct predictions made by the model over all kinds of predictions made.

Accuracy = (True Positives + True Negatives) / Total Outputs

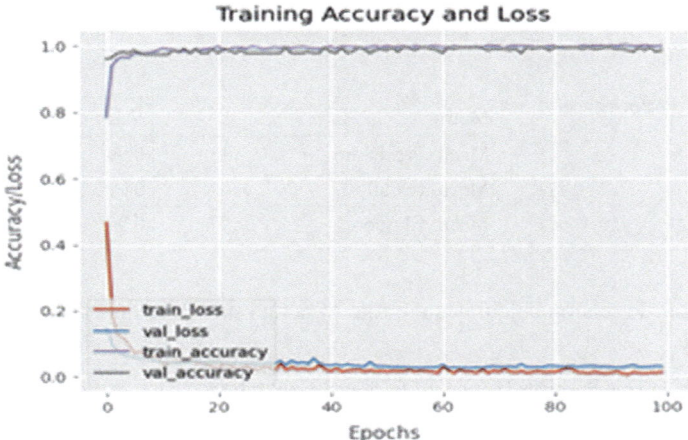

Fig. 45.5 Training accuracy/loss on RMFRD dataset

Fig. 45.6 Confusion matrix

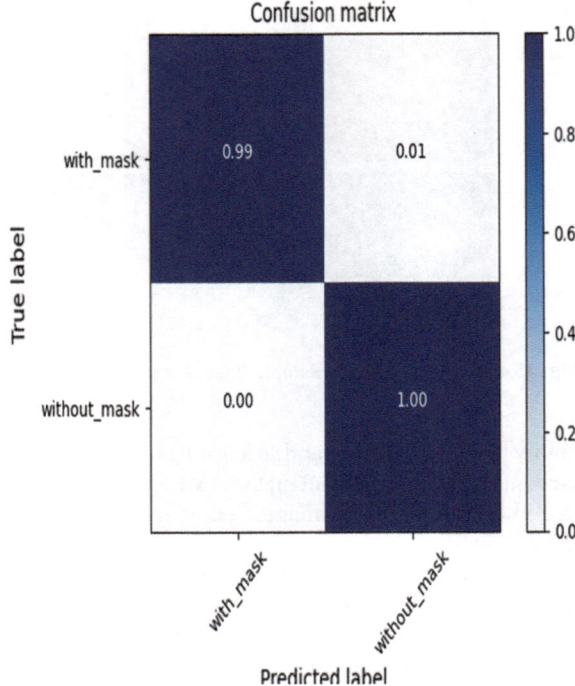

Table 45.1 Performance measure per class

	Mask	No Mask
Precision	0.98	0.99
Recall	0.99	0.99
F1-score	0.98	0.99

Table 45.2 Performance metrics

Metric	Value
Macro Precision	0.99
Macro Recall	0.99
Macro F1-score	0.99

- Precision: Precision calculates the rate of actual positives out of those predicted positive.

  Precision = True Positives / (True Positives + False Positives)

- Recall: Recall measures the rate of actual positives over the predicted values that are actually positive.

  Recall = True Positives / (True Positives + False Negatives)

- F1-score: f1-score is the harmonic mean of the precision and recall and conveys the balance between the precision and the recall.

  F1-score = 2 * [(Precision*Recall) / (Precision + Recall)]

45.5 Results

As shown in Fig. 45.7 Face Mask detector system is able to classify Mask indicated by green color box and NoMask with red color box. It also displays the summary about total number of faces detected and total number of faces out of the detected ones violating the constraint. System is also able to detect people wearing mask inappropriately as No Mask. Even the half occluded faces are been detected correctly. The system fails to detect in case the person's face is not clearly visible or not toward the camera angle.

45.6 Conclusion

The goal of the Face Mask Detection system using deep learning models is addressed in this work. The proposed model has a real-time detection rate, leading to a reduction in miss rate. Face Mask Detection system proposed in this work gives good results as evident from the experimental results. In the future, our objective will be to incorporate the proposed model in videos. We can collate the social distancing model to the existing work so that precautionary measures for Covid-19 are monitored. Further, we can add the Alerting model which will send alerts to the authority with the picture of the person not wearing a mask. The existing method works well with human faces clearly visible in the future it can be improved to work with complex occluded faces. Further experiments can be conducted in an organized manner to achieve the above-mentioned objectives.

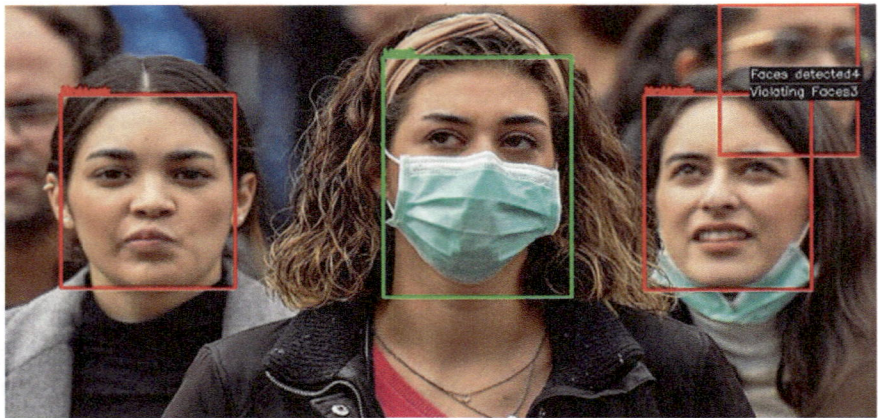

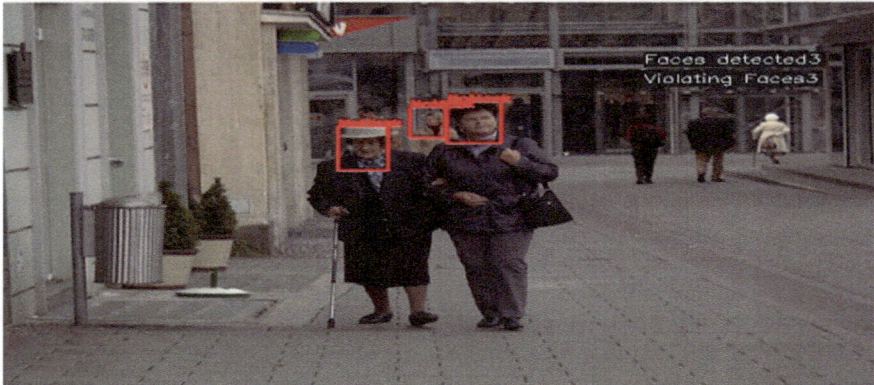

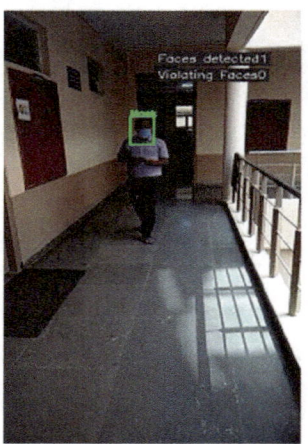

Fig. 45.7 Test results of the trained model

References

1. Chu, Derek K., Elie A. Akl, Stephanie Duda, Karla Solo, Sally Yaacoub, Holger J. Schünemann, Amena El-harakeh. "Physical distancing, face masks, and eye protection to prevent person-to-person transmission of SARS-CoV-2 and COVID-19: a systematic review and meta-analysis." The Lancet (2020).
2. WHO coronaviruses (COVID19), "https://www.who.int/emergencies/diseases/novel-coronavirus-2019", 2020, [Online; accessed June 15, 2020])
3. Howard, J., Huang, A., Li, Z., Tufekci, Z., Zdimal, V., van der Westhuizen, H.M., von Delft, A., Price, A., Fridman, L., Tang, L.H. and Tang, V., Face masks against COVID-19: an evidence review,2020.
4. West, R., Michie, S., Rubin, G.J. and Amlôt, R., Applying principles of behaviour change to reduce SARS-CoV-2 transmission. *Nature Human Behaviour*, pp.1–9, 2020
5. Ma, Q.X., Shan, H., Zhang, H.L., Li, G.M., Yang, R.M. and Chen, J.M., Potential utilities of mask-wearing and instant hand hygiene for fighting SARS-CoV-2. *Journal of medical virology*, 2020.
6. Eikenberry, S.E., Mancuso, M., Iboi, E., Phan, T., Eikenberry, K., Kuang, Y., Kostelich, E. and Gumel, A.B., To mask or not to mask: Modeling the potential for face mask use by the general public to curtail the COVID-19 pandemic. Infectious Disease Modelling, 2020.
7. Cheng, V.C., Wong, S.C., Chuang, V.W., So, S.Y., Chen, J.H., Sridhar, S., To, K.K., Chan, J.F., Hung, I.F., Ho, P.L. and Yuen, K.Y., The role of community-wide wearing of face mask for control of coronavirus disease 2019 (COVID-19) epidemic due to SARS-CoV-2. Journal of Infection, 2020.
8. Kelly, M.D.: Visual identification of people by computer. Stanford University, Department of Computer Science (1970)
9. Sirovich, L., Kirby, M.: Low-dimensional procedure for the characterization of human faces. Josa a **4**(3), 519–524 (1987)
10. Kirby, M., Sirovich, L.: Application of the Karhunen-Loeve procedure for the characterization of human faces. IEEE Trans. Pattern Anal. Mach. Intell. **12**(1), 103–108 (1990)
11. Schölkopf, B., Smola, A. and Müller, K.R., October. Kernel principal component analysis. In *International conference on artificial neural networks* (pp. 583–588). Springer, Berlin, Heidelberg, 1997.
12. Girshick, Ross. "Fast r-cnn." In *Proceedings of the IEEE international conference on computer vision*, pp. 1440–1448, 2015.
13. Ren, Shaoqing, Kaiming He, Ross Girshick, and Jian Sun. "Faster r-cnn: Towards real-time object detection with region proposal networks." In *Advances in neural information processing systems*, pp. 91–99, 2015.
14. Liu, Wei, Dragomir Anguelov, Dumitru Erhan, Christian Szegedy, Scott Reed, Cheng-Yang Fu, and Alexander C. Berg. "Ssd: Single shot multibox detector." In *European conference on computer vision*, pp. 21–37. Springer, Cham, 2016.
15. Redmon, J., Divvala, S., Girshick, R. and Farhadi, A., You only look once: Unified, real-time object detection. In *Proceedings of the IEEE conference on computer vision and pattern recognition* (pp. 779–788), 2016
16. Redmon, J., Farhadi, A.: Yolov3: An incremental improvement. ArXiv preprint arXiv **02767**, 2018 (1804)
17. Sandler, M., Howard, A., Zhu, M., Zhmoginov, A. and Chen, L.C., Mobilenetv2: Inverted residuals and linear bottlenecks. In *Proceedings of the IEEE conference on computer vision and pattern recognition* (pp. 4510–4520), 2018.
18. Wang, Z., Wang, G., Huang, B., Xiong, Z., Hong, Q., Wu, H., Yi, P., Jiang, K., Wang, N., Pei, Y., Chen, H.: Masked face recognition dataset and application. ArXiv preprint arXiv **09093**, 2020 (2003)

19. Howard, A.G., Zhu, M., Chen, B., Kalenichenko, D., Wang, W., Weyand, T., Andreetto, M. and Adam, H., Mobilenets: Efficient convolutional neural networks for mobile vision applications. ArXiv preprint arXiv: 1704.04861, 2017.
20. Yang, S., Luo, P., Loy, C.C. and Tang, X., Wider face: A face detection benchmark. In Proceedings of the IEEE conference on computer vision and pattern recognition (pp. 5525–5533), 2016.

Chapter 46
Underwater Image Segmentation Using Fuzzy-Based Contrast Improvement and Partition-Based Thresholding Technique

Pratima Sarkar, Sandeep Gurung, and Sourav De

Abstract Underwater images suffer from haziness due to the presence of particles and changes in water density in the marine environment. Underwater image segmentation is one of the challenging areas due to the blurriness of the images. Image preprocessing is necessary before image segmentation due to unclear image, so the work is divided into three parts: contrast improvement, sharpening, and segmentation of an image. The fuzzy-based contrast improvement technique is proposed with Contrast Limited Adaptive Histogram Equalization to enhance contrast. An unsharp mask is used to sharpen an image. The proposed partition-based thresholding segmentation is used to segment the image. In this method, the partition is made to calculate an appropriate threshold value to segment each partition. Quantitative and qualitative analysis has been shown in results and discussions part using entropy as a measurement parameter. 75.67% accuracy was received by the work. Also, the work is compared with existing work.

46.1 Introduction

In the marine environment investigation, one of the major tasks is object detection from an image, which consists of two phases: localization and classification of an object from an image. Image segmentation helps in the region of interest detection that is used to localize an object. Underwater images are suffered from low contrast and bluish color so before segmenting an image, preprocessing is necessary. Underwater images are unclear because of uneven water density, reflection, and refraction of lights, the existence of different particles, etc.

P. Sarkar (✉) · S. Gurung · S. De
Sikkim Manipal Institute of Technology, SMU, Majitar, India
e-mail: pratima.s@smit.smu.edu.in

S. Gurung
e-mail: sandeep.gu@smit.smu.edu.in

Coochbehar Government Engineering College, Cooch Behar, India

© The Author(s), under exclusive license to Springer Nature Singapore Pte Ltd. 2022   473
V. Bhateja et al. (eds.), *Evolution in Computational Intelligence*,
Smart Innovation, Systems and Technologies 267,
https://doi.org/10.1007/978-981-16-6616-2_46

Image contrast is one of the important features of an image to get better information from an image and sharpening gives better information about edges so that above-mentioned operations are selected for enhancement.

In the proposed work two novel techniques have been used to improve the contrast of the image using fuzzy logic-based Contrast Limited Adaptive Histogram Equalization (CLAHE) and partition-based thresholding segmentation approach. To improve the sharpness of the image, the Unsharp mask approach is employed. One of the most popular techniques to improve contrast is Histogram Equalization (HE) [1, 2]. For improving contrast, it is important to make darker pixels darker and lighter pixel lighter. HE gives a probability value for each intensity value that exists in an image, which helps us to identify the pixel intensity distribution. Fuzzy logic is selected to assign a much lower value to darker pixel and a higher value to lighter pixel. Calculation of HE is faster so here CLAHE is used. Adaptive Histogram Equalization (AHE) computes multiple histograms and redistributes the probability intensity values to improve contrast. The main disadvantage of AHE is that it over amplifies the noise [3]. So CLAHE with fuzzy logic is used to improve contrast.

Image sharpening is done by Unsharp mask, which is used to improve image edge information [4]. This mask usually generates blur images by using the Gaussian blur effect or generates negative of an image and subtract from the original image. It allows high-frequency components to improve edge information. The main disadvantage of the Unsharp mask generates unwanted edges [5]. The unwanted edges can be reduced by controlling the contrast, radius of the edges, and threshold of brightness change.

In earlier few years' different techniques are used to segment underwater images based on supervised learning, clustering-based technique, and some of the modelings like snake models, etc. One of the unsupervised techniques used is in [6] Co-saliency detection combined with K-Means clustering and saliency cues integration. The work generates a saliency map to highlight salient regions of an image. After highlighting the regions image segmentation was implemented using Gaussian modeling and bias correction technique. Sonar images are captured by using an echo signal of sound. The snake model is used to target the detection and localization of an object from a sonar image [7]. It modifies the existing snake model. It introduces a greedy approach to converge contour more efficient way rather than directly using energy minimizing contour. Poisson-Gaussian theory to describe the nature of noise exists in an image [8]. After that mean-shift algorithm is used to divide the images into different regions. Underwater image segmentation needs enhancement before segmentation is divided so some of the literature follows three phases de-noising, enhancement, and segmentation [8]. Semantics-based segmentation techniques also segment an underwater image [9]. Deep learning-based SUIN-Net model with fully connected convolution encoder-decoder techniques and the concept of residual skip block are used to learn semantics.

It has been observed that the histogram of underwater images contains multiple peaks and valleys so the global thresholding method is not suitable for the work. In the locally adaptive thresholding method, an image is divided into overlapping segments, then computes an array of threshold values to calculate the global threshold value [10]. The complexity of the adaptive thresholding technique is much higher.

Now, region-growing segmentation requires a starting point called seed point [11]. Neighbors of seed point are checked for homogeneity depending upon the threshold value, if neighbor's intensity values are within threshold values it keeps on merging neighbors. This procedure will be repeated until the condition dissatisfies. The main problem related to the region growing is critical to determine the seed point within the region of interest.

The performance of segmentation has been measured by the entropy of an image in many previous works [3, 10, 12]. It is a parameter that measures randomness in an image so a lesser value of entropy is a reflection of good segmentation. Mean Square Error and Pixel Signal to Noise ratio are used to analyze the quality of the reconstructed image [8]. The intersection over union is used as a performance measure parameter in the deep learning-based technique [9].

The contributory parts of this paper are given below:

- Here CLAHE applied on the fuzzy plane instead of directly applied on the original image.
- Fuzzy logic tries giving lesser value to the darks region and higher value to the light region so the contrast will improve automatically.
- The proposed work partition the entire image into sub-images. Partition is important to identify the proper threshold value for each segment.
- The modified thresholding technique takes less time than adaptive thresholding as it is not considering the overlapping area.

46.2 Proposed Methodology

The main objective of the proposed work is to segment the background from its foreground from an underwater debris image. To enhance the quality of the underwater image, contrast improvement and sharpening are employed in this approach. Fuzzy-based contrast limited adaptive histogram equalization is proposed for contrast improvement and Unsharp masking is used to enhance sharpness. The following steps are involved in the work (Fig. 46.1).

1. Contrast improvement by fuzzy-based CLAHE
2. Image Sharpening using Unsharp Mask
3. Partition-based image segmentation

46.2.1 Contrast Improvement

In the last few years' fuzzy logic is applied for the improvement of contrast, noise removal, etc. To improve the contrast of an image in this approach, it has been tried to make the darker pixels more darker, the lighter pixels more lighter and no changes for the average intensity pixels. To implement the above concept, the fuzzy logic

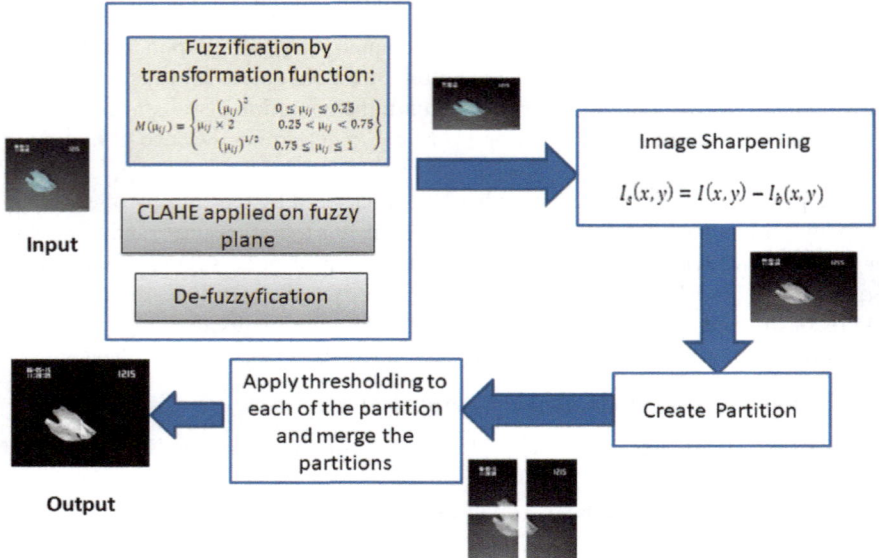

Fig. 46.1 Schematic diagram of the proposed solution

with Contrast Limited Adaptive Histogram Equalization (CLAHE) is proposed. The following steps are followed to improve contrast into an image:

1. Image intensity values are mapped into a fuzzy plane (Fuzzification)
2. CLAHE Applied on fuzzy plane
3. De-fuzzification to get transformed intensity values

While applied fuzzification on an image it transformed all intensity values ranging from 0 to 1 that is a fuzzy plane. In fuzzy plane values are assigned in such a way so that darker pixel assigned with lesser value and brighter pixel assigned with higher value. An image P with size $m \times n$ and its intensity value varies from (0 to L-1). For mapping original intensity value into fuzzy plane following expressions are used:

$$P_{mid} = \frac{I_{max} - I_{min}}{2} \tag{46.1}$$

P_{mid} is a middle-intensity value of the image, I_{max} and I_{min} are the maximum and minimum intensity values of the image.

$$\mu_{ij} = \frac{P_{ij}}{2 \times I_{mid}} \tag{46.2}$$

where P_{ij} is an intensity value of image P at location ij.

To assign darker pixel with lower value and brighter pixel with higher value the membership function is expressed as follows:

$$M(\mu_{ij}) = \begin{cases} (\mu_{ij})^2 0 \le \mu_{ij} \le 0.25 \\ \mu_{ij} \times 2 0.25 < \mu_{ij} < 0.75 \\ (\mu_{ij})^{1/2} 0.75 \le \mu_{ij} \le 1 \end{cases} \qquad (46.3)$$

$M(\mu_{ij})$ is the membership function used to map intensity values of an image into fuzzy plane f.

The CLAHE [13] is applied on a fuzzy plane expressed as follows:

$$HE = \left[I_{f\_max} - I_{f\_min} \right] \times P_k(I_{cu}) + I_{f\_min} \qquad (46.4)$$

where I_{f_max} and I_{f_min} are the minimum and maximum intensity values of the image. $P_k(I_{cu})$ is cumulative distribution function as per input image [13].

$$P_k(I_{cu}) = \sum_{j=0}^{k} P_I(I_j) \qquad (46.5)$$

$P_I(I_j)$ is probability density function given as follows:

$$P_I(I_j) = \frac{n_k}{N} \qquad (46.6)$$

where n_k is the number of pixels with intensity value k and N is total number of pixels. For de-fuzzification all fuzzy plane values are multiplied with level L-1. Result of de-fuzzification is contrast improved image.

46.2.2 Image Sharpening

Image sharpening is mostly used for improving texture and edge detail. In Unsharp mask, blur images of the same image are created by Gaussian blur effects. Expression to implement Unsharp mask is:

$$I_s(x, y) = I(x, y) - I_b(x, y) \qquad (46.7)$$

where $I_s(x, y)$ is a sharp image, $I(x, y)$ is high contrast image, and $I_b(x, y)$ is a blur image.

46.2.3 Image Segmentation

A partition-based thresholding technique is used to segment an image. A sharp image is taken as input for image segmentation. Threshold value-based image segmentation is chosen as it works well with high contrast images. The selection of threshold value is done from the histogram. An image is segmented properly if an image is bimodal, i.e., two peaks are separated by a valley of a histogram. Partition-based thresholding segmentation algorithm consists of the following steps:

Partition-based thresholding algorithm:

Step 1: Read sharp Image

Step 2: Divide an image into two equal nonoverlapping parts called sub-regions.

Step 3: Generate a histogram for each of the sub-regions

Step 3.1: If a valley between two peaks exists

then call *Calthresh* (Image)

Go to for each partition step 2.

Step 4: After segmenting each partition combine all to get segmented images.

Step 5: Applied dilation or erosion as per image requirement to get better segmentation.

Calthresh (Image)

Step 1: Let the intensity values for each pixel be f(i,j) where i and j are the Locations.

Step 2: Threshold value $T = \sum_{i=1,j=1}^{i=m,j=n} f(i,j)$, where the image size is $m \times n$.

Step 3: Based on the following formula segmented the sub-image

$$G(i, j) = \left\{ \begin{array}{l} 1 \ \ if \ \ f(i, j) > T) \\ 0 \ \ Otherwise \end{array} \right\}$$

The output of the partition-based thresholding is the segmented image.

46.3 Result and Discussions

In the work image segmentation, we have implemented underwater debris images. JAMSTEC Deep-sea Debris Database [14] is used to collect underwater images. We have collected set of videos from the above-mentioned website videos that are converted into images by extracting frames. MATLAB 2015 platform is used for implementing the work. In Fig. 46.2, the original and the resultant images after applying fuzzy contrast limited adaptive histogram equalization are shown.

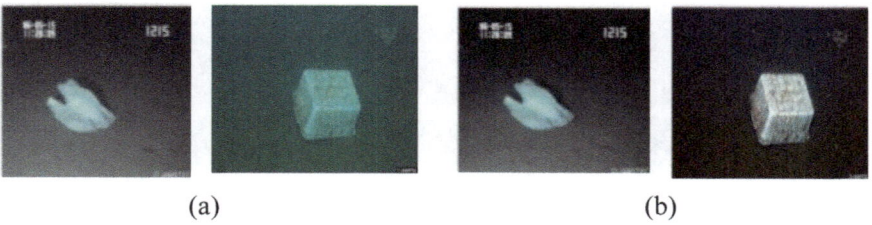

(a)   (b)

Fig. 46.2   **a** Original Image **b** Image after contrast improvement

Fig. 46.3 Sharp images
after applying Unsharp mask

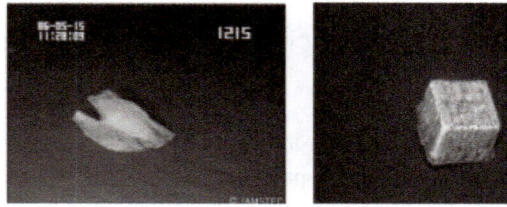

After contrast enhancement, to segment an image, it is important to enhance edge information in an image so image sharpening can be implemented. The Unsharp mask is employed for image sharpening and the results are shown in Fig. 46.3.

After that, the partition-based thresholding segmentation is applied. In this stage, the images are divided into different segments according to it is specified in the algorithm. The sharp images, those are in Fig. 46.3 are partitioned into two regions. If detection of the threshold values is not possible then each of the regions is subdivided into two equal parts. Finally, the threshold value for segmentation of four regions is determined and no further division is required. Here both the images are divided into four partitions as shown in Fig. 46.4.

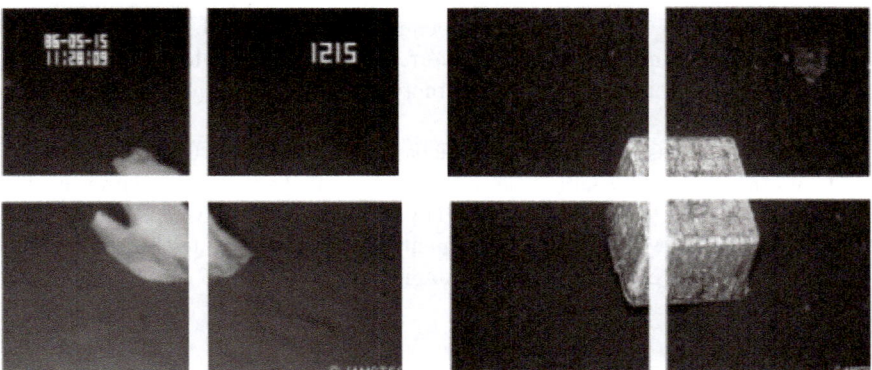

Fig. 46.4 Partition generated from sharp images

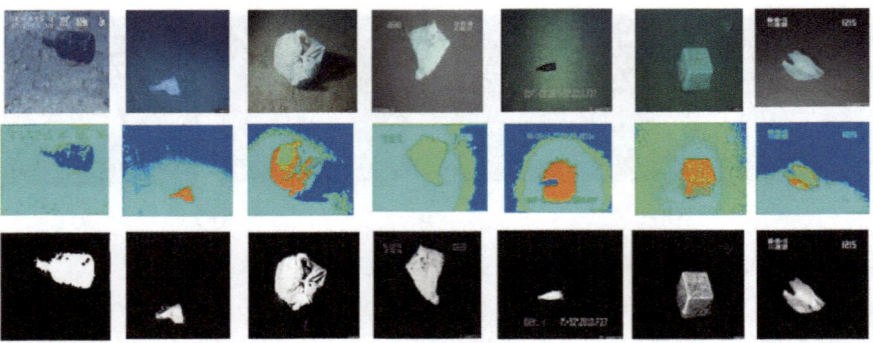

Fig. 46.5 1st row original image, 2nd row segmented image by K-Means, 3rd row segmented image by proposed work

Partition-based thresholding techniques are used to segment each partition. Then combined all segmented partitions of an image to regenerate the complete image. According to the requirement of the segmented image, we have performed dilation or erosion morphological operation to get better segmentation. Few of the segmentation results are shown in Fig. 46.5.

The goodness of image segmentation can be measured by maximum uniformity inside the partition and minimum uniformity between inter-partitions. Entropy is the property of an image that determines the randomness of an image [12, 15]. To measure disorder in a partition or to evaluate segmentation in the work entropy is used as a parameter. Let us assume the image is I and Shannon's entropy is E. it is expressed as [13]:

$$E(I) = -\sum_{k=0}^{L-1} p(k)log_2^{p(k)} \tag{46.8}$$

where I is the image used for calculating entropy. L is the number of gray levels in the image I. P(k) is the probability of occurrence of intensity value k in an image I. Lower the entropy value segmentation is more uniform, i.e., segmentation quality is better [15].

Quantitative assessment has been made based on entropy and the results are tabulated Table 46.1. From the table it has been observed that entropy value improved by 75.67% with respect to original image and by using modified K-Means [5] improved by 69.64%. For the evaluation of the segmentation we have selected 100 underwater images. Most of the entropy values are ranging between 0 and 2 and the maximum accuracy received was 75.67%.

Table 46.1 Entropy values before and after segmentation

Sl. No	Entropy of original Image	K-Means	% of improvement w.r.t original image in K-Means	Entropy of segmented Image	% of improvement w.r.t original image in proposed work
1	5.2186	2.6078	50%	1.3339	74.49%
2	5.1574	2.6512	48.54%	1.4336	72.20%
3	6.2319	3.2219	49.39%	1.5492	75.14%
4	5.9378	3.1198	47.45%	2.0059	66.22%
5	5.9748	2.0076	66.49%	1.4842	66.53%
6	7.1522	3.7890	48.15%	1.7304	75.82%
7	4.9085	2.0098	59.18%	1.9709	61.39%
8	5.6020	1.9841	66.07%	1.5069	73.25%
9	5.6538	1.7984	**69.64%**	1.4283	75.26%
10	7.4923	4.0304	46.22%	1.9556	**75.67%**

46.4   Conclusion

In the work, underwater images are segmented using the partition-based thresholding segmentation technique. The existing state of the work related to underwater image segmentation is very much data dependent as using deep learning techniques or accuracy of the segmentation is not measured so the work addressed the stated problems. For the work, we have selected underwater debris images. To enhance the contrast of an image fuzzy-based CLAHE is proposed and for sharpening existing an Unsharp mask is used. Then we have proposed partition-based segmentation novel technique to segment images.

After image segmentation qualitative and quantitative analysis was conducted using MATLAB. Here we have selected four types of objects for the experiment such as plastic, glass bottle, plastic bottle, and a plastic bucket. For quantitative analysis, entropy is selected as a parameter. From the above experiment, we can conclude that after segmentation, most of the entropy value lies between 0 and 2. The proposed work compared with the K-Means segmentation technique and the result shows our work gives almost 6–29% better accuracy. So this is a novel approach to segment underwater images.

References

1. Magudeeswaran, V., Singh, J.F.: Contrast limited fuzzy adaptive histogram equalization for enhancement of brain images, Int. J. Imag. Syst. Technol. **27**(1), 98–103, 17
2. De, S., Dey, S., Paul, S.: Underwater image enhancement using neighbourhood based two level contrast stretching and modified artificial bee colony. In: 7th IEEE Uttar Pradesh Section International Conference on Electrical, Electronics and Computer Engineering (UPCON 2020) (Accepted)
3. Magudeeswaran, V., Ravichandran, C. G.: Fuzzy Logic-Based Histogram Equalization for Image Contrast Enhancement"Hindawi Publishing Corporation Mathematical Problems in Engineering Volume 2013, Article ID 891864, 10 pages. https://doi.org/10.1155/2013/891864
4. Al-Ameen, Z., Muttar, A., Al-Badrani, G.: Improving the sharpness of digital image using an amended unsharp mask filter. Int J Image, Graphics Signal Process **11**, 1–9 (2019). https://doi.org/10.5815/ijigsp.2019.03.01.
5. Rajeev, A.A., Hiranwal, S., Sharma, V.K.: Improved segmentation technique for underwater images based on k-means and local adaptive thresholding. In: Mishra D., Nayak M., Joshi A. (eds) Information and Communication Technology for Sustainable Development. Lecture Notes in Networks and Systems, vol 10. Springer, Singapore (2018). https://doi.org/10.1007/978-981-10-3920-1_45
6. Zhu, Y., Hao, B., Jiang, B., Nian, R., He, B., Ren, X., Lendasse, A.: Underwater image segmentation with co-saliency detection and local statistical active contour model. 1–5 (2017). https://doi.org/10.1109/OCEANSE.2017.8084742
7. Wu, J., HaitaoGuo: Underwater sonar image segmentation based on snake model. Appl. Mech. Mater. 448–453, 3675–3678 (2013) Oct 31. https://doi.org/10.4028/www.scientific.net/AMM.448-453.3675. ISSN: 1662–7482. © 2014 Trans Tech Publications, Switzerland
8. Boudhane, M., Nsiri, B.: Underwater image processing method for fish localization and detectionin submarine environment. J. Vis. Commun. Image R. **39**, 226–238 (2016). https://doi.org/10.1016/j.jvcir.2016.05.017 1047–3203
9. Liu, F., Fang, M.: Semantic segmentation of underwater images based on improved deeplab. J. Mar. Sci. Eng. **8**, 188 (2020). https://doi.org/10.3390/jmse8030188
10. Rezaei, F., Izadi, H., Memarian, H., Baniassadi, M.: The effectiveness of different thresholding techniques in segmenting micro CT images of porous carbonates to estimate porosity. J. Petrol. Sci. Eng. (2019). https://doi.org/10.1016/j.petrol.2018.12.063
11. EL Allaoui, A., Mohammed, M.: Evolutionary region growing for image segmentation. Int. J. Appl. Eng. Res. **13**(5), 2084–2090 (2018). ISSN 0973–4562
12. De, S., Bhattacharyya, S., Dutta, P.: Automatic magnetic resonance image segmentation by fuzzy intercluster hostility index based genetic algorithm: an application. Appl. Soft Comput. **47**, 669–683 (2016)
13. César Mello Román, J., Luis VázquezNoguera, J., Legal-Ayala, H., Pinto-Roa, D.P., Gomez-Guerrero, S., García Torres, M.: Entropy and contrast enhancement of infrared thermal images using the multiscale top-hat transform. Entropy, 21(3), 244 (2019). https://doi.org/10.3390/e21030244
14. http://www.godac.jamstec.go.jp/catalog/dsdebris/e/index.html. Accessed 22 Jan 2021
15. Zhang, H., Fritts, J.E., Goldman, S.: An entropy-based objective evaluation method for image segmentation. In: Proceedings of SPIE Storage and Retrieval Methods and Applications for Multimedia, pp. 38–49 (2004)

Chapter 47
Classification of EEG Signals for Seizure Detection Using Feature Selection and Channel Selection

Saurav Suman, Vamsi Deekshit Kanakavety, Ajay Venkatesh kattoju, and Pradnya Ghare

Abstract In this paper, we have introduced a new method for the classification of electroencephalogram (EEG) signals for seizure detection using feature selection and channel selection. The channels are selected based on the method of variance. Further, features have been extracted from intrinsic mode functions obtained after decomposing the EEG signal with empirical mode decomposition. The feature selection is done using a one-way analysis of variance test with a predefined threshold probability value. The classification of seizure and non-seizure signals is done on selected features using Decision Tree and k-nearest neighbor algorithms. The proposed method for classification based on a combination of feature selection and channel selection provides an accuracy of 99.6% which is better as compared to the existing methods. Thus, optimizing the number of channels and using accurate features gives us a better way of analyzing the EEG signals for seizure detection.

47.1 Introduction

The human brain is a complex organ that controls all the functionalities of our body. The transmission of the information between two neurons takes place in terms of electric pulses. These impulses can be acquired using different modalities, electroencephalogram (EEG) is one such modality that helps in analyzing the brain and its behavior based on the received frequency of the signal. It measures the voltage fluctuation within the neurons in the brain. Epilepsy or seizure is one of the most commonly occurring brain disorders in which an individual behaves abnormally, shows seizure activities, and may lead to a lack of awareness. If seizures can be predicted reliably and the occurrence of seizures is found with high sensitivity and specificity, it could help significantly in reducing the physical damages [1].

S. Suman · V. D. Kanakavety (✉) · A. V. kattoju · P. Ghare
Department of Electronics and Communication, Visvesvaraya National Institute of Technology, Nagpur, India
e-mail: vamsikanakavety144@gmail.com

© The Author(s), under exclusive license to Springer Nature Singapore Pte Ltd. 2022   483
V. Bhateja et al. (eds.), *Evolution in Computational Intelligence*,
Smart Innovation, Systems and Technologies 267,
https://doi.org/10.1007/978-981-16-6616-2_47

Many nonlinear methods have come into existence for the extraction of information from EEG signals like the Lyapunov exponent which was predominantly useful in extracting the neuronal behavior [2]. The complexity changes of the EEG signals can be quantitatively described using the correlation function [3] and to find the irregularity, and the variable nature of the EEG signals fractal parameter was used [4]. The entropy index can also be used as a quantitative measure for EEG estimation and this can be used as a tool for classifying the signals [5].

Recently, new methods have been available for the analysis of nonstationary and nonlinear signals proposed in [6], which were based on the empirical mode decomposition (EMD) [7]. This method was found to be less computationally complex and the results of the classification were pretty much more accurate as compared to the earlier methods. V. Bajaj et al. [8] have used bandwidth parameters, such as amplitude modulation bandwidth and frequency modulation bandwidth, for classifying the EEG signals into seizure and non-seizure. The algorithms used for seizure detection ought to be accurate and precise. The developments until now have focused upon obtaining a better accuracy but at the same time computational complexity increases. In this paper, we have combined both feature selection and channel selection methods to obtain a better accuracy along with a low computational cost. To classify the EEG signals into seizure and non-seizure class, the raw data are first decomposed using the EMD method. Further, 12 different statistical features were extracted and a one-way analysis of variance (ANOVA) test was conducted, for selecting the most variant features. Out of the different channel selection algorithms, selection based on variance method was used. The classification has been done using the classification learner tool in MATLAB R2018b and the algorithms used are k-nearest neighbor and Decision trees. The experimental results provided better results compared to Liang et al. [9] and V. Bajaj et al. [8].

The course content of the paper is as follows: Sect. 47.2 consists of EEG data set, Channel selection, EMD method, Feature extraction, and Feature selection. Section 3 comprises of Results and Discussion of the classification of EEG signals into seizure and non-seizure. Further, Section 4 provides the Conclusion.

47.2 Methodology

47.2.1 Data Set

The data used in this paper are taken from the University of Bonn [10]. It consists of five subsets, namely Z, O, N, F, S where each subset consists of 100 single-channel EEG recordings where each channel is of 23.6 s duration. The data in subsets Z and O are recorded from five healthy persons extracranially, i.e., surface EEG recordings, where Z consists of data when eyes were open and O when eyes were closed. Whereas N, F, and S are obtained intracranially, taken out of the archive data from presurgical diagnosis. Further, N and F consist of data of five epileptic patients recorded during the seizure-free intervals. F from the epileptic focal zone and N from the opposite

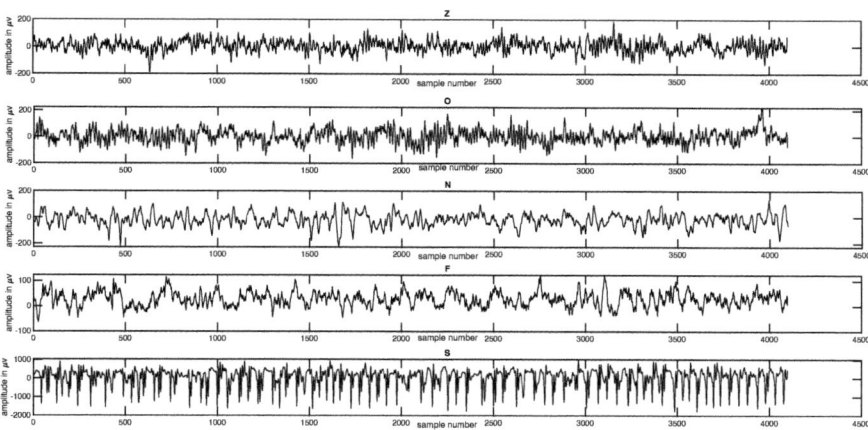

Fig. 47.1 One sample signal from each dataset (Z, O, N, F, S)

side of the brain where the hippocampus is formed. S subset consists of data taken from all the regions showing seizure activities. In a combined manner all these subsets are classified into two categories non-seizure (Z, O, N, F) and seizure (S). All the above recordings were done using a 128 channel electrode system. Further, for the discretization of the signals 12 bit analog to digital converter (ADC) was used and the data was sampled at the rate 173 Hz. The channels are recorded using a common average montage reference and data was randomized with reference to the recording region. To avoid the artifacts, data were filtered using a band pass filter with a cutoff frequency range of 0.5 40 Hz. The sample signal from each subset plotted in Fig. 47.1 shows the nature of the raw EEG data acquired from five different situations mentioned earlier.

47.2.2 Channel Selection

The EEG signals contain data of a large number of channels, this huge amount of data when used for classification for seizure detection accounts for a very high processing time and large computational complexity. The process can be optimized by the reduction in the number of channels processed, This is termed as channel selection. In cases, when more channels are used for classification, more features are available for the classifier to distinguish any complex pattern. This can lead to a larger overlapping of data among the channels, for the channels far away from the focus may contain the same data during the ictal and interictal periods. Thus while training these data we may encounter overfitting and hence channel selection becomes important to avoid that. Many algorithms are used for channel selection in literature [11].

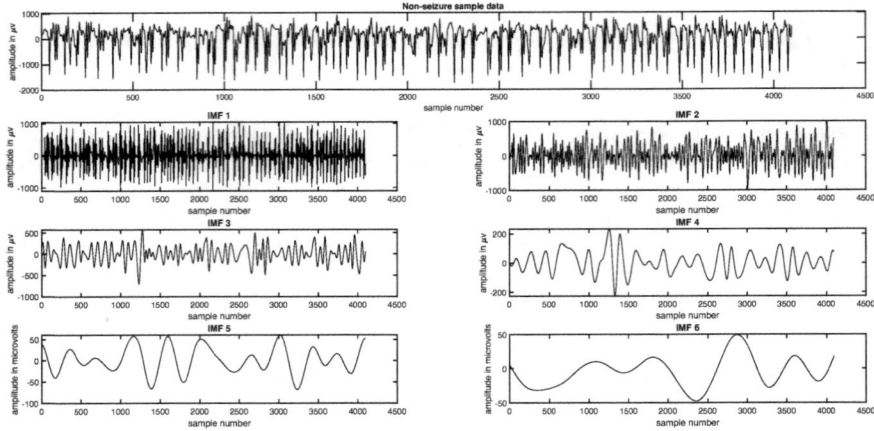

Fig. 47.2 IMFs obtained after EMD process on a sample signal

Out of the different methods for channel selection this paper uses the method of maximum variance, we have two cases for channel selection. Case (a) when the top 20 channels are selected based on maximum variance, from each subset. In this case, we have 80 (N, F, Z, O) channels of non-seizure data and 20 channels of seizure data. Case (b) When the top 50 channels are selected from each subset and thus leading to 200 channels of non-seizure and 50 channels of seizure data.

47.2.3 Empirical Mode Decomposition

Bajaj and Pachori [8] EMD method for signal processing, is quite useful in the analysis of nonlinear and nonstationary signals. It decomposes the signal into a set of Intrinsic Mode Functions (IMFs). In Fig. 47.2, we have shown 6 out of 8 IMFs generated from raw EEG signal. It can be considered as the most suitable method for natural signals like EEG, as it doesn't make any assumptions about the composition of signal and operates in both time and frequency domain. IMFs are intrinsic signals which are modulated both in amplitude and frequency domain. The condition of signal to be an IMF are described in [12].

47.2.4 Feature Extraction and Selection

Out of many popular methods used for feature extraction (wavelet analysis [13], Fourier transform [14], etc.), we have chosen EMD as it gives us intrinsic functions which are both frequency and amplitude modulated and thus making it easier to

Impulse factor: $x_{IF} = \dfrac{x_p}{\dfrac{1}{N}\displaystyle\sum_{i=1}^{N}\|x_i\|}$ X_o is peak value	Crest factor: $x_{crest} = \dfrac{x_p}{\sqrt{\dfrac{1}{N}\displaystyle\sum_{i=1}^{N} x_i^2}}$ X_o is peak value
Kurtosis: $x_{kurt} = \dfrac{\dfrac{1}{N}\displaystyle\sum_{i=1}^{N}(x_i - \bar{x})^4}{\left[\dfrac{1}{N}\displaystyle\sum_{i=1}^{N}(x_i - \bar{x})^2\right]^2}$	Skewness: $x_{skew} = \dfrac{\dfrac{1}{N}\displaystyle\sum_{i=1}^{N}(x_i - \bar{x})^3}{\left[\dfrac{1}{N}\displaystyle\sum_{i=1}^{N}(x_i - \bar{x})^2\right]^{3/2}}$

Fig. 47.3 Mathematical formulation of some extracted features

extract required features in the time domain itself. All the 12 features extracted from IMFs are—Mean, Median, Maximum, Minimum, Kurtosis, Skewness, Root Mean Square, Variance, Shape Factor, Standard deviation, Impulse Factor, Crest Factor. Figure 47.3 shows some standard mathematical formulations. It is very important to figure out which features will describe the signals accurately while classification, as not every feature might be useful for a particular application. Also, the computational overload is reduced when we are able to optimize the number of features used for further training and testing of data. In this paper, we are using a one-way ANOVA test, which is a means to gather information about the relationship of variables [15]. A statistically significant result is obtained when a probability (p-value) is less than the predefined threshold. In this paper, we have run a one-way ANOVA test on the EEG signals for all the 12 features mentioned earlier for each of the IMFs generated after empirical mode decomposition of all five datasets. The results showing p-values are summed up in Table 47.1. The significance value or threshold of probability is taken as 0.05.

47.3 Results and Discussion

The results of the ANOVA test are tabulated in Table 47.1, it can be clearly seen that with the threshold of probability value 0.05, mean, median, and skewness can be rejected as a feature used for classification as their p-value are greater than the threshold. For each of the two cases of channel selection, we have considered only the first three IMFs out of eight IMFs obtained after empirical mode decomposition. The reason being that other IMFs contribute very little toward the signal and would not show desired results during classification. DT and KNN algorithms are used for

Table 47.1 p-value table

Features/IMF no.	IMF 1	IMF 2	IMF 3	IMF 4	IMF 5	IMF 6
Mean	3.77E-08	1.15E-06	**0.0905**	**0.1992**	**0.2696**	**0.3683**
Median	0.0012	5.44E-05	**0.1896**	**0.8523**	**0.1496**	**0.2199**
Variance	1.71E-60	7.30E-79	1.03E-48	8.40E-25	1.15E-21	1.66E-16
Standard deviation	5.26E-128	2.30E-143	1.79E-99	8.29E-63	4.76E-49	8.80E-47
Maximum	4.11E-141	3.41E-132	2.06E-104	7.92E-63	1.97E-42	3.64E-38
Minimum	6.03E-143	9.31E-135	8.74E-97	9.43E-65	2.78E-39	3.32E-39
Skewness	**0.7804**	**0.9732**	**0.6028**	5.04E-06	**0.428**	**0.7193**
Kurtosis	3.90E-32	1.94E-31	2.29E-12	7.93E-04	0.0043	0.0264
Root mean square	2.48E-127	3.27E-143	6.42E-100	8.67E-63	3.57E-49	1.92E-46
Shape factor	8.87E-20	6.22E-29	3.93E-10	0.0208	0.0131	0.0348
Impulse factor	1.53E-53	3.57E-53	4.91E-18	0.0309	**0.1989**	**0.553**
Crest factor	2.98E-68	1.10E-63	2.00E-19	**0.0554**	**0.8196**	**0.5145**

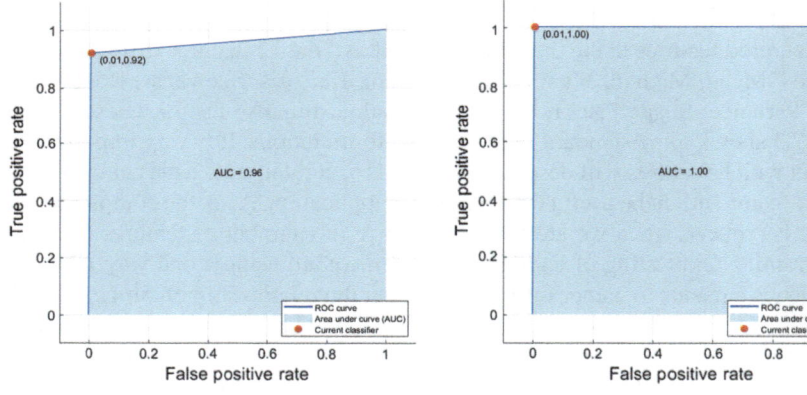

(a) ROC curve without channel selection.

(b) ROC curve when top 50 channels were selected from each dataset.

Fig. 47.4 ROC curves when KNN algorithm was used for classification

classifying the EEG signals into seizure and non-seizure data. DT starts from the root node and then compares the values with trained data, takes the decision and moves toward the leaf nodes and continues the same process making a tree-like structure by taking a decision at every node. KNN is a supervised method that uses the trained data to classify new data points into class based on the similarity. However, KNN is preferred over DT, as in DT algorithm even a small change in the data leads to larger changes in its structure and provides relatively inaccurate results. The performance of a classifier is judged by the computation of sensitivity (SEN), specificity (SPE), and accuracy (A).

Table 47.2 Classification results

Methods	Algorithm	IMF1			IMF2			IMF3		
		SEN	SPE	A	SEN	SPE	A	SEN	SPE	A
W/O channel	DT	97.2	92.7	96.4	98.02	96.84	97.8	94.29	79.3	91.4
Selection	KNN	98.02	96.8	97.8	98.02	96.8	97.6	93.3	88.88	92.6
Top 20	DT	100	95.23	99	96.25	85	94	90.3	70.5	87
Channels	KNN	**100**	95.23	**99**	97.53	94.73	97	96.25	85	94
Top 50	DT	99	96	98.4	98.5	95.9	98	94.66	88.63	93.6
Channels	KNN	**100**	98	**99.6**	99	97.95	98.8	94.66	88.63	93.6

The performance of different cases is tabulated in Table 47.2. In the case when we applied only feature selection and no channel selection we got an accuracy of 97.8% and sensitivity of 98% using the KNN algorithm and the corresponding receiver operating characteristics (ROC) curve is shown in Fig. 47.4a. For the remaining two cases, both feature selection and channel selection were applied. In the second case, the top 20 channels were selected and classification was applied to them and we got an accuracy of 99% and 100% sensitivity with the same KNN algorithm. For the final case, we have selected the top 50 channels and applied the classification then we got an accuracy of 99.6% and again 100% sensitivity using the same KNN algorithm and the corresponding ROC curve is shown in Fig. 47.4b. We have compared our results with that of Liang et al. [4] since they have worked on the same EEG dataset. Reference [4] randomly selects 60% features from the dataset for classification and got an accuracy of 97.82–98.51% which was better than methods used by others for classifying seizure and non-seizure signals by that time. Our proposed method provides an accuracy of 99.60% which is much better as compared to previously used methods. These precise results along with low computational load make the proposed method a very efficient approach to achieve real-time implementation of EEG data for seizure detection.

47.4 Conclusion

EMD process is used in this paper to decompose highly complex EEG signals. The features were extracted from the IMFs generated by EMD process. All the features obtained in this paper are of time domain, and further features are selected based on a one-way ANOVA test. Further, the method of maximum variance is used for channel selection. Out of 100 channels from each data set, the top 20 and top 50 most variant channels from each data set in two separate cases are considered for classification. Only feature selection doesn't provide sufficiently accurate results, this motivated us to optimize the number of channels processed as well. Selection of most variant channels removes redundant channels, reduces the computational time and avoids overfitting of data during classification. Thus a unique combination of feature selection and channel selection is used in this paper for the classification of

EEG signals into seizure and non-seizure. The results obtained indicate that using accurate features and avoiding redundant channels has led to an improvement in the accuracy as it avoids overfitting of channel data during classification. Further, the case in which 50 channels were selected gives slightly better results than the selection of 20 channels, as in the former case more information has been fed to the classifiers. Thus even without losing much information the signals are predicted with higher accuracy. The proposed method using the KNN classifier has provided a better classification accuracy, and could achieve real-time performance due its low computational complexity. This method can further be used for other nonstationary signals as well.

References

1. Iasemidis, L.D., Shiau, D.S., Chaovalitwongse, W., Sackellares, J.C., Pardalos, P.N., Principe, J.C., Carney, P.R., Prasad, A., Veeramani, B., Tsakalis, K.: Adaptive epileptic seizure prediction system. IEEE Trans. Biomed. Eng. **50**(5), 616–627 (2003). May
2. Lehnertz, K., Elger, C.E.: Spatio-temporal dynamics of the primary epileptogenic area in temporal lobe epilepsy characterized by neuronal complexity loss. Electroencephalogram. Clin. Neurophysiology. **95**(2), 108–117 (1995)
3. A. Accardo, M. Affinito, M. Carrozzi, and F. Bouquet, "Use of the fractal dimension for the analysis of electroencephalographic time series," Biol Cybern., vol. 77, pp. 339 350, 1997
4. Srinivasan, V., Eswaran, C., Sriraam, N.: Approximate entropy-based epileptic EEG detection using artificial neural networks. IEEE Trans. Inf. Technol. Biomed. **11**(3), 288–295 (2007). May
5. Lay-Ekuakille, Aimé, et al. "Entropy index in quantitative EEG measurement for diagnosis accuracy." IEEE Transactions on Instrumentation and Measurement 63.6 (2013): 1440-1450
6. Oweis, R.J., Abdulhay, E.W.: Seizure classification in EEG signals utilizing Hilbert-Huang transform. Biomed. Eng. OnLine **10**, 38 (2011)
7. Pachori, R.B., Bajaj, V.: Analysis of normal and epileptic seizure EEG signals using empirical mode decomposition. Comput. Methods Progr Biomed. **104**(3), 373–381 (2011)
8. V. Bajaj and R. B. Pachori 2012 Classification of Seizure and non-seizure EEG Signals Using Empirical Mode Decomposition, IEEE TRANSACTIONS ON INFORMATION TECHNOLOGY IN BIOMEDICINE, VOL. 16, NO. 6
9. Liang, S.F., Wang, H.C., Chang, W.L.: Combination of EEG complexity and spectral analysis for epilepsy diagnosis and seizure detection. EURASIP J. Adv. Signal Process. **2010**, 853434 (2010)
10. Andrzejak, R.G., Lehnertz, K., Mormann, F., Rieke, C., David, P., Elger, C.E.: Indications of nonlinear deterministics and finite-dimensional structures in time series of brain electrical activity: Dependence on recording region and brain state. Phys. Rev. E **64**(6), 061907 (2001)
11. Alotaiby, T., El-Samie, F.E.A., Alshebeili, S.A., et al.: A review of channel selection algorithms for EEG signal processing. EURASIP J. Adv. Signal Process. **2015**, 66 (2015). https://doi.org/10.1186/s13634-015-0251-9
12. Flandrin, P., Rilling, G., Goncalves, P.: Empirical mode decomposition ' as a filter bank. IEEE Signal Process. Lett. **11**(2), 112–114 (2004). Feb.
13. Schuyler, R., White, A., Staley, K., Cios, K.J.: Epileptic seizure detection. IEEE Eng. Med. Biol. Mag. **26**(2), 74–81 (2007)
14. Dastidar, S.G., Adeli, H., Dadmehr, N.: Mixed-band wavelet-chaos neural network methodology for epilepsy and epileptic seizure detection. IEEE Trans. Biomed. Eng. **54**(9), 1545–1551 (2007). Sep.
15. Ostertagová, Eva, Ostertag, Oskar: Methodology and Application of One-way ANOVA. American Journal of Mechanical Engineering **1**(7), 256–261 (2013). https://doi.org/10.12691/ajme-1-7-21

Chapter 48
Index-Based Improved High Capacity Data Hiding Technique

Pratap Chandra Mandal and Imon Mukherjee

Abstract Steganography is the art of secret communication to avoid detection by an unauthorized person. The pixel value differencing (PVD) technique hides the confidential message into the difference between two adjacent pixels. The PVD-based techniques have less embedding capacity due to unidirectional embedding. An index-based data hiding is one of the solutions to enhance the embedding capacity. It has been noticed that index-based data hiding fails to recover accurate data for some cases. This identified problem has been resolved in the proposed technique. It has achieved an average embedding capacity (AEC) of 2.49 bpp which is better than the index-based data hiding technique. The proposed technique withstands steganalysis attacks.

48.1 Introduction

The progress of recent digital technology has made data transmission quicker. On the other hand, it has also made the intrusion of data easier. Thus, protecting the privacy of data during communication is a severe problem. Data hiding [6, 11, 14–17] techniques hide messages into cover media imperceptibly. Steganography [2, 3, 8, 10, 12, 13] is a data hiding technique that hides the confidential data invisibly.

 Smooth regions of an image contain pixels with similar features. Massive alteration in the smooth area can easily be noticed by the human eye. An edge area can be used for embedding massive bits. Based on this concept, PVD [18] method embeds data on the difference between the two adjacent pixels. The drawback of such a steganography method is the low payload due to unidirectional embedding. To enhance the payload of the PVD method, an index-based steganographic scheme

P. C. Mandal (✉)
B.P. Poddar Institute of Management & Technology, Kolkata, India
e-mail: pcmandal9@gmail.com

I. Mukherjee
Indian Institute of Information Technology Kalyani, Kalyani, India
e-mail: imon@iiitkalyani.ac.in

© The Author(s), under exclusive license to Springer Nature Singapore Pte Ltd. 2022   491
V. Bhateja et al. (eds.), *Evolution in Computational Intelligence*,
Smart Innovation, Systems and Technologies 267,
https://doi.org/10.1007/978-981-16-6616-2_48

is presented in the work [7], but the technique fails to recover the exact message for some cases. The work [6] offers higher EC using PVD and LSB methods concurrently in the bit plane. Later, Li et al. [8] mixes PVD, MF, and particle swarm optimization (PSO) techniques for achieving improved capacity. The work [9] presented a data hiding technique combining PVD and side match methods to enhance the embedding capacity. Liu et al. [11] have presented an RDH technique in the encrypted image (RDHEI) by redundant space transfer (RST) scheme. The work [14] also has presented an RDHEI scheme by redundancy transfer and sparse block encoding. The limitation of both schemes is the lower embedding capacity.

48.1.1 Motivation and Challenges

The index-based data hiding technique [7] has been presented to improve the embedding capacity of the PVD technique. It has been noticed that the scheme [7] has not taken all the cases between the basic pixel and three other pixels during the embedding process. Therefore, that scheme fails to recover accurate data for some cases. This identified problem has been resolved in the proposed technique.

The challenge we have faced here is to preserve the trade-off between the payload and the quality of the images. To maintain a decent visual quality, we have chosen a threshold value that is set at 5. The overflow/underflow problems in this technique are overcome by the pixel value adjustment strategy.

48.1.2 Contribution

- The proposed technique resolves the limitation of the index-based data hiding technique [7].
- Without compromising on the quality of the images, the AEC of all the test images becomes 2.49 bpp.
- It withstands steganalysis attacks.

Index-based data hiding

In index-based steganographic approach [7], images are partitioned into non-overlapping 2×2 blocks. A basic pixel is calculated using a basic pixel selection function. The number of bits to be embedded is determined by the difference between the basic pixel and three other pixels. The weakness of the scheme [7] is that during embedding it has not taken into account all the cases for three-pixel pairs as shown in Step 7 and Step 8 in Algorithm 48.1. Therefore, the scheme fails to recover the accurate data for some cases.

Basic pixel selection function

This function determines the basic pixel. Images are divided into $K \times K$ blocks. Value of K is chosen here as 2. Basic pixel selection function $F\ (e, f)$ can be defined using Eq. (48.1) as

$$F(e, f) = [(K \times (e \bmod K) + (K - 1) \times (f \bmod K)] \bmod K^2, \quad (48.1)$$

where e, f are the index values of row and column, respectively.

Assume, the secret information be "011100110000011..." which is to be embedded in a 2×2 blocks containing the pixels 26, 61, 76, and 94. Applying the embedding process of the work [7] the stego pixels become 30, 69, 78, and 91 and the secret information which is retrieved in the extraction process from four stego pixels becomes "101011011101". Both are not the same. Hence, the algorithm fails to retrieve accurate data in this case. To overcome the limitation of work [7], an improved data hiding technique has been proposed by considering all the cases that may occur during the embedding process and it retrieves the accurate data.

The rest of the paper is designed as follows: Sect. 48.2 presents the proposed work and Sect. 48.3 shows the experimental results. The work has been concluded in Sect. 48.4.

48.2   Proposed Work

In this technique, cover images are divided into non-overlapping 2×2 blocks. Blocks are selected one by one until the last block is reached.

48.2.1   Embedding Process

Algorithm 48.1 embeds n bits message using PVD technique in three-pixel pairs in horizontal, vertical, and diagonal directions within a block.

48.2.2   Extraction Process

In the receiver end, the image is partitioned into 2×2 blocks. The blocks are selected one by one after extracting data from a block. Algorithm 48.2 presents the extraction process for extracting data from a block containing four stego pixels.

Algorithm 48.1: *Embedding*

#This algorithm embeds data into a block containing four pixels and produces corresponding stego pixels as:

1. Determine the basic pixel using the basic pixel selection function.
2. Compute the differences $d_{i,i \in \{1,2,3\}}$ from each pixel pair containing the basic pixel using Eq. (48.2). Let us consider $P(u + 1, v + 1)$ as the basic pixel,

$$d_{i,i \in \{1,2,3\}} \leftarrow \begin{cases} |P(u + 1, v + 1) - P(u + 1, v)|, & \text{for } i = 1. \\ |P(u + 1, v + 1) - P(u, v + 1)|, & \text{for } i = 2. \\ |P(u + 1, v + 1) - P(u, v)|, & \text{for } i = 3. \end{cases} \quad (48.2)$$

3. Compute the number of bits to be embedded for each d_i using the range table. A threshold value T limits the number of embedding bits as:

$$n_{i,i \in \{1,2,3\}} \leftarrow min(\lfloor \log_2 (u_i^{(a)} - l_i^{(a)} + 1) \rfloor, T), 1 \le a \le 6. \quad (48.3)$$

4. Read n_i bits from secret bit array S for each d_i using Eq. (48.3), and convert to decimal B_i.
5. The new differences d_i' are calculated by adding B_i with the lower bound $l_i^{(a)}$ using Eq. (48.4) as:

$$d_{i,i \in \{1,2,3\}}' \leftarrow \left| l_i^{(a)} + B_i \right|, \qquad 1 \le a \le 6. \quad (48.4)$$

6. Calculate the difference between d_i and d_i' as m_i using Eq. (48.5),

$$m_{i,i \in \{1,2,3\}} \leftarrow \left| d_i - d_i' \right|, \quad (48.5)$$

7. Let us consider, $P_0 \leftarrow P(u + 1, v + 1)$, $P_1 \leftarrow P(u + 1, v)$, $P_2 \leftarrow P(u, v + 1)$ and $P_3 \leftarrow P(u, v)$. Compute the new pixel pair P_0' and P_1' for the basic pixel pair as:

if $((P_0 \ge P_1 \ \& \ d_1 \ge d_1') \ || \ (P_0 < P_1 \ \& \ d_1 < d_1'))$ **then**
$\quad \lfloor \ (P_0', P_1') \leftarrow (P_0 - \lceil \frac{m_1}{2} \rceil, P_1 + \lfloor \frac{m_1}{2} \rfloor)$
if $((P_0 \ge P_1 \ \& \ d_1 < d_1') \ || \ (P_0 < P_1 \ \& \ d_1 \ge d_1'))$ **then**
$\quad \lfloor \ (P_0', P_1') \leftarrow (P_0 + \lceil \frac{m_1}{2} \rceil, P_1 - \lfloor \frac{m_1}{2} \rfloor)$

8. Remaining two pixel pairs are modified to new pixel values for a given value $k, k \in \{2, 3\}$ as:

if $((P_0 \ge P_k \ \& \ d_k \ge d_k') \ || \ (P_0 < P_k \ \& \ d_0 < d_k'))$ **then**
$\quad \lfloor \ (P_0', P_k') \leftarrow (P_0 - \lceil \frac{m_1}{2} \rceil, P_k - \lceil \frac{m_1}{2} \rceil + m_k)$
if $((P_0 \ge P_k \ \& \ d_k < d_k') \ || \ (P_0 < P_k \ \& \ d_k \ge d_k'))$ **then**
$\quad \lfloor \ (P_0', P_k') \leftarrow (P_0 + \lceil \frac{m_1}{2} \rceil, P_k + \lceil \frac{m_1}{2} \rceil - m_k)$

9. The basic pixel is adjusted to maintain the identical value for all pixel pairs.

Algorithm 48.2: *Extraction*

#This algorithm extracts n secret bits from three pixel pairs within a block containing four stego pixels as:

1. Identify the basic pixel using the basic pixel selection function.
2. Compute the difference values, $D'_{i,i\in\{1,2,3\}}$ between basic pixel $P'(u+1, v+1)$ and three other pixels using Eq. (48.6),

$$
D'_{i,i\in\{1,2,3\}} \leftarrow
\begin{cases}
\left| P'(u+1, v+1) - P'(u+1, v) \right|, & \text{for } i = 1. \\
\left| P'(u+1, v+1) - P'(u, v+1) \right|, & \text{for } i = 2. \\
\left| P'(u+1, v+1) - P'(u, v) \right|, & \text{for } i = 3.
\end{cases}
\tag{48.6}
$$

3. Determine the lower bound $l_i^{(a)}$ from the range table for each D'_i, where $1 \le a \le 6$.
4. Compute the decimal value B'_i of the secret bits as:

$$
B'_{i,i\in\{1,2,3\}} \leftarrow \left| D'_i - l_i^{(a)} \right|, \text{ where } 1 \le a \le 6.
\tag{48.7}
$$

5. The number of embedded bits n'_i from pixel pair can be computed using the threshold value T as:

$$
n'_{i,i\in\{1,2,3\}} \leftarrow min(\lfloor \log_2(u_i^{(a)} - l_i^{(a)} + 1) \rfloor, T), \text{ where } 1 \le a \le 6.
\tag{48.8}
$$

6. Convert each decimal value B'_i to n'_i binary bits using Eqs. (48.7) and (48.8).
7. Concatenate all the bits in proper order.

Sample example

A sample example is taken to make the problem clear. The threshold value T is set at 5. Figure 48.1 presents the embedding of data in a block containing four pixels, and Fig. 48.2 presents the extraction of data from that block. From the example it is confirmed that the proposed technique can extract accurate data.

48.2.3 Overflow/Underflow Problem

During embedding, the boundary pixels may generate overflow/underflow problems. Let us take four stego pixels as BS, A, B, and C. The overflow problems have been resolved in this technique by pixel value adjustment strategy as follows:

If the value of the pixel BS surpasses 255, adjust the other pixels by subtracting $|BS - 255|$ from them and then update the value BS to 255, i.e.,

$BS > 255$: $A = A - |BS - 255|$, $B = B - |BS - 255|$, $C = C - |BS - 255|$, $BS = 255$.

Same strategy is taken for the other pixels.

In case of underflow problem, if pixel BS becomes less than 0, adjust the other pixels by adding the absolute value of BS with them and set BS to 0 as

$BS < 0$: $A = A + |BS|$, $B = B + |BS|$, $C = C + |BS|$, $BS = 0$.

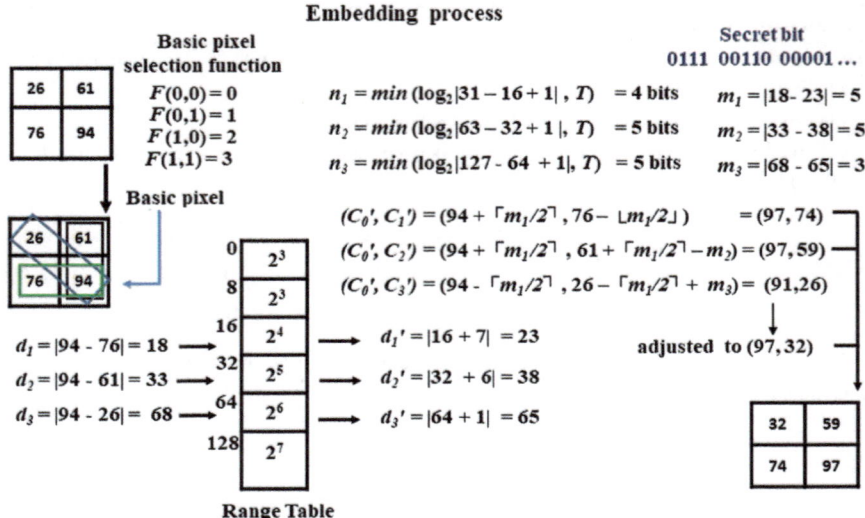

Fig. 48.1 Example of embedding process within a block

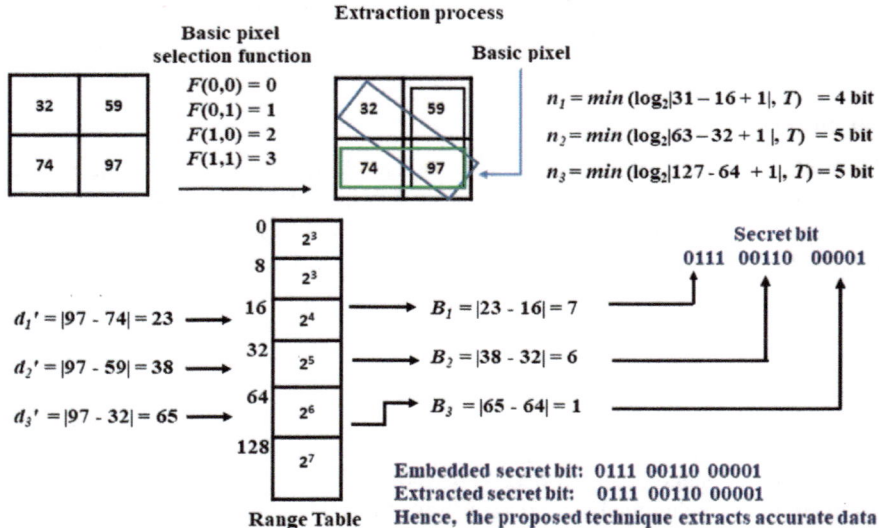

Fig. 48.2 Example of extraction process

Same strategy is taken for other pixels.

Theorem 48.1 *In the proposed technique, the maximum possible distortion in a pixel is $2^n - 1$, where $3 \leq n \leq 5$.*

Proof Each block has three-pixel pairs including the basic pixel pair. In the case of the basic pixel pair, the highest distortion in any pixel is 2^{n-1}, where $3 \leq n \leq 5$. The threshold value is selected at 5. The highest distortion in the pixels in other pixel pairs except basic pixel is $2^n - 1$, where $3 \leq n \leq 5$. Therefore, the maximum pixel distortion in this technique $= max\ (2^{n-1}, 2^n - 1) = 2^n - 1$. Hence proved. □

48.3 Experimental Results

All the experiments are performed using 500 grayscale images of size 512×512 [4, 5]. The whole image is used for embedding the confidential data. Algorithms are programmed using MATLAB R2015a. The performance of this technique is compared with the recent state-of-the-art works [6–9, 11, 14].

48.3.1 Embedding Capacity

The embedding capacity of the proposed technique is measured by bits per pixel (bpp). The AEC of 500 images is 2.49 bpp.

48.3.2 Visual Quality

The proposed work is measured by the quality assessment metrics like, PSNR, MSSIM index, UIQI, and NCC. The average PSNR and MSSIM of all the test images is 38.87 dB and 0.994. The average UIQI of the images is 0.997. The average NCC value of all the test images is $0.996 \approx 1$.

48.3.3 Resistance Against Different Attacks

Security of the technique has been verified concerning the resistance against steganalysis attack, viz., StegExpose.

Steganalysis using StegExpose

Table 48.1 Steganalysis results of StegExpose

File name	Stego?	Primary Set	χ^2	Sample pairs	RS analysis	Fusion (mean)
Lena	FALSE	0.02887	0.00129	0.02909	0.01718	0.01911
Lena-st	FALSE	0.03765	0.00005	0.02205	0.00441	0.01616
Baboon	FALSE	0.02899	0.03328	0.04561	0.01451	0.03060
Baboon-st	FALSE	0.07751	0.00219	0.08135	0.05370	0.06619
Couple	FALSE	0.01462	0.00211	0.00372	0.00482	0.00632
Couple-st	FALSE	0.03259	0.00150	0.02517	0.00979	0.01726
Elaine	FALSE	0.00828	0.00007	0.00153	0.00669	0.00430
Elaine-st	FALSE	0.05238	0.00225	0.02956	0.02861	0.02820
Barbara	FALSE	0.06531	0.00614	0.02778	0.01992	0.02979
Barbara-st	FALSE	0.05148	0.00007	0.02490	0.02551	0.02566
Peppers	FALSE	0.01126	0.00303	0.01077	0.01278	0.00946
Peppers-st	FALSE	0.05120	0.00132	0.04800	0.05542	0.06398

To confirm the security of the proposed technique, steganalysis tests such as Primary sets, χ^2, Sample pairs, RS analysis, and Fusion detection are conducted using a standard steganalysis tool, viz., "StegExpose" [1].

Table 48.1 presents the output after analyzing the images by StegExpose with the threshold value at 0.1. An image is regarded as a stego if fusion (mean) crosses the threshold value. Six sample cover images and their corresponding stego versions are presented in Table 48.1. All the values are near to zero and the difference between the analyzed value of the cover image and their corresponding stego form is negligible. It confirms the security of the technique.

48.3.4 Comparative Analysis with Existing Algorithms

The proposed technique is compared with the recent state-of-the-art works [6–9, 11, 14]. All the techniques have used 512×512 images. The comparison of the proposed technique with works [6, 7, 11, 14] is shown in the Table 48.2. It outperforms the index-based data hiding [7] method with AEC and better image quality. The work [6] offers higher AEC using PVD and LSB methods concurrently in the bit plane, but the image quality of the proposed technique is higher than the work [6]. Both the schemes [11, 14] have achieved higher PSNR value with lower AEC compared to the proposed technique.

The proposed work also compared with the techniques [8, 9] and presented in Table 48.3. Our scheme outperforms the work [8] in terms of AEC and the work [9] in terms of visual quality of the images.

Table 48.2 Comparison of AEC and PSNR with the recent state-of-the-art works

Schemes	Metrics	Airplane	Baboon	Peppers	Lena	Man	Lake	City	Average
Jung [6]	AEC	4.00	4.02	4.00	4.00	4.01	4.00	4.00	**4.00**
	PSNR	33.13	31.74	33.55	33.21	32.72	33.84	32.06	**32.89**
Jung et al. [7]	AEC	2.35	2.62	2.33	2.35	2.43	2.42	2.42	**2.42**
	PSNR	30.66	25.96	30.4	31.94	28.63	28.99	30.45	**29.58**
Liu et al. [11]	AEC	1.71	0.62	1.61	1.61	1.29	–	–	**1.37**
	PSNR	41.00	47.74	41.70	41.49	43.21	–	–	**43.03**
Qin et al. [14]	AEC	2.08	0.61	1.81	1.84	1.58	–	–	**1.58**
	PSNR	43.79	53.31	45.56	45.46	46.49	–	–	**46.96**
Proposed scheme	AEC	2.36	2.78	2.33	2.34	2.38	2.44	2.44	**2.44**
	PSNR	38.92	38.40	38.96	38.95	38.88	38.81	38.85	**38.82**

Table 48.3 Comparison of AEC and PSNR with the recent state-of-the-art work

Schemes	Metrics	Lena	Baboon	Peppers	Tiffany	Jet	Average
Li et al. [8]	AEC	2.14	2.64	2.14	2.12	2.11	**2.23**
	PSNR	42.74	36.63	42.45	42.96	43.23	**41.60**
Liu et al. [9]	AEC	2.71	3.08	2.72	2.71	2.74	**2.79**
	PSNR	36.70	32.04	34.83	35.91	36.19	**35.14**
Proposed scheme	AEC	2.34	2.78	2.33	2.50	2.48	**2.49**
	PSNR	38.95	38.40	38.96	38.83	38.94	**38.82**

48.4 Conclusion

The proposed technique has resolved the limitation of the previous index-based steganographic technique. The embedding capacity of the PVD method is utilized to embed the data in horizontal, vertical, and diagonal directions. The proposed method achieves high embedding capacity while maintaining good visual quality by using a threshold value. It has attained an AEC of 2.49 bpp for all the test images. The proposed work withstands steganalysis attacks. In future, we will try to implement the technique in the frequency domain.

DECLARATION: We have taken permission from competent authorities to use the images/data as given in the paper. In case of any dispute in the future, we shall be wholly responsible.

References

1. Boehmm, E.: Stegexpose: a tool for detecting LSB steganography (2014). https://github.com/b3dk7/StegExpose
2. Das, S., Muhammad, K., Bakshi, S., Mukherjee, I., Sa, P.K., Sangaiah, A.K., Bruno, A.: Lip biometric template security framework using spatial steganography. Pattern Recognit. Lett. **126**, 102–110 (2019)
3. Grajeda Marín, I.R., Montes Venegas, H.A., Marcial Romero, J.R., Hernández Servín, J.A., Muñoz Jiménez, V., Luna, G.D.I.: A new optimization strategy for solving the fall-off boundary value problem in pixel-value differencing steganography. Int. J. Pattern Recognit. Artif. Intell. **32**(01), 1860010–17 (2018)
4. Images. The BOSSbase-1.01 Database. Binghamton University. Available from: http://dde.binghamton.edu/download/
5. Images. University of Southern California. The USC-SIPI Image Database (2019). Available from: http://sipi.usc.edu/database/database.php
6. Jung, K.H.: Data hiding scheme improving embedding capacity using mixed PVD and LSB on bit plane. J. Real-Time Image Process. **14**(1), 127–136 (2018)
7. Jung, K.H., Yoo, K.Y.: High-capacity index based data hiding method. Multimed. Tools Appl. **74**(6), 2179–2193 (2015)
8. Li, Z., He, Y.: Steganography with pixel-value differencing and modulus function based on PSO. J. Inf. Secur. Appl. **43**, 47–52 (2018)
9. Liu, H.H., Lin, Y.C., Lee, C.M.: A digital data hiding scheme based on pixel-value differencing and side match method. Multimed. Tools Appl. **78**(9), 12157–12181 (2019)
10. Liu, W., Yin, X., Wei, L., Zhang, J., Zeng, J., Shi, S., Mao, M.: Secure halftone image steganography with minimizing the distortion on pair swapping. Signal Process. **167**, 1–10 (2020)
11. Liu, Z.L., Pun, C.M.: Reversible data-hiding in encrypted images by redundant space transfer. Inf. Sci. **433**, 188–203 (2018)
12. Mandal, P.C., Mukherjee, I., Chatterji, B.N.: High capacity reversible and secured data hiding in images using interpolation and difference expansion technique. Multimed. Tools Appl. 1–22 (2021)
13. Paul, G., Davidson, I., Mukherjee, I., Ravi, S.S.: Keyless dynamic optimal multi-bit image steganography using energetic pixels. Multimed. Tools Appl. **76**(5), 7445–7471 (2017)
14. Qin, C., Qian, X., Hong, W., Zhang, X.: An efficient coding scheme for reversible data hiding in encrypted image with redundancy transfer. Inf. Sci. **487**, 176–192 (2019)
15. Shen, S.Y., Huang, L.H.: A data hiding scheme using pixel value differencing and improving exploiting modification directions. Comput. Secur. **48**, 131–141 (2015)
16. Singh, A.K.: Data hiding: current trends, innovation and potential challenges. ACM Trans. Multimed. Comput. Commun. Appl. (TOMM) (2020)
17. Weng, S., Zhang, C., Zhang, T., Chen, K.: High capacity reversible data hiding in encrypted images using SIBRW and GCC. J. Vis. Commun. Image Represent. 102932 (2020)
18. Wu, D.C., Tsai, W.H.: A steganographic method for images by pixel-value differencing. Pattern Recognit. Lett. **24**(9–10), 1613–1626 (2003)

Chapter 49
IRIS Position-Based Wheelchair Maneuver Using Semantic Segmentation

Hrithik Aditya, Vishal Chawla, Rohit Maheswari, and A. G. Keskar

Abstract The paper presents a novel system for wheelchair maneuver to aid people suffering from limb-related paralysis. It features Iris Center Localization using Hough Transform for tracking the direction of eye gaze. The limitations observed with Hough transform in previous research works were resolved using Semantic Segmentation for improved detection of the Iris against the Sclera background. Hence, we have demonstrated how Deep Learning models can perform Semantic Segmentation for Object Detection. Our approach gave an accuracy of 99.2% and allowed the wheelchair to move with more precision due to the scope of calibration of the motor torque/speed with respect to the determined eye gaze estimate. Moreover, different methods to avoid object collision were studied and implemented. Further, we provide an abstract design for the proposed wheelchair.

49.1 Introduction

The project aims to develop a wheelchair system capable of being maneuvered by a physically challenged human being either disabled by incapacitation or otherwise impaired from any limb-related paralysis. For this purpose, the direction of eye gaze is used as an indicator of the direction in which the user wants to navigate. Our project work aggregates several challenges confronted while developing such a system that effectively detects as well as tracks the location of the iris on a real-time basis. As a part of our survey, we have investigated several methods before presenting a Deep Learning model for Iris detection based on Semantic Segmentation and Iris Center Localization using Hough Transform. Our project further incorporates another Deep Learning model for detecting obstacles such as road traffic and pedestrians in the path of the wheelchair. The work carried out bequeaths an abstract design of the wheelchair and the study of various techniques related to image processing using Deep learning to effectuate Semantic Segmentation and their application.

H. Aditya (✉) · V. Chawla · R. Maheswari · A. G. Keskar
Department of Electronics and Communication Engineering, Visvesvaraya National Institute of Technology, Nagpur, India

501
V. Bhateja et al. (eds.), *Evolution in Computational Intelligence*,
Smart Innovation, Systems and Technologies 267,
https://doi.org/10.1007/978-981-16-6616-2_49

49.2   Literature Review

Several attempts have been made to develop an efficient wheelchair system. Through our survey, we have evaluated some of those projects and the technologies which they adopted. We critique the following drawbacks regarding the same.

Eye-controlled wheelchair using Hough transform [1] was a project which aimed at tracking the movement of the eye using the Hough transform. A circle can be described by

$$(x - a)^2 + (y - b)^2 = r^2 \tag{49.1}$$

In order to determine the center coordinates (a, b) and the radius, r we must consider three points on the same arc such that

$$
\begin{aligned}
(x_1-a)^2 + (y_1-b)^2 &= r^2 \\
(x_2-a)^2 + (y_2-b)^2 &= r^2 \\
(x_3-a)^2 + (y_3-b)^2 &= r^2
\end{aligned}
\tag{49.2}
$$

The Hough transform algorithm can then transform the image domain, containing the edge point set (x_n, y_n) into the parameter domain, which contains the corresponding parameter set (a_n, b_n). The Hough transform offers good anti-jamming performance as it is capable enough to extract comprehensive description parameters on the basis of local metrics. Because of its robustness, we shall apply the Hough transform algorithm for our objectives and resolve low accuracy issues by proper separation of the iris from the rest of the eye background(sclera) using Semantic Segmentation.

Smart wheelchair with object detection using deep learning [2] was a project that aimed at using deep learning for eye gaze tracking but there were unsolved issues with classification. In Eyeball Gesture Controlled Automatic Wheelchair Using Deep Learning [3] the position of the eye gaze was detected using deep learning but the eyeball gestures were limited to simply left, right, or straight. Also, the accuracy achieved was moderate. Therefore, accuracy had to be improved and better classification for the model was needed.

The person in the wheelchair might wear spectacles and hence reflected light or glare from the spectacles might impair proper detection of the eye and cause difficulty in any further computer vision tasks. The issue can be addressed by separating reflection layers from the desired background. A speculative study about eyeglass reflection properties and reflection layer separation has been done in [4–7] which can be utilized in our system as a part of preprocessing.

Voice-controlled wheelchair [8] was a project which aimed at using the voice commands to act as input for the wheelchair and based on the command the wheelchair would move to the left, right, or straight. The speech recognition was

done by a voice recognition module, connected with Arduino. The wheelchair oper-
ated on a real analogous voice signal of the user. The limitations associated with this
project were low accuracy due to noise and disturbances and thus we learned that it
was very important to control the noise to achieve superior accuracy.

EOG-driven wheelchair [9] was a project that used an Electrooculography(EOG)
system to measure the electrical potential of one's own body action to track the
movement of the retina. Signals were generated and sent to the wheelchair as input.
The issue with such a project was that it was extremely costly and required a lot
of human effort thus we understood that it was important to make a cost-effective
solution for our problem.

49.3 Design Specifications and Implementation

The wheelchair system needs to be equipped with two cameras interfaced to a micro-
controller; one to capture the image of the eye and another to monitor traffic in front
of the wheelchair for object detection (Fig. 49.1). The camera feed is given to the
system that performs video preprocessing to improve the clarity of each frame. The
preprocessed images are fed as input to our deep learning model for the segmentation
of the eye into the sclera and iris, of which the sclera class is disregarded as it is
unnecessary. The iris class being elliptical or nearly circular in geometry undergoes

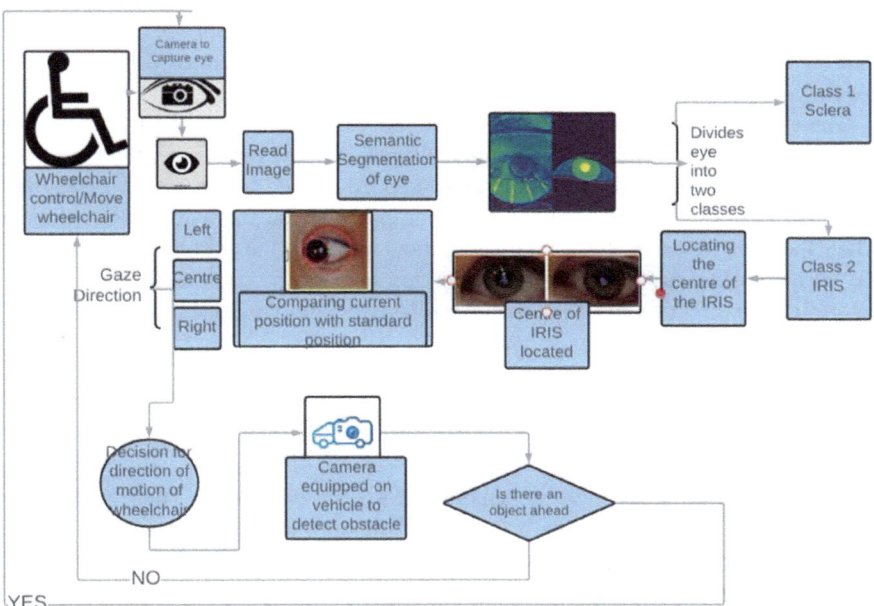

Fig. 49.1 Abstract wheelchair system

the process of centralization by the use of Hough Transform. Next, we estimate the direction of gaze based on the localized coordinates of the center of the iris. The estimate contains the direction of gaze as well as the extremity or the degree to which the user is looking to the left or to the right from the central forward-looking position of the eye. The wheelchair must contain motors capable of being calibrated with this estimate. The wheelchair should be interfaced with sensors that detect immediate obstacles in the path of the wheelchair so that if there is an obstacle present in front of the wheelchair there should not be any collision. Once these objects are identified, through the use of sensors like IR Sensors, Ultrasonic Sensors, we can calculate the proximity of such obstacles in the path of the wheelchair. However, the scope of the project is limited to software implementation of iris detection using segmentation and eye-tracking using eye center localization and object detection.

49.3.1 Iris Detection and Centralization

For segmentation, we have applied the concept of transfer learning. We have used the COCO model as our pretrained model which has been trained on the COCO (Common Objects in Context) dataset which is a popular open-source object recognition database. It is a Deep Learning R-CNN (Region-based Convolutional Neural Network) model. The model has been reused for our task of iris segmentation by training it with our own open eye dataset containing 250 images containing 200 images for training and 50 for testing. This optimization through the use of transfer learning allows the improved performance of the model for our specific task. The training has been done using ResNet-50 as backbone architecture with a total of 9 epochs and batch size as 1.

The center coordinates once localized after application of Hough Transform over a particular frame are stored. When the feed moves to the next frame the new coordinates of the center of the IRIS are computed using the same procedure. The new coordinates along the horizontal axis are compared to the standard iris coordinates when the user looks in the forward direction which is the middle position along the horizontal x-axis. The difference between the two coordinates is then used to determine the direction in which the user is looking and wishes to move.

49.3.2 Object Detection for Traffic Monitoring

The Haar Cascade Classifier implemented for detecting cars contains 13 stages having a total of 250 features with 4, 7, 12, 22 features in the first 4 stages. Another classifier for detecting a human being contains more than 1000 features with around 1400 decision nodes present in 30 stages.

For Segmentation of obstacles in the path of movement open-source Torchvision FCN (Fully Convolutional Network) Segmentation model has been used with

ResNet-101 as backbone trained on Pascal VOC dataset. It contains 20 classes of objects which include outdoor objects such as cars, bikes, buses, pedestrians, and indoor objects such as sofas, chairs, dining tables, potted plants. The complete study of the model architecture can be found in [10].

49.4 Result and Evaluation

49.4.1 Iris Detection and Tracking

Hough transform circumscribes the object and returns its center and radius. It is an extension of canny edge detection which allows the extraction of circular objects. Therefore its performance depends on the edges present in the image or in this case the border of the circle being distinguishable. Hence, image clarity, brightness, and contrast are important factors that affect the efficiency of Hough Transform. Noise and distortion in the images may be because of the inherent bias of the parameter space, statistical noise due to irrelevant structures, background noise as well as due to feature distortion [11]. If the difference between the pixel intensities between the edge and the region surrounding it is insufficient or if the image is blurred and the outline of the circle is not clear the algorithm fails to detect that circle.

For our experiment, we captured high-resolution images (225×225) from a 5MP camera. As we can observe the algorithm fails to detect the iris in Fig. 49.2 and detects multiple nonexistent circles in Fig. 49.3. Hence, we must find a way to improve the distinguishability of the iris to improve performance. However, in our current scope, we assume that the voting and peak detection mechanisms of Hough Transform are reliable. Hence, we apply a deep learning model for the semantic segmentation of the eye which distinguishes the iris highlighting it so that Hough Transform can now be applied subsequently.

Fig. 49.2 Missed detection in several frames

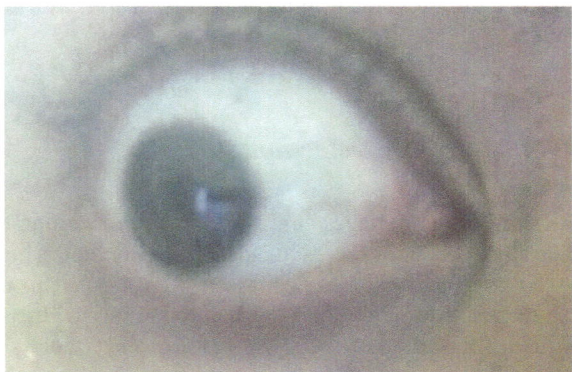

Fig. 49.3 Undesired circles
detected

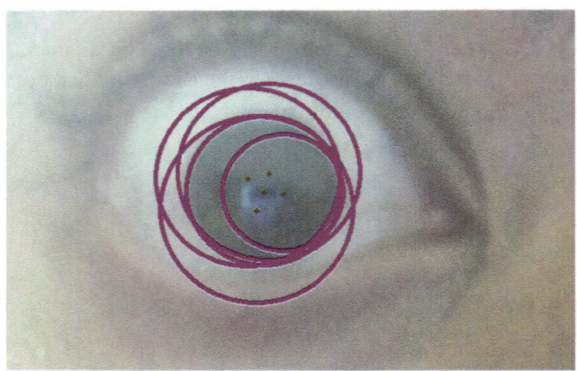

Fig. 49.4 Semantic
segmentation of eye to detect
the IRIS

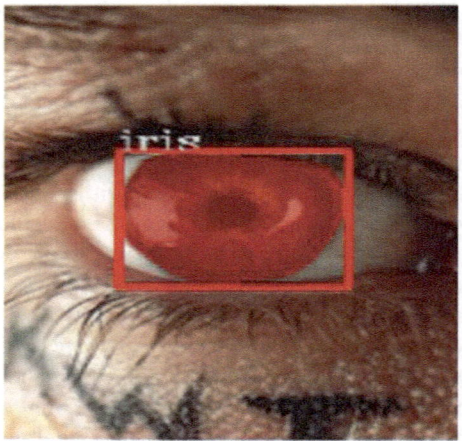

   The Deep Learning model offers a Mean Average Precision(mAP) of 0.992 which
shows an improvement in the accuracy of iris detection as compared to the models
reviewed in [3, 12].

49.4.2 Object Detection

As shown in the images below our machine learning model based on Haar Cascade
Classifier produces several misdetections including both False Positives (FP) as well
as False Negatives (FN). This can be attributed to uneven brightness or contrast
of images. Although this may be moderately addressed by performing Histogram
Equalization, the lack of sharp features nevertheless degrades the performance of
these classifiers. As the classifier is trained with positive and negative images of a
single object, the scope of using Haar Cascade is limited as the weights of features

Fig. 49.5 Prediction of
multiple false positives

Fig. 49.6 Missed detection;
FN prediction

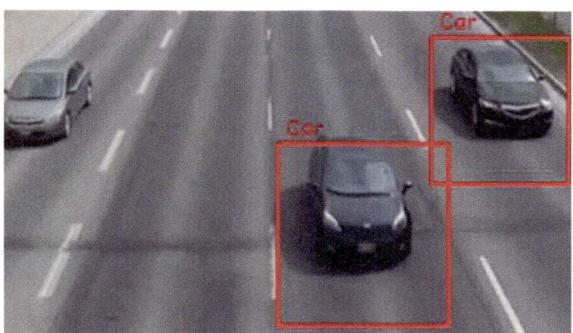

referring to different classes of objects cannot be combined in a single classifier and
hence the detection of each object will require a classifier of its own.

Therefore we augment a deep learning model into our wheelchair system for the
purpose of object detection. The model accurately detects 20 categories of obstacles
each of which is color mapped to be distinguishable. The model size is around
360 MB and the inference time is around 0.0175 s on an NVIDIA GTX 1080 TI
GPU. The fcn_resnet101 model structure offers 63.7% mean IoU (Intersection over
Union) and 91.9% Global Pixelwise Accuracy. More details about the model are
provided in [10].

49.4.3   Conclusion and Future Work

As per our objectives, we have suggested an abstract design for the entire wheelchair
system. Through our experiments, we have evaluated the limitations of the standard Hough Transform. We overcame the problems related to noisy images having
distorted features by first segmenting the iris using our deep learning model and
subsequently applying Hough Transform upon it. Further, we have provided a study
of two methods for Object Detection based on the use of Haar Cascade Classifiers
and semantic segmentation using Deep Learning FCN model. It can be concluded

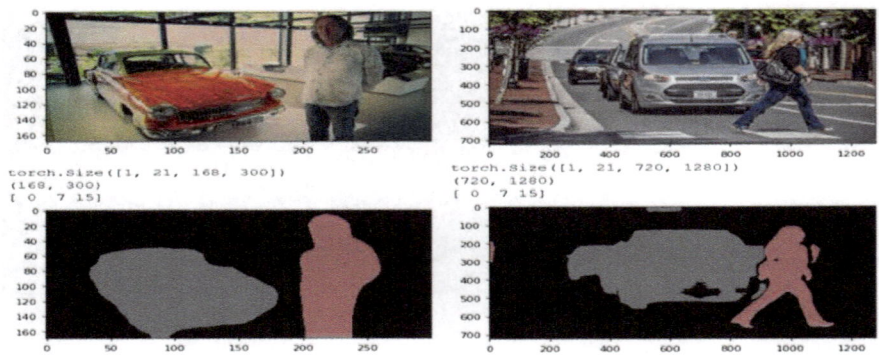

Fig. 49.7 Semantic segmentation; representation of different classes with different colors

Table 49.1 Comparison table between our wheelchair system versus other systems

	Xiong et al. [12]	Adarsh and Megha [3]	Anagha et al. [1]	Our
IRIS DETECTION	87% using Hough Transform	99.11% using Deep Learning	Haar Cascade and Hough Transform	99.2% using Hough Transform and Semantic Segmentation
OBJECT DETECTION	✕	✕	Using ultrasonic sensors (absolute accuracy 1–3%)	63.7% mean IoU, 91.9% Global Pixelwise Accuracy
Eyeball gestures detected	15 Gestures	3 Gestures	5 Gestures	Measures deviation from central position

that the FCN model is more reliable and it provides a wider scope of detection of 20 classes by the use of a single model. We have also expanded the scope of previous projects which only considered three eyeball gestures; left, right, or center. Through the use of Hough Transform to monitor relative changes in the position of the IRIS, we have increased the resolution as well as the freedom for detecting the direction of eye gaze.

The principle of our project is reflection. We believe in the process of lifelong learning and acknowledge any further work that may be undertaken on the project. Hence, the future scope of the project includes spectacle glare or reflected light removal for improved iris detection[4], calibrating the torque/speed of the wheelchair motors with the movement of iris, and the application of more efficient methods in place of Standard Hough Transform. Additionally, the eye gaze might be momentarily misdirected in a manner that misguides the wheelchair. Hence, we suggest the use of PID controllers to check the overshoot and rise time when such a scenario arises.

References

1. Anagha Dwajan, B., Sowmya, M.S., Bhavani, S., Priya Reddy, N., Usha, M.R.: 2020, Eye gaze controlled wheelchair. Int. J. Eng. Res. Technol. (IJERT) **9**(5) (May 2020)
2. Sayali, G., Miss Samiksha, B., Miss. Nikeeta, S.: Smart wheelchair with object detection using deep learning. Int. Res. J. Eng. Technol. (IRJET) **6**(12) Dec (2019)
3. Rajesh, A., Mantur, M.: Eyeball gesture controlled automatic wheelchair using deep learning. In: 2017 IEEE Region 10 Humanitarian Technology Conference (R10-HTC), Dhaka, Bangladesh, pp. 387–391 (2017). https://doi.org/10.1109/R10-HTC.2017.8288981
4. Sandhan, T., Choi, J.Y.: Anti-Glare: tightly constrained optimization for eyeglass reflection removal. In: 2017 IEEE Conference on Computer Vision and Pattern Recognition (CVPR), Honolulu, HI, USA, 2017, pp. 1675–1684. https://doi.org/10.1109/CVPR.2017.182
5. Li, Y., Brown, M.S.: Single image layer separation using relative smoothness. In: 2014 IEEE Conference on Computer Vision and Pattern Recognition, Columbus, OH, USA, 2014, pp. 2752–2759. https://doi.org/10.1109/CVPR.2014.346
6. Chang Shih, Y., Krishnan, D., Durand, F., Freeman, W.T.: Reflection removal using ghosting cues. In: 2015 IEEE Conference on Computer Vision and Pattern Recognition (CVPR), Boston, MA, USA, 2015, pp 3193–3201. https://doi.org/10.1109/CVPR.2015.7298939
7. Levin, A., Weiss, Y.: User assisted separation of reflections from a single image using a sparsity prior. IEEE Trans. Pattern Anal. Mach. Intell. **29**(9), 1647–1654 (2007). https://doi.org/10.1109/TPAMI.2007.1106. Sept
8. Bramhe, M.V., Vijay, N., Bhagyashree Rao, K.: Voice controlled wheelchair for physically disabled person. Int. J. Adv. Res. Electr. Electron. Instrument. Eng. **6**(2) February (2017)
9. Naga Rajesh, A., Chandralingam, S., Anjaneyulu, T., Satyanarayana, K.: EOG Controlled Motorized Wheelchair for Disabled Persons. World Acad Sci. Eng. Technol. Int. J. Medical, Health, Biomed. Pharmaceut. Eng. **8**(5) (2014)
10. Long, J., Shelhamer, E., Darrell, T.: Fully convolutional networks for semantic segmentation. IEEE Conference on Computer Vision and Pattern Recognition (CVPR) **2015**, 3431–3440 (2015). https://doi.org/10.1109/CVPR.2015.7298965
11. Mukhopadhyay, P., Chaudhuri, B.B.: A survey of hough transform. Pattern Recogn. **48** (3), 993–1010 (2015). ISSN 0031–3203
12. Xiong, J., Xu, W., Liao, W., Wang, Q., Liu, J., Liang, Q.: Eye control system base on ameliorated hough transform algorithm. IEEE Sens. J. **13**(9), 3421–3429 (2013). https://doi.org/10.1109/JSEN.2013.2262934. Sept

Chapter 50
COVID-19 Social Distancing Surveillance System

Akarsh Raj, Surbhi Mahajan, Sachi Bundele, and Punitkumar Bhavsar

Abstract The pandemic from COVID-19 impinged our day-to-day lives and wreaked havoc upon many sectors in our society. This worldwide pandemic, which had its onset in January 2020, has forced us to reconsider our perception of what "normal" should be. While there's no official cure yet, various vaccines have been rolled out and are expected to take effect soon. However, the efficacy of vaccines has been a debatable issue. Thus, the most effective way to battle this situation would be to strictly follow the precautionary measures advised by the governing authorities. Wearing mask and following the social distancing norms are considered as one of the most effective ways to control the spread of infection [1]. However, this new normal becomes difficult to implement as many people tend not to follow social distancing. While it is difficult to check whether people are following social distancing, we propose a solution which would come in handy in such circumstances and would hasten the process of contact tracing in comparison to manual inspection. In this study, we strive to present a video surveillance model, which would allow the detection of social distancing between people based on object detection and tracking algorithms. The specific algorithm used in our study for object detection is the YOLO algorithm and monitoring the distance between any two persons is done using a technique called Perspective Transformation. The proposed method shows promising results which could be implemented as a surveillance system for monitoring social distancing.

A. Raj · S. Mahajan · S. Bundele · P. Bhavsar (✉)
Department of Electronics and Communication Engineering,
Visvesvaraya National Institute of Technology Nagpur, Nagpur, India
e-mail: punitbhavsar@ece.vnit.ac.in

S. Bundele
e-mail: sachisb@students.vnit.ac.in

© The Author(s), under exclusive license to Springer Nature Singapore Pte Ltd. 2022   511
V. Bhateja et al. (eds.), *Evolution in Computational Intelligence*,
Smart Innovation, Systems and Technologies 267,
https://doi.org/10.1007/978-981-16-6616-2_50

50.1   Introduction

Infectious diseases and epidemics are truly one of the greatest threats to humanity. The Black Death, which was a global devastating epidemic of Bubonic Plague that struck Europe around the 1300s, and the Spanish Flu Pandemic of 1918 are only a few examples of these deadly disasters which have affected large human populations and caused havoc over the period of their evolution [2].

Coronavirus disease (COVID-19) is the newly discovered contagious disease. It was first discovered in Wuhan region of China and has been spreading worldwide thereafter [3]. The World Health Organisation has declared it as a pandemic. Its general symptoms include cough, fever, fatigue, breathing difficulties, and loss of smell and taste. Symptoms begin from one to fourteen days after exposure to the virus. The infection spreads due to discharge in the air with droplets of saliva when a person coughs or sneezes. Prevention can be taken by social distancing, quarantining of the patients, ventilation of indoor spaces, wearing masks, washing hands at regular intervals, and keeping hands away from face [4].

COVID-19 is spreading rapidly throughout the world and the situation is worsening day by day. The total number of cases seen worldwide as of May 2021 is 169 million and the total number of deaths is around 3 million. Out of these cases, India has reported around 27.7 million cases and around 3 lac deaths [5]. The increasing number of cases poses a grave situation which must be dealt with swiftly. Vaccinating people would be one option to stop the number of infections. Around 1.7 billion people have been vaccinated worldwide. In India, around 42 million people have been fully vaccinated, which constitutes only 3.1% of the total population [6]. Thus, vaccination might be the most effective method, but vaccinating the entire population would take longer. Thus, in the meanwhile, it is imperative that people follow the social distancing norms to curb the spread of the virus.

50.1.1   Importance of Social Distancing

COVID-19 spreads through air transmission. When people sneeze, cough, sing, talk, or breathe it transmits respiratory droplets which act as carriers of this virus. So, people who are in the vicinity of the infected person (within 6 feet) are at high risk to catch the disease [7]. The situation is becoming extremely dire nowadays and many countries are experiencing the second wave of this deadly disease. The massive rise in cases attributes to the fact that many people ignore the social distancing norms. This, in turn, leads to an unprecedented rise in clusters i.e. a group of people who become infected at once. If people were to follow the norms, the number of cases can be significantly reduced. Thus, it becomes imperative for us to adhere to strict social distancing as it significantly reduces the chances of getting infected [8].

50.1.2 Comparison Between Various Object Detection Algorithms

The task of labelling an object with a class comes under object classification in computer vision, whereas the finding the location of the object inside an image is called object localisation. Object detection is a technique in which both the above tasks are amalgamated together to achieve the process of identifying the class and also finding the location of the object by drawing a bounding box around it as a single objective.

There are many object detection methods like traditional sliding window technique and also deep learning methods such as the family of R-CNNs, SSD, and YOLO, the latter being more talked about for real-time applications [9].

R-CNN or Region-Based Convolutional Neural Network is a region proposal-based object detection algorithm that was first introduced in a paper by Ross Girshik et al. from UC Berkeley titled "Rich feature hierarchies for object detection and semantic segmentation" in 2014 [10]. Its major drawback is that its extremely slow speed owing to the fact that the number of these region proposals is very high and each of such proposals is sent to the R-CNN for feature extraction.

SSD stands for Single Shot Multibox Detector. It was proposed by Szegedy et al. in November 2016 [11]. This method uses a single pass for object detection and classification. It uses a regression-based algorithm which eliminates the need of object proposals. The network makes predictions from feature maps with different resolutions to fit the most accurate bounding box.

The YOLO algorithm, which stands for "You Only Look Once" is one of the fastest object detection algorithms, proposed by Szegedy et al. in November 2016 [12]. It proves to be an excellent choice for achieving real-time detection due to a relatively faster algorithm with around 30–45 FPS per second [13].

Each of the above algorithms has its own merits and limitations and depending on the application and GPU limitations. In surveillance system requires to process the images almost in real time which requires the fast processing time with relatively acceptable level of performance for the other parameters. YOLO is selected for this purpose as evidenced from Fig. 50.1.

Fig. 50.1 Numerical comparison between R-CNN, SSD, and YOLO [14]

Method	mAP	FPS	batch size	# Boxes	Input resolution
Faster R-CNN (VGG16)	73.2	7	1	~ 6000	~ 1000 × 600
Fast YOLO	52.7	155	1	98	448 × 448
YOLO (VGG16)	66.4	21	1	98	448 × 448
SSD300	74.3	46	1	8732	300 × 300
SSD512	76.8	19	1	24564	512 × 512
SSD300	74.3	59	8	8732	300 × 300
SSD512	76.8	22	8	24564	512 × 512

50.1.3 Architecture of YOLO v3

The YOLO v2 version though slightly faster fails in terms of accuracy in front of algorithms such as SSD and RetinaNet. Thus jeopardising just a bit in speed and with significantly high accuracy, YOLO v3 is selected for the application in this study.

YOLO v3 uses a variant of Darknet, where 53 more layers are stacked onto the original 53 layer network trained on Imagenet which gives us a 106 layer fully convolutional underlying architecture. Figure 50.2 shows the architecture of YOLO v3. This version performs detection by applying 1×1 direction kernels on feature maps of three sizes at three different locations in the network and detects objects at 3 different scales. These layers are 82nd, 94th, and 106th layer.

YOLO v3 typically takes an input image of dimensions 416×416 and it predicts at three scales, which are accurately determined by downsampling the input image's dimensions by 32, 16, and 8 pixels, respectively. The 82nd layer performs the first detection. The network down samples the picture for the first 81 layers, resulting in a stride of 32 for the 81st layer. The resulting feature map would be 13×13 if the selected image is of the size 416×416. The 1×1 detector kernel is used here to make a detection related function map of $13 \times 13 \times 255$.

The function map from layer 79 is then passed through a few convolutional layers and is then upsampled by two times to 26×26 dimensions. The function map from layer 61 is then depth concatenated with this one. After that, some 1×1 convolutional layers are applied to the fully combined feature maps to fuse the features from the 61st layer. The 94th layer then performs the second detection, resulting in a detection function map with dimensions of $26 \times 26 \times 255$.

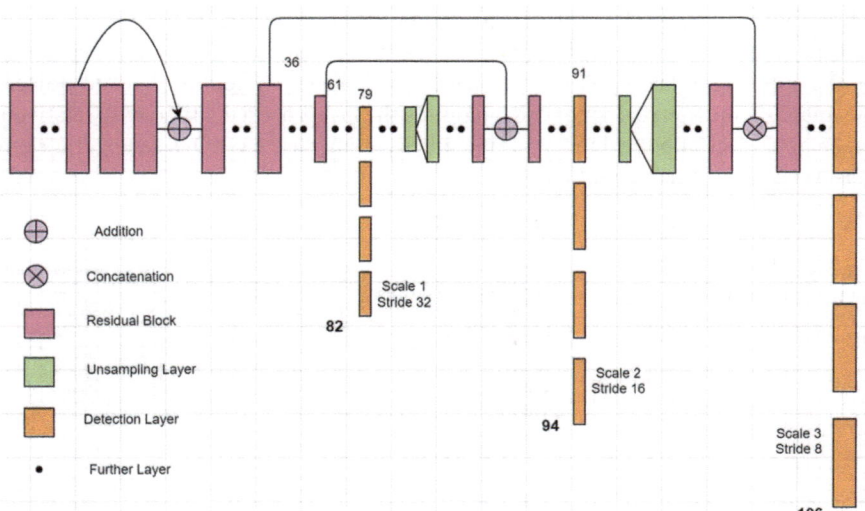

Fig. 50.2 Architecture of YOLO v3 [15]

The feature map obtained from layer 91 is subjected to some convolutional layers before being concatenated in depth with a feature map obtained from layer 36, following a similar method. Following that, some 1×1 convolutional layers then fuse the information obtained from the 36th layer, as before. The final of these three is generated at the 106th layer, yielding a feature map with dimensions of $52 \times 52 \times 255$ [16].

These 3 layers obtained by OpenCV function getUnconnectedOutLayers() that are sent through the forward pass, help in the object detection step.

50.2   Method

50.2.1   Object Detection Using YOLO v3

YOLO (You Only Look Once) is a single stage object detection algorithm in which predicts bounding boxes and classes of the image at once. Each bounding box uses 4 descriptors—centre coordinates, width, and height of the bounding box.

YOLO predicts every class's confidence score inside each bounding box using regression. Since for this complete process, the image is passed through CNN only once, that is why YOLO is a very fast object detector and therefore used for real-time applications. YOLO v3 algorithm along with the COCO dataset was used for "person" class only object detection.

50.2.2   Distance Measurement

We utilised the concept of perspective transformation. Here we worked in an environment having still camera and moving object. So we took the coordinates of 4 points, namely, points "a", "b", "c", and "d" in the background that formed a rectangle in reality (rect. a'b'c'd' as shown in Fig. 50.3) with the assumption that every person in the image is walking on a flat surface.

Along with these four points, we also took three more points "e", "f", and "g". Distance between points "e" and "f" is used to define social distancing threshold (in our case 6 feet) in horizontal direction and points e and g are used for the vertical direction in real world.

Taking the bottom centre of each bounding box, we mapped them to the transformed domain and using this transformed space, we calculated the pairwise distance between each such point.

Using distances $\overline{ef}$ and $\overline{eg}$, if the distance is less than the calculated social distancing violating threshold, we made the red bounding box otherwise green bounding boxes were selected.

Demonstration of Perspective Transformation

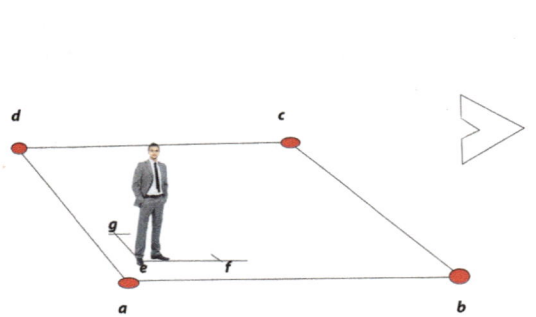
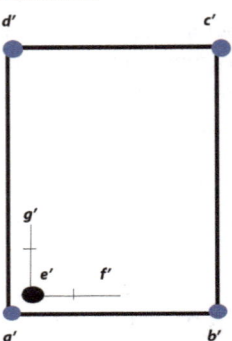

Fig. 50.3 Demonstration of perspective transformation

50.3   Result

The result is analysed based on the performance evaluation parameters as shown in Figs. 50.5 and 50.6. Apart from this, the designed system also provides an alert if the number of people in the frame exceeds a certain manually provided threshold.

Performance evaluation of classification technique for social distance norm

1. TPR stands for true positive rate. In our application, true positives are defined as the number of instances where the system accurately detects instances of social distancing violations. In the Fig. 50.4, it is represented by **Person C** and **Person D**. TPR is the ratio of true positives to the total predictions made by the system.
2. FPR stands for false positive rate. In our application, false positives are defined as the number of instances where our system detects non-violations in place of social distancing violations and non-violations in place of violations. Here, it is **Person B**. TPR is the ratio of false positives to the total predictions made by the system.
3. FNR stands for false negative rate. Here, it is defined as the number of instances where our system predicted social distancing violations in place of non-violations as shown by **Person A** in Fig. 50.4. FNR is the ratio of false negatives to the total predictions made by the system.

Using the above parameters, we can define Precision and Recall as:

– Precision—It is defined as the parameter that gives us the proportion of positive identifications which are actually correct. It tells us how accurate our predictions are. Mathematically, precision is defined as follows:

$$Precision = \frac{TPR}{TPR + FPR}$$

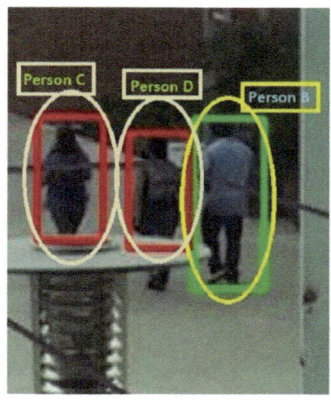 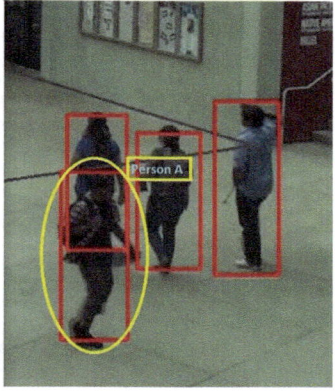

Fig. 50.4   Estimating TP, FP, and FN in the data

– Recall—It is the proportion of actual positives which are identified correctly. It
  accounts for whether we are able to detect all objects present in the image or not.
  Mathematically, recall is defined as follows:

$$Recall = \frac{TPR}{TPR + FNR}.$$

We now calculate precision and recall for different values of IoUs from our observations and try to calculate the average precision (AP) and Average Recall (AR).

– Intersection over Union (IoU)—If we wish to evaluate the intersection over union
  whilst doing object detection, we would need the ground-truth bounding boxes
  (testing set) and the actual predicted bounding boxes. IoU is needed because
  the parameters of our model such as the sliding window size, feature extraction
  method, constantly vary, and thus complete matching between the ground truth
  and predicted boxes is difficult.

To calculate Average Precision and Average Recall, we have considered a sample
of 60 frames for each value of IoU and calculated the values of TPR, FPR, and FNR
for each observation. The following table was obtained after all the observations
were made (Table 50.1).

Thus, using the above table as a reference, we get the Average Precision as
90.354% and Average Recall as 89.548%.

In the result, people following the social distancing rule can be seen in green
rectangles and those violating it can be seen in red boxes. The result also provides
details about the total count of people in the test area and number of people following
and violating social distancing norms.

Figure 50.5 shows one of the output frames from the video surveillance. Here,
minimum number of people allowed on the street was set to 15 and since the total
number of pedestrians is 16 that's why an alert message is displayed as "**Threshold**

Table 50.1 Average precision and recall values

IOU	TPR	FPR	FNR	Precision tp/(tp+fp)	Recall tp/(tp+fn)
0.5	0.8124	0.0872	0.1003	0.9030	0.8901
0.6	0.8310	0.0850	0.0838	0.9072	0.9083
0.7	0.8066	0.0860	0.1073	0.9036	0.8825
0.8	0.8043	0.0886	0.1069	0.9007	0.8826
0.9	0.8317	0.0891	0.0792	0.9032	0.9130

Fig. 50.5 Sample result showing threshold exceeding case

Exceeding". In contrast, Fig. 50.6 shows less number of people on street compared to the set threshold of 15 and the corresponding message displayed is "**Below Threshold**".

50.4 Limitations

1. The proposed system works well only when we are dealing with a single storey landscape. Multiple storeys or levels within our landscape may not provide good accuracy with our proposed method.

Fig. 50.6 Sample result showing below threshold case

2. By the inherent property of YOLO v3, FNR and FPR may increase if the person to be detected appears inherently small when seen from the camera, or is comparatively far away from the camera.

50.5 Conclusions

The coronavirus pandemic has affected our day-to-day lives and people are expected to adhere to the new normal by following the guidelines for social distancing norms and wearing masks in public places. These precautionary steps certainly help to reduce the spread of the disease, however, adherence to the guidelines is difficult due to the lack of awareness among people. Hence, this study exploited technological advancement to monitor and identify the violation of social distancing norm in a predefined area. This study adopted the use of YOLO algorithm for person detection. The bounding boxes and corresponding coordinates as identified by the algorithm are then used for calculating the distance between two persons. Perspective transformation is then used to map the camera view to plane in two-dimensional representation for calculation of the distances. The results show the effectiveness of this method for monitoring distancing norms in public places like online exam centres, banks, schools, hotels, restaurants, etc. Thus, the proposed system can be used as a video surveillance system for social distancing norms. The future studies will focus on methodologies to overcome the above-mentioned limitations.

References

1. World Health Organisation COVID-19 precautions and effects. https://www.who.int/emergencies/diseases/novel-coronavirus-2019
2. Huremović, D.: Brief history of pandemics (pandemics throughout history). Psychiatry of Pandemics, pp. 7–35 (2019). Published 2019 May 16. https://doi.org/10.1007/978-3-030-15346-5_2
3. The Guardian Report.: How did coronavirus start and where did it come from? Was it really Wuhan's animal market? https://www.theguardian.com/world/2020/apr/28/how-did-the-coronavirus-start-where-did-it-come-from-how-did-it-spread-humans-was-it-really-bats-pangolins-wuhan-animal-market. Accessed 25 Feb 2021
4. Transmission of virus: WHO precautions. https://www.who.int/vietnam/news/detail/14-07-2020-q-a-how-is-covid-19-transmitted
5. https://www.worldometers.info/coronavirus/?utm_campaign=homeAdvegas1?
6. https://ourworldindata.org/covid-vaccinations?country=OWID_WRL
7. The Medical News Today.: Coronavirus cause: origin and how it spreads. Accessed 25 Feb 2021
8. After effects of COVID -19. https://www.who.int/docs/default-source/coronaviruse/risk-comms-updates/update-36-long-term-symptoms.pdf?sfvrsn=5d3789a6_2
9. Lee, A.: Comparing deep neural networks and traditional vision algorithms in mobile robotics. https://www.cs.swarthmore.edu/~meeden/cs81/f15/papers/Andy.pdf
10. Girshick, R., Donahue, J., Darrell, T., Mali, J.: Rich feature hierarchies for accurate object detection and semantic segmentation. arXiv:1311.2524
11. Liu, W., Anguelov, D., Erhan, D., Szegedy, C., Reed, S., Fu, C.Y., Berg, A.C.: SSD: single shot multibox detector. In: European Conference on Computer Vision, pp. 21–37. Springer, Cham (2016). arXiv:1512.02325
12. You only look once: unified, real-time object detection. arXiv:1506.02640
13. https://towardsdatascience.com/r-cnn-fast-r-cnn-faster-r-cnn-yolo-object-detection-algorithms-36d53571365e#:~:text=YOLO
14. Numerical Comparison between RCNN, YOLO, SSD. https://jonathan-hui.medium.com/object-detection-speed-and-accuracy-comparison-faster-r-cnn-r-fcn-ssd-and-yolo-5425656ae359
15. Architecture of Yolo v3. https://miro.medium.com/max/3802/1*d4Eg17IVJ0L41e7CTWLLSg.png
16. Kathuria, A.: What's new in Yolo v3? https://towardsdatascience.com/yolo-v3-object-detection-53fb7d3bfe6b
17. Object detection. https://towardsdatascience.com/object-detection-simplified-e07aa3830954

Chapter 51
Secure and Efficient Text Encryption Using Elliptic Curve Cryptography

Ningthoukhongjam Tutu Raja⬤ **and Khumanthem Manglem Singh**⬤

Abstract As data is being generated and shared in very large amounts, there is a need to secure the data during transmission, processing, or even storing. Elliptic Curve Cryptography has been in wide use in recent times to secure the data transmission as it provides higher level of security with significantly smaller sized keys than other methods of encryption and decryption. In this paper, we propose a simple but efficient way to protect data with the use of ECC. The proposed techniques are compared with some other similar techniques.

51.1 Introduction

Cryptography is the process of securing information from unwanted access by making the data indecipherable. Public key cryptosystems use key pairs namely public and private key. Public key may be known to others but the private key is only known to the owner. Elliptic Curve Cryptography is a type of public key cryptography where some parameters used in the process of encryption are declared as public information. However, even with this information, it is very hard to get any information about the original data from the encrypted data. Researchers have also used ECC to encrypt image and make it unreadable to any unauthorized persons. ECC is widely used as it is hard to crack as it deals with the elliptic curve discrete logarithmic problem. Elliptic curves were first used to design public key cryptosystem by Neal Koblitz [1] and Victor Miller [2], independently in 1985. Since then ECC has been used in many applications due to the advantageous features it possesses. Many organizations have also standardized ECC [3–6]. It is gradually starting to replace other existing public key cryptosystems. Many forms of elliptic curves have been proposed [7–9] by many authors. In this paper, the curve equation of the form,

$$y^2 = x^3 + \alpha x + \beta \tag{51.1}$$

N. T. Raja (✉) · K. M. Singh
National Institute of Technology, Manipur 795004, India

© The Author(s), under exclusive license to Springer Nature Singapore Pte Ltd. 2022   521
V. Bhateja et al. (eds.), *Evolution in Computational Intelligence*,
Smart Innovation, Systems and Technologies 267,
https://doi.org/10.1007/978-981-16-6616-2_51

is used in order to encrypt and decrypt the data. This equation is also known as Weierstrass equation, where α and β are known constants which satisfy the equation,

$$4\alpha^3 + 27\beta^2 \neq 0 \tag{51.2}$$

51.1.1 Modular Arithmetic on Elliptic Curve (E)

In elliptic curves, operations on the points are done using some modular arithmetic. Functions used to perform these operations are given below:

1. Point Negation [P_negation(Point)]:
   This returns the negation of a point. ie:
   for a point (x,y), negation is given by:
   (x, -y modulus p)
2. Inverse Modulo[Inv_mod(k,p)]:
   This function returns the inverse of k modulo p. ie:
   for a non-zero integer k, inverse modulo returns the only integer x such that:
   (x * k) mod p == 1, where p is a prime.
3. Point Addition [Point_add(Point1,point2)]
   This function results in the addition of two points on the elliptic curve according to the group law.
4. Scalar Multiplication [S_mult(k,point)]
   This function returns k * p1. This is computed using the double and point addition algorithm.
5. On the Curve[Is_on_curve(Point)]
   It is always a good practice to check whether the resultant point obtained after any operation lies on the curve that we are using. This function asserts the same.
   i.e.,
   We check the point p1 using the curve Eq. (51.1), whether p1 is on the curve or not.

51.2 Literature Review

There are many encryption algorithms proposed in the past decade. Some of the techniques focused on the storage limitations of existing methods, some focused on the security that is provided by the method. But the most common parameter that is being focused on by researchers to improve is observed to be the time taken by the process.

In a paper [10], by Laiphrakpam Dolendro Singh and Khumanthem Manglem Singh, a new technique is proposed which does not require mapping each character to affine points on the Elliptic curve. The encryption is done by grouping the characters and by forming large integers from the groups.

Similarly, Keerthi and Surendiran developed a method [11] that requires no mapping of characters. The characters are grouped according to their hexadecimal values. This method is fast compared to others but it is only limited to characters having ASCII values less than 256.

In another paper by Binay Kumar et al. [12], a method is proposed where the encryption is done using ECC followed by a modified Arnold transformation on the encrypted points. This technique is very fast for smaller messages however as the message becomes large, the computation time increases exponentially and is not very efficient. Also this method requires a lookup table and the size of overhead of the transmission is also large.

In [13], the author proposed a new encryption technique using ECC through compression with artificial intelligence. The original text is compressed using a substitution technique by searching for repeating group of characters and then replacing them with a binary number. This process is then followed by encryption of the compressed text using ECC. This method provides a better security but the time taken for computing is very high.

In a paper [14] by Prasenjit Das and Chandan Giri, two techniques are introduced to map the characters before encryption and after decryption of the text messages. The first method grouped the characters by comparing the largest ASCII value to a base. This decreases the number of groups. The second method maps the characters according to the largest ASCII value. Both the methods presented in this paper are good for characters with lower ASCII values.

In their paper [15], Sujatha Kota et al suggested an interesting method to select private key for ECC using Particle Swarm Optimization and Cuckoo Search Algorithm. Their method increases the accuracy of ECC, of which the Cuckoo Search Algorithm performed the best. Since there is a new process of selecting the private key, the time taken increases in this method.

A new authenticated encryption scheme is proposed in this paper [16]. Here, the characters are encoded by dividing the message into blocks and perform some XOR operations. Then the resultant is mapped to the elliptic curve. This method is very secure and relatively fast.

51.3   Proposed Method

We have introduced a method to encrypt text messages using Elliptic curve cryptography. The detailed methodology is presented in Sects. 51.3.1 and 51.3.2. In this paper we will be using the NIST 192 curve and the already defined parameters. The following parameters are used:

$\alpha = 6277101735386680763835789423207666416083908700390324961276.$

$\beta = 2455155546008943817740293915197451784769108058161191238065.$

$p = 6277101735386680763835789423207666416083908700390324961279.$

g = (60204628237568865675821348058752611191669897663688468488818,
17405033229362203140485755228021941036402348892738665041)
n = 62771017353866807638357894231760590137671947731828422840081.
Using these parameters, first we calculate the keys for sender and receiver.
At receiver's end:
Private key (Pr$_R$) = random number between 1 and n
Public key (Pu$_R$) = s_mult(Pr$_R$, g)
Pu$_R$ is made public for any user who wants to communicate.
At sender's end:
Private key (Pr$_S$) = random number between 1 and n
Public key (Pu$_S$) = s_mult(Pr$_S$, g)
Since Pu$_R$ is public, use it to generate the shared secret key.
Shared key (Sk) = s_mult(Pr$_S$, Pu$_R$)

51.3.1 Encryption Process

1. Take the text message and list them as their ASCII values.
2. Find the number of digits (dig) in the largest value of the ASCII values.
3. Convert the ASCII values to string form and pad '0'(s) to all the converted ASCII values in such a way that the length of all the strings matches the largest value.
4. Form string(s) of length 16*dig by concatenating the padded values
5. If the number of strings is not even, pad the value with 8*dig'32's.
6. Convert the strings to integers and form pairs.
   Take encryption matrix, $M_e = \begin{pmatrix} x_k & 0 \\ 0 & y_k \end{pmatrix}$.
7. Perform matrix multiplication between M_e and all the converted pairs. The resultant pairs are the cipher texts (C_{xi}, C_{yi}).
8. Pu$_S$ and [dig, (C_{xi}, C_{yi})] are sent to the receiver's end.

51.3.2 Decryption Process

1. After receiving Pu$_S$ and (C_{xi}, C_{yi}), the shared key is calculated as:
2. Shared key (Sk) = s_mult(Pr$_R$, Pu$_S$) = (x_k, y_k)
3. Take decryption matrix, $M_d = \begin{pmatrix} 1/x_k & 0 \\ 0 & 1/y_k \end{pmatrix}$
4. Perform matrix multiplication between M_d and all the cipher pairs (C_{xi}, C_{yi}). Convert these values to strings
5. If the last string is a string '32's of the length 8*dig, remove the string.
6. If the string is not of the length 16*d, append '0' in front of the string to make the length of the string 16*dig.

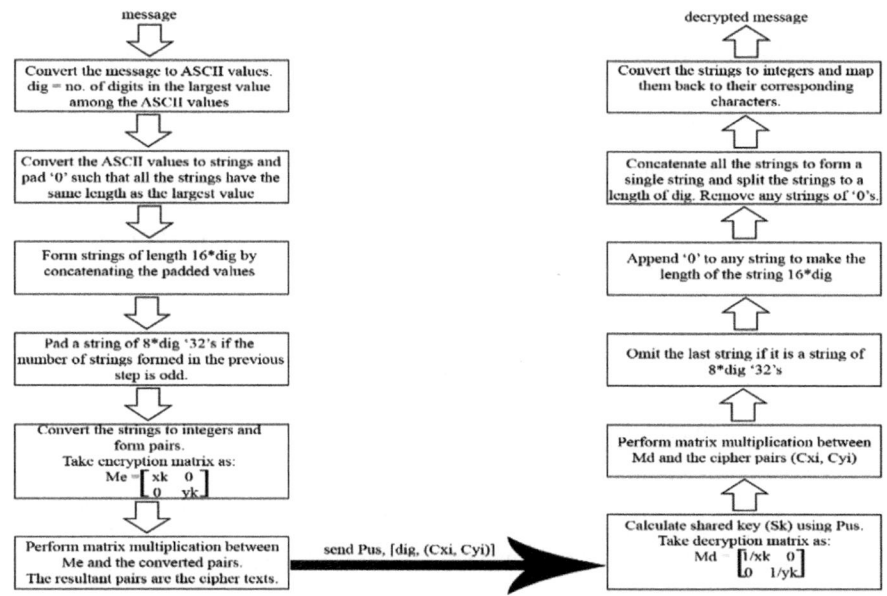

Fig. 51.1 Proposed Method

7. Concatenate all the strings to form a single string and split the string into strings of length dig. Note: omit all the '000'strings.
8. Convert the strings to integers and map the ASCII values to their corresponding characters to get the original text (Fig. 51.1).

51.4 Example

Elliptic curve parameters used are stated previously.
Message to be encrypted = 'National Institute of Technology, Manipur, 795004'.
Sk = shared key = (xk, yk) =
(1798224889846244383113448495972364923480460346610550498372,
402942287075564456621833000944951367047819934 1583180139646)

51.4.1 Encryption Process

ascii_vaues=[78,97,116,105,111,110,97,108,32,73,110,115,116,105,116,117,116,1
1. 01,32,111,102,32,84,101,99,104,110,111,108,111,103,121, 44, 32, 77, 97, 110,
105, 112, 117, 114, 44, 32, 55, 57, 53, 48, 48, 49]
2. dig = 3, since the largest value is 117.

string converted values =

3. '0780971161051111100971080320731101151161051161171161010320791020320
84101099104110111108111103121044032055057053048048049'

   i.e., all the 2 length strings are appended with '0' to make it a string of length 3(=dig). Therefore, '78' becomes '078' and so on.

   String of 16*dig length =

4. '078097116105111110097108032073110115116105116117','11610103207910203
2084101099104110111108111103121','044032055057053048048049'

5. Since there are only 3 strings, pad and form pairs of integers:

   pair1 = (780971161051111100971080320731101151161051161 17,
   11610103207910203208410109910411011110811110312 1),
   pair2 = (44032055057053048048049,
   32323232323232323232323232323232323232323232323232)

6. Encryption matrix, $M_e = \begin{pmatrix} x_k & 0 \\ 0 & y_k \end{pmatrix}$

7. After multiplication we get, (Cx1, Cy1) =
   (14043617800542278412834798506815314689937582683958291950831629844 7
   1156347109579993771162657707732794 61524,467820153977868491166064204
   2570750796829365987094327247875098866897722437359216934258340521 82
   387886435166),
   (Cx2, Cy2) =
   (79179537354672985262287836556046613641219650843098016550174328544 3
   31978752276228,
   13024397157998043042321874778018630045990139285912406357253315088 2
   7080943835329450126714117384538770055872)

8 Pu$_S$ and [dig, (C$_{xi}$, C$_{yi}$)] are sent to the receiver's end.

51.4.2 Decryption Process

   Sk = (x$_k$, y$_k$)
   = (calculated using Pu$_S$)
1.
   = (1798224889846244383113448495972364923480460346610550498372,
   40294228707556445662183300094495136704781993415831801396 46)

2. Decryption matrix, $M_d = \begin{pmatrix} 1/x_k & 0 \\ 0 & 1/y_k \end{pmatrix}$

3. Performing matrix multiplication on (Cxi, Cyi) we get,
   ('780971161051111100971080320731101151161051161 17',
   '116101032079102032084101099104110111108111103121',
   '44032055057053048048049',
   '32323232323232323232323232323232323232323232323232')

4. Since the last string is 8*dig '32's, remove the string.

5. Make all the string length of 16*dig:
   ('078097116105111110097108032073110115116105116117',
   '116101032079102032084101099104110111108111103121',
   '000000000000000000000000000044032055057053048048049')

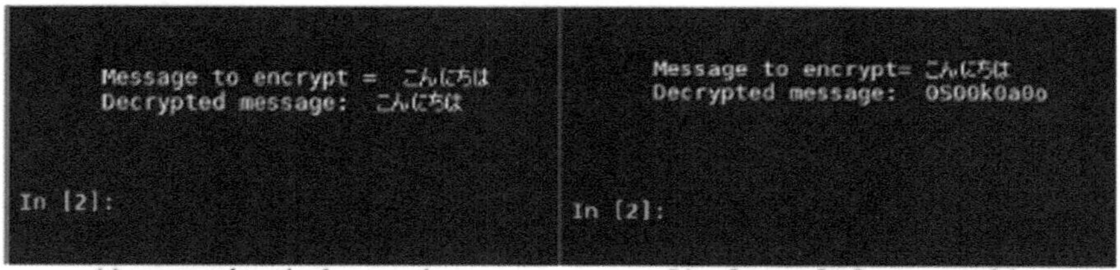

(a) proposed method encrypting   (b) reference [11] unsuccessful
and decrypting japanese characters   in decrypting japanese characters
successfully.   (ASCII value > 255).

Fig. 51.2 Encryption of characters with higher ASCII value (here > 12371)

6. Divide string into length of dig and omit all the '000's (if any):
   ['078','097','116','105','111','110','097','108','032','073','110','115','116','105',   '116',
   '117', '116', '101', '032', '079', '102', '032', '084', '101', '099', '104','110','111', '108',
   '111', '103', '121', '044', '032', '055', '057', '053', '048', '048', '049']
7. Convert to integers and map to their corresponding ASCII value. The decrypted
   message is:
   'National Institute of Technology, Manipur, 795004'.

51.5   Results and Analysis

Our method is executed and compared with other text encryption methods of recent
past. The results shown are implemented a desktop Processor with 3.2 GHz Intel i7
CPU, 8 GB RAM, Window 10, 64 bit operation system using Python 3.7.

Among all the methods compared in the study, reference [16] and [11] can only
be used for encrypting and successfully decrypting characters having ASCII values
less than 256. In the paper [12], only the characters defined in the lookup table can
be processed by the proposed method. Whereas, our method, [10] can be used to
encrypt any character having a defined ASCII value and then successfully decrypt it
to get the original message (Fig. 51.2).

51.6   Conclusion

In this paper, we have put forward an efficient method to encrypt text messages. The
converted ASCII values are mapped using the shared secret key and then transmitted
with the length of the largest digit. Compared to previously proposed methods (shown
in Table 51.1), our method has shown improvements in various aspects. The proposed
method is secure and also efficient in terms of storage and time. The cipher size of our
method is also significantly smaller compared to existing techniques. This method is
also applicable to all the characters with defined ASCII value. Our method also does

Table 51.1 Comparison of performance with other methods

Algorithms	Proposed method	Reference[10]	Reference[11]	Reference[12]	Reference[16]
Encryption time (in secs)	**0.0467**	0.0518	0.0699	0.0747	0.0637
Decryption time (in secs)	**0.0155**	0.0169	0.0159	0.0781	0.0458
Cipher size (kB)	0.328	0.348	**0.3**	25.584	0.554

(Length of message = number of characters = 504)

not require any dedicated lookup table and hence the process is faster and consumes less storage.

In future research, our method can also be implemented on images by grouping the pixels and mapping them accordingly to hide the data in images.

References

1. Koblitz, N.: Elliptic curve cryptosystems. Math. Comput. **48**(177), 203–209 (1987)
2. Miller, V.: Use of elliptic curve in cryptograhy. Adva. Cryptogr. CRYPTO'85, LNCS 218, 417–126 (1986)
3. ANSI X9.63: Public Key Cryptography for the Financial Services Industry: Key Agree- ment and Key Transport Using Elliptic Curve Cryptography. American National Standards Institute (2001)
4. IEEE 1363–2000: Standard Specifications for Public Key Cryptography (2000)
5. ISO/IEC 15946-3: Information Technology–Security Techniques—Cryptographic Tech-niquesBased on Elliptic Curves—Part 3: Key Establishment. (2002)
6. NIST: Digital Signature standard. FIPS publication, 186-2(2000)
7. Peter, L.: Montgomery: speeding the pollard and elliptic curve methods of factorization. American Mathematical Society (1987). http://www.jstor.org/stable/pdfplus/2007888.pdf
8. N. P. Smart: The Hessian form of an elliptic curve. Springer-Verlag Berlin Heidelberg, ISBN 978-3-540-42521-2 (2001)
9. Billet, O., Joye, M.: The Jacobi model of an Elliptic Curve and the Side-channel Analysis. Springer-Verlag Berlin Heidelberg. ISBN 978-3-540-40111-7 (2003)
10. Singh, L.D., Singh, K.M.: Implementation of text encryption using elliptic curve cryptography. Procedia Comput. Sci. **54**(1), 73–82 (2015)
11. Keerthi, K., Surendiran, B.: Elliptic curve cryptography for secured text encryption. In: International Conference on Circuit, Power and Computing Technologies (ICCPCT), pp. 1–5 (2017)
12. Singh, B.K., Tsegaye, A., Singh, J.: Probabilistic data encryption using elliptic curve cryptography and Arnold transformation. In: International Conference on I-SMAC (IoT in Social, Mobile, Analytics and Cloud) (I-SMAC), pp. 644–651(2017)
13. Som, S.: Encryption technique using elliptic curve cryptography through compression and artificial intelligence. In: Bokhari, M., Agrawal, N., Saini, D. (eds) Cyber Security. Advances in Intelligent Systems and Computing, vol. 729. Springer, Singapore, (2018)
14. Das, P., Giri, C.: An efficient method for text encryption using elliptic curve cryptography. In: 8th International Advance Computing Conference (IACC), pp. 96–101 (2018)

15. Kota, S., Padmanabhuni, V.N., Budda, K., et al.: Authentication and encryption using modified elliptic curve cryptography with particle swarm optimization and cuckoo search algorithm. J. Inst. Eng. India Ser. B **99**, 343–351 (2018)
16. Almajed, H. N., Almogren, A.S.: SE-Enc: A secure and efficient encoding scheme using elliptic curve cryptography. IEEE Access **7**, 175865–175878 (2019)

Chapter 52
Masked Face Detection Using Transfer Learning

Sourav Mohanty and M. A. Lakshmi

Abstract COVID-19 pandemic has created chaos all over the world thereby reducing productivity and affecting day to day life of people. Wearing a face mask has become the need of the hour. In the future, it is expected that masks will become an integral part of human lifestyle and people have to wear masks to avail certain services. Therefore, the detection of face mask has become a global problem to avoid the spread of this virus. This paper presents a deep learning approach to detect faces with masks from an image or from a video stream, which can be used in surveillance tasks. The method attains accuracy up to 99% on Simulated Face Mask Dataset. We have manually tested this model on webcam, and it detects the presence of masks correctly without overfitting.

52.1 Introduction

COVID-19 pandemic has made a large impact worldwide. In the early 2020, SARS-COV-2 was categorized as a new type of virus by the World Health Organization (WHO). The disease soon expanded over the world, affecting many countries and resulting in both economic and human loss of life. Although many medical professionals and researchers have come up with solutions to stop its spread, no cure for this virus has been reported till date. Because this virus spreads through the air and through touch, wearing a mask is an important precaution to take during the pandemic. As a result, Face Mask Detection has become an important application that requires users to wear face masks, maintain physical distance, and wash their hands using hand sanitizers. Face mask detection has been a prominent problem in the field of image, video processing, and computer vision. Due to rapid improvements in the domain of machine learning and neural networks, this problem is well

S. Mohanty (✉) · M. A. Lakshmi
Department of CSE, GITAM University Hyderabad, Hyderabad, Telangana, India

M. A. Lakshmi
e-mail: amuddana@gitam.edu

V. Bhateja et al. (eds.), *Evolution in Computational Intelligence*,
Smart Innovation, Systems and Technologies 267,
https://doi.org/10.1007/978-981-16-6616-2_52

addressed until now. Nowadays, this technology has a large use case because it can not only detect from static images but can also detect from real-time video streams. With the help of object detection and image classification, very high accuracy can be easily achieved, which makes this trustworthy for detecting and penalizing those who don't follow the rules.

This paper aims at creating a model that can detect masks on a person's face so that the corresponding authority can be informed in a smart city network. First, photos from real-time video footage of public locations are captured using CCTV cameras. This is then given as input to our model. These facial pictures are retrieved and utilized to detect the presence of a mask on the face. When the architecture detects people who are not wearing a mask, the information can be passed on to the authorities, who can then take the appropriate action. On data obtained from various sources, the proposed system produced promising results.

52.2 Literature Review

Since 2012, science has made significant progress in constructing deep learning models for image classification almost every year. The ImageNet challenge has been the key benchmark for measuring progress because of its huge scale and challenging data, and it has resulted in several pretrained models such as VGG-16, MobileNet V2, Inception V2, Res-Net, Xception, and others. So, before we started, we studied a few research papers.

Bosheng Qin and Dongxiao Li [1] developed a face mask detection system based on the SRCNet classification network, which quantified a three-category classification problem based on unconstrained 2D facial image images and classified them as "Correct Facemask," "Incorrect Facemask," and "No Face Mask" with an accuracy of 98.7%. The dataset that they used was medical mask dataset from kaggle consisting of 678 annotated images. Md. Sabbir Ejaz et al. [2] designed the Principal Component Analysis (PCA) algorithm for Masked Face Detection. Their dataset was a mixture of images from ORL Database and their own images. A total of 500 photos were used in this experiment. On different test scenarios, 300 photos were used as training images and 80–200 photos were used as test images. PCA was found to be effective in detecting faces without a mask, with an accuracy of 96.25%, but in identifying faces with a mask, its accuracy drops to 68.75%. Li et al. [3] used an object detection algorithm named YOLOv3 for face detection, which is built on the darknet-53 deep learning network architecture, where the WIDER FACE and Celebi databases were utilized for training, while the FDDB database was used for evaluation. The dataset included 32203 images and 393703 annotated faces, which were divided into three sets: training (40%), validation (10%), and testing (50%). The accuracy of this model was 93.9%. In light of the foregoing, we can deduce that a small number of research articles on Face Mask Detection have been published. As a result, we chose Transfer Learning Models like Mobile Net v2 for future enhancements.

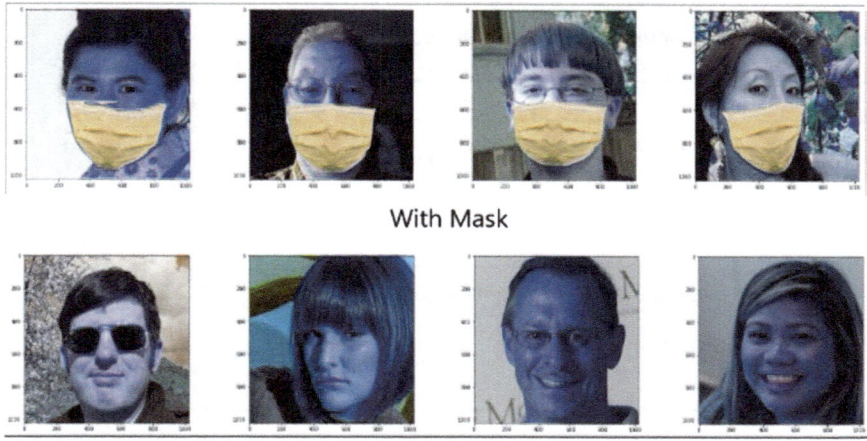

With Mask

Without Mask

Fig. 52.1 Sample images from dataset

52.3   Proposed System

52.3.1   Dataset Description

A Simulated Masked Face Dataset (SMFD) [4] named Mask Face-Net is used. Based on the dataset Flickr-Faces-HQ (FFHQ), Masked Face-Net is a collection of human faces with a correctly or poorly worn mask consisting of 137,016 images. Out of which 5869 images are taken for dataset construction. Respective folders for train and test images along with two folders inside each for two classes and stored 80% images for training and 20% images for testing are created. Using data input pipelines, train images are split into both train (70%) and validation batches (30%) (Fig. 52.1).

52.3.2   Proposed Approach

So, to make sure everyone is following the rules and to keep a track of people who are not wearing masks, the power of deep learning and computer vision is put together to create a model by fine-tuning a pretrained model named MobileNet V2 using transfer learning that can detect faces with and without masks from real-time video generators such as webcam or from an image or a saved video file. This model can have various applications in offices, railway stations, airports, and other public places where surveillance cameras are used for face detection. The suggested system integrates OpenCV, Tensor Flow, and Keras to recognize a person wearing a face

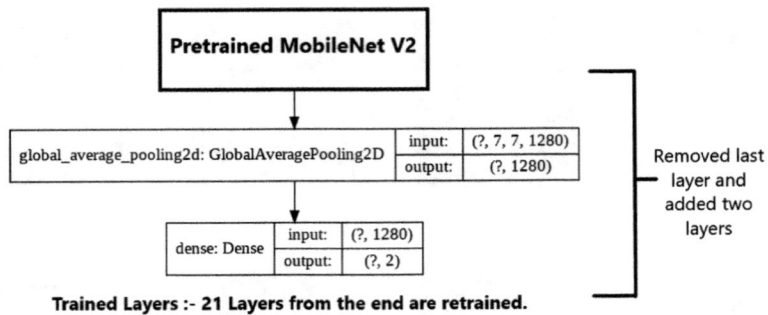

Fig. 52.2 Block diagram of our model

```
Conv_1 (Conv2D)                   (None, 7, 7, 1280)   409600   block_16_project_BN[0][0]

Conv_1_bn (BatchNormalization)    (None, 7, 7, 1280)   5120     Conv_1[0][0]

out_relu (ReLU)                   (None, 7, 7, 1280)   0        Conv_1_bn[0][0]

global_average_pooling2d (Globa   (None, 1280)         0        out_relu[0][0]

dense (Dense)                     (None, 2)            2562     global_average_pooling2d[0][0]
================================================================================================
Total params: 2,260,546
Trainable params: 1,208,642
Non-trainable params: 1,051,904
```

Fig. 52.3 Architecture of our model

mask in an image/video stream using computer vision and deep learning algorithms (Figs. 52.2 and 52.3).

1. Generating Data input pipelines using Image Data Generator
2. Train a predefined deep learning model (MobileNet V2)
3. Using Computer Vision to classify masked and without masked face.

52.3.2.1 Transfer Learning

Transfer learning is a technique through which we can inherit features from a model that is trained on a large dataset and use it to solve real-time problems that do not have enough data or have a small dataset. It is particularly popular in solving deep learning problems currently as it can train deep convolutional neural networks with a small amount of data. This is mostly valuable in the field of data science and object detection, as most real-world situations do not have enough data points to train complicated models. Even when the new model is trained on a limited dataset, transfer learning improves its performance. InceptionV3, Xception, VGG16, MobileNet, MobileNetV2, ResNet50, and other well-known pretrained models were trained using 14 million photos from the ImageNet dataset.

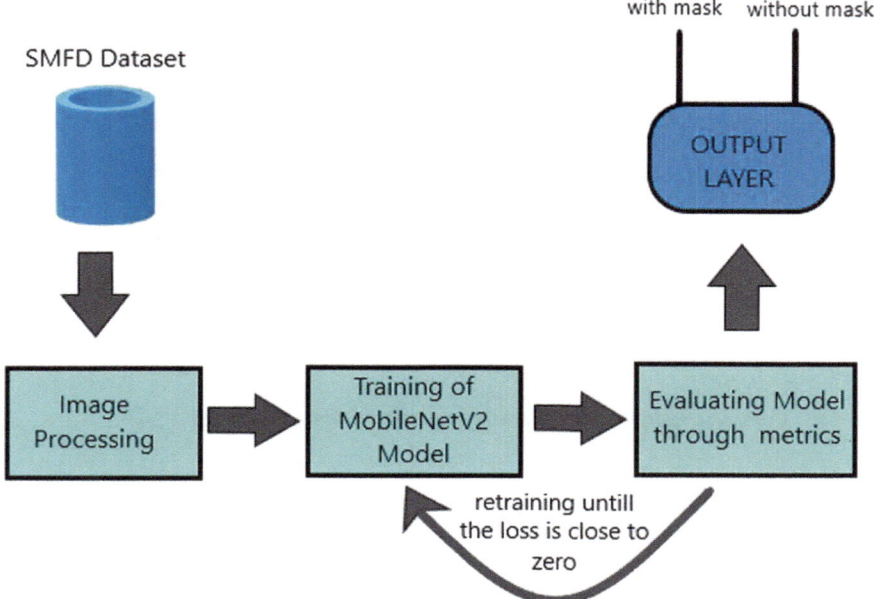

Fig. 52.4 Schematic representation of the proposed model

In this research, an approach that is based on transfer learning is implemented using a pretrained model named as MobileNet V2 to classify images. We removed the last layer of the pretrained model with 1000 units and added a dense layer of two units with an activation function of SoftMax. All the predefined weights of the model were frozen except the last 23 layers, which gave us a pretty decently performing model. This model is trained for 15 epochs with a batch size of 32 (Fig. 52.4).

52.3.2.2   Image Processing

Image Processing uses computer algorithms to process digital images. It can be easily implemented by using Keras predefined libraries. Since we had a good amount of images to work with, we only used the preprocessing_function of MobileNet V2 named preprocess_input() to adequate our image to the format the model requires. After that, the dataset is rescaled to 224 × 224 pixels and is categorized into train, test and validation batches. The class mode is set to categorical which will convert the labels to one hot encoded vector. Figure 52.5 shows how images are stored in our train batches.

During prediction, each frame from the output source is passed through a face classifier named as Haar Cascade Classifier to detect the bounding box coordinates of the face. Paul Viola and Michael Jones developed the Haar Cascade classifier as an

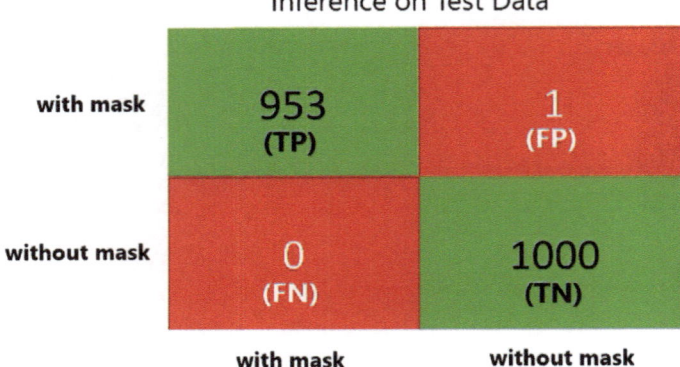

Fig. 52.5 Confusion matrix on test data without normalization

effective object identification approach in their work "Rapid Object Detection with a Boosted Cascade of Simple Features" in 2001[5]. The detectMultiScale() function of OpenCV is used to find the coordinates of the faces. It features parameters such as scaleFactor, which determines how much the image size is decreased, and minNeighbors, which determines how many neighbors each candidate rectangle should keep. With experimentation, a scale factor of 1.1 and minNeighbors of 5 gave a decent output. This function returns an array of face coordinates consisting of four values X-coordinate, Y-coordinate, width, and height of the detected feature of the face. These faces are then cropped and resized to 224 × 224. Keras works with images in batches. As a result, the number of samples (or images) is counted using the first dimension. An additional dimension is required to make a batch of images (samples, size1, size2, channels). Therefore, the images are converted into arrays using img_to_array() function, and an additional dimension is added to the image array using expand_dims() function from NumPy library. This resultant sliced image is then passed through the preprocess_input() function of MobilenetV2 provided by Keras. The preprocess input function is used to convert an image to the format required by the model. After preprocessing, the image is sent to the model for prediction and the class with the highest probability is displayed on the output screen.

52.4 Results

For compiling our model, we used ADAM optimizer as proposed by Jagadeeswari and Uday Theja [6] with a learning rate of 0.0001 loss function of "categorical cross-entropy" and metric set to accuracy. We also created a model checkpoint function that monitors validation loss and saves only the best weights. The model performed exceptionally well by giving a training accuracy of 100% on the 12th epoch and testing accuracy of 99.99%. Accuracy, Precision, Recall, and F1 score are some of

Table 52.1 Performance of the proposed model

	Precision	Recall	F1-Score	Support
With mask	1.00	1.00	1.00	954
Without mask	1.00	1.00	1.00	1000
Accuracy			1.00	1954
Macro average	1.00	1.00	1.00	1954
Weighted average	1.00	1.00	1.00	1954

the performance metrics used to evaluate the transfer learning model's performance. Table 52.1 shows the values of the performance metrics when our mode was evaluated on test datasets (Figs. 52.6 and 52.7).

Fig. 52.6 Training versus testing loss

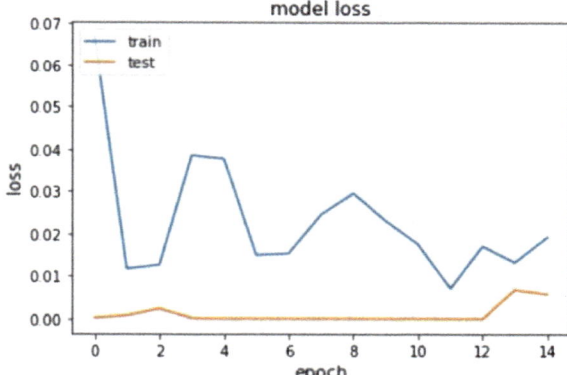

Fig. 52.7 Training versus testing accuracy

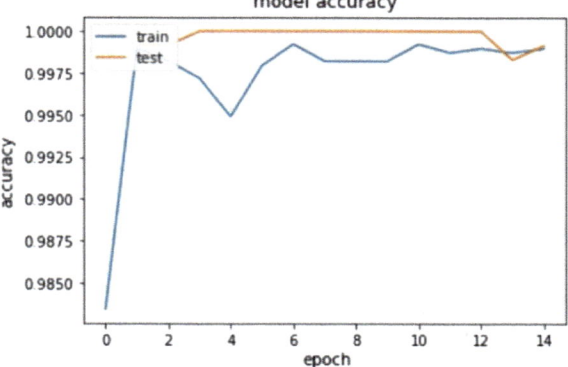

Table 52.2 Comparison of the proposed model with previous models

	Dataset	Model proposed	Accuracy
Ref. [2]	Medical mask dataset from kaggle	SRCNet classification network	98.7%
Ref. [3]	Mixture of ORL Database and own images	Principal Component Analysis (PCA)	96.25% (without mask) 68.75% (with mask)
Ref. [4]	WIDER FACE and Celebi Databases	YOLO V3	93.9%
Ref. [1]	MaskedFace-Net	Transfer learning of MobileNet V2	100%

52.5   Output

The proposed model can be used in conjunction with surveillance cameras to prevent COVID-19 transmission by detecting people who aren't wearing face masks. To execute this, we need computer vision. In this research, Haar Cascade Face Classifier is implemented for detecting the coordinates of faces in the video streams. The sliced image is then further pre-processed with the preprocess_input() function provided by Keras, and the output is displayed on the screen with the help of cv2 (Table 52.2).

For commercial use, this application can be integrated with surveillance cameras in public places. After integration opencv is used for working with the video data. The camera source is connected to the program using the VideoCapture() function provided by opencv. Now the entire video is looped by each frame, and face coordinates are present in each frame are stored in a numpy array. The face images are then further sliced from each frame, preprocessed, and sent to the model for prediction. Using OpenCV, a rectangle is drawn along the face coordinates present in each frame, and a text containing the class and confidence score of the class is displayed on the top of the rectangle. This process is repeated for all frames present in the video. The video output on webcam generated about 20–25 fps. A python Tkinter App is developed to automate the above process.

52.6   Conclusion and Future Scope

Because of COVID-19 pandemic, the world is facing a huge crisis. It has been proved through experiment that the spread of COVID occurs more in crowded areas like in public places. Since it transmits through contact and air, scientists have shown that wearing a mask can limit the transmission of this disease to a greater extent. As a result, ruling parties of various nations have made it essential to wear masks in public areas, reducing the virus's transmission rate. Since it is difficult to monitor crowd individually, we have developed a deep learning model in this paper that is built on transfer learning of MobileNetV2. In the proposed work, the proposed model has achieved an accuracy of 99.99% on SMFD (Simulated Mask Face Dataset). The same work can be further improved by using various state of the art object detection models like Fast RCNN, EfficientNet, Retina Net, YOLO (You Look Only Once) for detecting the bounding boxes of faces instead of using Haar cascade Classifier. This will increase the accuracy of detecting face coordinates and reduce undetected faces in a single frame. Furthermore, this model can be trained on a larger dataset with real-face masked images to provide better accuracy and output.

References

1. Qin, B., Li, D.: Identifying facemask-wearing condition using image superresolution with classification network to prevent covid-19 (2020)
2. Ejaz, M.S., Islam, M.R., Sifatullah, M., Sarker, A.: Implementation of principal component analysis on masked and non-masked face recognition. In: 2019 1st International Conference on Advances in Science, Engineering and Robotics Technology (ICASERT). pp. 1–5 (2019)
3. Li, C., Wang, R., Li, J., Fei, L.: Face detection based on yolov3. In: Recent Trends in Intelligent Computing, Communication and Devices, pp. 277–284. Springer (2020)
4. Dataset, Cabani, A., Hammoudi, K., Benhabiles, H., Melkemi, M.: MaskedFace-Net - A dataset of correctly/incorrectly masked face images in the context of COVID-19. Smart Health. ISSN 2352–6483, Elsevier (2020)
5. Viola, P., Jones, M.: Rapid object detection using a boosted cascade of simple features. In: Proceedings of the 2001 IEEE Computer Society Conference on Computer Vision and Pattern Recognition. CVPR 2001, pp. I-I (2001). https://doi.org/10.1109/CVPR.2001.990517
6. Jagadeeswari, C., Uday Theja, M.: Performance evaluation of intelligent face mask detection system with various deep learning classifiers. Int. J. Adv. Sci. Technol. **29**(11s), 3074–3082 (2020)

Chapter 53
Feature Selection Technique-Based Approach for Suggestion Mining

A. Ramesh, K. Pradeep Reddy, M. Sreenivas, and Para Upendar

Abstract The consumers share their opinions toward commercial entities like services, brands and products in discussion forums, blogs, online reviews and social media platforms. Opinion mining or sentiment analysis is one popular technique to extract positive, negative or neutral opinion of consumers on different entities. Suggestion mining is one type of opinion mining technique to extract the advices, tips or wishes that are hidden in the text. Suggestion mining is used in different applications like marketing, economics, politics and advertising. The application of suggestion mining provides the motivation for the SemEval 2019 Task 9, which introduced two subtasks such as subtask A and subtask B to differentiate the sentences into suggestions or non-suggestion sentences. Subtask A contains training and test datasets from same domain whereas subtask B contains training and test datasets from different domains. In this work, the subtask A dataset is used for classifying the given sentences into suggestion sentences or not. In this work, the experiment was conducted with different feature selection techniques such as Chi-square (CHI2), Document Frequency Difference (DFD) and Multivariate Relative Discriminative Criterion (MRDC). The feature selection techniques select the important features from the large feature set. The extracted features are used to represent the sentence vectors. These vectors are forwarded to two classification algorithms such as Support Vector Machine and Random Forest. The proposed system attained good accuracy of 83.47% for suggestion mining.

A. Ramesh (✉)
Vardhaman College of Engineering, Hyderabad, Telangana, India

K. P. Reddy
CMR Institute of Technology, Hyderabad, Telangana, India

M. Sreenivas
Sreenidhi Institute of Science and Technology, Hyderabad, Telangana, India

P. Upendar
MLR Institute of Technology, Hyderabad, Telangana, India

© The Author(s), under exclusive license to Springer Nature Singapore Pte Ltd. 2022   541
V. Bhateja et al. (eds.), *Evolution in Computational Intelligence*,
Smart Innovation, Systems and Technologies 267,
https://doi.org/10.1007/978-981-16-6616-2_53

53.1 Introduction

Online forums and review-based portals contain plethora of user-generated text. These texts often contain many different opinions, which are the subject of research in area of opinion mining. Sentiment analysis or opinion mining is a process of computationally identifying and categorizing the opinions from unstructured data. This is used to identify a user's perspective like positive, negative or neutral on a product. On the other hand, there are also different types of information that exist within these texts such as suggestions. Unlike opinions, suggestions appear in different parts of a text and also appear more sparsely. Suggestion mining is a relatively young field of research when compared to sentiment analysis and is realized as a standard text classification with two classes such as suggestion and non-suggestion.

Opinion mining is used to identify whether the product is successful in the market or not. Suggestion mining finds out ways to enhance the product to satisfy the customers. Review texts are mainly used to identify the sentiments of the user. Besides sentiments, review texts also contain valuable information such as advice, recommendations, tips and suggestions on a variety of points of interest [1]. These suggestions will help other customers to make their choices and the sellers can improve their products' quality. While mining for suggestions, the propositional aspects like mood, modality, sarcasm and compound statements have to be considered. Several applications such as customer to customer suggestions, enhancement of sentiment detection, product improvement, recommender systems and suggestion summarization used the techniques of suggestion mining [2].

The researchers proposed different systems for suggestion mining based on deep learning and machine learning techniques. In this work, the experiment was conducted with the terms used by the authors used in their texts. The experiment started with selection of terms based on the frequency of the terms used in the corpus. It was observed that the achieved results are not good for suggestion mining. Later, the experiment continued with the terms, which are selected by the feature selection algorithms. Three feature selection algorithms are used in this work to identify the relevant features. The training dataset of suggestion mining is represented as vectors with these identified features. Support Vector Machine (SVM) and Random Forest (RF) classification algorithms are used to evaluate the performance of this system. The subtask A dataset of SemEval 2019 competition is used for this experiment, and accuracy measure is used to evaluate the performance.

This paper is planned in six sections. The existing proposals of different researchers for suggestion mining are discussed in Sect. 2. The features of dataset and evaluation measures are explained in Sect. 3. The feature selection techniques and the proposed system are explained in Sect. 4. The experimental results of the proposed system are presented in Sect. 5. The conclusions of this work are specified in Sect. 6 with future scope for suggestion mining.

53.2 Literature Survey

Suggestion mining is a trending research domain that focuses on the extraction of tips, advice, and recommendations from unstructured text. To better recognize suggestions, instead of only matching feature words, one must have the ability to understand long and complex sentences. SemEval-2019 Task 9 provides the suggestion mining task [3]. The task can be recognized as a text classification task, given a sentence collected from user feedback, participating systems are required to give a binary output by marking it as suggestion or non-suggestion.

Samuel Pecar et al. proposed [4] neural model architecture for suggestion mining. For subtask A, they employed bidirectional LSTM encoder, which consists of two stacked layers followed by self-attention. For subtask B, better performance was proved by only one layer in one direction to reduce learning process and over-fitting. The proposed model attained test evaluation score of 0.6816 and 0.6850 for subtask A and subtask B, respectively.

Rajalakshmi et al. developed [5] a system named as suggestion mining task in SemEval-2019. This system used SMOTE technique for data augmentation to balance the imbalance in the data, linguistic rule-based method for feature extraction and deep learning technique (Convolutional Neural Network (CNN)) for classification. The extracted Bag of Words (BOW) features are used in MLP, RF and CNN classifier for suggestion mining. They compared the results of CNN with Multi-Layer Perceptron (MLP) model and Random Forest classifier (RF). They found that the CNN model performs better compared with RF and MLP classifiers for both the subtasks.

Yimeng Zhuang developed [6] a Self-Attention Network (SAN) model by concentrating on modeling long-term dependency. They adopted various techniques such as back-translation, linguistic features, contextualized embedding and auxiliary loss to augment the system. Their model achieved an F1 score of 76.3 and obtained fourth rank in the competition. Yunxia Ding et al. described [7] a deep-learning system named as Stacked Bidirectional Long-Short Memory Network (SBiLSTM). They used Word2Vec model to learn the distributed representations from sentences. The ensemble technique is performed well to improve the effectiveness of their model. The proposed model achieved an F1-score of 0.5659.

Ping Yue et al. used [8] bidirectional encoder representation learned from transformers (BERT) to address the problem of domain-specific suggestion mining. The BERT is also used to extract feature vectors and perform fine-tuning. They applied an ensemble model to combine the BiLSTM, CNN and GRU models. They observed that these models achieved good performance in the final evaluation phase. Qimin Zhou et al. used [9] pre-trained BERT model to learn the word or sentence embeddings. BERT stands for Bidirectional Encoder Representation from Transformers is the latest breakthrough in the field of NLP provided by Google Research. The embedding vectors that are extracted from BERT as the input to the CNN layer. The proposed system obtained an F1 score of 0.715.

Table 53.1 The Dataset characteristics

Dataset	Total number of suggestions	Total number of Non-suggestions	Total number of sentences
Training data	2085	6415	8500
Test data	296	296	592

53.3 Dataset Characteristics

"Task 9—Suggestion mining from online reviews and forums" is introduced in SemEval 2019 competition, which has two subtasks [3]. Subtask A is to classify a sentence into a suggestion or a non-suggestion. The features of dataset are represented in Table 53.1. The training dataset is imbalanced, which means there is a huge variation in the number of sentences of suggestion class and non-suggestion class whereas the test dataset is balanced. It was observed that the imbalance of data reduces the accuracy of our proposed approach.

Different evaluation measures such as recall, precision, F1-score and accuracy are used by the researchers for evaluating the solutions of suggestion mining. In this work, accuracy measure is used to evaluate the performance of our proposed model. Accuracy is the number of test sentences that are correctly predicted their class label from a given set of test sentences [10]. Because of complexity, we presented only accuracy results in the paper.

53.4 Feature Selection Technique-Based Approach for Suggestion Mining

The proposed model is represented in Fig. 53.1.

In this model, first apply different preprocessing techniques such as tokenization, removal of emoticons, hashtags, retweets, @mentions, stopword removal and stemming for removing the irrelevant information in the text. Stopwords are words that are frequently used in the text but they don't have any discriminative power. Prepositions, conjunctions, articles etc. are the examples of stopwords. Stemming is a technique of transforming different forms of same word into its root form. After cleaning the data in the training dataset, the next step is extracting all the terms. Next, feature selection techniques are used to compute the ranks of the terms. The top-ranked terms are used as features and the training dataset sentences are represented as vectors with these features. The vectors are forwarded to machine learning algorithms to generate the classification model. The test sentences vectors are passed to the model that predicts the class label of test sentence as suggestion sentence or non-suggestion sentence.

In this proposed model, the feature selection technique step is important to improve the accuracy of suggestion mining. Feature selection is an important step in the

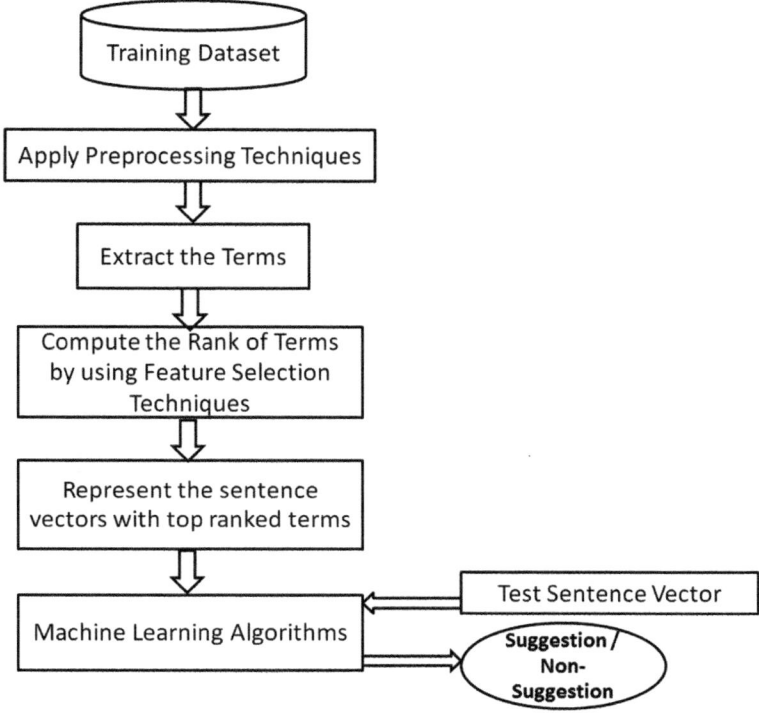

Fig. 53.1 The proposed model

machine learning algorithms. Feature selection techniques recognize the informative and relevant features from a set of features.

53.4.1 Chi-Square (CHI2)

Chi-square measure determines the dependency score among the class and a feature [11]. The high score indicates that the feature is more relevant to the given class, and low score indicates that the feature is less informative to the class. Eq. (53.1) designates the CHI2 measure for term t in a specific class C.

$$\chi^2(t, C) = \frac{N(AD - BC)^2}{(A + B)(A + C)(B + D)(C + D)} \tag{53.1}$$

In this measure, A is number of class C documents contains term t, B is the number of other than class C documents contain the term t, C is the number of class C documents doesn't contain term t and D is the number of other than class C documents doesn't contain the term t, N is the total document count. Eq. (53.2) is used to compute the CHI2 of a term across all classes.

$$\chi^2(t) = \sum_{i=1}^{m} P(Ci)\chi^2(t, Ci) \tag{53.2}$$

where, m is number of classes, $P(C_i)$ is the proportion of documents in class C_i relative to the total number of documents in the training dataset.

53.4.2 Document Frequency Difference (DFD)

DFD measure is proposed by Nicholls and Song [12] based on the observation of the sentiment word distribution in positive and negative class of documents. Equation (53.3) is used to calculate the score for features by using DFD measure.

$$DFD(t) = \frac{DFt, pos - DFt, neg}{N} \tag{53.3}$$

where, $DF_{t, pos}$ is the number of positive class documents that contain the term t, $DF_{t, neg}$ is the number of negative class documents that contain the term t, and N is the total document count in the dataset. DFD measure normalizes the scores of terms among 0 and 1.

53.4.3 Multivariate Relative Discriminative Criterion (MRDC)

Multivariate relative discriminative criterion (MRDC) measure is used to compute the rank of a feature based on the relative discriminative criterion (RDC) and correlation between features [13]. Equation (53.4) is used to compute MRDC.

$$MRDC(fi) = RDC(fi) - \sum_{fi \neq fj, fj \in S} correlation(fi, fj) \tag{53.4}$$

Equation (53.5) is used to compute the relative discriminative criterion (RDC) of a feature. RDC measure works based on the difference among the document counts of a feature in positive and negative class of documents.

$$RDC(wi, tcj(wi)) = \left(\frac{|df_{pos}(wi) - df_{neg}(wi)|}{\min(df_{pos}(wi), df_{neg}(wi)) \times tcj(wi)} \right) \tag{53.5}$$

where, $df_{pos}(w_i)$ and $df_{neg}(w_i)$ are the document counts of a feature w_i in positive class and negative class of documents, respectively. $tc_j(w_i)$ is the number of times the feature w_i occurred in jth class. The feature w_i may repeat several times in a specific class of documents.

The correlation among features f_i and f_j is calculated by using Eq. (53.6).

$$correlation(fi, fj) = \left| \frac{\sum_{D=1}^{N} (fi, D - \overline{f}i)(fj, D - \overline{f}j)}{\sqrt{\sum_{D=1}^{N} (fi, D - \overline{f}i)^2} \sqrt{\sum_{D=1}^{N} (fj, D - \overline{f}j)^2}} \right|$$

(53.6)

Where, $f_{i, D}$, $f_{j, D}$ are the number of times feature f_i and f_j occurred in a document D respectively N is the total document count in dataset. $\overline{f}i$ and $\overline{f}j$ are the mean values of feature f_i and f_j, respectively.

53.5 Experimental Results

In this work, the experiment was carried out with three feature selection algorithms such as CHI2, DFD and MRDC and two machine learning algorithms such as SVM and RF. The feature selection algorithms select the best informative features by computing the ranks of the features. The machine learning algorithms build the classification model by using sentence vectors of training dataset. The classification model is used to predict the class label of test documents. Table 53.2 shows the accuracies of suggestion mining when experimented with feature selection algorithms and machine learning algorithms.

The experiment was performed with highest-ranked features of 4000 terms. The experiment starts with top-ranked 1000 features and then incremented by 1000 in next iterations. It was found that the accuracy was reduced when experiment continued with after 4000 features. Thus, the experiment stopped with top ranked 4000 features.

It was observed that the MRDC feature selection algorithm identified best informative features when compared with CHI2 and DFD algorithms. The RF classifier achieved the best accuracy of 83.47 for suggestion mining when compared with

Table 53.2 The accuracies of suggestion mining

Number of features/Machine learning algorithms—Feature selection techniques	Support vector machines			Random forest		
	CHI2	DFD	MRDC	CHI2	DFD	MRDC
1000	74.20	73.97	77.87	77.87	77.33	79.46
2000	75.39	74.56	78.56	79.53	78.25	80.79
3000	76.87	75.12	80.37	80.41	79.46	82.18
4000	78.25	76.33	81.29	81.64	80.21	83.47

SVM classifier. The high ranked 4000 terms achieved best accuracy for suggestion mining in both classifiers. It was identified that the accuracy of suggestion mining is increased when the number of high-ranked features is increased in the experiment. The SVM classifier with MRDC feature selection algorithms obtained good accuracy of 81.29 for suggestion mining when experimented with top-ranked 4000 features.

53.6 Conclusion and Future Plan

Suggestion mining is used to extract advice from text such as that provided in blogs, online reviews, discussion forums and social media platforms where consumers share their opinions toward commercial entities like services, brands and products. In this work, feature selection-based approach is proposed for suggestion mining. The experiment was performed with three feature selection algorithms such as CHI2, DFD and MRDC and two machine learning algorithms such as SVM and RF. The random forest classifier performance is good when compared with SVM. The MRDC measure recognized the best features when compared with CHI2 and DFD. The RF achieved an accuracy of 83.37 for suggestion mining when experimented with high-ranked features of 4000, which are identified by MRDC measure.

In future work, we will propose a new feature selection algorithm to identify best informative features as well as to avoid the redundant features and also plan to apply deep learning techniques to improve the accuracy of suggestion mining.

References

1. Negi, S., Buitelaar, P.: Suggestion mining from opinionated text. In: Sentiment Analysis in Social Networks, pp. 129–139 (2017)
2. Negi, S., Asooja, K., Mehrotra, S., Buitelaar, P.: A study of suggestions in opinionated texts and their automatic detection. In: Proceedings of the Fifth Joint Conference on Lexical and Computational Semantics, pp. 170–178 (2016)
3. Negi, S., Daudert, T., Buitelaar, P.: Semeval-2019 task 9: Suggestion mining from online reviews and forums. In: Proceedings of the 13th International Workshop on Semantic Evaluation (SemEval-2019)
4. Pecar, S., Simko, M., Bielikova, M.: NL-FIIT at SemEval-2019 Task 9: neural model ensemble for suggestion mining. In: Proceedings of the 13th International Workshop on Semantic Evaluation (SemEval-2019), pp. 1218–1223 Minneapolis, Minnesota, USA, June 6–7, 2019
5. Rajalakshmi, S., Angel Deborah, S., Milton Rajendram, S., Mirnalinee, T.T.: SSN-SPARKS at SemEval-2019 Task 9: mining suggestions from online reviews using deep learning techniques on augmented data. In: Proceedings of the 13th International Workshop on Semantic Evaluation (SemEval-2019), pp. 1237–1241 Minneapolis, Minnesota, USA, June 6–7, 2019
6. Zhuang, Y.: Yimmon at SemEval-2019 Task 9: suggestion mining with hybrid augmented approaches. In: Proceedings of the 13th International Workshop on Semantic Evaluation (SemEval-2019), pp. 1267–1271 Minneapolis, Minnesota, USA, June 6–7, 2019
7. Ding, Y., Zhou, X., Zhang, X.: YNU DYX at SemEval-2019 Task 9: a stacked BiLSTM model for suggestion mining classification. In: Proceedings of the 13th International Workshop on

Semantic Evaluation (SemEval-2019), pp. 1272–1276 Minneapolis, Minnesota, USA, June 6–7, 2019

8. Yue, P., Wang, J., Zhang, X.: YNU-HPCC at SemEval-2019 Task 9: using a BERT and CNN-BiLSTM-GRU model for suggestion mining. In: Proceedings of the 13th International Workshop on Semantic Evaluation (SemEval-2019), pp. 1277–1281 Minneapolis, Minnesota, USA, June 6–7, 2019

9. Zhou, Q., Zhang, Z., Wu, H., Wang, L.: ZQM at SemEval-2019 Task9: a single layer CNN based on pre-trained model for suggestion mining. In: Proceedings of the 13th International Workshop on Semantic Evaluation (SemEval-2019), pp. 1287–1291 Minneapolis, Minnesota, USA, June 6–7, 2019

10. Raghunadha Reddy, T., Vishnu Vardhan, B., Vijaypal Reddy, P.: A survey on authorship profiling techniques. Int. J. Appl. Eng. Res. **11**(5), 3092–3102 (2016)

11. Yang, Y., Pedersen, J.O.: A comparative study on feature selection in text categorization. In: Fourteenth International Conference on Machine Learning, Morgan Kaufmann Publishers Inc, pp. 412–420 (1997)

12. Nicholls, C., Song, F.: Comparison of feature selection methods for sentiment analysis. In: Advances in artificial intelligence, pp. 286–289. Springer, Berlin (2010)

13. Labani, M., Moradi, P., Ahmadizar, F., Jalili, M.: A novel multivariate filter method for feature selection in text classification problems. Eng. Appl. Artif. Intell. **70**, 25–37 (2018)

Chapter 54
Anti-Jamming Wireless Communication Using Chaos-Based Code Selection Spread Spectrum Technique

Balamurugan Gopalakrishnan(ID) **and M. A. Bhagyaveni**(ID)

Abstract Physical layer performance of wireless communication, which is subjected to reactive jamming attack, is analyzed in this paper. The main problem in conventional direct sequence spread spectrum (DSSS) is the requirement of pre-shared key between sender and receiver. Many researchers focused to remove pre-shared keys for anti-jamming communication, but in those techniques, it requires enough time to de-spread the data. So to overcome the problem of pre-shared key and to enhance the security of wireless communication with minimum de-spreading time, chaos-based code selection (CCS) DSSS has been proposed in this paper. Matlab simulation results show that our proposed technique CCS DSSS indicates much better performance when compared with other unshared secret spreading keys in terms of security and de-spreading time. The de-spreading time required to extract the original message of size 512 bits is almost 0.05888 ms, which is minimum compared with other unshared secret spreading key spread spectra.

54.1 Introduction

Secured communication plays an important role in transferring confidential information over a wireless network in the presence of eavesdroppers and jammer still remains a challenging task [1]. Due to the broadcasting nature of wireless communication, wireless networks are more difficult to secure by traditional cryptography schemes. The cryptography techniques are generally implemented in higher layers; physical layer security is now emerging as a promising technique to realize wireless secrecy in communications. To overcome jamming attack, traditionally spread

B. Gopalakrishnan (✉)
MIT Campus, Anna University, Chennai, India

M. A. Bhagyaveni
College of Engineering Guindy, Anna University, Chennai, India

© The Author(s), under exclusive license to Springer Nature Singapore Pte Ltd. 2022   551
V. Bhateja et al. (eds.), *Evolution in Computational Intelligence*,
Smart Innovation, Systems and Technologies 267,
https://doi.org/10.1007/978-981-16-6616-2_54

spectrum technique is commonly used. Spread spectrum techniques such as direct sequence spread spectrum [1–4] (DSSS) and frequency hopping spread spectrum [1, 2, 5, 6] (FHSS) have been commonly adopted techniques to defend against jamming attack.

Many researchers have focused to remove the requirement of a pre-shared key for anti-jamming communication. The authors of [2–4] presented an uncoordinated DSSS (UDSSS) and delayed seed disclosure (DSD) DSSS technique, respectively, to overcome the reactive jamming attack without a pre-shared secret key. In these techniques, the communication efficiency decreases and also receiver requires enough time to de-spread the message. Nguyen Xuan Quyen et al. [7] discussed a chaos-based DSSS; the message bit duration is varied according to a chaotic behaviour, and these data bits are spread with a pseudo-noise (PN) sequence. The receiver requires an identical regeneration of both PN sequence and also the chaotic behaviour to improve the data security significantly. Arash Tayebi et al. [8] analyzed the performance of synchronization in direct sequence spread spectrum under jamming attacks using the chaotic sequence. In this, the amount of power required for each jammer to collapse the synchronization was analyzed.

Yao Liu et al. [9] developed Randomized Differential DSSS (RD-DSSS) to overcome jamming without a pre-shared key. RD-DSSS encodes each bit with the correlation of unpredictable spreading codes. The bit "0" is encoded with two different spreading codes and a bit "1"is encoded using two identical spreading codes. Hence, this technique generates high overhead. Balamurugan et al. [10] developed Random Code key Selection using Codebook (RCSC) DSSS, which provides the solution to anti-jamming. The transmitter randomly selects a key from codebook, and the next spreading key is disclosed at the end of the current message transmission. At first, the receiver guesses the key used by the transmitter, and thereafter next key is extracted from the current de-spread message. Alagil et al. [11] proposed Randomized Positioning DSSS (RP-DSSS) scheme as an extension to improve the security of RD-DSSS. The vulnerability of index codes roots raises from the fact that they are located at an end of a spread message. To protect index codes from adversaries, they randomly relocated the index codes for each message. In [12] the author proposed a technique to identify the location of random seed based on multiple messages to recover spreading code.

In our approach, the legitimate node selects the codekey from the codebook with the help of chaotic discrete values. Here, instead of using static chaotic initial value, it can be made as dynamic initial value by measuring the variations of received signal strength (RSS) on the wireless channel between sensor nodes and PDA device. Due to this, it is very difficult for the intruder to predict the codekey used by the legitimate nodes. From the above literature survey, it has been found that many authors are working in spread spectrum communication without a pre-shared key. But no one considers to minimizing the time required for de-spreading data by the legitimate node. This is the first paper to discuss about minimizing the de-spreading time without pre-shared codekey with the help of chaos signal in spread spectrum communication.

54.2 System Description

In this section, we described Chaos-based Code Selection DSSS communication system structure and operation. The CCS DSSS does not require pre-sharing of keys like traditional DSSS. The keys are selected from the codebook by the chaotic sequence, which is generated from discrete chaotic function. Based on the initial condition of the chaos signal, codekeys are extracted from the codebook. Figure 54.1a, b illustrates the transmitter and receiver process in CCS DSSS technique. The codebook generation, chaos initial seed generation, and chaos-based code selection blocks are common to both transmitter and receiver.

The codebook generator consists of two blocks, namely, PN sequence and group creator with indexed codebook. The PN sequence is generated by linear feedback shift registers and it is denoted by pn. The codekeys are created by permutation and combination of pn sequence with codekey length (a) and it is created by $G = pnC_a$; $a \in (4, 5, 6)$. For each generated codekey (k), a unique index (i) is assigned to form a codebook (A). It is shown in Eq. 54.1; y is the length of index and it is calculated as $\lceil log_2 G \rceil$. The dimension of codebook (A) is $[G \times (y + a)]$ where each bit $i_{(a,b)} \in (1, 0)$ and $k_{(a,b)} \in (1, 0)$.

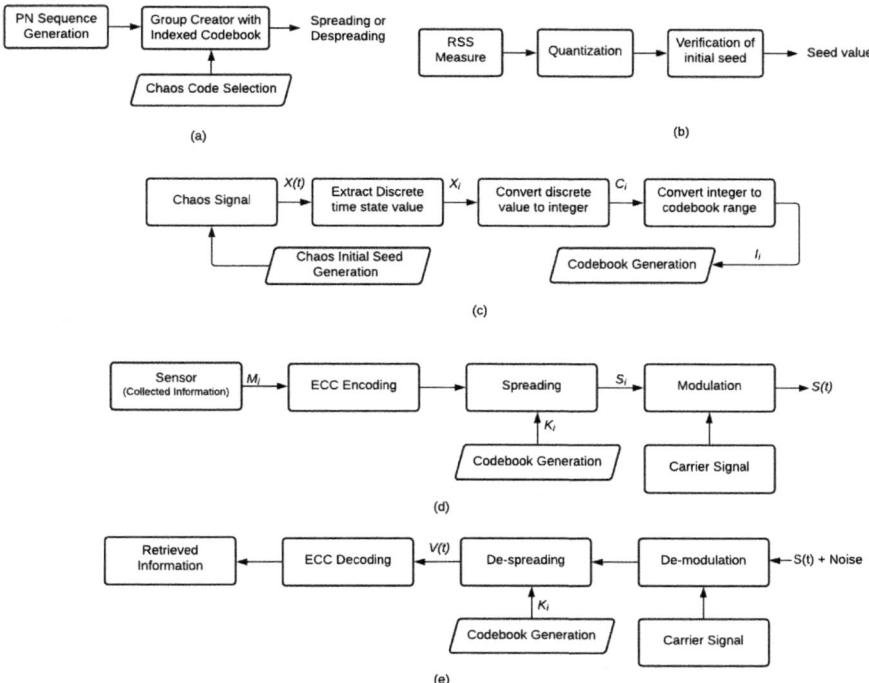

Fig. 54.1 **a** Codebook Generation **b** Chaos Initial Seed Generation **c** Chaos-based Code Selection **d** CCS DSSS transmitter block **e** CCS DSSS receiver block

$$A = \begin{vmatrix} i_{(1,1)} & i_{(1,2)} & . & i_{(1,y)} & k_{(1,1)} & k_{(1,2)} & k_{(1,3)} & k_{(1,4)} & k_{(1,a)} \\ i_{(2,1)} & i_{(2,2)} & . & i_{(2,y)} & k_{(2,1)} & k_{(2,2)} & k_{(2,3)} & k_{(2,4)} & k_{(2,a)} \\ i_{(3,1)} & i_{(3,2)} & . & i_{(3,y)} & k_{(3,1)} & k_{(3,2)} & k_{(3,3)} & k_{(3,4)} & k_{(3,a)} \\ i_{(4,1)} & i_{(4,2)} & . & i_{(4,y)} & k_{(4,1)} & k_{(4,2)} & k_{(4,3)} & k_{(4,4)} & k_{(4,a)} \\ & . & & & . & . & . & . & . \\ & . & & & . & . & . & . & . \\ i_{(G,1)} & i_{(G,2)} & . & i_{(G,y)} & k_{(G,1)} & k_{(G,2)} & k_{(G,3)} & k_{(G,4)} & k_{(G,a)} \end{vmatrix} \qquad (54.1)$$

Chaotic sequences are generated by discrete chaotic functions. We used logistic map to select codekey from the codebook for spreading and de-spreading the data. Logistic map is expressed in Eq. 54.2. It maps the current state x_i to the next state x_{i+1}. The logistic map created by varying the bifurcation parameter r from 2.5 to 4.0 makes the sequence a-periodic.

$$x_{i+1} = \left(\frac{r}{2}\right)\left(1 - x_i^2\right) - 1 \qquad (54.2)$$

The initial seed selection for chaos signal is based on the received signal strength between transmitter and receiver. The chaotic sequence is highly dependent upon the initial seed value and any small discrepancies in the initial seed value result in complete change in chaotic sequence, and thus high data security is achieved. The complete steps to select the chaos-based codekey selection are described as follows.

Step 1: Multiple packets are exchanged between two devices and channel radio signal strength (RSS) is measured by $P_r = P_o - 20\log\left(\frac{d}{d_0}\right)$ where P_r is the received power expressed in dBm, and P_o is the power of the signal separated at a distance d_0 away from the transmitter and d is the distance between two antennas.

Step 2: The extracted radio signal measurement is encoded into binary bit-based quantization. The upper level quantization is represented as 1 if $Q^+ = \mu + \alpha\sigma$, lower level quantization is represented as 0 if $Q^- = \mu - \alpha\sigma$ and if the RSS value lies between Q^+ and Q^- then it is dropped. Where μ is mean value, σ is standard deviation and α is constant term and should be greater than zero.

Step 3: The extracted bits stream after quantization is usually not identical on both sides due to bidirectional channel characteristics. To correct the mismatching bits, error correction code (ECC) is used. The authorized node permutes the bit stream randomly, divides into small blocks, and sends the parity information of each block to another node. Similarly, another node also computes the parity information and check for parity mismatch. The node performs a binary search on the message block and converts the mismatched bits. The computed matched value is converted to decimal fraction format, and it is used as chaos initial seed value (x_0).

Step 4: The obtained chaos initial seed value (x_0) is used in logistic map expressed in Eq. 54.2 to extract discrete chaotic values(x_{i+1}). By iterating the chaotic equation, different discrete chaotic values are generated. Convert the generated chaotic sequence x_i to chaotic integer (C_i) value by $C_i = int[(x_i * 10^\vartheta)]$, where ϑ is an integer chosen to set precision value and it should be $10^\vartheta > G$. Modulus operation is performed by $C_i mod G$ to obtain chaotic index (I_i) value. Corresponding to the generated index (I_i), codekey (K_i) is obtained from the codebook (A).

The CCS DSSS system spreading and de-spreading process is shown in Fig. 1a, b. In this system, the initial value of chaos is generated by chaos initial seed generation by exchanging RSS signal between sensor and PDA device. It is then encoded into bit streams by using quantization, and identical bits are obtained by exchanging parity bits. The obtained bits are set as initial value (x_i) of the chaos signal. Based on the initial condition, discrete chaotic values (x_{i+1}) are generated. The chaotic discrete (x_{i+1}) values are converted to chaotic integer value (C_i).

The obtained chaotic integer value (C_i) is used as an index I_i.

$$C_i \in I_i, where I_i \in (1, \ldots, G) \tag{54.3}$$

C_i chaotic index is used for selecting the key K_i from the codebook (A).

$$K_i = A_{C_i}, where \ C_i \in G \tag{54.4}$$

The selected codekey K_i is spread with physiological signal M_i, where codekey is generated by chaotic logistic map at that instant. It is then passed through the modulation. During transmission, noise interferes with the message signal.

$$S(t) = (M_i.K_i)\sin(2\pi f_c t) + N_i(t) \tag{54.5}$$

On the receiving side, the received signal is demodulated and then it is de-spread by multiplying with the codekey K_i. The codekey K_i is selected based on chaotic signal generated at receiver. Here, both the sender and receiver initial chaos values are selected by exchanging RSS signal. Then, de-spread signal enters the correlator.

$$\widehat{m}(t) = \int_{t_n}^{t_{n+1}} S(t).\sin(2\pi f_c t).dt \tag{54.6}$$

The output of coherent detector $\widehat{m}(t)$ is multiplied with the codekey K_i and then integrated over one bit period T_b, where $T_b = aT_c, a \in (4, 5, 6)$

$$v(t) = \int_0^{T_b} \widehat{m}(t).K_i dt \tag{54.7}$$

Finally, the decision device constructs the received message depending on the polarity of the output of the correlator $v(t)$.

$$b_n = \begin{cases} 1, v(t) \geq 0 \\ 0, v(t) < 0 \end{cases} \qquad (54.8)$$

54.3 Simulation Results

In this section, employing MATLAB computer simulation, the performance of Chaos-based Code Selection DSSS was evaluated. For evaluation, we used Intel I5 core processor, 2.5 GHz, and 8 GB RAM.

54.3.1 Probability of Message Being Jammed

In our simulation, we consider the tolerable ECC error is 3%. We performed 500 trials and RS (255, 223) code is used. In each trial, we spread the message by picking a random code sequence from codebook. We also generated random codes for the reactive jammer. To calculate the probability that the message gets jammed, we count the number of error bits that are affected by the jammer. If this number is larger than [(ECC) x (length of message)] then the trial is considered as failed. The probability that the message jammed is calculated by using Eq. 54.9.

$$\text{Probability of message being jammed} = \frac{\text{Number of failed trials}}{\text{Total number of trials}} \qquad (54.9)$$

Figure 54.2 shows that the probability of message being jammed for codebook size 500 is tested under different SNR values. The codekey length is set to five chips, and 1000 blocks of information are transmitted for testing. It is inferred that probability of message being jammed for DSSS is high when compared with UDSSS, DSD-DSSS, and CCS-DSSS techniques. In DSSS technique, the spreading key is pre-shared in the presence of jammer, and only one codekey is used to de-spread the entire message. So if the jammer identifies the secret key, then it can replicate the secret code and jam the entire network. Whereas in UDSSS, DSD-DSSS, and CCS-DSSS, the secret key is not pre-shared, and also the spreading codekey varies dynamically so it is very difficult for the intruder to jam the message. The probability of message being jammed (SNR = 0 dB) is about 0.206 for DSSS, whereas UDSSS, CCS-DSSS, and DSD-DSSS, the probability of message being jammed (SNR = 0 dB) is almost 0.06.

Fig. 54.2 Probability of message being jammed for codebook size 500

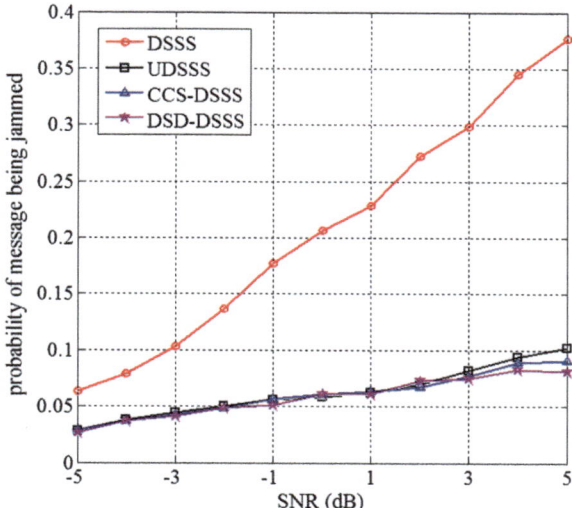

54.3.2   De-Spreading Time

The de-spreading message is tested with payload size of 512 bits. The de-spreading time of a receiver is measured for an average of 10 rounds. We tested UDSSS, DSD DSSS, and CCS DSSS techniques using similar codebook size ranging from 50 to 400 for fixed codekey length of five chips.

When compared to all other techniques, DSSS takes minimum de-spreading time because DSSS technique uses only one code sequence for spreading and de-spreading the entire message transmission, but the probability of jamming the message is high. The UDSSS and DSD DSSS require more time to de-spread the data because they have to check all the code sequences used in the codebook. Whereas in CCS DSSS, both the transmitter and receiver generate the initial seed value for chaotic signal by exchanging RSS signal, and then based on RSS signal, initial seed value is computed by both sensor and PDA device. The codekey is selected from the codebook based on chaotic value and since their behaviours are unpredictable so it is very difficult for the intruder to de-spread the data. For the codebook size 300, the de-spreading times are 0.0007 ms, 0.1098 ms, 0.0705 ms, and 0.05888 ms for DSSS, UDSSS, DSD DSSS, and CCS DSSS, respectively (Fig. 54.3).

54.3.3   Computational Complexity

The computation complexity analysis is shown in Table 54.1. From analysis, we noticed that de-spreading time of CCS-DSSS technique is low compared with without

Fig. 54.3 De-spreading time for different codebook sizes

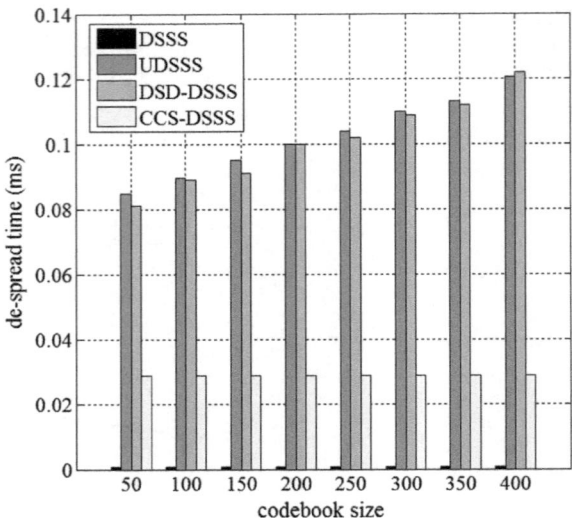

Table 54.1 Computational complexity

Spreading technique	Storage memory	De-spreading time (ms)	Key pre-shared
DSSS	15 bits	0.0007	Yes
UDSSS	33 KB	0.1098	No
DSD-DSSS	33 KB	0.0705	No
CCS-DSSS	33 KB	0.0588	No

pre-shared spreading techniques because, if the initial seed value is correctly identified, then based on that codekeys are calculated independently by using chaotic map. The memory required to store codekeys without pre-shared spreading techniques is determined by using Eq. 54.10.

$$\text{Memory requirement} = (G * a * pn) + (G * I_n) \qquad (54.10)$$

where G is the created groups, a is the length of codekey, pn is the pseudorandom sequence length, and I_n is the index assigned for each codekey; it is determined as $\lceil log_2 G \rceil$.

54.4 Conclusion

A secure spread-spectrum Chaos-based Code Selection DSSS is presented and analyzed. The use of chaos in spreading techniques provides additional security to the communication system. The probability of message being jammed (SNR = 0 dB) is 0.06 and 0.20 for CCS-DSSS and DSSS, respectively. This indicates that the intruder finds it very difficult to eavesdrop and jam the message. The de-spreading time required to extract the original message is almost 0.05888 ms irrespective of codebook size, which is minimum when compared with DSD-DSSS and UDSSS under larger codebook size. The obtained results show that CCS-DSSS is more secure, and the receiver de-spreads the message with minimum de-spreading time compared with others without pre-shared spreading techniques.

References

1. Pickholtz, R., Schilling, D., Milstein, L.: Theory of spread-spectrum communications—A tutorial. IEEE Trans. Commun. **30**(5), 855–884 (1982)
2. Popper, C., Strasser, M., Capkun, S.: Anti-jamming broadcast communication using uncoordinated spread spectrum techniques. IEEE J. Sel. Areas Commun. **28**(5), 703–715 (2010)
3. Liu, A., Ning, P., Dai, H., Liu, Y., Wang C.: Defending DSSS-based broadcast communication against insider jammers via delayed seed disclosure. In Proceedings of the 26th Annual Computer Security Applications Conference (ACSAC '10). ACM, New York, NY, USA, pp. 367–376 (2010)
4. Cassola, A., Jin, T., Noubir, G., Thapa, B.: Efficient spread spectrum communication without preshared secrets. IEEE Trans. Mob. Comput. **12**(8), 1669–1680 (2013)
5. Gopalakrishnan, B., Bhagyaveni, M.A.: Anti-jamming communication for body area network using chaotic frequency hopping. Healthcare Technol. Lett. **4**(6), 233–237 (2017)
6. Dhivyadharshini, Gopalakrishnan, B.: Comparative analysis of FH and CFH spread spectrum under different jammers. In: 2020 International Conference on Communication and Signal Processing (ICCSP), Chennai, India, pp. 1361–1365 (2020)
7. Xuan Quyen, N., Van Yem, V., Manh Hoang, T.: A chaos-based secure direct-sequence/spread-spectrum communication system. Abstract Appl. Anals. Article ID 764341, 1–11 (2013)
8. Tayebi, A., Berber, S., Swain, A.: Performance analysis of chaotic DSSS-CDMA synchronization under jamming attack. Circuits, Syst. Signal Process. **35**(12), 4350–4371 (2016)
9. Liu, Y., Ning, P., Dai, H., Liu A.: Randomized differential DSSS: Jamming-Resistant wireless broadcast communication. In: Proceedings of Conference IEEE INFOCOM. San Diego, CA. pp. 1-9 (2010)
10. Gopalakrishnan, B., Bhagyaveni, M.A.: Random codekey selection using codebook without pre-shared keys for anti-jamming in WBAN. Comput. Electr. Eng. **51**, 89–103 (2016)
11. Alagil A., Alotaibi, M., Liu Y.: Randomized positioning DSSS for anti-jamming wireless communications. In: Proceedings of International Conference on Computing, Networking and Communications (ICNC), Kauai, HI, pp. 1–6 (2016)
12. Alagil A., Liu Y.: Random allocation seed-DSSS broadcast communication against jamming attacks. In: Security and Privacy in Communication Networks Secure Comm 2019. Lecture Notes of the Institute for Computer Sciences, Social Informatics and Telecommunications Engineering, Vol. 304. Springer, Cham (2019)

Author Index

Printed by Printforce, the Netherlands